Biohydrogen Production

Rapid accumulation of pollutants and waste has resulted in a persistent demand for environmental clean-up through sustainable microbial technologies that combine biodegradation with bioenergy generation. Biohydrogen has emerged as a popular biofuel because it produces zero-carbon emissions and is environmentally friendly.

This volume in the *Wastewater Treatment and Research* series delves into the various ways in which biohydrogen can be produced. This includes photosynthetic biohydrogen production, where cyanobacteria decompose water to hydrogen under anaerobic conditions using photosynthesis, anaerobic dark fermentation where hydrogen is produced by decomposition of organic matter by anaerobic microorganisms, and light and dark fermentation. With special emphasis on cell immobilization strategies, membrane processes, and nanomaterials, this book also highlights recent innovations in biohydrogen production such as culture enrichment, pretreatment of wastes, metabolic and genetic engineering, and the coupling process of light and dark fermentation. By providing an ecofriendly roadmap to turning waste into bioenergy, this book serves as an instrumental guide to averting one of the most pressing environmental crises of our times.

Professors, researchers, students, and science enthusiasts will appreciate this holistic overview of biohydrogen production, especially in terms of its long-term viability, cost, and environmental implications.

Wastewater Treatment and Research

Series Editor: Maulin P. Shah

Wastewater Treatment: Molecular Tools, Techniques, and Applications
Edited by Maulin P. Shah

Advanced Oxidation Processes for Wastewater Treatment: An Innovative Approach
Edited by Maulin P. Shah, Sweta Parimita Bera, and Gunay Yildiz Tore

Emerging Technologies in Wastewater Treatment
Edited by Maulin P. Shah

Bio-Nano Filtration in Industrial Effluent Treatment: Advanced and Innovative Approaches
Edited by Maulin P. Shah

Membrane and Membrane-Based Processes for Wastewater Treatment
Edited by Maulin P. Shah

Phycoremediation Processes in Industrial Wastewater Treatment
Edited by Maulin P. Shah

Recent Trends in Industrial Wastewater Treatment
Edited by Tripti Malik and Maulin P. Shah

Algal-based Waste Treatment Technologies in a Circular Economy
Edited by Günay Yıldız Töre and Maulin P. Shah

Biohydrogen Production: An Overview of Diverse Feedstock and Modern Technologies
Edited by Venkatesh Chaturvedi, Pradeep Verma and Maulin P. Shah

For more information about this series, please visit: www.routledge.com/Wastewater-Treatment-and-Research/book-series/WASTEWATER

Biohydrogen Production

An Overview of Diverse Feedstock and Modern Technologies

Edited by
Venkatesh Chaturvedi, Pradeep Verma
and Maulin P. Shah

CRC Press
Taylor & Francis Group
Boca Raton London New York

CRC Press is an imprint of the
Taylor & Francis Group, an **informa** business

Designed cover image: Shutterstock

First published 2026
by CRC Press
4 Park Square, Milton Park, Abingdon, Oxon, OX14 4RN

and by CRC Press
2385 NW Executive Center Drive, Suite 320, Boca Raton FL 33431

CRC Press is an imprint of the Taylor & Francis Group, an informa business

British Library Cataloguing-in-Publication Data
A catalogue record for this book is available from the British Library

ISBN: 9781032941110 (hbk)
ISBN: 9781032940335 (pbk)
ISBN: 9781003568612 (ebk)

DOI: 10.1201/9781003568612

Typeset in Times
by codeMantra

Contents

Preface

Owing to rapid urbanization and industrialization, pollutants and wastes are rapidly accumulating in the environment leading to serious consequences. Thus, environmental clean-up through sustainable microbial technologies, where biodegradation is coupled to bioenergy generation, has become a prerequisite. Biohydrogen is considered as one of the most sought-after biofuels due to its zero-carbon emission. It can be used as an alternate biofuel to alleviate stresses due to carbon dioxide emissions and depletion of fossil fuels. Biohydrogen is mainly produced via three major routes, i.e., photosynthetic biohydrogen production where algae/cyanobacteria decompose water to hydrogen under anaerobic conditions through photosynthesis, anaerobic dark fermentation where hydrogen is produced by decomposition of organic matter by anaerobic microorganisms, and light and dark fermentation, which is a recent innovation where light and dark fermentation processes are combined together. A vast array of wastes, such as household waste, industrial waste from food processing industries, lignocellulose from agriculture, and livestock residues, has been tested for hydrogen production with promising results. This approach is environmentally sustainable as it couples the degradation of wastes by microorganisms with hydrogen production.

An immense amount of research has been carried out to increase hydrogen production through innovative developments and enhancement strategies for improving the biorefinery route of conversion of industrial/household/agriculture wastes and livestock residues to biohydrogen. Recent innovations for biohydrogen production involve culture enrichment, pretreatment of wastes, metabolic and genetic engineering, and the coupling process of light and dark fermentation for biohydrogen. Recently, cell immobilization strategies, membrane processes, and nanomaterials have also been employed to enhance biohydrogen production. Its long-term viability and sustainability have been assessed.

This book describes a clean, ecofriendly route for the conversion of wastes into bioenergy, abating the environmental crisis and leading to a sustainable future. Researchers, students, and science enthusiasts alike will appreciate this holistic view of biohydrogen production through recent innovative technologies that enhance biohydrogen production so as to bring down its cost and make it a sustainable source of energy for coming generations.

About the Editors

Venkatesh Chaturvedi earned his M.Sc. degree in Biochemistry in 2004 and Ph.D. degree in Biotechnology in 2011, from Banaras Hindu University, Varanasi, Uttar Pradesh, India. He has more than 12 years of teaching experience in government/private teaching institutions. He is currently working as Assistant Professor at the School of Biotechnology, Institute of Science, Banaras Hindu University since 2019. He is a Group Leader of the Bioremediation and Environmental Biotechnology Laboratory at the School of Biotechnology, Banaras Hindu University. His area of expertise involves environmental biotechnology, bioremediation, and industrial biotechnology. He has undertaken an extramural research grant in the area of bioremediation. He has published more than 40 research/review articles in renowned scientific journals such as *International Journal of Biological Macromolecules, International Biodeterioration and Biodegradation, Process Biochemistry, World Journal of Microbiology and Biotechnology, Fuel Processing Technology*, and *Frontiers in Microbiology.* He is a member of various national and international societies/academies. He has also published around ten book chapters in books published by national/international publishers. He has also served as a reviewer for many journals from different publishers.

Pradeep Verma earned his Ph.D. from Sardar Patel University, Gujarat, India, in 2002. In the same year, he was selected as a UNESCO fellow and joined the Czech Academy of Sciences, Prague, Czech Republic. He later moved to Charles University, Prague, to work as a postdoctoral fellow. In 2004, he worked as a visiting scientist at the UFZ Centre for Environmental Research, Halle, Germany. He was awarded a DFG fellowship that provided him with another opportunity to work as a postdoctoral fellow at Gottingen University, Germany. Later in 2007, he returned to India and joined Reliance Life Sciences, Mumbai, and worked extensively on biobutanol production, which attributed a few patents to his name. Later, he was awarded JSPS Postdoctoral Fellowship Program and joined the Laboratory of Biomass Conversion, Research Institute of Sustainable Humanosphere (RISH), Kyoto University, Japan. He has also been the recipient of various prestigious awards, such as the Ron Cockcroft award by the Swedish Society, and the UNESCO Fellow ASCR Prague. Recently, for his contribution to the area of fungal microbiology, industrial biotechnology, and environmental bioremediations, he has been awarded the prestigious Fellow Award from the International Academy of Sciences (IAWS), 2024; Association of Fungal Biologist (AFB), 2024; National Academy of Biological Sciences (NABS), 2023; Mycological Society of India (2020), P.C. Jain Memorial Award (MSI), and Biotech Research Society of India (2021). In 2020, he was also awarded Fellow of the Biotechnology Research Society of India (BRSI) and Fellow of the Academy of Sciences of AMI India 2021 (FAMSc). Furthermore, he has also been awarded the JSPS Bridge Fellow award in 2022 and a short-term visit to Kyoto University, Kyoto, Japan, to strengthen ties between the two laboratories. He is currently working as a professor at the Department of Microbiology, CURAJ (Central University of Rajasthan). His area of expertise involves microbial diversity, bioremediation, bioprocess development, lignocellulosic, and algal biomass-based biorefinery. He holds several international patents in the field of microwave-assisted biomass pretreatment and biobutanol production. He has more than 106 research articles in peer-reviewed international journals and has contributed to 66 published book chapters in different edited books. He has also edited 26 books for various international publishers. He is a guest editor for several journals, such as *Biomass Conversion and Biorefinery*, *Agriculture*, *Frontier in Nanotechnology*, and *International Journal of Environmental Research and Public Health*. He is also an editorial board member for the *Journal of Current Nanomedicine* and an acting reviewer for more than 70 journals of different publishers.

Maulin P. Shah is an active researcher and scientific writer in his field for over 20 years. He earned a B.Sc. degree (1999) in Microbiology from Gujarat University, Godhra (Gujarat), India. He also earned his Ph.D. degree (2005) in Environmental Microbiology from Sardar Patel University, Vallabh Vidyanagar (Gujarat), India. He is the Chief Scientist and Head of the Industrial Wastewater Research Lab, Division of Applied and Environmental Microbiology Lab at Enviro Technology Ltd., Ankleshwar, Gujarat, India. His work focuses on the impact of industrial pollution on the microbial diversity of wastewater and genetically engineering high-impact microbes for the degradation of hazardous materials. His research interests include biological wastewater treatment, environmental microbiology, biodegradation, bioremediation, and phytoremediation of environmental pollutants from industrial wastewaters. Majority of his work focuses on isolation, screening, identification, and genetic engineering of high-impact microbes for the degradation of hazardous materials. He has published more than 250 research papers in national and international journals of repute on various aspects of microbial biodegradation and bioremediation of environmental pollutants. He is the editor of more than 50 books of international repute and is an active editorial board member in top-rated journals.

Contributors

Thilini U. Ariyadasa
Department of Chemical and Process Engineering, Faculty of Engineering
University of Moratuwa
Moratuwa, Sri Lanka

Janhavee Asthana
School of Biotechnology, Institute of Science
Banaras Hindu University
Varanasi, Uttar Pradesh, India

Janani Balaji
Department of Genetic Engineering
SRM Institute of Science and Technology
Kattankulathur, Chengalpattu, Tamil Nadu, India

Aditi Baldev
Department of Genetic Engineering
SRM Institute of Science and Technology
Kattankulathur, Chengalpattu, Tamil Nadu, India

Sneha Banerjee
Department of Biosciences
JIS University
Kolkata, West Bengal, India

Bindu Battan
Department of Biotechnology
Kurukshetra University
Kurukshetra, Haryana, India

Debalina Bhattacharya
Department of Microbiology
Maulana Azad College
Kolkata, West Bengal, India

Arghadeep Bhowmick
Department of Biosciences
JIS University
Kolkata, West Bengal, India

Sulekha Chahal
Department of Biotechnology
Kurukshetra University
Kurukshetra, Haryana, India

Kajal Chauhan
Department of Biotechnology
Kurukshetra University
Kurukshetra, Haryana, India

Rajveer Singh Chauhan
Department of Botany
Deen Dayal Upadhyaya Gorakhpur University
Gorakhpur, Uttar Pradesh, India

Deepak Choudhary
Plant Functional Genomics Lab, Biotechnology Unit, Department of Botany (UGC-Centre for Advance Studies)
Jai Narain Vyas University
Jodhpur, Rajasthan, India

Taranjit Kaur Gargeria
School of Biotechnology, Institute of Science
Banaras Hindu University
Varanasi, Uttar Pradesh, India

Chaitanya Gatkal
School of Biotechnology, Institute of Science
Banaras Hindu University
Varanasi, Uttar Pradesh, India

Monalisa Ghara
Amity Institute of Biotechnology
AUUP
Noida, Uttar Pradesh, India

Biva Ghosh
Department of Biosciences
JIS University
Kolkata, West Bengal, India

Deepshikha Gupta
Department of Zoology
Deen Dayal Upadhyaya Gorakhpur University
Gorakhpur, Uttar Pradesh, India

Nikita Gupta
Department of Botany
Deen Dayal Upadhyaya Gorakhpur University
Gorakhpur, Uttar Pradesh, India

Noorunnisa Mohamed Hanifa
Department of Chemical and Process Engineering, Faculty of Engineering
University of Moratuwa
Moratuwa, Sri Lanka

Asim Kumar Jana
Department of Biotechnology
Dr B R Ambedkar National Institute of Technology
Jalandhar, Punjab, India

Shweta Jha
Plant Functional Genomics Lab, Biotechnology Unit, Department of Botany (UGC-Centre for Advance Studies)
Jai Narain Vyas University
Jodhpur, Rajasthan, India

Angelaalincy Maria Joseph
Vinayaka Mission's Kirupananda Medical College and Hospitals
Vinayaka Missions Research Foundation (Deemed to be University)
Salem, Tamil Nadu, India

Keerthana Karthick
Department of Genetic Engineering
SRM Institute of Science and Technology
Kattankulathur, Chengalpattu, Tamil Nadu, India

Deepak Kumar
Institute of Environmental Studies
Kurukshetra University
Kurukshetra, Haryana, India

Manoj Kumar
Department of Biotechnology
Dr B R Ambedkar National Institute of Technology
Jalandhar, Punjab, India

Prasann Kumar
Department of Agronomy, School of Agriculture
Lovely Professional University
Phagwara, Punjab, India

Sakshi Kumari
Amity Institute of Biotechnology
AUUP
Noida, Uttar Pradesh, India

Kanagaraj Kuppusamy
Center for Supramolecular Chemistry and Catalysis and Department of Chemistry
College of Science, Shanghai University
Shanghai, China

Ambuj Kumar Kushwaha
School of Biotechnology
Institute of Science, Banaras Hindu University
Varanasi, Uttar Pradesh, India

Prerana Limbani
Department of Genetic Engineering
SRM Institute of Science and Technology
Kattankulathur, Chengalpattu, Tamil Nadu, India

Vishal Malhotra
Amity Institute of Biotechnology
AUUP
Noida, Uttar Pradesh, India

Hendrina Marian
Department of Genetic Engineering
SRM Institute of Science and Technology
Kattankulathur, Chengalpattu, Tamil Nadu, India

Mainak Mukhopadhyay
Department of Biosciences
JIS University
Kolkata, West Bengal, India

Senthilkumar Nangan
Center for Research and Development (CRD), Vinayaka Mission's Kirupananda Variyar Arts and Science College
Vinayaka Missions Research Foundation (Deemed to be University)
Salem , Tamil Nadu, India

Nikhil Gauravarapu Navlur
Department of Biotechnology
Dr B R Ambedkar National Institute of Technology
Jalandhar, Punjab, India

Pemaththu Hewa Viraj Nimarshana
Department of Mechanical Engineering, Faculty of Engineering
University of Moratuwa
Moratuwa, Sri Lanka

Debarati Paul
Amity Institute of Biotechnology
AUUP
Noida, Uttar Pradesh, India

Aadharshini Pradeep
Department of Genetic Engineering
SRM Institute of Science and Technology
Kattankulathur, Chengalpattu, Tamil Nadu, India

Monika Prakash Rai
Department of Biotechnology
Motilal Nehru National Institute of Technology Allahabad
Prayagraj, Uttar Pradesh, India

Mohandass Ramya
Department of Genetic Engineering
SRM Institute of Science and Technology
Kattankulathur, Chengalpattu, Tamil Nadu, India

Pasupathi Rathinasabapathi
Department of Genetic Engineering
SRM Institute of Science and Technology
Kattankulathur, Chengalpattu, Tamil Nadu, India

Galbadage Don Shehan Sandeepa
Department of Chemical and Process Engineering, Faculty of Engineering
University of Moratuwa
Moratuwa, Sri Lanka

Aditi Sarker
Department of Biotechnology
Dr B R Ambedkar National Institute of Technology
Jalandhar, Punjab, India

Nancy Sikodia
Department of Biotechnology
Kurukshetra University
Kurukshetra, Haryana, India

Pradeep Kumar Singh
Department of Biochemistry
Dr. Rammanohar Lohia Avadh University
Ayodhya, Uttar Pradesh, India

Samraggi Singha
Department of Genetic Engineering
SRM Institute of Science and Technology
Kattankulathur, Chengalpattu, Tamil Nadu, India

Sheetal Soni
Plant Functional Genomics Lab, Biotechnology Unit, Department of Botany (UGC-Centre for Advance Studies)
Jai Narain Vyas University
Jodhpur, Rajasthan, India

Ananya Sood
Department of Genetic Engineering
SRM Institute of Science and Technology
Kattankulathur, Chengalpattu, Tamil Nadu, India

Meenakshi Suhag
Institute of Environmental Studies
Kurukshetra University
Kurukshetra, Haryana, India

Surabhi Sirohi
Institute of Environmental Studies
Kurukshetra University
Kurukshetra, Haryana, India

Shilpi Tyagi
Plant Functional Genomics Lab, Biotechnology Unit, Department of Botany (UGC-Centre for Advance Studies)
Jai Narain Vyas University
Jodhpur, Rajasthan, India

Shrasti Vasistha
Department of Biosciences
IMS University Courses Campus
Ghaziabad, Uttar Pradesh, India

Deepti Verma
School of Biotechnology
Institute of Science, Banaras Hindu University
Varanasi, Uttar Pradesh, India

Wanni Arachchige Jalitha Wasath
Department of Chemical and Process
Engineering, Faculty of Engineering
University of Moratuwa
Moratuwa, Sri Lanka

Kandasamy Yazhini
Department of Genetic Engineering
SRM Institute of Science and Technology
Kattankulathur, Chengalpattu, Tamil Nadu, India

Valorization of Microalgal Biomass for Biohydrogen Production

1

Noorunnisa Mohamed Hanifa, Galbadage Don Shehan Sandeepa, Wanni Arachchige Jalitha Wasath, Pemaththu Hewa Viraj Nimarshana, and Thilini U. Ariyadasa

1.1 INTRODUCTION

The expanding global population has resulted in a notable surge in energy demand, thereby intensifying the challenges associated with energy scarcity. From 2012 to 2020, global energy consumption increased from 579.2 EJ to 663.6 EJ, with projections indicating a rise to 859.9 EJ by 2040, marking a substantial 48% growth (Jiménez-Llanos et al., 2020; Rady et al., 2024). Fossil fuels continue to serve as the dominant global energy source and their extensive use has greatly contributed to environmental challenges, particularly through the emission of greenhouse gases (GHGs) (Wijayasekera et al., 2020). GHG emissions, particularly carbon dioxide (CO_2), are recognized as primary contributors to global warming and climate change. In response to the environmental challenges posed by fossil fuel consumption, increasing attention is being directed toward the development of alternative renewable energy sources, including solar, wind, geothermal, and biomass, as sustainable strategies for reducing GHG emissions (Premaratne et al., 2022). Among these, hydrogen (H_2) production from renewable biomass has garnered considerable attention as a promising sustainable energy solution (Ahmed et al., 2022).

H_2 has emerged as a key renewable energy carrier due to its sustainability, high-energy potential, and efficient conversion capabilities (S. Li et al., 2022). Moreover, H_2 with a lower heating value of 120,000 kJ/kg offers high energy density, clean combustion, and versatility, enabling its application across a wide range of sectors, including power generation, transportation, and industrial processes (Pathy et al., 2022;

DOI: 10.1201/9781003568612-1

Yong & Mat Isa, 2023). Global H_2 production is estimated at ~500 billion m^3 annually, with an annual growth rate of around 10%. Specifically, about 40% of H_2 is produced from natural gas, 30% from heavy oil and naphtha, and 18% from coal, while electrolysis and biomass account for just 4% and 1%, respectively (Jiménez-Llanos et al., 2020). Conventional H_2 production methods, including steam methane reforming, partial oxidation, and coal gasification, pose significant challenges due to their high carbon emissions and energy-intensive processes (Kumar et al., 2021). These limitations have driven increasing interest in sustainable and renewable H_2 production through biological processes.

The production of biohydrogen through the valorization of microalgal biomass offers a technically viable and environmentally sustainable pathway for renewable energy generation, capable of addressing the growing global energy demand while mitigating ecological impacts (Thevarajah et al., 2025). As a third-generation feedstock, microalgae is an efficient feedstock for biohydrogen production due to a 5- to 10-fold higher growth rate than terrestrial plants, efficient CO_2 sequestration, adaptability to various water sources such as freshwater, seawater, and wastewater, and no reliance on arable land for cultivation (Ahmed et al., 2022; Hanifa et al., 2025; Hidalgo et al., 2023; Liyanaarachchi et al., 2021). Furthermore, the biochemical composition of microalgae, consisting of 20%–30% carbohydrates, 40%–60% proteins, and 10%–20% lipids, renders it a highly favorable feedstock for biohydrogen production, as these macromolecules serve as essential substrates for microbial and enzymatic conversion processes (Piyatilleke et al., 2024; Thevarajah et al., 2024). Notably, *Chlamydomonas reinhardtii* and *Chlorella* sp. have been extensively studied for biohydrogen production, attributed to their relatively simple cellular organization and high adaptability to both photoautotrophic and mixotrophic growth conditions (Goria et al., 2024). Furthermore, biohydrogen production from microalgae could be achieved through various biological pathways, including direct and indirect bio-photolysis, dark fermentation, and photo-fermentation (Ayodele et al., 2023). Additionally, the research focuses on developing integrated hybrid systems such as photo-dark coupled fermentation and combined fermentation with microbial electrolysis cells (MECs) for biohydrogen production to address the limitations of conventional biohydrogen generation techniques and maximize overall yield (Rady et al., 2024).

However, a comprehensive understanding of biochemical pathways and optimization strategies for biohydrogen production is essential to effectively address the technical and economic challenges associated with the process. Accordingly, this chapter aims to scrutinize the prospects of valorizing microalgal biomass for biohydrogen production through an in-depth analysis of optimal strain selection, cultivation and harvesting methodologies, and pretreatment techniques that enhance the efficiency of biohydrogen production. Moreover, the key challenges encountered in biohydrogen production and proposed future research directions aimed at advancing biohydrogen production will also be addressed.

1.2 MICROALGAE FOR BIOHYDROGEN PRODUCTION

Microalgae offer significant potential for biohydrogen production, particularly when cultivated under controlled conditions, optimizing parameters such as thermal conditions, pH levels, illumination strength, and nutrient concentration available. Moreover, efficient biomass utilization for biohydrogen production depends on effective cultivation practices, innovative reactor designs, and optimized harvesting and pretreatment techniques, positioning this approach as a sustainable pathway for renewable energy generation. This section discusses microalgal strains with the potential for biohydrogen production, alongside a detailed review of the cultivation, harvesting, and pretreatment techniques essential for optimizing the biohydrogen yield (Figure 1.1).

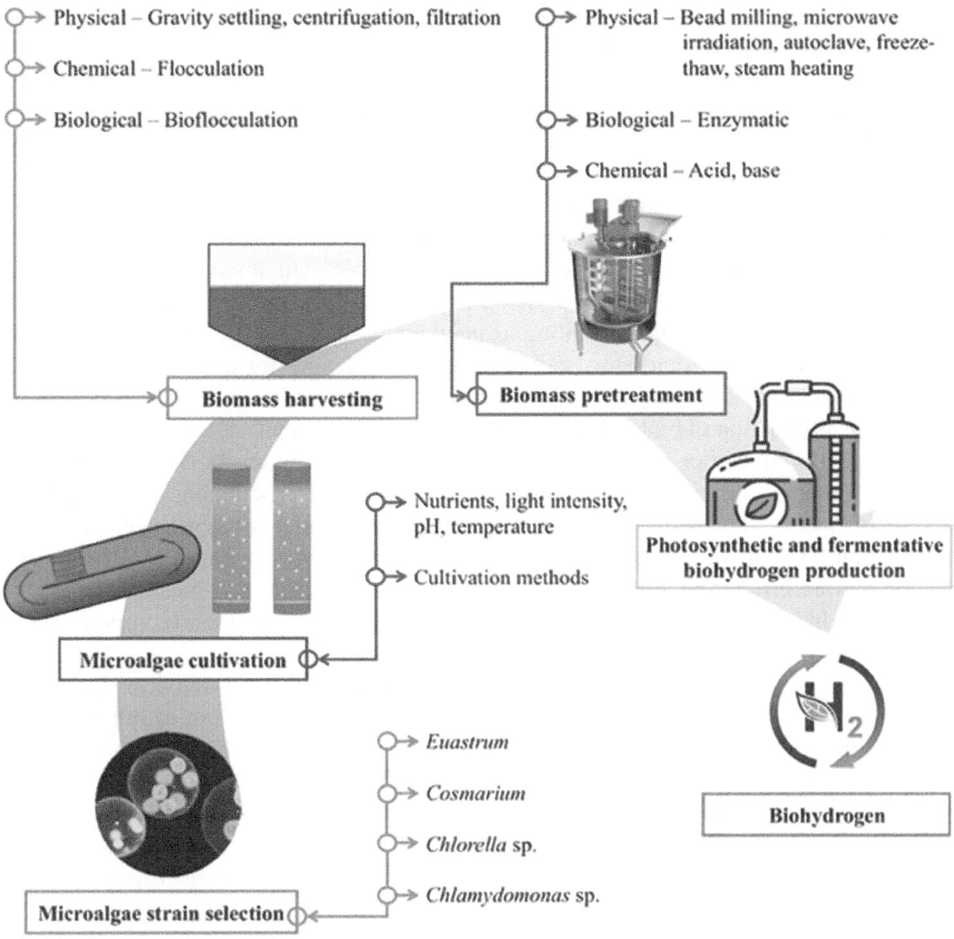

FIGURE 1.1 Schematic diagram of biohydrogen production from microalgae.

1.2.1 Microalgal Strains

Biohydrogen production from microalgae is influenced by specific physiological and environmental conditions that optimize metabolic pathways for efficient H_2 generation (Jiao et al., 2024). Recent studies have focused on a range of microalgal species to evaluate their biohydrogen production capabilities, highlighting the diversity and potential of microalgae. Numerous microalgal species, including *Euastrum*, *Cosmarium*, *Chlorella* sp., *Ulothrix*, *Selenastrum*, *Chlamydomonas* sp., *Closterium*, *Micractinium* sp., *Pandorina*, *Stigeoclonium*, and *Staurastrum*, have been identified as capable of biohydrogen synthesis (Y. Wang et al., 2020).

For instance, the lipid-extracted residual biomass of *Scenedesmus obliquus*, when subjected to processing through a thermal-alkaline approach at a temperature of 100°C, demonstrated a substantial enhancement in H_2 yield, reaching a maximum of 45.54 mL/g volatile solids (VS) (Yang et al., 2010). Moreover, *Anabaena siamensis* TISTR 8012, a cyanobacterium capable of nitrogen fixation, exhibited a peak rate of H_2 generation of 32 μmol H_2/mg chlorophyll/h when cultured in BG110 medium with 0.5% fructose under steady illumination level of 200 μE/m²/s for 12 hours, surpassing the rates of other strains such as *Anabaena* PCC 7120 and *Synechocystis* PCC 6803 (Taikhao & Phunpruch, 2017). Similarly, *Planktosphaeria subcordiformis* showed promising results with a H_2 concentration of 63.2% in biogas, and production rates ranging from 0.53 to 0.70 cm³/h in natural water-based media (Dudek et al., 2022).

Furthermore, *Tetraspora* sp. CU2551 produced 12.8±0.9 mL H_2/25 mL medium, achieving a H_2 production rate of 182±20 nmol/mg dry weight/h using calcium alginate-immobilized cells under aerobic conditions (Maswanna et al., 2020). Moreover, *C. reinhardtii* with codon-optimized genes for ferrochelatase and leghemoglobin achieved a 6.8-fold increase in protein expression, enhancing H_2 yield to 23.5 µmol/mg chlorophyll/h (Wu et al., 2011). According to a study by Taikhao et al. (2013), the halotolerant cyanobacterium demonstrated high H_2 production rates, with an optimal rate of 13.804 µmol H_2/mg chlorophyll/h under nitrogen deprivation and specific conditions of 0.4 µM ferric ions (Fe^{3+}) (Taikhao et al., 2013). Additionally, *C. reinhardtii* stands out as a model organism due to its extensively characterized H_2 metabolism and the application of advanced genetic engineering tools, which have significantly enhanced H_2 yields in controlled laboratory environments (King et al., 2022). Despite its prominence in research, the inherent limitations of *C. reinhardtii*, including its heightened sensitivity to solar radiation, low lipid levels, and difficulties in growing outdoors, pose significant challenges for commercial-scale biohydrogen generation (Scranton et al., 2015).

Moreover, the production of biohydrogen by microalgae relies on physiological conditions, especially in environments with reduced oxygen (O_2) levels and elevated light intensity (Bahadar et al., 2022). The activity of hydrogenase, a crucial enzyme involved in biohydrogen production, is extremely sensitive to O_2, inhibiting its function during direct bio-photolysis and compromising biohydrogen yield, as discussed in detail in Section 1.3 (Bahadar et al., 2022). Consequently, strains with enhanced O_2 tolerance or anaerobic adaptation are preferred, and advancements in genetic engineering have enabled the development of oxygen-tolerant hydrogenase variants, reducing the inhibitory effects of O_2 on H_2 production (Jiao et al., 2024).

Maintaining O_2 concentrations at or below 1% is critical for preserving hydrogenase activity in *C. reinhardtii*, which is achieved by using a sulfur-deprived culture medium to inhibit photosystem II (PSII) activity, thereby limiting O_2 production and promoting an anaerobic environment conducive to hydrogenase function (Vijayaraghavan, Karthik, & Kamala Nalini, 2009). *Chlorella vulgaris*, among other species, is extensively utilized for biohydrogen generation without pretreatment, achieving production rates of H_2 spanning from 0.37 to 19 mL H_2/g VS (Wieczorek, Kucuker, & Kuchta, 2014). Therefore, the maximum H_2 yield reported for *C. vulgaris* underscores the potential of microalgal biomass as a robust and versatile substrate for sustainable biohydrogen production.

1.2.2 Cultivation Systems of Microalgae

Microalgal cultivation systems are essential for optimizing biohydrogen production, leveraging specialized reactor configurations and designs to enhance the efficiency of microbial metabolic processes. Microalgae-based bioreactors are crucial for optimizing biohydrogen production by enhancing light absorption and microbial metabolic efficiency. Reactor designs, including photobioreactors (PBRs), continuous stirred tank reactors (CSTRs), fixed-bed reactors, and multi-stage systems, offer distinct advantages for biohydrogen production depending on the specific microbial requirements (Jiao et al., 2024; Zarei et al., 2024). PBRs, such as tubular, vertical column, and flat panel types, are designed to maximize light penetration and the surface-area-to-volume ratio, essential for optimal biohydrogen synthesis. Tubular PBRs, having diameters from 3 to 6 cm and extending up to 100 m in length, provide effective gas exchange and light penetration, achieving H_2 yield of 2.5 L H_2/L reactor volume (Huang et al., 2017). Vertical column PBRs, which use water jackets for temperature regulation, facilitate deep light penetration, with light intensities reaching up to 2,000 µmol/m²/s, thus supporting denser cultures and higher biomass productivity (Jiao et al., 2024). Flat panel PBRs, typically applied at smaller scales (below 10 m³), offer precise control over pH, temperature, and optical density, although they require more energy for mixing and aeration (Jiao et al., 2024). Advances in light optimization, such as the use of planar optical guides with light-dispersing nanoparticles, have enhanced light distribution by 10.3 times, increasing biomass productivity by 220% compared to conventional systems (Sun et al., 2016). Although CSTRs are effective for maintaining good substrate-microbe contact and mass transfer, they suffer from biomass washout due to high mixing rates and short hydraulic retention times, which limit H_2 production rates (Luo et al., 2011). Fixed-bed reactors, with lower turbulence compared to CSTRs, offer improved mass transfer but encounter challenges in maintaining uniform microbial activity due to pH gradients, requiring recirculation to improve efficiency (A. Sharma & Arya, 2017; A. Wang et al., 2011).

Additionally, recent advancements focus on integrating multiple reactor types, such as MECs, dark fermenters, and photo-fermenters, into multi-reactor configurations for the enhancement of biohydrogen yield and overall process efficiency (Bora et al., 2022). Multi-reactor systems have demonstrated significant improvements, achieving H_2 production rates as high as 4.52 m^3 $H_2/m^3/d$ with MEC configurations using butyrate-adapted biofilms (Show et al., 2019). Moreover, hybrid bioreactors, comprising a combination of dark and photo-fermentation mechanisms, improve overall H_2 yields by maximizing substrate utilization, with H_2 and organic acids being fermented in separate stages. For instance, four-stage systems utilizing an integration of photolysis, photo-fermentation, and dark fermentation have been developed, significantly boosting energy recovery through organic acids and reducing the requirement for external substrates (Rai & Singh, 2016). These advancements highlight the potential of integrated reactor designs to optimize biohydrogen production, although challenges related to system complexity, cost, and scalability remain. However, continued research is essential to refine these technologies, making microalgal biohydrogen production a commercially viable renewable energy source.

1.2.3 Harvesting Methods

Harvesting and dewatering are crucial steps in microalgal biomass production, focusing on efficiently segregating microalgal cells from the growth medium and preparing the biomass for subsequent applications in biohydrogen production. Efficient biomass harvesting is critical in biohydrogen production, where achieving high cell densities is essential for maximizing biohydrogen yield and optimizing process efficiency. The efficiency of various harvesting methods is influenced by several factors, including the microalgal strain, cell size, and the nature of the intended end products (de Morais et al., 2023). Conventional microalgal harvesting methods, such as gravity settling, filtration, centrifugation, and flocculation, each present unique advantages and limitations in terms of efficiency, cost-effectiveness, and scalability (Deepa, Sowndhararajan, & Kim, 2023).

Filtration utilizes membranes with varying pore sizes to separate microalgal strains. For instance, microfiltration membranes exhibiting pore sizes between 0.1 and 10 μm are employed to retain microalgal cells. However, the filtration method is often plagued by issues such as membrane fouling and clogging, which could significantly impair its efficiency (Castro-Muñoz & García-Depraect, 2021). In contrast, centrifugation offers enhanced efficiency by subjecting microalgal suspension to centrifugal forces, which promote the sedimentation of microalgal cells, thereby facilitating separation from the culture medium. Commercial-scale centrifuges utilized in microalgal processing include decanter centrifuges, disk stack centrifuges, perforated basket centrifuges, and cyclones, each distinguished by specific operational parameters such as centrifugal force, throughput capacity, and separation efficiency (Kim et al., 2015).

Flocculation is widely considered cost-effective and energy-efficient for microalgal harvesting, making it particularly ideal for large-scale implementation due to its simplicity and scalability. The process facilitates the aggregation of microalgal cells by employing various flocculants, including inorganic, organic, and polymeric agents, or via mechanisms such as auto-flocculation, electro-flocculation, and microbial flocculation (Mubarak et al., 2019). Flocculation is highly versatile, with options for the selection of suitable flocculants based on the specific harvesting requirements. For instance, X. Wang et al. (2024) demonstrated that the use of inorganic flocculants, such as polymeric aluminum chloride (PAC), and organic flocculants, including cationic polyacrylamide (CPAM), effectively promotes the aggregation of microalgal cells. The process enhances photobiological biohydrogen production in both the prokaryotic strain *Synechocystis* sp. and the eukaryotic microalgae *C. reinhardtii*. The combined PAC-CPAM flocculant achieved flocculation efficiencies surpassing 92.2% for both strains, with H_2 production improving by 65.9% in *Synechocystis* and 80.2% in *C. reinhardtii* compared to the controls. Additionally, hydrogenase activity was significantly enhanced, reaching 1.14 μmol H_2/mg chlorophyll/h in *Synechocystis* and 28.2 μmol H_2/mg chlorophyll/h in *C. reinhardtii* (X. Wang et al., 2024). Therefore, the study underscores the potential of flocculation techniques to optimize microalgal biohydrogen production, further advancing the viability of microalgae-based biohydrogen as a sustainable renewable energy source.

Gravity settling is another economically efficient method for harvesting microalgal cells from the liquid using gravitational forces (Mathimani & Mallick, 2018). The settling rate depends on factors such as cell size, density, and shape. Microalgae with larger or denser cells settle more efficiently, whereas gravity settling is less effective for small-sized algae or cells with low cell density (Milledge & Heaven, 2013). To improve settling, coagulants or flocculants could be added, and lamella separators or sedimentation tanks could be utilized (Molina Grima et al., 2003; Pragya et al., 2013). Despite these enhancements, challenges such as low slurry output, the requirement for thickening, and limited efficiency persist, restricting the scalability and reliability of the process (Barros et al., 2015).

Moreover, integrating multiple harvesting methods can optimize overall efficiency and enhance performance, despite requiring meticulous coordination of each method's operational efficiency. A detailed evaluation that includes cost analysis, energy consumption, and resource recovery is essential for selecting the most appropriate harvesting and dewatering system, tailored to the specific characteristics of each microalgal strain.

1.2.4 Pretreatment Methods

Microalgal biomass pretreatment is essential for enhancing biohydrogen synthesis, as the rigid cell walls, composed of complex carbohydrates, proteins, and lipids, protect intracellular components such as fermentable sugars. Disruption of cell wall barriers is essential to enhance substrate accessibility, thereby improving efficiency during fermentation processes. Pretreatment methods, including mechanical, chemical, and biological approaches, offer distinct advantages and challenges, influencing their effectiveness and suitability for large-scale applications. Table 1.1 provides a comparison of biohydrogen yields obtained from different microalgal species subjected to various pretreatment conditions.

Mechanical pretreatment techniques for microalgal cell disruption are widely applied in bioprocessing to promote the release of intracellular components critical for biohydrogen production. Common techniques include bead milling, high-pressure homogenization (HPH), ultrasonication, microwave irradiation, hydrodynamic cavitation, and pulsed electric fields (PEFs), each leveraging different physical

TABLE 1.1 Biohydrogen yields from various microalgal species following different pretreatment methods

PRETREATMENT METHOD		*SPECIES*	*BIOHYDROGEN YIELD*	*REFERENCE*
Mechanical	Sonication	*Chlorella sorokiniana*	52 mL/g VS	Roy et al. (2014)
	Steam heating/ Microwave irradiation	*Chlorella pyrenoidosa*	8.8–12.7 mL/g TVS	Bhatia et al. (2023)
	Thermal	*C. vulgaris*	190.90 mL/g VS	Stanislaus et al. (2018)
Chemical	Alkaline	*C. vulgaris*	1178.4 mL/L	H. Liu et al. (2022)
	H_2O_2	*C. sorokiniana*	63 mL/g VS	Roy et al. (2014)
	NaOH	*Scenedesmus sp.* (lipid extracted)	16.89 mL/g VS	Yang et al. (2010)
Biological	Enzymatic hydrolysis	*C. vulgaris*	528 mL/L	H. Liu et al. (2022)
Combined	Steaming/Microwaving with diluted HCl	*C. pyrenoidosa*	75.6–83.3 mL/g TVS	Bhatia et al. (2023)

TVS, Total Volatile Solids; VS, Volatile Solids; H_2O_2, Hydrogen peroxide; NaOH, Sodium hydroxide; HCl, Hydrochloric acid.

forces to disrupt microalgal cells efficiently (Nagarajan et al., 2020). Bead milling, a pioneering technique in its field, involves agitating microalgal biomass involving steel, ceramic, glass, or quartz beads to generate shear forces, resulting in cell disruption through mechanical agitation and bead collisions (Onumaegbu et al., 2018). Studies have shown that smaller beads of sizes between 0.3 and 0.4 mm significantly improve cell disruption efficiency at higher agitator speeds of 9–10 m/s, achieving nearly 97% disruption in under 1 minute (Postma et al., 2017). Simultaneously, HPH directs microalgal suspensions through a high-pressure nozzle, inducing cell disruption through rapid pressure drops and impact forces, rendering the process viable for large-scale applications (Samarasinghe et al., 2012). Despite being highly effective, HPH often requires multiple passes to reach optimal cell disruption. Moreover, other methods, including ultrasonication and microwave irradiation, depend on parameters such as frequency, exposure time, and temperature control to ensure efficient disruption and energy management (Kurokawa et al., 2016; Xia et al., 2013). Moreover, mechanical processing techniques play a crucial role in microalgal pretreatment but are frequently being integrated with additional methods to optimize energy efficiency and maximize biohydrogen production.

Furthermore, thermal pretreatment, a type of mechanical pretreatment, employs elevated temperatures to disrupt microalgal cell structures, thereby facilitating the extraction of valuable compounds. For example, autoclaving utilizes heat at 121°C and pressures between 1.1 and 1.5 MPa to enhance carbohydrate hydrolysis when coupled with chemical treatments (Onumaegbu et al., 2018). Steam explosion exposes microalgal biomass to temperatures of 180°C–220°C and pressures of 1.5–2.5 MPa followed by rapid decompression, which could effectively rupture cell walls but is highly energy-intensive and less practical for large-scale applications (Nurra et al., 2014). Freeze-thaw cycles involve alternating between freezing at –20°C and thawing at 30°C to mechanically disrupt cells, though this method could be energy-demanding and impractical for industrial applications (Nagarajan et al., 2020). Despite achieving ~30% higher H_2 yields compared to untreated biomass, the process faces limitations in economic viability due to high energy costs associated with the pretreatment (Nagarajan et al., 2020).

Chemical pretreatment of microalgal biomass, typically utilizing strong acids or alkalis, is a common method for promoting cell disruption and enhancing carbohydrate hydrolysis. For instance, sulfuric acid facilitates hydrolysis by breaking down cell wall polymers, releasing valuable intracellular compounds, while sodium hydroxide saponifies the membrane and cell wall lipids, disintegrating the cellular structure (Nagarajan et al., 2020).

The aforementioned treatments often combine with hydrothermal processes to enhance sugar recovery, as demonstrated with *Nannochloropsis oceanica*, a mild sulfuric acid pretreatment that increased fermentability and subsequent H_2 yield in dark fermentation (Xia et al., 2015). Despite being effective, chemical pretreatment could produce inhibitory compounds such as furfural, 5-hydroxymethylfurfural (HMF), acetate, and propionate, potentially impeding downstream fermentation processes. For instance, *Clostridium tyrobutyricum* undergoes inhibition at furfural concentrations as low as 0.3 g/L and HMF at 0.6 g/L, reducing the potential of the production of oxidoreductases such as butanol (Y. Liu et al., 2018). However, certain strains depict resilience; *Thermoanaerobacterium thermosaccharolyticum* can tolerate up to 725 mg/L of inhibitors, maintaining H_2 production from sugarcane bagasse hydrolysate (Hu & Zhu, 2017).

Biological pretreatment employs enzymes to degrade microalgal cell walls, by avoiding harmful by-products and preserving fermentable sugars (Bhushan et al., 2023). Enzymatic hydrolysis, utilizing cellulases, amylases, and other enzymes, could break down cell wall polymers, enhancing sugar release (Bhushan et al., 2023). For instance, enzymatic pretreatment of *C. vulgaris* biomass with a mixture of cellulases and amylases achieved a H_2 generation of 135 mL H_2/g VS, signifying a 7-fold increase compared to untreated biomass (Wieczorek et al., 2014). Although enzymatic methods are ecofriendly and highly specific, they are often constrained by high costs and the requirement for tailored enzyme combinations (S. Sharma et al., 2022). Using bacterial or fungal crude enzymes could offer a cost-effective alternative, although optimization for specific microalgal strains is necessary (Prajapati et al., 2015).

Therefore, the efficiency of biohydrogen synthesis from microalgal biomass could be significantly enhanced by combining various pretreatment methods. For instance, integrating mechanical or chemical pretreatment with subsequent enzymatic treatment could improve carbohydrate release and H_2 yields, addressing the challenges posed by complex cell walls (Nagarajan et al., 2020). However, selecting an appropriate pretreatment strategy requires a careful evaluation of several critical factors such as cell disruption efficiency, carbohydrate release, energy demands, and overall cost-effectiveness. Despite their potential effectiveness, high-energy methods may prove economically unsustainable due to the substantial energy consumption associated with their implementation (Krishnamoorthy et al., 2022). Therefore, assessing operational and maintenance costs, as well as considering the impact on environmental effects, is essential for determining the long-term feasibility and sustainability of the system. The optimal pretreatment approach must align with the specific characteristics of the microalgal biomass while meeting the fermentation process requirements to enhance biohydrogen production and achieve sustainable outcomes. Current research continues to refine suitable pretreatment strategies, aiming to enhance the efficiency and feasibility of biohydrogen synthesis from microalgal feedstocks. However, sustained research is vital to further optimize the strategies, addressing challenges and ensuring the practical application on a larger scale.

1.3 MECHANISMS OF BIOHYDROGEN PRODUCTION BY MICROALGAE

Understanding the mechanisms of biohydrogen production from microalgae is crucial for developing strategies to improve biohydrogen generation efficiency, improve energy yields, and address challenges related to the commercial viability of renewable energy applications (Anwar et al., 2019). The production of biohydrogen by microalgae can be achieved through two main approaches, involving direct H_2 production using microalgae and utilizing microalgal biomass as a feedstock for microbial fermentation-driven H_2 generation. Direct biohydrogen production is facilitated through the incorporation of photosynthetic microalgae to convert sunlight and water directly into molecular biohydrogen through enzyme activity, while indirect biohydrogen production involves the conversion of organic substrates into biohydrogen through metabolic pathways that are not directly linked to photosynthetic reactions (Jiao et al., 2024). Therefore, this section aims to discuss the biomechanisms involved in biohydrogen production in microalgae, emphasizing specific enzymes and their roles. Furthermore, a detailed comparison of the advantages and disadvantages associated with the production of biohydrogen is provided in Table 1.2, highlighting their respective benefits and challenges.

1.3.1 Photosynthetic Pathways for the Production of Biohydrogen

The key mechanism of H_2 production in microalgae is linked to photosynthesis, the photochemical conversion of light energy into stored chemical energy by microalgae. The microalgal photosynthesis process comprises two main stages: the light-dependent reactions, primarily associated with biohydrogen production, and the light-independent reactions, commonly known as the Calvin cycle (J. Zhang et al., 2023; Q. Zhang, 2021).

Direct Photolysis: In the process of direct photolysis, chlorophyll molecules within the photosystems integrate into the thylakoid membranes of chloroplasts, absorbing light energy. The two main photosystems involved are: PSII and photosystem I (PSI) (Suresh et al., 2023). PSII is primarily responsible for the process that splits water molecules into O_2, protons, and electrons, which is known as photolysis. The electrons generated are relayed through a series of

TABLE 1.2 Advantages and disadvantages of various biohydrogen production mechanisms

TYPE OF MECHANISM	*ADVANTAGES*	*DISADVANTAGES*	*REFERENCES*
Direct photolysis	• Readily accessible resources such as water and sunlight • Requires low light intensity • Simple process without many intermediary steps	• Hydrogenase enzyme is highly oxygen-sensitive, limiting H_2 yield • Difficult to maintain optimal conditions for high H_2 production	Penconi et al. (2015), Suresh et al. (2023), and Xuan et al. (2023)
Indirect photolysis	• Separates O_2 and H_2 production, reducing O_2 inhibition of hydrogenase • Potential for higher H_2 yields than direct photolysis (~1 compared to that of ~10^{-3} mol H_2/mol substrate)	• Requires a two-step process, increasing complexity • Relies on carbohydrate synthesis and breakdown, making the process less direct and slower	Show et al. (2012) and J. Zhang et al. (2023)
Photo-fermentation	• Utilizes solar energy efficiently in the presence of light • Able to fully convert organic compounds to H_2 • High H_2 conversion yield (0.5–2 mol H_2/mol substrate) by nitrogenase enzyme in bacteria	• Necessity of pretreatment • Light dependency can restrict operation times	Ahmed et al. (2021) and Sarangi and Nanda (2020)
Dark fermentation	• Less sensitive to O_2 than photosynthetic pathways • Low-energy requirements • High yields of H_2 production (2–4 mol H_2/mol substrate) • Economically feasible	• Thermodynamic limitations • Incomplete degradation of organic substrates • Inefficient substrate conversion	Fakhimi et al. (2020), Iqbal et al. (2022), and Jiao et al. (2024)

H_2, Hydrogen; O_2, Oxygen.

carriers in the electron transport chain (ETC) to PSI, re-energizing the electrons through light absorption (Vinyard & Brudvig, 2017). The energized electrons from PSI are subsequently utilized to reduce ferredoxin, a small iron-sulfur protein typically found in the stroma of chloroplasts. Consequently, the production of biohydrogen in microalgae occurs when electrons from reduced ferredoxin are transferred to protons by [FeFe]-hydrogenase enzymes, resulting in biohydrogen formation. Similarly, [NiFe]-hydrogenases can also catalyze H_2 production, though they generally exhibit lower activity in microalgae and are more sensitive to O_2, which limits their efficiency in biohydrogen generation under aerobic conditions. The overall reaction for direct photolysis is denoted in Equation 1.1 (Sung et al., 2022).

$$2H_2O + \text{Light energy} \rightarrow 2H_2 + O_2 \tag{1.1}$$

The photolysis of water is a critical step in the production of biohydrogen where light energy is absorbed by PSII, which in turn excites electrons in the reaction (Penconi et al., 2015). The high-energy electrons generated are transmitted through the ETC to PSI, while the photolysis of water replaces the electrons lost by PSII. The oxygen-evolving complex (OEC) within the PSII catalyzes the splitting of water, producing molecular O_2, electrons, and protons as denoted in Equation 1.2 (Pisciotta et al., 2011).

$$2H_2O + \text{Sunlight} \rightarrow 4H^+ + 4e^- + O_2 \quad (1.2)$$

The generated electrons are used to reduce ferredoxin, which subsequently transfers electrons to [FeFe]-hydrogenase enzymes, enabling the production of biohydrogen. Simultaneously, the resulting protons participate in biohydrogen gas formation, as illustrated in Equation 1.3 (Pisciotta et al., 2011).

$$4H^+ + 4e^- \rightarrow 2H_2 \quad (1.3)$$

The direct photolytic pathway is theoretically highly efficient because it enables the direct conversion of light energy into H_2, allowing for maximum energy transfer without intermediate reactions or energy carriers. However, the presence of O_2, a by-product of water photolysis, can inhibit hydrogenase enzyme activity, thereby restricting H_2 production (Xuan et al., 2023). An in-depth explanation of production by microalgae through the direct bio-photolysis pathway is provided in Figure 1.2.

Indirect Photolysis: In the process of indirect photolysis, the electrons generated through water splitting are not directly employed in H_2 production, as biohydrogen production is temporarily dissociated from the photosynthetic ETC. The electrons are initially utilized for the synthesis of carbohydrates and the evolution of O_2 via photosynthesis. Subsequently, the synthesized carbohydrates are metabolized through cellular catabolic pathways to generate reducing equivalents to produce H_2 (J. Zhang et al., 2023). The process primarily comprises two steps: the fixation of CO_2 by photosynthesis and the decomposition of carbohydrates. Initially, microalgae convert solar energy into organic compounds, such as lipids and carbohydrates, producing O_2 as a by-product and increasing the amount of endogenous substrate. In the second phase, metabolic pathways oxidize the fixed carbon stored in carbohydrates, releasing a significant pool of electrons for downstream biohydrogen production (Show et al., 2019).

One of the main advantages of indirect photolysis is its capability to mitigate O_2 sensitivity issues, as O_2 generation and H_2 production take place in separate phases of the process. The sequence of reaction

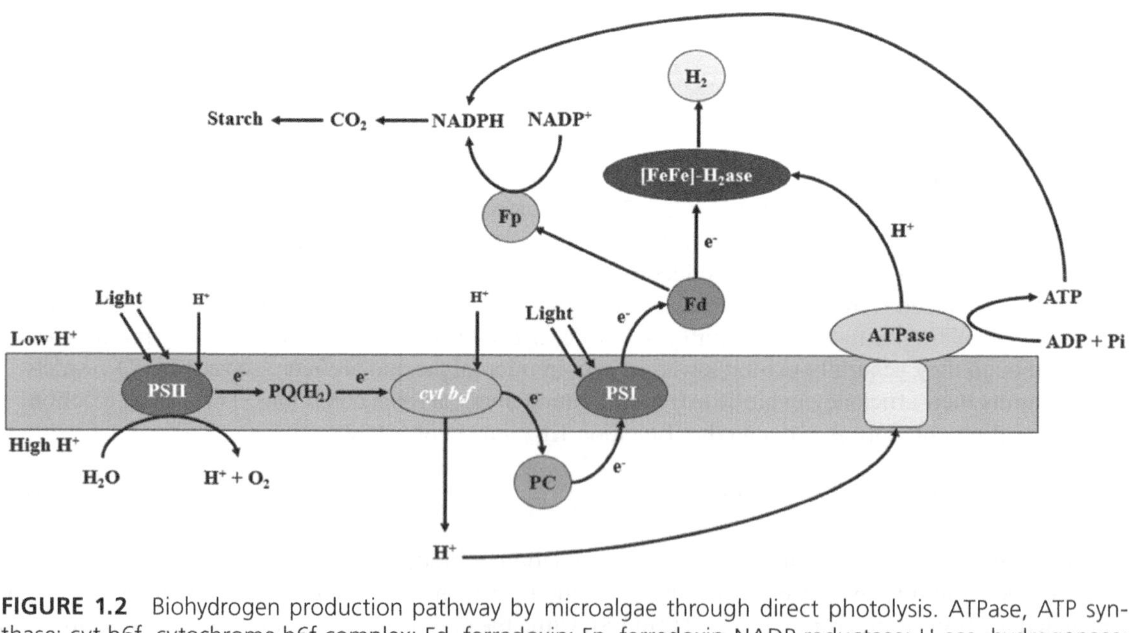

FIGURE 1.2 Biohydrogen production pathway by microalgae through direct photolysis. ATPase, ATP synthase; cyt b6f, cytochrome b6f complex; Fd, ferredoxin; Fp, ferredoxin-NADP reductase; H_2ase, hydrogenase; PC, plastocyanin; PQ, plastoquinones; PQ (H_2), reduced plastoquinones; PS, photosystem. (Modified from Lam et al. (2019).)

steps that facilitate biohydrogen production through indirect photolysis is detailed in Equations 1.4 and 1.5 (Ahmed et al., 2021).

$$\text{Aerobic reaction: } 12H_2O + 6CO_2 + \text{Solar energy} \rightarrow C_6H_{12}O_6 + 6O_2 + 6H_2O \quad (1.4)$$

$$\text{Anaerobic reaction: } C_6H_{12}O_6 + 12H_2O + \text{Solar energy} \rightarrow +6CO_2 + 12H_2 + 6H_2O \quad (1.5)$$

Key enzymes drive the metabolic shift, notably fermentative enzymes such as pyruvate-ferredoxin oxidoreductase (PFOR), which facilitate electron transfer crucial for activating hydrogenase activity in anaerobic environments. The ability to adjust the process for high H_2 yields, coupled with metabolic flexibility, is pivotal for optimizing biohydrogen production on an industrial scale. Moreover, leveraging microalgae's ability to function under both aerobic and anaerobic conditions offers a flexible and promising system for sustainable biohydrogen production. Additionally, the generation of valuable co-products, including O_2 and bioactive compounds, during the aerobic phase may further improve the economic feasibility of the process (Anwar et al., 2019).

1.3.2 Fermentative Pathways for Indirect Biohydrogen Production

In addition to the direct photosynthetic pathways, biohydrogen can also be produced through the fermentation of microalgae under anaerobic conditions, as a form of indirect biohydrogen production. During fermentative biohydrogen generation, microalgae utilize stored organic compounds as substrates, where production occurs predominantly in the dark. Furthermore, fermentative biohydrogen generation provides notable benefits, such as promoting environmental sustainability, reducing pollution, and lowering dependence on fossil fuels (Ahmed et al., 2021). Moreover, the fermentation processes, primarily categorized into dark fermentation and photo-fermentation, offer significant potential for biohydrogen production as a sustainable alternative to fossil fuels.

Dark Fermentation: During dark fermentation, microalgae metabolize stored carbohydrates, such as starch, when light is not available. Biohydrogen production through dark fermentation is affected by numerous variables, which include the rate of substrate disintegration, anaerobic strains, and the microbial utilization of the substrate (Sarangi & Nanda, 2020). Moreover, the breakdown of the organic compounds will result in the generation of pyruvate via glycolysis, followed by further metabolism that leads to biohydrogen production. Moreover, fermentative enzymes, including hydrogenase and PFOR, play a key role in influencing the process (Ahmed et al., 2021). Pyruvate is oxidatively decarboxylated by PFOR to generate acetyl-CoA, CO_2, and reduced ferredoxin. The ferredoxin, which was reduced, then transfers electrons to hydrogenase, resulting in the production of H_2 gas as depicted in Equation 1.6 (Sarangi & Nanda, 2020).

$$\text{Pyruvate} + \text{CoA} + 2\text{Fd}_{ox} \rightarrow \text{Acetyl-CoA} + CO_2 + 2\text{Fd}_{red} \quad (1.6)$$

Generally, dark fermentation exhibits lower sensitivity to O_2 compared to the photosynthetic pathway, enabling it to proceed effectively in anoxic or low-light environments, where light-dependent processes are not feasible (Zarei et al., 2024). The primary challenge in industrial H_2 production lies in the suboptimal biohydrogen yields, which are mainly attributed to the inherent limitations associated with microalgal biology. H_2 production via dark fermentation can be enhanced by stimulating microbial metabolic activity and improving electron transfer efficiency during the fermentation process (J. Zhang et al., 2023).

The integration of genetic engineering, metabolic engineering, and molecular biology is crucial to overcome the limitations and achieve improved H_2 production.

Photo-Fermentation: Photo-fermentation, facilitated by nitrogenase enzymes and primarily performed by purple non-sulfur photosynthetic bacteria, involves microalgae harnessing solar energy for the conversion of organic carbon molecules to H_2 and CO_2 in the presence of light (Jiao et al., 2024). Additionally, photo-fermentation is considered the most efficient for H_2 production due to its ability to achieve a high conversion yield from substrates (Iqbal et al., 2022).

Microalgae are regarded as a promising candidate for photo-fermentation due to their rapid growth rate, capacity for CO_2 fixation, and potential to produce a diverse array of organic compounds. In contrast to other forms of fermentation, photo-fermentation is not subjected to the inhibitory effects of O_2 on the nitrogenase enzyme (Jiao et al., 2024). Moreover, integrating photosynthetic organisms that fix carbon with H_2-producing bacteria facilitates the development of synergistic systems, where the photosynthetic organisms capture CO_2, while the bacteria concurrently produce H_2, thereby achieving both carbon sequestration and renewable H_2 production in a unified process. This collaboration provides a sustainable pathway to achieve biohydrogen yields of up to 3–4 mol H_2 per mol of water in photo-fermentation processes, while simultaneously mitigating carbon emissions and enhancing renewable energy production (Fakhimi et al., 2020).

1.3.3 Role of Hydrogenase and Nitrogenase Enzymes in the Biohydrogen Production

Hydrogenase and nitrogenase enzymes are considered to be the key cellular machinery responsible for catalyzing both direct and indirect biohydrogen production pathways in microalgae. Hydrogenase is primarily found in green microalgae and cyanobacteria, whereas nitrogenase, capable of producing significant amounts of H_2 under anaerobic conditions, is found exclusively in cyanobacteria (Khetkorn et al., 2017). Furthermore, hydrogenases and nitrogenases facilitate the proton reduction process, leading to the generation of molecular H_2.

Hydrogenase is the primary enzyme responsible for catalyzing biohydrogen in microalgae. The two primary classes of hydrogenases relevant to biohydrogen production in microalgae are [FeFe]-hydrogenases and [NiFe]-hydrogenases (Xuan et al., 2023). [FeFe]-hydrogenases are predominantly related with the production of biohydrogen in green microalgae such as *C. reinhardtii*, where the enzymes play a crucial role in photobiological H_2 production. These enzymes primarily function in the chloroplasts under anaerobic or sulfur-deprived conditions, directly coupling to the photosynthetic ETC. Notably, [FeFe]-hydrogenases have an activity nearly 10–100 times more than [NiFe]-hydrogenases and exceed 1,000 times than nitrogenases (S. Li et al., 2022). Moreover, [FeFe]-hydrogenases catalyze the activation of H_2 through the reversible reaction denoted in Equation 1.7.

$$\text{Activation of hydrogen: } H_2 \leftrightarrow 2H^+ + 2e^- \tag{1.7}$$

Consequently, electrons derived from the splitting of water molecules are employed during photosynthesis, thereby directly linking the production of biohydrogen to solar energy conversion (Xuan et al., 2023). However, the primary limitation of [FeFe]-hydrogenases is their extreme sensitivity to O_2, which is a by-product of the photosynthetic process. Since microalgae are oxygenic photosynthetic organisms, the simultaneous production of O_2 during water splitting inhibits the function of the hydrogenase enzymes. Therefore, the sensitivity to O_2 poses a significant challenge to the sustained generation of biohydrogen in microalgae under normal growth conditions (Hemschemeier et al., 2023).

Despite being less prevalent in microalgae, [NiFe]-hydrogenases are widely studied in cyanobacteria, due to their mutual photosynthetic characteristics with microalgae. Compared to [FeFe]-hydrogenases, [NiFe]-hydrogenases are more tolerant to O_2, despite being less efficient in the production of biohydrogen. Although the role of [NiFe]-hydrogenase enzymes in microalgal biohydrogen production is limited, [NiFe]-hydrogenases offer the potential for developing more resilient biohydrogen production systems (Hemschemeier et al., 2023).

Nitrogenase enzymes, primarily known for their role in nitrogen (N_2) fixation, contribute to the production of biohydrogen as a by-product of N_2 reduction. Nitrogenase enzymes are commonly found in diazotrophic bacteria such as *Azotobacter* sp. and certain cyanobacteria (Esteves-Ferreira et al., 2017). Moreover, nitrogenase enzymes catalyze the reduction of atmospheric N_2 to ammonia (NH_3), with the integrated production of H_2 as depicted in Equation 1.8 (Rutledge & Tezcan, 2020).

$$\text{Nitrogen reduction}: N_2 + 8H^+ + 8e^- + 16ATP \rightarrow 2NH_3 + H_2 \tag{1.8}$$

The reaction is mediated by a complex enzyme system comprising the Fe protein and the MoFe protein. Despite being less efficient at producing biohydrogen compared to hydrogenase enzymes, nitrogenase enzymes are of significant interest, primarily due to the ability to function under aerobic conditions and the high environmental adaptability exhibited, including tolerance to varying temperatures, pH levels, and nutrient availability (Xuan et al., 2023).

1.4 FACTORS AFFECTING BIOHYDROGEN PRODUCTION FROM MICROALGAE

Biohydrogen generation from the valorization of microalgal biomass is a multi-parameter process, which is impacted by numerous factors throughout the production cycle. The following section discusses how light intensity, pH, temperature, substrate, nutrients, and H_2 partial pressure (HPP) influence the microalgal growth and metabolic activity to enhance the biohydrogen yield. Furthermore, Figure 1.3 provides an overview of the key factors that must be considered and optimized to achieve cost-effective and sustainable biohydrogen production.

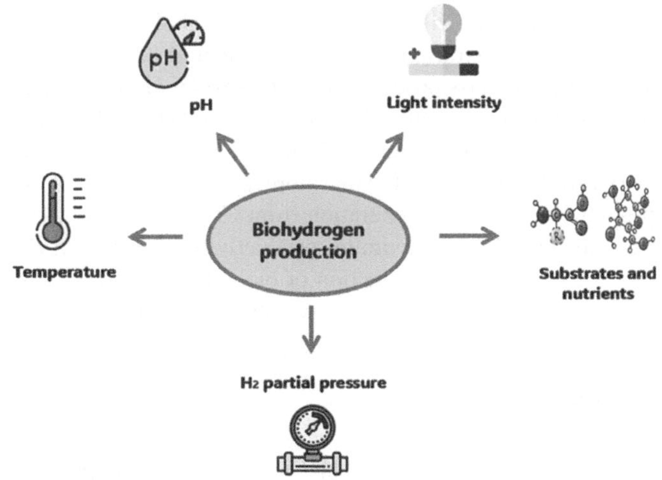

FIGURE 1.3 Factors affecting biohydrogen production from microalgae.

1.4.1 Light Intensity

Light intensity plays a critical role in biohydrogen production from microalgae by directly influencing photosynthesis and metabolic pathways. For most microalgae, the rate of photosynthesis is maximized when the light intensity reaches 100–500 μE/m²/s with maximum productivity occurring at average irradiance levels between 50 and 100 μE/m²/s (Chowdury et al., 2020). When light intensity exceeds optimal levels, microalgae absorb photons at a rate that surpasses photosynthetic capacity, resulting in a light saturation effect. Consequently, photosynthetic efficiency decreases due to the accumulation of surplus photons, which cannot be effectively utilized. Moreover, prolonged exposure to higher light intensity can damage the photosynthetic system, particularly PSII, through photoinhibition. This results in reduced photosynthetic efficiency and limits the growth rate of microalgae, ultimately restricting biohydrogen production (Pathy et al., 2022). Hwang and Lee (2021) investigated the influence of light intensity on microalgal biohydrogen production, reporting that *C. reinhardtii* and *Chlorella sorokiniana* achieved mean hydrogen yields of 108±4 μmol/L and 88±7 μmol/L, respectively, under an illumination of 100 μmol/m² s. Notably, at a lower light intensity of 50 μmol/m² s, *C. reinhardtii* exhibited a maximum light-to-hydrogen energy conversion efficiency of 1.61%, while *C. sorokiniana* recorded a peak efficiency of 1.06%, highlighting the strain-specific optimization potential for photobiological hydrogen production systems (Hwang & Lee, 2021). Therefore, the optimal lighting conditions for biohydrogen production could be achieved by adjusting light intensity during different growth phases, using light modulation techniques such as light/dark cycle or varying light exposure, selecting light-tolerant strains, and monitoring light conditions in real time to prevent photoinhibition.

1.4.2 pH

pH is a physical parameter affecting both growth and metabolic pathways related to biohydrogen production using microalgae. A slight change in the pH of the medium influences the activities of enzymes such as hydrogenase, involves biohydrogen production and shifts metabolic pathways, reducing H_2 yield, particularly under acidic conditions (Chowdury et al., 2020). For instance, at low pH, typically below 6, the activity of hydrogenase is inhibited, leading to the accumulation of by-products such as acetate and ATP, reducing biohydrogen production efficiency. On the other hand, pH values greater than 9 hinder photosynthesis due to the formation of carbonate ions, which are scarcely utilized by microalgae for growth (Min Woon et al., 2023).

Microalgal growth and biohydrogen production are generally optimized within a pH range of 6–9, although the exact pH requirements may differ based on the metabolic characteristics and environmental adaptability of the specific microalgal strain. For instance, Qiu et al. (2017) reported that a pH of 6 in the culture medium is optimal for the biomass productivity of *C. sorokiniana*, a species with significant potential for biohydrogen production (Qiu et al., 2017). Moreover, Rashid et al. (2013) reported that *Chlorella* sp. exhibited an optimal biohydrogen production rate of 24 mL/L/h at pH 8 when fructose was utilized as the primary carbon substrate (Rashid et al., 2013). Furthermore, Leong et al. (2024) implemented an automatic pH control strategy to promote biohydrogen generation from *C. vulgaris* via dark fermentation, achieving a biohydrogen yield of 221 mL/g VS and maximum productivity of 391.1 mL/L/h (Leong et al., 2024). Consequently, the media varies during the production of biohydrogen and understanding both initial and operational pH is critical, as it depends on the microalgae strain and characteristics of the substrate used.

1.4.3 Temperature

Temperature is a critical factor, as it directly affects both the biomass production rate of microalgae and the activity of hydrogenase enzymes responsible for biohydrogen generation. The most commonly used microalgae in biohydrogen production, including *Chlorella* sp., *C. reinhardtii*, and *Scenedesmus* sp., exhibit optimal growth within a temperature range of 15°C –35°C (Dolganyuk et al., 2020). However, exceeding the optimal temperature range leads to enzyme denaturation, reduced metabolic efficiency, and inhibited microalgal growth (Ayub et al., 2024). For instance, ribulose-1,5-bisphosphate carboxylase (Rubisco), a key enzyme in photosynthesis, becomes less effective at elevated temperatures, resulting in a decline in biomass productivity, inducing a reduction in the biohydrogen yield (Min Woon et al., 2023).

In addition to the biomass growth rate, hydrogenase enzymes involved in biohydrogen production exhibit an optimal temperature range at which their catalytic activity is maximized. Xiao et al. (2013) investigated the effect of temperature on biohydrogen production by *Klebsiella pneumoniae* ECU-15 and the results demonstrated that cell growth, biohydrogen production, and specific H_2 productivity increased up to 37°C with the maximum value of 5.5 mg/L, 5363.8 mL/L, 8.9 mmol/g cell/h, respectively. Exceeding 37°C resulted in a decline in these parameters, highlighting the importance of maintaining the temperature within the optimal range for each specific microalga to maximize biohydrogen yield (Xiao et al., 2013).

1.4.4 Substrates and Nutrients

A substrate functions as a source of organic or inorganic compounds that supply the necessary nutrients and energy to sustain cellular growth and drive metabolic pathways critical for biohydrogen production. The substrate requirements for microalgal cultures are governed by the specific growth mode, such as heterotrophy, mixotrophy, or autotrophy. For instance, Jiménez-Llanos et al. (2020) demonstrated that *Chlorella* sp. is capable of H_2 production by utilizing both endogenous and exogenous substrates via heterotrophic mode (Jiménez-Llanos et al., 2020). Starch, synthesized during the light phase of microalgal photosynthesis, serves as the primary endogenous substrate and is utilized to supply electrons for biohydrogen production. Additionally, exogenous carbon sources, including glucose, fructose, sucrose, and acetate, can significantly augment microalgal growth and optimize the metabolic pathways involved in biohydrogen synthesis. Moreover, industrial wastewater and organic waste materials, including agricultural residues such as corn stalks and sugarcane molasses, offer a sustainable and cost-effective alternative to conventional carbon sources. For example, a study by Min Woon et al. (2023) reported that the *C. vulgaris* MSU 01 strain generated 220 mL/L of H_2 by utilizing corn stalk at the concentration of 4 g/L as the substrate (Min Woon et al., 2023). The utilization of wastewater and organic residues facilitates biomass valorization by converting waste into energy, while simultaneously contributing to waste remediation.

Additionally, the availability of key nutrients, particularly nitrogen and phosphorus, is critical for maintaining robust growth and supporting optimal metabolic activity, directly influencing the efficiency of biohydrogen production. Nitrogen facilitates protein synthesis, while phosphorus is critical for energy transfer and cellular processes (Singh & Das, 2020). Moreover, micronutrients, particularly metal ions including iron (Fe), magnesium (Mg), zinc (Zn), and nickel (Ni), facilitate the activation of enzymes and co-enzymes related to microalgal growth and biohydrogen production. For instance, Fe and Ni activate co-elements of [FeFe]-hydrogenase and [NiFe]-hydrogenase enzymes, responsible for catalyzing the oxidative production of H_2. Furthermore, Mg and Zn are involved in the transportation of molecules across the membrane and act as co-factors for enzymes in biohydrogen production (Jiao et al. 2024; Goria et al., 2024). Additionally, sulfur deprivation is another critical factor that could stress the microalgal culture during bio-photolysis, leading to reduced photosynthetic activity and carbon fixation resulting in lower H_2 production rates (Min Woon et al., 2023).

1.4.5 H_2 Partial Pressure

In fermentative biohydrogen production, HPP greatly influences H_2 yield, as the accumulation of H_2 in the culture medium inhibits [FeFe]-hydrogenase activity. Ayodele et al. (2023) reported the influence of HPP on biohydrogen yield using a custom-designed anaerobic bio-disc reactor with a selected strain of chemotrophic *Clostridium* spp. The study indicated that increasing the total pressure by 0.18 bar resulted in a 19.5% reduction in H_2 yield, whereas a 0.11 bar decrease in pressure led to a 7% increase in H_2 yield (Ayodele et al., 2023).

Additionally, the reduction of HPP could be achieved by purging the bioreactor headspace with gases such as CO_2. Ayodele et al. (2023) reported that purging the headspace with CO_2 at a controlled flow rate of 300 mL/min maximized H_2 yield, reaching an optimal value of 1.68 mol H_2/mol of the substrate. The CO_2 purging strategy effectively reduced the HPP, facilitating a more stable and efficient fermentation process conducive to continuous H_2 production (Ayodele et al., 2023). Therefore, purging the headspace with CO_2 reduces HPP, alleviates product inhibition on key metabolic enzymes, and enables microorganisms to maintain efficient biohydrogen production, resulting in maximized biohydrogen yield.

1.5 CHALLENGES AND FUTURE PERSPECTIVES

Biohydrogen production from the valorization of microalgal biomass has emerged as a promising and sustainable solution for renewable energy generation. However, the production process is met with several technical, economic, and environmental challenges that must be addressed to achieve viability for large-scale commercialization (Salakkam et al., 2021). Although microalgae offer several advantages, the process of converting microalgal biomass into biohydrogen on an industrial scale remains complex. From issues related to the composition of the microalgal biomass to the energy demands of pretreatment processes, these barriers hinder the efficiency, scalability, and economic feasibility of biohydrogen production systems (Velmozhina et al., 2023). Therefore, the aim of this section is to explore key technical challenges and outline potential advancements to address limitations, aiming to improve the viability of biohydrogen as a sustainable energy source. Table 1.3 summarizes the challenges and potential solutions for microalgae-based biohydrogen production.

1.5.1 Economic Viability

One of the primary challenges in biohydrogen production from microalgae is the economic feasibility of the technologies required for large-scale implementation. The high costs associated with the upstream and downstream processing of microalgal biomass for biohydrogen production pose a significant barrier to commercialization (Jiao et al., 2024). Several factors contribute to these high costs, including the energy and resource demands of cultivation, harvesting, and pretreatment, as well as the costs of bioreactor construction and maintenance. Furthermore, downstream processes such as H_2 purification, storage, and distribution significantly increase the economic costs, limiting its viability as a widely accessible energy solution (Jiao et al., 2024). As a result, the overall cost of the biohydrogen production process remains economically unfeasible in comparison to other H_2 production methods, such as steam methane reforming or renewable energy-powered electrolysis (Mullen et al., 2023). The estimated cost of biohydrogen production through direct photolysis is approximately USD 7.24/kg, while the cost of indirect photolysis and fermentation processes is slightly higher, at USD 7.54/kg and USD 7.61/kg, respectively (Nirmala et al., 2023). In comparison, the cost of producing H_2 is USD 2.4/kg from fossil fuels, USD 2/kg from carbon capture and storage, USD 4/kg from electrolyzed water, USD 3.6/kg from solar thermochemical methods, and USD 3.7/kg from wind. Therefore, the notable discrepancy in production costs indicates the need for in-depth research and targeted efforts to enhance the cost-effectiveness of biohydrogen production from microalgae.

One approach to progress the economic viability of biohydrogen production is to develop integrated systems that generate multiple products from microalgal biomass. In addition to H_2, microalgae can be used to produce high-value co-products such as bio-oils, pigments, or biofertilizers, which create

TABLE 1.3 Challenges and potential solutions for microalgae-based biohydrogen production

CHALLENGES	*POTENTIAL SOLUTIONS*	*REFERENCES*
High production costs of biohydrogen from microalgae	• Develop integrated biorefinery systems to produce high-value co-products (bio-oils, pigments, biofertilizers) to offset costs • Optimize cultivation and pretreatment processes • Innovate in bioreactor design to lower costs	Nobre et al. (2013)
Energy-intensive cultivation and processing of microalgal biomass	• Use open pond systems to reduce energy demands • Invest in energy-efficient harvesting techniques (flocculation, membrane filtration) • Optimize downstream processing to reduce energy consumption	Novoveská et al. (2023)
Inconsistent H_2 production due to O_2 inhibition in photolysis	• Engineer algal strains with enhanced O_2 tolerance • Modify metabolic pathways to disrupt O_2, H_2 cycles • Explore bacterial co-cultures to consume O_2 and maintain anaerobic conditions	Ban et al. (2018)
Metabolic limitations of microalgae for sustained H_2 production	• Employ genetic engineering to introduce oxygen-tolerant hydrogenase enzymes • Optimize carbon storage pathways • Enhance overall photosynthetic efficiency • Use synthetic biology to improve H_2 yield	J. Zhang et al. (2023)
Technological challenges in scaling up production	• Develop hybrid or multi-stage photobioreactors with advanced light management (LED lighting, optical light guides) • Improve fluid dynamics for better light and nutrient distribution • Design reactors to maximize energy efficiency	Benner et al. (2022)
Low energy return on investment (EROI)	• Investigate energy-efficient conversion methods (biochemical fermentation) • Explore integrated systems like IBHR to maximize energy output while minimizing input energy	Redwood et al. (2012)

IBHR, Integrated Biohydrogen Refineries.

additional revenue streams that help offset H_2 production costs. For example, Nobre et al. (2013) evaluated the potential of biohydrogen production from carotenoids-extracted *Nannochloropsis* biomass through dark fermentation. The study resulted in the generation of 60.6 mL H_2/g residual biomass denoting the potential of improving the economic viability of biohydrogen production through integrated biorefinery approaches that capitalize on the simultaneous extraction of valuable co-products (Nobre et al., 2013).

Furthermore, intensive work on the optimization of H_2 yield needs to be conducted with regard to biohydrogen facility, operation, and materials costs. For example, the economic feasibility and performance of a scaled-up pilot modular photoreactor were evaluated over a 6-month field operation to assess its viability for large-scale biohydrogen production. The design analysis provided useful insights into the comparative costs of key components such as chemicals supplied as nutrients, materials, labor, water consumption, land leasing, energy, and other operational expenses for a commercial-scale photoreactor. Moreover, the cost analysis revealed that out of the total construction cost of USD 0.75/m², fabrication materials and nutrient-enriched chemicals constituted the largest expenditures, accounting for 84% of the total costs (Show et al., 2012). Additionally, conducting techno-economic analysis (TEA) studies of various H_2 generation processes aids in identifying the most cost-effective biohydrogen production methods. For instance, a TEA conducted by Ganeshan et al. (2023) reported that, among various biohydrogen production techniques, gasification proved to be the most economical at a cost of USD 1.2/kg of H_2, followed closely by anaerobic digestion at USD 1.25/kg of H_2 (Ganeshan et al., 2023). Ultimately, addressing the current cost challenges through technological improvements and integrated biorefineries will be critical in transforming microalgal biohydrogen production into a commercially feasible renewable energy source.

1.5.2 Metabolic Limitations

A key characteristic of direct photolysis is its utilization of water as the primary substrate and solar energy as the energy source for the reaction. As a result, direct photolysis becomes an attractive method for biohydrogen production, as both essential reactants including water and sunlight are abundant. Nevertheless, the process of direct photolysis faces major challenges in practical applications, H_2 generation primarily being inconsistent (Show et al., 2019). As discussed in Section 1.3, alongside H_2, photosynthetic microorganisms also generate molecular O_2, which is an effective inhibitor of the [Fe]-hydrogenase enzyme. Although initial biohydrogen production rates remain high under light exposure for approximately 60 hours, O_2 accumulation progressively suppresses hydrogenase activity, leading to a decline in H_2 yield and eventual cessation of production (Jiao et al., 2024). Therefore, biohydrogen from direct photolysis is generated transiently with low (<1.5%) photochemical conversion efficiency. One approach to overcome the challenges of O_2 buildup in biohydrogen production involves engineering microalgal strains to disrupt the O_2-H_2 cycles or modifying the photosynthesis-respiration balance, thereby reducing O_2 sensitivity (Schumann et al., 2023).

Additionally, to address the metabolic constraints, genetic engineering strategies have been explored, including the introduction of alternative hydrogenase enzymes with higher O_2 tolerance. The enhancement of carbon storage pathways to decrease competition for reducing equivalents and the optimization of photosynthetic efficiency increase the overall energy available for biohydrogen production. Furthermore, gene manipulation of hydrogenase can improve enzyme activity and is suitable for utilization with both light and dark fermentation (J. Zhang et al., 2023). Biohydrogen production by microalgae is transient under natural conditions, primarily controlled by the cell's metabolic network, which dynamically responds to environmental factors and internal metabolic requirements. For example, a study on algal-bacterial co-cultures aimed at enhancing photolysis-mediated H_2 production by green algae was conducted by Ban et al. (2018). The bacterium *Pseudomonas* sp. strain D formed a syntrophic relationship with *C. reinhardtii*, where the bacteria rapidly consumed O_2, creating anaerobic conditions favorable for H_2 production (Ban et al., 2018). Furthermore, the algal-bacterial interaction also prevented chlorophyll degradation, increased starch accumulation, and maintained protein integrity, all of which contributed to improved H_2 yields. Moreover, *Pseudomonas* sp. strain D showed similar beneficial relationships with other algae species, including *Chlorella* and *Scenedesmus*. Despite these efforts, significant improvements remain necessary, requiring further research to optimize the metabolic aspects of direct photolysis for industrial-scale biohydrogen production.

1.5.3 Energy Assessment

The energy balance of the process is a critical factor in determining the feasibility of biohydrogen production from microalgal biomass. Achieving a positive energy return on investment (EROI) requires that the energy output from H_2 production exceeds the energy input throughout the entire process, from cultivation to bioresource conversion (Ahmed et al., 2021). Therefore, a thorough energy assessment must include all stages, comprising pretreatment, cultivation, harvesting, biomass processing, and the H_2 production phase itself.

Although pretreatment is crucial for optimizing downstream processes, such as extraction and conversion, by facilitating the disruption of complex biomass structures, it can be energy-intensive, adding to the overall energy consumption of the process (González-Gloria et al., 2021). Moreover, the commercial cultivation of microalgae, particularly in closed PBRs, is energy-intensive as artificial lighting, temperature regulation, mixing, and CO_2 supply contribute to substantial energy demands, often making the process less energy-efficient (Thevarajah et al., 2025). Although open pond systems reduce energy requirements, challenges including lower productivity, land use, and a greater risk of contamination remain prevalent (Novoveská et al., 2023). Moreover, the harvesting and processing of microalgae are considered to be the most energy-intensive stages due to the typically low biomass concentration achieved in traditional autotrophic cultivation (Coelho et al., 2014). Conventional methods such as centrifugation, filtration, and flocculation consume significant amounts of energy, negatively impacting the overall energy balance (Thevarajah et al., 2022). Additionally, the conversion of microalgae biomass to biohydrogen is a major factor affecting energy efficiency. Thermochemical methods including pyrolysis and gasification are energy-intensive due to the high temperatures required, whereas biochemical approaches, such as fermentation, are less energy-demanding but may deliver lower H_2 yields (Das et al., 2021).

Therefore, identifying conversion pathways that strike a balance between high H_2 yield and minimal energy consumption is essential for ensuring a favorable energy balance. To improve the energy efficiency of biomass pretreatment for biohydrogen production, hybrid methods combining mechanical and chemical pretreatment processes can be explored. For instance, integrating low-energy mechanical treatments with mild acid/alkali-based thermal pretreatment could reduce energy consumption while minimizing the generation of fermentation inhibitors, improving both extraction efficiency and sustainability of the process (Nagarajan et al., 2020). Moreover, minimizing energy consumption can be accomplished through the implementation of open pond systems to decrease energy requirements, alongside the adoption of energy-efficient harvesting methods such as flocculation and membrane filtration (Zhao et al., 2020). A study by Redwood et al. (2012) explored energy-efficient methods for biohydrogen production through an Integrated Biohydrogen Refinery (IBHR) system, which converted biomass into electricity using hydrothermal hydrolysis, extractive biohydrogen fermentation, and photobiological H_2 fermentation, for electricity generation in a fuel cell (Redwood et al., 2012). The selective separation of organic acids from waste-fed fermentations provided a suitable substrate for photo-fermentative H_2 production, significantly boosting total energy output up to 11 times. Furthermore, the energy balance of biohydrogen production through waste-fed fermentation outperformed anaerobic decomposition, solar, and wind power, excluding additional H_2 from electrochemical production. Despite these advancements, further research and developments are essential to fully realize the potential of enhancing energy efficiency and scalability of biohydrogen production using microalgae.

1.5.4 Technological Challenges

Scaling up biohydrogen production from microalgae presents major technological challenges, as the optimized conditions of lab-scale systems such as controlled light, nutrient, and CO_2 delivery often lead to reduced efficiencies when applied at industrial scales (Thevarajah et al., 2025). Large-scale PBRs pose issues including light attenuation, uneven nutrient distribution, and difficulties in temperature regulation (Abdur Razzak et al., 2024). Moreover, inefficient fluid dynamics in reactors can lead to issues such as variations in light intensity and nutrient concentration gradients, which hinder microalgal growth and reduce biohydrogen yields, posing a significant challenge to the overall scalability and viability of biohydrogen production.

Innovations in reactor design, such as hybrid or multi-stage PBRs, offer a promising solution, for enhancing microalgal growth while addressing the associated challenges. Advanced light management technologies, specifically LED lighting and optical light guides, can improve light distribution by channeling light to deeper regions of the culture, ensuring more consistent illumination throughout the reactor (Benner et al., 2022).

For example, Hu et al. (2017) developed a PBR for the mass culturing of *Dunaliella tertiolecta*, utilizing optimized internal LED lighting to enhance growth. The study resulted in achieving a maximum cell density of 1.88×10^3 cells/L, 67% of the cell density in ideal conditions, suggesting that adjusting LED spacing can further increase growth rates (J. Y. Hu & Sato, 2017). Moreover, enhanced fluid dynamics, achieved through optimized flow patterns and mixing strategies, ensures uniform distribution of nutrients and gases, preventing the establishment of nutrient imbalances and pH variations. Although these innovations in reactor design present viable solutions to current limitations in scale-up procedures, further research is essential to refine these technologies, ensuring they are cost-effective, energy-efficient, and capable of sustaining high biohydrogen yields at an industrial scale.

1.6 CONCLUSION

The production of biohydrogen from the valorization of microalgae biomass gained significant attention due to its compatibility with other renewable energy sources. Maximizing biohydrogen yield requires the implementation of advanced cultivation techniques, efficient harvesting strategies, effective pretreatment methods, and optimizing critical parameters. However, achieving improved biohydrogen yields requires overcoming challenges associated with the valorization of microalgal biomass, which necessitates technological advancements and collaborative efforts across multiple disciplines. The development of integrated systems for producing high-value co-products has the potential to address issues associated with high production costs with microalgae-based biohydrogen production. Moreover, employing genetic engineering approaches to introduce oxygen-tolerant hydrogenase enzymes enhances the metabolic pathways involved in biohydrogen generation, effectively addressing and eliminating existing metabolic limitations. Furthermore, the integration of open pond systems for the cultivation of microalgae is essential to reduce the energy demand of cultivation. Hence, overcoming these challenges is imperative for the successful valorization of microalgae biomass to produce biohydrogen at a commercially viable scale and contribute to a sustainable energy future.

REFERENCES

Abdur Razzak, S., Bahar, K., Islam, K. M. O., Haniffa, A. K., Faruque, M. O., Hossain, S. M. Z., & Hossain, M. M. (2024). Microalgae cultivation in photobioreactors: Sustainable solutions for a greener future. *Green Chemical Engineering*, *5*(4), 418–439. https://doi.org/10.1016/J.GCE.2023.10.004

Ahmed, S. F., Mofijur, M., Nahrin, M., Chowdhury, S. N., Nuzhat, S., Alherek, M., Rafa, N., Ong, H. C., Nghiem, L. D., & Mahlia, T. M. I. (2022). Biohydrogen production from wastewater-based microalgae: Progresses and challenges. *International Journal of Hydrogen Energy*, *47*(88), 37321–37342. https://doi.org/10.1016/j.ijhydene.2021.09.178

Ahmed, S. F., Rafa, N., Mofijur, M., Badruddin, I. A., Inayat, A., Ali, M. S., Farrok, O., & Yunus Khan, T. M. (2021). Biohydrogen production from biomass sources: Metabolic pathways and economic analysis. *Frontiers in Energy Research*, *9*, 753878. https://doi.org/10.3389/FENRG.2021.753878/

Anwar, M., Lou, S., Chen, L., Li, H., & Hu, Z. (2019). Recent advancement and strategy on bio-hydrogen production from photosynthetic microalgae. *Bioresource Technology*, *292*, 121972. https://doi.org/10.1016/J.BIORTECH.2019.121972

Ayodele, D. T., Ogunbiyi, O. D., Akamo, D. O., Otun, K. O., Akinpelu, D. A., Adegoke, J. A., Fapojuwo, D. P., & Oladoye, P. O. (2023). Factors affecting biohydrogen production: Overview and perspectives. *International Journal of Hydrogen Energy*, *48*(71), 27513–27539. https://doi.org/10.1016/j.ijhydene.2023.04.001

Ayub, H. M. U., Nizami, M., Qyyum, M. A., Iqbal, N., Al-Muhtaseb, A. H., & Hasan, M. (2024). Sustainable hydrogen production via microalgae: Technological advancements, economic indicators, environmental aspects, challenges, and policy implications. *Environmental Research*, *244*, 117815. https://doi.org/10.1016/J.ENVRES.2023.117815

Bahadar, A., Kanthasamy, R., Sait, H. H., Zwawi, M., Algarni, M., Ayodele, B. V., Cheng, C. K., & Wei, L. J. (2022). Elucidating the effect of process parameters on the production of hydrogen-rich syngas by biomass and coal Co-gasification techniques: A multi-criteria modeling approach. *Chemosphere*, *287*, 132052. https://doi.org/10.1016/J.CHEMOSPHERE.2021.132052

Ban, S., Lin, W., Wu, F., & Luo, J. (2018). Algal-bacterial cooperation improves algal photolysis-mediated hydrogen production. *Bioresource Technology*, *251*, 350–357. https://doi.org/10.1016/J.BIORTECH.2017.12.072

Barros, A. I., Gonçalves, A. L., Simões, M., & Pires, J. C. M. (2015). Harvesting techniques applied to microalgae: A review. *Renewable and Sustainable Energy Reviews*, *41*, 1489–1500. https://doi.org/10.1016/J.RSER.2014.09.037

Benner, P., Meier, L., Pfeffer, A., Krüger, K., Oropeza Vargas, J. E., & Weuster-Botz, D. (2022). Lab-scale photobioreactor systems: Principles, applications, and scalability. *Bioprocess and Biosystems Engineering*, *45*(5), 791. https://doi.org/10.1007/S00449-022-02711-1

Bhatia, S. K., Rajesh Banu, J., Singh, V., Kumar, G., & Yang, Y.-H. (2023). Algal biomass to biohydrogen: Pretreatment, influencing factors, and conversion strategies. *Bioresource Technology*, *368*, 128332. https://doi.org/10.1016/j.biortech.2022.128332

Bhushan, S., Jayakrishnan, U., Shree, B., Bhatt, P., Eshkabilov, S., & Simsek, H. (2023). Biological pretreatment for algal biomass feedstock for biofuel production. *Journal of Environmental Chemical Engineering*, *11*(3), 109870. https://doi.org/10.1016/J.JECE.2023.109870

Bora, A., Mohanrasu, K., Angelin Swetha, T., Ananthi, V., Sindhu, R., Chi, N. T. L., Pugazhendhi, A., Arun, A., & Mathimani, T. (2022). Microbial electrolysis cell (MEC): Reactor configurations, recent advances and strategies in biohydrogen production. *Fuel*, *328*, 125269. https://doi.org/10.1016/J.FUEL.2022.125269

Castro-Muñoz, R., & García-Depraect, O. (2021). Membrane-based harvesting processes for microalgae and their valuable-related molecules: A review. *Membranes*, *11*(8), 585. https://doi.org/10.3390/MEMBRANES11080585

Chowdury, K. H., Nahar, N., Deb, U. K., Chowdury, K. H., Nahar, N., & Deb, U. K. (2020). The growth factors involved in microalgae cultivation for biofuel production: A review. *Computational Water, Energy, and Environmental Engineering*, *9*(4), 185–215. https://doi.org/10.4236/CWEEE.2020.94012

Coelho, R. S., Vidotti, A. D. S., Reis, É. M., & Franco, T. T. (2014). High cell density cultures of microalgae under fed-batch and continuous growth. *Chemical Engineering Transactions*, *38*, 313–318. https://doi.org/10.3303/CET1438053

Das, P., V.P., C., Mathimani, T., & Pugazhendhi, A. (2021). Recent advances in thermochemical methods for the conversion of algal biomass to energy. *Science of the Total Environment*, *766*, 144608. https://doi.org/10.1016/J.SCITOTENV.2020.144608

de Morais, E. G., Sampaio, I. C. F., Gonzalez-Flo, E., Ferrer, I., Uggetti, E., & García, J. (2023). Microalgae harvesting for wastewater treatment and resources recovery: A review. *New Biotechnology*, *78*, 84–94. https://doi.org/10.1016/J.NBT.2023.10.002

Deepa, P., Sowndhararajan, K., & Kim, S. (2023). A review of the harvesting techniques of microalgae. *Water*, *15*(17), 3074. https://doi.org/10.3390/W15173074

Dolganyuk, V., Belova, D., Babich, O., Prosekov, A., Ivanova, S., Katserov, D., Patyukov, N., & Sukhikh, S. (2020). Microalgae: A promising source of valuable bioproducts. *Biomolecules*, 10(8), 1153. https://doi.org/10.3390/biom10081153

Dudek, M., Dębowski, M., Nowicka, A., Kazimierowicz, J., & Zieliński, M. (2022). The effect of autotrophic cultivation of Platymonas subcordiformis in waters from the natural aquatic reservoir on hydrogen yield. *Resources*, *11*(3), 31. https://doi.org/10.3390/RESOURCES11030031

Esteves-Ferreira, A. A., Cavalcanti, J. H. F., Vaz, M. G. M. V., Alvarenga, L. V., Nunes-Nesi, A., & Araújo, W. L. (2017). Cyanobacterial nitrogenases: Phylogenetic diversity, regulation and functional predictions. *Genetics and Molecular Biology*, *40*(1), 261–275. https://doi.org/10.1590/1678-4685-GMB-2016-0050

Fakhimi, N., Gonzalez-Ballester, D., Fernández, E., Galván, A., & Dubini, A. (2020). Algae-bacteria consortia as a strategy to enhance H_2 production. *Cells*, 9(6), 1353. https://doi.org/10.3390/CELLS9061353

Ganeshan, P., Vigneswaran, V. S., Gowd, S. C., Kondusamy, D., Sanjay Kumar, C., Krishnamoorthy, N., Kumar, D., Juneja, A., Paramasivan, B., Raju, N. N., Rajendran, K., & Pugazhendhi, A. (2023). How does techno-economic analysis and lifecycle assessment help in commercializing the biohydrogen supply chain? *Fuel*, 341, 127601. https://doi.org/10.1016/J.FUEL.2023.127601

González-Gloria, K. D., Rodríguez-Jasso, R. M., Shiva, Aparicio, E., Chávez González, M. L., Kostas, E. T., & Ruiz, H. A. (2021). Macroalgal biomass in terms of third-generation biorefinery concept: Current status and techno-economic analysis – A review. *Bioresource Technology Reports*, *16*, 100863. https://doi.org/10.1016/J.BITEB.2021.100863

Goria, K., Singh, H. M., Singh, A., Kothari, R., & Tyagi, V. V. (2024). Insights into biohydrogen production from algal biomass: Challenges, recent advancements and future directions. *International Journal of Hydrogen Energy*, *52*, 127–151. https://doi.org/10.1016/j.ijhydene.2023.03.174

Hanifa, N. M., Thevarajah, B., Liyanaarachchi, V. C., Nimarshana, P. H. V., Boopathy, R., & Ariyadasa, T. U. (2025). Simultaneous bioremediation, resource recovery and natural dye production via textile wet processing wastewater-integrated microalgae cultivation: a review. Current Pollution Reports, 11(1). https://doi.org/10.1007/s40726-025-00363-9

Hemschemeier, A., Posewitz, M. C., & Happe, T. (2023). Hydrogenases and hydrogen production. In The Chlamydomonas Sourcebook: Volume 2: Organellar and Metabolic Processes (pp. 343–367). Academic Press. https://doi.org/10.1016/B978-0-12-821430-5.00008-0

Hidalgo, D., Martín-Marroquín, J. M., & Corona, F. (2023). Metal-based nanoadditives for increasing biomass and biohydrogen production in microalgal cultures: A review. *Sustainable Chemistry and Pharmacy*, *33*, 101065. https://doi.org/10.1016/j.scp.2023.101065

Hu, B., & Zhu, M. (2017). Direct hydrogen production from dilute-acid pretreated sugarcane bagasse hydrolysate using the newly isolated Thermoanaerobacterium thermosaccharolyticum MJ1. Microbial Cell Factories, 16(1). https://doi.org/10.1186/s12934-017-0692-y

Hu, J. Y., & Sato, T. (2017). A photobioreactor for microalgae cultivation with internal illumination considering flashing light effect and optimized light-source arrangement. *Energy Conversion and Management*, *133*, 558–565. https://doi.org/10.1016/J.ENCONMAN.2016.11.008

Huang, Q., Jiang, F., Wang, L., & Yang, C. (2017). Design of photobioreactors for mass cultivation of photosynthetic organisms. *Engineering*, *3*(3), 318–329. https://doi.org/10.1016/J.ENG.2017.03.020

Hwang, J.-H., & Lee, W. H. (2021). Continuous photosynthetic biohydrogen production from acetate-rich wastewater: Influence of light intensity. *International Journal of Hydrogen Energy*, *46*(42), 21812–21821. https://doi.org/10.1016/j.ijhydene.2021.04.052

Iqbal, K., Saxena, A., Pande, P., Tiwari, A., Chandra Joshi, N., Varma, A., & Mishra, A. (2022). Microalgae-bacterial granular consortium: Striding towards sustainable production of biohydrogen coupled with wastewater treatment. *Bioresource Technology*, *354*, 127203. https://doi.org/10.1016/J.BIORTECH.2022.127203

Jiao, H., Tsigkou, K., Elsamahy, T., Pispas, K., Sun, J., Manthos, G., Schagerl, M., Sventzouri, E., Al-Tohamy, R., Kornaros, M., & Ali, S. S. (2024). Recent advances in sustainable hydrogen production from microalgae: Mechanisms, challenges, and future perspectives. *Ecotoxicology and Environmental Safety*, *270*, 115908. https://doi.org/10.1016/J.ECOENV.2023.115908

Jiménez-Llanos, J., Ramírez-Carmona, M., Rendón-Castrillón, L., & Ocampo-López, C. (2020). Sustainable biohydrogen production by Chlorella sp. microalgae: A review. *International Journal of Hydrogen Energy*, *45*(15), 8310–8328. https://doi.org/10.1016/j.ijhydene.2020.01.059

Khetkorn, W., Rastogi, R. P., Incharoensakdi, A., Lindblad, P., Madamwar, D., Pandey, A., & Larroche, C. (2017). Microalgal hydrogen production – A review. *Bioresource Technology*, *243*, 1194–1206. https://doi.org/10.1016/J.BIORTECH.2017.07.085

Kim, K., Shin, H., Moon, M., Ryu, B. G., Han, J. I., Yang, J. W., & Chang, Y. K. (2015). Evaluation of various harvesting methods for high-density microalgae, Aurantiochytrium sp. KRS101. *Bioresource Technology*, *198*, 828–835. https://doi.org/10.1016/J.BIORTECH.2015.09.103

King, S. J., Jerkovic, A., Brown, L. J., Petroll, K., & Willows, R. D. (2022). Synthetic biology for improved hydrogen production in Chlamydomonas reinhardtii. *Microbial Biotechnology*, *15*(7), 1946–1965. https://doi.org/10.1111/1751-7915.14024

Krishnamoorthy, A., Rodriguez, C., Durrant, A., Krishnamoorthy, A., Rodriguez, C., & Durrant, A. (2022). Sustainable approaches to microalgal pre-treatment techniques for biodiesel production: A review. *Sustainability*, *14*(16), 9953. https://doi.org/10.3390/SU14169953

Kumar, R., Kumar, A., & Pal, A. (2021). An overview of conventional and non-conventional hydrogen production methods. *Materials Today: Proceedings*, *46*, 5353–5359. https://doi.org/10.1016/j.matpr.2020.08.793

Kurokawa, M., King, P. M., Wu, X., Joyce, E. M., Mason, T. J., & Yamamoto, K. (2016). Effect of sonication frequency on the disruption of algae. *Ultrasonics Sonochemistry*, *31*, 157–162. https://doi.org/10.1016/J.ULTSONCH.2015.12.011

Lam, M. K., Loy, A. C. M., Yusup, S., & Lee, K. T. (2019). Biohydrogen production from algae. In Biomass, Biofuels, Biochemicals: B*iohydrogen*, Second Edition (pp. 219–245). Elsevier. https://doi.org/10.1016/B978-0-444-64203-5.00009-5

Leong, Y. K., Su, R.-H., Varjani, S., Lee, D.-J., & Chang, J.-S. (2024). Integrated biohydrogen production and dairy manure wastewater treatment via a microalgae platform. *International Journal of Hydrogen Energy*, *52*, 404–417. https://doi.org/10.1016/j.ijhydene.2023.06.308

Li, S., Li, F., Zhu, X., Liao, Q., Chang, J. S., & Ho, S. H. (2022). Biohydrogen production from microalgae for environmental sustainability. *Chemosphere*, *291*, 132717. https://doi.org/10.1016/J.CHEMOSPHERE.2021.132717

Liu, H., Zhang, Z., Lu, C., Wang, J., Wang, K., Guo, S., & Zhang, Q. (2022). Effects of enzymatic hydrolysis and alkalization pretreatment on biohydrogen production by chlorella photosynthesis. *Bioresource Technology*, *349*, 126859. https://doi.org/10.1016/j.biortech.2022.126859

Liu, Y., Geng, Y., Zhou, Q., & Yuan, W. (2018). The effect of furfural and 5-hydroxymethyl furfural on butyric acid fermentation by Clostridium tyrobutyricum. *Journal of Chemical Technology & Biotechnology*, *93*(3), 849–854. https://doi.org/10.1002/JCTB.5439

Liyanaarachchi, V. C., Premaratne, M., Ariyadasa, T. U., Nimarshana, P. H. V., & Malik, A. (2021). Two-stage cultivation of microalgae for production of high-value compounds and biofuels: A review. *Algal Research*, *57*, 102353. https://doi.org/10.1016/j.algal.2021.102353

Luo, G., Talebnia, F., Karakashev, D., Xie, L., Zhou, Q., & Angelidaki, I. (2011). Enhanced bioenergy recovery from rapeseed plant in a biorefinery concept. *Bioresource Technology*, *102*(2), 1433–1439. https://doi.org/10.1016/J.BIORTECH.2010.09.071

Maswanna, T., Lindblad, P., & Maneeruttanarungroj, C. (2020). Improved biohydrogen production by immobilized cells of the green alga Tetraspora sp. CU2551 incubated under aerobic condition. *Journal of Applied Phycology*, *32*(5), 2937–2945. https://doi.org/10.1007/S10811-020-02184-3

Mathimani, T., & Mallick, N. (2018). A comprehensive review on harvesting of microalgae for biodiesel – Key challenges and future directions. *Renewable and Sustainable Energy Reviews*, *91*, 1103–1120. https://doi.org/10.1016/J.RSER.2018.04.083

Milledge, J. J., & Heaven, S. (2013). A review of the harvesting of micro-algae for biofuel production. *Reviews in Environmental Science and Biotechnology*, *12*(2), 165–178. https://doi.org/10.1007/s11157-012-9301-z

Min Woon, J., Shiong Khoo, K., Akermi, M., Alanazi, M. M., Wei Lim, J., Jing Chan, Y., Sean Goh, P., Silas Chidi, B., Kee Lam, M., Zaini, J., Roil Bilad, M., Zhou, Y., Tasnim Sahrin, N., & Musa Ardo, F. (2023). Reviewing biohydrogen production from microalgal cells through fundamental mechanisms, enzymes and factors that engendering new challenges and prospects. *Fuel*, 346, 128312. https://doi.org/10.1016/j.fuel.2023.128312

Molina Grima, E., Belarbi, E. H., Acién Fernández, F. G., Robles Medina, A., & Chisti, Y. (2003). Recovery of microalgal biomass and metabolites: Process options and economics. *Biotechnology Advances*, *20*(7–8), 491–515. https://doi.org/10.1016/S0734-9750(02)00050-2

Mubarak, M., Shaija, A., & Suchithra, T. V. (2019). Flocculation: An effective way to harvest microalgae for biodiesel production. *Journal of Environmental Chemical Engineering*, *7*(4), 103221. https://doi.org/10.1016/J.JECE.2019.103221

Mullen, D., Herraiz, L., Gibbins, J., & Lucquiaud, M. (2023). On the cost of zero carbon hydrogen: A techno-economic analysis of steam methane reforming with carbon capture and storage. *International Journal of Greenhouse Gas Control*, *126*, 103904. https://doi.org/10.1016/J.IJGGC.2023.103904

Nagarajan, D., Chang, J. S., & Lee, D. J. (2020). Pretreatment of microalgal biomass for efficient biohydrogen production – Recent insights and future perspectives. *Bioresource Technology*, *302*, 122871. https://doi.org/10.1016/J.BIORTECH.2020.122871

Nirmala, N., Praveen, G., AmitKumar, S., SundarRajan, P. S., Baskaran, A., Priyadharsini, P., SanjayKumar, S. P., Dawn, S. S., Pavithra, K. G., Arun, J., & Pugazhendhi, A. (2023). A review on biological biohydrogen production: Outlook on genetic strain enhancements, reactor model and techno-economics analysis. *Science of the Total Environment*, *896*, 165143. https://doi.org/10.1016/J.SCITOTENV.2023.165143

Nobre, B. P., Villalobos, F., Barragán, B. E., Oliveira, A. C., Batista, A. P., Marques, P. A. S. S., Mendes, R. L., Sovová, H., Palavra, A. F., & Gouveia, L. (2013). A biorefinery from Nannochloropsis sp. microalga – Extraction of oils and pigments. Production of biohydrogen from the leftover biomass. *Bioresource Technology*, *135*, 128–136. https://doi.org/10.1016/J.BIORTECH.2012.11.084

Novoveská, L., Nielsen, S. L., Eroldoğan, O. T., Haznedaroglu, B. Z., Rinkevich, B., Fazi, S., Robbens, J., Vasquez, M., & Einarsson, H. (2023). Overview and challenges of large-scale cultivation of photosynthetic microalgae and cyanobacteria. *Marine Drugs*, 21(8), 445. https://doi.org/10.3390/MD21080445

Nurra, C., Torras, C., Clavero, E., Ríos, S., Rey, M., Lorente, E., Farriol, X., & Salvadó, J. (2014). Biorefinery concept in a microalgae pilot plant. Culturing, dynamic filtration and steam explosion fractionation. *Bioresource Technology*, *163*, 136–142. https://doi.org/10.1016/J.BIORTECH.2014.04.009

Onumaegbu, C., Mooney, J., Alaswad, A., & Olabi, A. G. (2018). Pre-treatment methods for production of biofuel from microalgae biomass. *Renewable and Sustainable Energy Reviews*, *93*, 16–26. https://doi.org/10.1016/J.RSER.2018.04.015

Pathy, A., Nageshwari, K., Ramaraj, R., Pragas Maniam, G., Govindan, N., & Balasubramanian, P. (2022). Biohydrogen production using algae: Potentiality, economics and challenges. *Bioresource Technology, 360*, 127514. https://doi.org/10.1016/j.biortech.2022.127514

Penconi, M., Rossi, F., Ortica, F., Elisei, F., & Gentili, P. L. (2015). Hydrogen production from water by photolysis, sonolysis and sonophotolysis with solid solutions of rare earth, gallium and indium oxides as heterogeneous catalysts. *Sustainability, 7*(7), 9310–9325. https://doi.org/10.3390/SU7079310

Pisciotta, J. M., Zou, Y., & Baskakov, I. V. (2011). Role of the photosynthetic electron transfer chain in electrogenic activity of cyanobacteria. *Applied Microbiology and Biotechnology, 91*(2), 377–385. https://doi.org/10.1007/s00253-011-3239-x

Piyatilleke, S., Thevarajah, B., Nimarshana, P., & Ariyadasa, T. U. (2024). Microalgal biofuels: Challenges and prospective in the framework of circular bioeconomy. *Energy Nexus*, 100338. https://doi.org/10.1016/j.nexus.2024.100338

Postma, P. R., Suarez-Garcia, E., Safi, C., Olivieri, G., Olivieri, G., Wijffels, R. H., & Wijffels, R. H. (2017). Energy efficient bead milling of microalgae: Effect of bead size on disintegration and release of proteins and carbohydrates. *Bioresource Technology, 224*, 670–679. https://doi.org/10.1016/J.BIORTECH.2016.11.071

Pragya, N., Pandey, K. K., & Sahoo, P. K. (2013). A review on harvesting, oil extraction and biofuels production technologies from microalgae. *Renewable and Sustainable Energy Reviews, 24*, 159–171. https://doi.org/10.1016/J.RSER.2013.03.034

Prajapati, S. K., Bhattacharya, A., Malik, A., & Vijay, V. K. (2015). Pretreatment of algal biomass using fungal crude enzymes. *Algal Research, 8*, 8–14. https://doi.org/10.1016/J.ALGAL.2014.12.011

Premaratne, M., Nishshanka, G. K. S. H., Anthonio, R. A. D. P., Liyanaarachchi, V. C., Thevarajah, B., Nimarshana, P. H. V., Malik, A., & Ariyadasa, T. U. (2022). Resource recovery from waste streams for production of microalgae biomass: A sustainable approach towards high-value biorefineries. *Bioresource Technology Reports, 18*, 101070. https://doi.org/10.1016/j.biteb.2022.101070

Qiu, R., Gao, S., Lopez, P. A., & Ogden, K. L. (2017). Effects of pH on cell growth, lipid production and CO2 addition of microalgae Chlorella sorokiniana. *Algal Research, 28*, 192–199. https://doi.org/10.1016/j.algal.2017.11.004

Rady, H. A., Ali, S. S., & El-Sheekh, M. M. (2024). Strategies to enhance biohydrogen production from microalgae: A comprehensive review. *Journal of Environmental Management, 356*. https://doi.org/10.1016/j.jenvman.2024.120611

Rai, P. K., & Singh, S. P. (2016). Integrated dark- and photo-fermentation: Recent advances and provisions for improvement. *International Journal of Hydrogen Energy, 41*(44), 19957–19971. https://doi.org/10.1016/J.IJHYDENE.2016.08.084

Rashid, N., Lee, K., Han, J., & Gross, M. (2013). Hydrogen production by immobilized Chlorella vulgaris: Optimizing pH, carbon source and light. *Bioprocess and Biosystems Engineering, 36*(7), 867–872. https://doi.org/10.1007/s00449-012-0819-9

Redwood, M. D., Orozco, R. L., Majewski, A. J., & Macaskie, L. E. (2012). An integrated biohydrogen refinery: Synergy of photofermentation, extractive fermentation and hydrothermal hydrolysis of food wastes. *Bioresource Technology, 119*, 384–392. https://doi.org/10.1016/J.BIORTECH.2012.05.040

Roy, S., Kumar, K., Ghosh, S., & Das, D. (2014). Thermophilic biohydrogen production using pre-treated algal biomass as substrate. *Biomass and Bioenergy, 61*, 157–166. https://doi.org/10.1016/j.biombioe.2013.12.006

Rutledge, H. L., & Tezcan, F. A. (2020). Electron transfer in nitrogenase. *Chemical Reviews, 120*(12), 5158. https://doi.org/10.1021/ACS.CHEMREV.9B00663

Salakkam, A., Sittijunda, S., Mamimin, C., Phanduang, O., & Reungsang, A. (2021). Valorization of microalgal biomass for biohydrogen generation: A review. *Bioresource Technology, 322*, 124533. https://doi.org/10.1016/J.BIORTECH.2020.124533

Samarasinghe, N., Fernando, S., Lacey, R., & Faulkner, W. B. (2012). Algal cell rupture using high pressure homogenization as a prelude to oil extraction. *Renewable Energy, 48*, 300–308. https://doi.org/10.1016/J.RENENE.2012.04.039

Sarangi, P. K., & Nanda, S. (2020). Biohydrogen production through dark fermentation. *Chemical Engineering & Technology, 43*(4), 601–612. https://doi.org/10.1002/CEAT.201900452

Schumann, C., Fernández Méndez, J., Berggren, G., & Lindblad, P. (2023). Novel concepts and engineering strategies for heterologous expression of efficient hydrogenases in photosynthetic microorganisms. *Frontiers in Microbiology*, 14, 1179607. https://doi.org/10.3389/FMICB.2023.1179607

Scranton, M. A., Ostrand, J. T., Fields, F. J., & Mayfield, S. P. (2015). Chlamydomonas as a model for biofuels and bio-products production. *The Plant Journal : For Cell and Molecular Biology, 82*(3), 523. https://doi.org/10.1111/TPJ.12780

Sharma, A., & Arya, S. K. (2017). Hydrogen from algal biomass: A review of production process. *Biotechnology Reports, 15*, 63–69. https://doi.org/10.1016/J.BTRE.2017.06.001

Sharma, S., Tsai, M. L., Sharma, V., Sun, P. P., Nargotra, P., Bajaj, B. K., Chen, C. W., & Dong, C. Di. (2022). Environment friendly pretreatment approaches for the bioconversion of lignocellulosic biomass into biofuels and value-added products. *Environments*, *10*(1), 6. https://doi.org/10.3390/ENVIRONMENTS10010006

Show, K. Y., Lee, D. J., Tay, J. H., Lin, C. Y., & Chang, J. S. (2012). Biohydrogen production: Current perspectives and the way forward. *International Journal of Hydrogen Energy*, *37*(20), 15616–15631. https://doi.org/10.1016/J.IJHYDENE.2012.04.109

Show, K. Y., Yan, Y., Zong, C., Guo, N., Chang, J. S., & Lee, D. J. (2019). State of the art and challenges of biohydrogen from microalgae. *Bioresource Technology*, *289*, 121747. https://doi.org/10.1016/J.BIORTECH.2019.121747

Singh, H., & Das, D. (2020). Biohydrogen from microalgae. In Handbook of Microalgae-Based Processes and Products: Fundamentals and Advances in Energy, Food, Feed, Fertilizer, and Bioactive Compounds (pp. 391–418). Academic Press. https://doi.org/10.1016/B978-0-12-818536-0.00015-4

Stanislaus, M. S., Zhang, N., Yuan, Y., Zheng, H., Zhao, C., Hu, X., Zhu, Q., & Yang, Y. (2018). Improvement of biohydrogen production by optimization of pretreatment method and substrate to inoculum ratio from microalgal biomass and digested sludge. *Renewable Energy*, *127*, 670–677. https://doi.org/10.1016/j.renene.2018.05.022

Sun, Y., Liao, Q., Huang, Y., Xia, A., Fu, Q., Zhu, X., & Zheng, Y. (2016). Integrating planar waveguides doped with light scattering nanoparticles into a flat-plate photobioreactor to improve light distribution and microalgae growth. *Bioresource Technology*, *220*, 215–224. https://doi.org/10.1016/J.BIORTECH.2016.08.063

Sung, Y. J., Yu, B. S., Yang, H. E., Kim, D. H., Lee, J. Y., & Sim, S. J. (2022). Microalgae-derived hydrogen production towards low carbon emissions via large-scale outdoor systems. *Bioresource Technology*, *364*, 128134. https://doi.org/10.1016/J.BIORTECH.2022.128134

Suresh, G., Kumari, P., & Venkata Mohan, S. (2023). Light-dependent biohydrogen production: Progress and perspectives. *Bioresource Technology*, *380*, 129007. https://doi.org/10.1016/J.BIORTECH.2023.129007

Taikhao, S., Junyapoon, S., Incharoensakdi, A., & Phunpruch, S. (2013). Factors affecting biohydrogen production by unicellular halotolerant cyanobacterium Aphanothece halophytica. *Journal of Applied Phycology*, *25*(2), 575–585. https://doi.org/10.1007/S10811-012-9892-3

Taikhao, S., & Phunpruch, S. (2017). Increasing hydrogen production efficiency of N_2-fixing cyanobacterium Anabaena siamensis TISTR 8012 by cell immobilization. *Energy Procedia*, *138*, 366–371. https://doi.org/10.1016/J.EGYPRO.2017.10.170

Thevarajah, B., Nimarshana, P. H. V., Chang, J., & Ariyadasa, T. U. (2025). Cost-effective large-scale production of Chlorella vulgaris-based biodiesel and bioethanol: A comparative assessment. Journal of the Taiwan Institute of Chemical Engineers, *179*, 106395. https://doi.org/10.1016/j.jtice.2025.106395

Thevarajah, B., Nimarshana, P. H. V., Sandani, W. P., Wasath, W. A. J., & Ariyadasa, T. U. (2025). *s* (pp. 307–324). Elsevier eBooks. https://doi.org/10.1016/b978-0-443-23967-0.00024-3

Thevarajah, B., Nimarshana, P. H. V., Sandeepa, G. D. S., Vivekanand, V., & Ariyadasa, T. U. (2025). Transformation of residual biomass into biofuels at commercial microalgae facilities: A perspective on sustainable energy recovery. Journal of Environmental Chemical Engineering, 117551. https://doi.org/10.1016/j.jece.2025.117551

Thevarajah, B., Nimarshana, P. H. V., Shehan Sandeepa, G. D., Boopathy, R., & Ariyadasa, T. U. (2024). Upcycling food processing industrial wastes in Spirulina cultivation: A perspective on the advancement of Sustainable Development Goal 12. *Trends in Food Science & Technology*, *149*, 104537. https://doi.org/10.1016/J.TIFS.2024.104537

Thevarajah, B., Nishshanka, G. K. S. H., Premaratne, M., Nimarshana, P. H. V., Nagarajan, D., Chang, J. S., & Ariyadasa, T. U. (2022). Large-scale production of Spirulina-based proteins and c-phycocyanin: A biorefinery approach. *Biochemical Engineering Journal*, *185*, 108541. https://doi.org/10.1016/J.BEJ.2022.108541

Velmozhina, K., Shinkevich, P., Zhazhkov, V., Politaeva, N., Korablev, V., Vladimirov, I., & Morales, T. C. (2023). Production of biohydrogen from microalgae biomass after wastewater treatment and air purification from CO_2. *Processes*, *11*(10), 2978. https://doi.org/10.3390/PR11102978

Vijayaraghavan, K., Karthik, R., & Kamala Nalini, S. P. (2009). Hydrogen production by Chlamydomonas reinhardtii under light driven sulfur deprived condition. *International Journal of Hydrogen Energy*, *34*(19), 7964–7970. https://doi.org/10.1016/J.IJHYDENE.2009.08.010

Vinyard, D. J., & Brudvig, G. W. (2017). Progress toward a molecular mechanism of water oxidation in photosystem II. *Annual Review of Physical Chemistry*, *68*, 101–116. https://doi.org/10.1146/annurev-physchem-052516-044820

Wang, A., Sun, D., Cao, G., Wang, H., Ren, N., Wu, W. M., & Logan, B. E. (2011). Integrated hydrogen production process from cellulose by combining dark fermentation, microbial fuel cells, and a microbial electrolysis cell. *Bioresource Technology*, *102*(5), 4137–4143. https://doi.org/10.1016/J.BIORTECH.2010.10.137

Wang, X., Chen, Y. P., Guo, J. S., Fang, F., & Yan, P. (2024). Flocculation-enhanced photobiological hydrogen production by microalgae: Flocculant composition, hydrogenase activity and response mechanism. *Chemical Engineering Journal*, *485*, 150065. https://doi.org/10.1016/J.CEJ.2024.150065

Wang, Y., Yang, H., Zhang, X., Han, F., Tu, W., & Yang, W. (2020). Microalgal hydrogen production. *Small Methods*, *4*(3), 1900514. https://doi.org/10.1002/SMTD.201900514

Wieczorek, N., Kucuker, M. A., & Kuchta, K. (2014). Fermentative hydrogen and methane production from microalgal biomass (Chlorella vulgaris) in a two-stage combined process. *Applied Energy*, *132*, 108–117. https://doi.org/10.1016/J.APENERGY.2014.07.003

Wijayasekera, S. C. N., Cooray, B. Y., Premaratne, M., & Ariyadasa, T. U. (2020). Assessment of the potential of CO_2 sequestration from cement flue gas using locally isolated microalgae. *2020 Moratuwa Engineering Research Conference (MERCon), Moratuwa, Sri Lanka, 2020*, pp. 124–129. https://doi.org/10.1109/MERCon50084.2020.9185234

Wu, S., Xu, L., Huang, R., & Wang, Q. (2011). Improved biohydrogen production with an expression of codon-optimized hemH and lba genes in the chloroplast of Chlamydomonas reinhardtii. *Bioresource Technology*, *102*(3), 2610–2616. https://doi.org/10.1016/J.BIORTECH.2010.09.123

Xia, A., Cheng, J., Lin, R., Lu, H., Zhou, J., & Cen, K. (2013). Comparison in dark hydrogen fermentation followed by photo hydrogen fermentation and methanogenesis between protein and carbohydrate compositions in Nannochloropsis oceanica biomass. *Bioresource Technology*, *138*, 204–213. https://doi.org/10.1016/J.BIORTECH.2013.03.171

Xia, A., Cheng, J., Song, W., Su, H., Ding, L., Lin, R., Lu, H., Liu, J., Zhou, J., & Cen, K. (2015). Fermentative hydrogen production using algal biomass as feedstock. *Renewable and Sustainable Energy Reviews*, *51*, 209–230. https://doi.org/10.1016/J.RSER.2015.05.076

Xiao, Y., Zhang, X., Zhu, M., & Tan, W. (2013). Effect of the culture media optimization, pH and temperature on the biohydrogen production and the hydrogenase activities by Klebsiella pneumoniae ECU-15. *Bioresource Technology*, *137*, 9–17. https://doi.org/10.1016/j.biortech.2013.03.109

Xuan, J., He, L., Wen, W., & Feng, Y. (2023). Hydrogenase and nitrogenase: Key catalysts in biohydrogen production. *Molecules*, 28(3), 1392. https://doi.org/10.3390/MOLECULES28031392

Yang, Z., Guo, R., Xu, X., Fan, X., & Li, X. (2010). Enhanced hydrogen production from lipid-extracted microalgal biomass residues through pretreatment. *International Journal of Hydrogen Energy*, *35*(18), 9618–9623. https://doi.org/10.1016/j.ijhydene.2010.07.017

Yong, K. S., & Isa, N. M. (2023). Investigation of the Hydrogen Gas Production by Using Solar Energy. Journal of Automotive Powertrain and Transportation Technology, 3(2), 28–34.

Zarei, Z., Malekshahi, P., Morowvat, M. H., & Trzcinski, A. P. (2024). A review of bioreactor configurations for hydrogen production by cyanobacteria and microalgae. *International Journal of Hydrogen Energy*, *49*, 472–495. https://doi.org/10.1016/J.IJHYDENE.2023.09.108

Zhang, J., Xue, D., Wang, C., Fang, D., Cao, L., & Gong, C. (2023). Genetic engineering for biohydrogen production from microalgae. *IScience*, *26*(8), 107255. https://doi.org/10.1016/J.ISCI.2023.107255

Zhang, Q. (2021). Waste to biohydrogen: Progress, challenges, and prospects. In Waste to Renewable Biohydrogen: Volume 1: Advances in Theory and Experiments (pp. 55–70). Academic Press. https://doi.org/10.1016/B978-0-12-821659-0.00002-2

Zhao, Z., Li, Y., Muylaert, K., & Vankelecom, I. F. J. (2020). Synergy between membrane filtration and flocculation for harvesting microalgae. *Separation and Purification Technology*, *240*, 116603. https://doi.org/10.1016/J.SEPPUR.2020.116603

Biohydrogen Production from Organic Wastes

2

Types of Microorganisms, Diverse Pathways, Enhanced Production, and Future Prospects

Deepak Kumar, Surabhi Sirohi, and Meenakshi Suhag

2.1 INTRODUCTION

Worldwide energy demands are rising day by day because of the immediate growth of the world population and rapid industrialization to fulfill their energy demands. Due to hasty industrialization and the increase in the expansion of the worldwide market, energy demands will increase by nearly 50% by 2050 (Nagarajan et al., 2020; Paramesh et al., 2018). Due to this, the depletion of non-renewable resources has been increasing, and their combustion results in increasing global warming and greenhouse gases (GHGs). Rising CO_2 concentrations, a known GHG emission, linked to the burning of non-renewable resources and found to be more than 409 ppm (Fawzy et al., 2020; Zhang et al., 2020; Osman et al., 2020), resulted in increasing global temperature by 2.8°C by the end of this century (UNEP, 2022). Avoiding a temperature increase beyond 2°C is the main concern, which is accomplished by the production of fuels with minimal or no carbon emissions (Rawat et al., 2023). Combustion of fossil fuel results in the high contribution of CO and CO_2 to the environment. A report estimates that there has been a 28.6% increase

DOI: 10.1201/9781003568612-2

in atmospheric CO_2 (Rawat et al., 2023). Furthermore, for sustainable and economic development, the globe will soon be calling for increased clean energy (Osman et al., 2020). Finding alternative renewable energy sources is therefore required to meet this energy requirement. At present, some alternative energy sources are being used including hydrogen, solar, wind, and bioethanol as well as biodiesel, biogas, and bioethanol (Sampath et al., 2020). The energy content of some commonly used fossil fuel sources is shown in Table 2.1 (Sampath et al., 2020).

In this context, biohydrogen is regarded as a fuel of the future since it is a renewable substitute for fossil fuels that are not renewable and a sustainable fuel with a high energy content. It is a non-toxic, zero-carbon emission fuel, and has an approximately 2.75-fold higher thermal efficiency and a high calorific value of nearly 122 kJ/g compared to hydrocarbon fuels (Patel et al., 2014; Goswami et al., 2021). It also has a high energy density, i.e. 142 MJ/kg, which is significantly higher than other fossil fuels such as gasoline (47 MJ/kg), methane (56 MJ/kg), and natural gas (54 MJ/kg) (Acar & Dincer, 2019). Biohydrogen is the most reliable energy source because it has high conversion efficiency, i.e. 99% in fuel cell systems (Nikolaidis & Poullikkas, 2017).

The molecules of hydrogen do not exist naturally in the earth's crust. So, it is necessary to produce molecular hydrogen for practical application. At the present time, the world produces nearly 368 trillion cubic meters of hydrogen annually (Pandu & Joseph, 2012). Out of the total worldwide production, methane contributes 48%, almost 30% by oil/naphtha refineries or the chemical industry, contribution of coal gasification is 18%, 3.9% and 0.1% come from electrolysis of water and various sources, respectively (Baghchehsaraee et al., 2010; Singh et al., 2023). According to this data, it is estimated that 96% of hydrogen is generated from fossil fuels (natural gas, coal, crude oil, oil shale, etc.). For the production of hydrogen, various fossil fuel-based traditional methods such as thermal decomposition, steam gasification, pyrolysis, catalyzed oxidation, water gas shift reaction (WGSR), auto-thermal reforming, etc. (Chandrasekhar et al., 2015; Osman et al., 2018; Osman et al., 2020) are used but these are non-sustainable, costly, and result in the huge emission of GHG at optimum conditions, i.e. high atmospheric pressure and temperature (Mishra et al., 2019; Rawat et al., 2023). However, these traditional methods are not eco-friendly and need a lot of energy (Balachandar et al., 2013). In contrast, biohydrogen production from renewable resources (water, organic wastes, and biomass) looks more promising than the above-discussed methods because it can use renewable energy resources, can typically operate at room temperature and atmospheric pressure, and does not produce CO_2 during the production process. Therefore, eco-friendly, renewable, and sustainable methods have been used for the production of biohydrogen.

TABLE 2.1 Energy content of various conventionally used fossil fuel sources

FUEL TYPE	*ENERGY CONTENT (MJ/KG)*
Hydrogen	120
Liquefied natural gas	54.4
Propane	49.6
Aviation gasoline	46.8
Automotive gasoline	46.4
Automotive diesel	45.6
Ethanol	29.6
Methanol	9.7
Coke	27
Wood (dry)	16.2
Bagasse	9.6

Source: Adopted from Sampath et al. (2020).

2.2 BIOHYDROGEN PRODUCTION FROM ORGANIC WASTE

More focus has been given in recent years to biohydrogen production from residues. The problem of scrap disposal and energy creation is simultaneously resolved by biohydrogen production from various carbon-based wastes. Organic wastes are produced from the trash of plants and animals which can easily decompose and their decomposition results in the production of carbon dioxide, methane, or simple organic molecules (Kadir et al., 2016). For biohydrogen production, multiple organic wastes are used such as industrial leftovers, agriculture residues, municipal sewage sludge, poultry waste, solid waste, manure, etc. (Dhanya et al., 2020). Various techniques using organic waste material to produce biohydrogen include: (1) bioelectrochemical system comprising microbial electrolysis cells (MECs), (2) biophotolysis (direct and indirect), (3) biofermentation (dark and light fermentation), and (4) gasification (Zhang & Angelidaki, 2014; Lin et al., 2018). Figure 2.1 shows the various biohydrogen production processes which are discussed below. From the above-discussed processes dark fermentation, MECs, and biomass gasification use biomass as a substrate to achieve high process efficiency except for the biophotolysis process (Rawat et al., 2023). Not all but biophotolysis and photo-fermentation are dependent on sunlight for the production of biohydrogen (Ghimire et al., 2015).

Different biohydrogen processes have different energy conversion efficiency which is highlighted in Table 2.2. The usable output of the energy conversion machine divided by the input energy is known as energy conversion efficiency (η).

2.3 FERMENTATION

The fermentation process is used to produce energy using an acceptor of electrons from the oxidation of multiple types of garbage-consuming microbes. The output of the fermentation process can be affected by the used catalyst (enzyme or microorganism), types of organic substrate (primarily a protein or carbohydrate), and other process parameters like pH, temperature, and medium composition

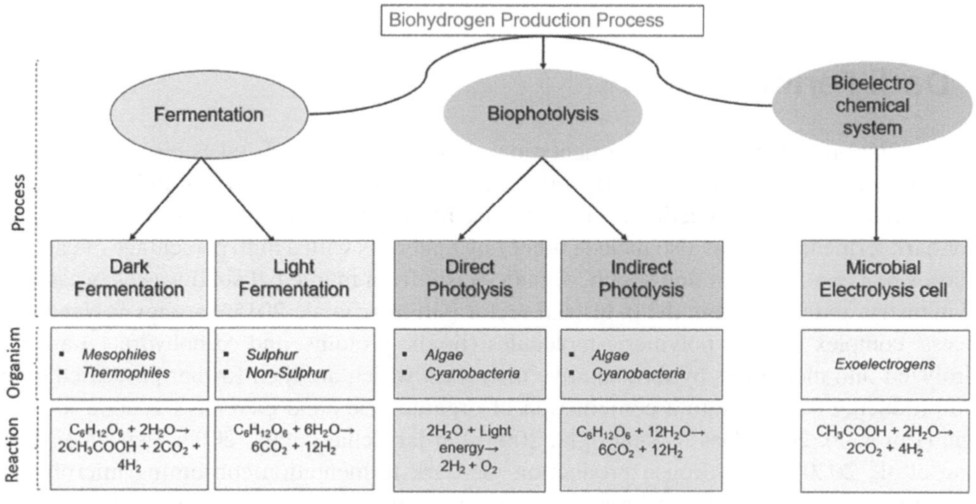

FIGURE 2.1 Various processes for biohydrogen production. (Edited from Osman et al. (2020).)

TABLE 2.2 Energy conversion efficiency of biohydrogen production process

BIOHYDROGEN PRODUCTION PROCESS	*ENERGY CONVERSION EFFICIENCY (%)*	*MICROORGANISMS USED*	*DESCRIPTION*	*REFERENCES*
Dark fermentation	30%–40%	***Clostridium*** sp., ***Enterobacter*** sp., ***E. coli***	Anaerobic breakdown of organic substrates to produce hydrogen and VFA	Das and Veziroğlu (2001)
Photo-fermentation	10%–20%	***Rhodobacter*** sp., ***Rhodopseudomonas*** sp., ***Rhodospirillum*** sp.	Utilizes photosynthetic bacteria under light to convert organic acids in biohydrogen	Hallenbeck and Benemann (2002)
Biophotolysis	1%–5%	Cyanobacteria (***Anabaena*** sp., ***Nostoc*** sp.), green algae (***Chlamydomonas reinhardtii***)	Direct hydrogen production through water splitting by photosynthetic microbes like cyanobacteria and green algae	Hitam and Jalil (2023) Melis (2002)
Microbial electrolysis cells	60%–80%	***Geobacter sulfurreducens***, ***Shewanella oneidensis***, ***Clostridium*** sp.	In MECs, microorganisms oxidize organic matter to generate biohydrogen with an external voltage	Rozendal et al. (2008)
Hybrid processes	40%–70%	Combinations of dark fermentation and photo-fermentative bacteria	Combines multiple processes to enhance hydrogen yield	Kapdan and Kargi (2006)

(Sampath et al., 2020). Two enzymes are responsible for the generation of hydrogen, i.e. hydrogenase and nitrogenase (Sarma et al., 2019). It can be aerobic or anaerobic (Tomasik & Horton, 2012). One of the effective methods of producing biohydrogen and further derivatives like alcohols and organic acids is by using microbes to ferment organic matter in anaerobic environments (Sampath et al., 2020). The fermentation process can be divided into two types depending on the utilization of light by microbes, i.e. (1) dark fermentation and (2) light fermentation. The process of dark fermentation proceeds in anaerobic conditions in which the breakdown of cellulosic organic biomass takes place which produces biohydrogen and organic acids or alcohols as a byproduct in the absence of light (Rizwan et al., 2019). In contrast to dark fermentation, photo-fermentation also takes place in anaerobic conditions in which organic molecules break into carbon dioxide and hydrogen in the presence of light-dependent bacteria (Mona et al., 2020).

2.3.1 Dark Fermentation

In dark fermentation or heterotrophic fermentation, the production of H_2 takes place in the omission of O_2, water, and sunlight using H_2-generating microbes like facultative anaerobes and obligate anaerobes. Organic substrates such as lignocellulosic biomass are necessary for dark fermentation. Alternative substrates for dark fermentation can also include other raw materials with a high percentage of carbohydrates, such as crops that contain sugar and starch, organic waste from municipal solid wastes, wastewater from the food industry, and waste from the industrial sector (Ghimire et al., 2015; Sarangi & Nanda, 2020). In this process, complex organic polymeric molecules (lipids, proteins, and carbohydrates) are pretreated and hydrolyzed into monomers by fermentative microbes, which are then further processed by essential hydrogen-producing bacteria into a combination of organic acids and alcohols having a lower molecular weight (Lin & Jo, 2003; Das & Veziroğlu, 2008; Vardar-Schara et al., 2008; Sharma & Arya, 2017; Nagarajan et al., 2020). Biohydrogen production via dark fermentation-consuming microbial communities is increasing. Mainly mesophilic and thermophilic microbial species are involved in converting

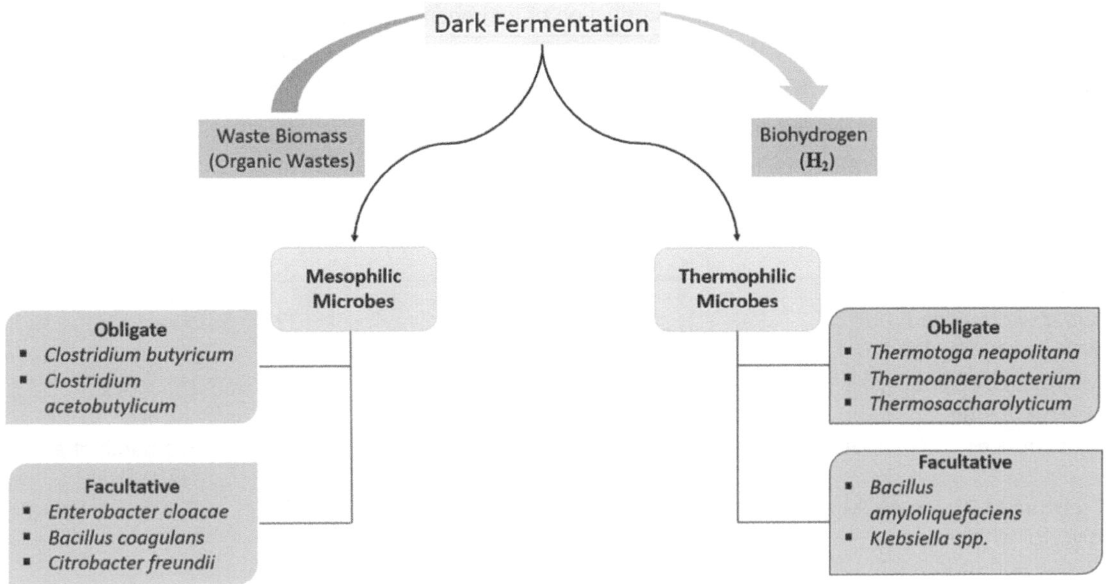

FIGURE 2.2 Microbial communities used in dark fermentation to produce biohydrogen.

waste biomass into hydrogen through dark fermentation. These microbes are either facultative or obligate (Sarangi & Nanda, 2020) as shown in Figure 2.2.

The primary advantage of this process is to produce hydrogen using various carbon sources in the absence of light and valuable byproducts including lactic acid, butyric acid, acetic acid, etc. (Sharma & Arya, 2017). Significantly, lower hydrogen production is a drawback. Moreover, carbon dioxide in the resulting gas combination needs to be removed (Saifuddin & Priatharsini, 2016). In the following chemical equation, hydrogenase enzyme acts as the key catalytic agent, i.e. $2H^+ + 2e^- \leftrightharpoons H_2$. Enzyme hydrogenases exist in three different forms, i.e. FeFe (iron-iron), NiFe (nickel-iron), and Fe-hydrogenases (metal-free) (Yagi & Higuchi, 2013). Metal-free hydrogenase has an iron-containing cofactor. Nevertheless, in the process of producing hydrogen through dark fermentation, either (NiFe) hydrogenase or (FeFe) hydrogenase can be very important (Barrozo & Orio, 2019). Different microorganisms are used in dark fermentation which is represented in Table 2.3.

2.3.2 Metabolic Pathways of Dark Fermentative Bacteria

No photon is required during dark fermentation, which takes place via glycolysis of sugars to pyruvate. Conversion of pyruvate depends on two pathways, i.e. whether facultative or obligate anaerobes during fermentation under anaerobic conditions. In the presence of oxygen, obligate anaerobes cannot endure, whereas facultative anaerobes can live with or without it. When the enzyme pyruvate formate lyase (facultative anaerobes) is present, pyruvate is converted to formate and acetyl-CoA. In the presence of enzyme formate hydrogen lyase (FHL), formate is then converted to carbon dioxide and hydrogen as shown in Figure 2.3.

Due to the obligate anaerobes, pyruvate is oxidized to acetyl-CoA and carbon dioxide in the presence of an enzyme pyruvate ferredoxin oxidoreductase (PFOR). Transfer of electrons takes place from PFOR which reduces the ferredoxin. The function of ferredoxin is to supply these electrons to the hydrogenase enzyme where H^+ is reduced to hydrogen which is supplied by redox mediator nicotinamide adenine dinucleotide (NADH; facilitated by NADH hydrogenase) as shown in Figure 2.4.

In dark fermentation the production of hydrogen occurs in two stages: (1) generation of reduced ferredoxin (Fd_{red}) by decarboxylation of pyruvate into acetyl-CoA (Equations 2.1 and 2.2). This reduced

TABLE 2.3 Numerous wastes used in the production of hydrogen in the dark fermentation process

SUBSTRATE USED	*TYPES OF MICROORGANISMS*	*PH*	*TEMP. (°C)*	*H_2 PRODUCE (mL/g VS)*	*REFERENCES*
Dairy compost	Varied culture	5.0	36.0±1	31.5	Xing et al. (2010)
Sewage slush	Mixed culture	7.0	37.0	11.2	Yang and Wang (2017)
A mixture of poplar leaves and sewage sludge				20.8	
Flower waste added with sewage sludge				32.0	
Ryegrass mixed with sewage sludge				51.0	
Waste of rice	Stimulated sewage slush	4.0–5.5	35.0	14.5±0.3	Alemahdi et al. (2015)
A mixture of 3% crude glycerol with sewage sludge and food waste	Mixed culture	5.5	35.0	179.3	Silva et al. (2018)
Distillery wastewater	Mixed culture	5.0	37.0	1.6±0.3	Laurinavichene et al. (2018)
Cassava wastewater	Mixed culture	5.5	37.0	39.8	Intanoo et al. (2016)
Rice straw	***Clostridium pasteurianum***	7.5	37.0±2	2.6	Srivastava et al. (2017)
Hydrolyzed food waste	***Aspergillus awamori**, A. oryzae*	4.0–4.6	37.0	219.9	Han et al. (2015)
Food waste+sludge	–	5.5±0.1	37.0	13.8	Kim et al. (2013)
Sugarcane bagasse	***Enterobacter aerogenes***	6.8	30.0	1000.0	Rai et al. (2014)
Wheat straw	***Caldicellulosiruptor saccharolyticus*** + *C. owensensis*	6.5±0.1	70.0	134.0	Soto et al. (2019)

ferredoxin donates the electron to the hydrogenase enzyme and produces hydrogen using obligate microbes (*Clostridium* sp.) (Khanna & Das, 2013).

$$\text{Pyruvate} + 2\text{Fd}(\text{oxi}) + \text{CoA} \rightarrow \text{CO}_2 + \text{Acetyl CoA} + 2\text{Fd}(\text{red}) \tag{2.1}$$

$$2\text{H}^+ + 2\text{Fd}(\text{red}) \rightarrow \text{H}_2 + 2\text{Fd}(\text{ox}) \tag{2.2}$$

(2) In this pyruvate and CoA are changed into acetyl CoA enzymes and formate in the presence of FHL which is the key mechanism for hydrogen production catalyzed by facultative anaerobes (*Klebsiella* sp.) (Equations 2.3 and 2.4).

$$\text{Pyruvate} + \text{CoA} \rightarrow \text{Formate} + \text{Acetyl CoA} \tag{2.3}$$

$$\text{Formate} \rightarrow \text{CO}_2 + \text{H}_2 \tag{2.4}$$

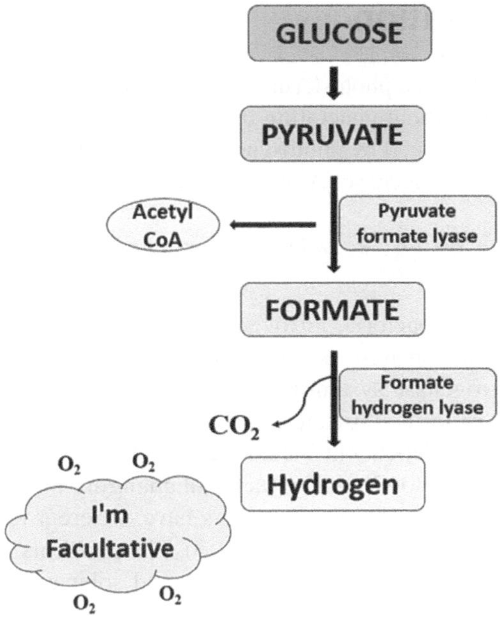

FIGURE 2.3 Production of biohydrogen via facultative anaerobes.

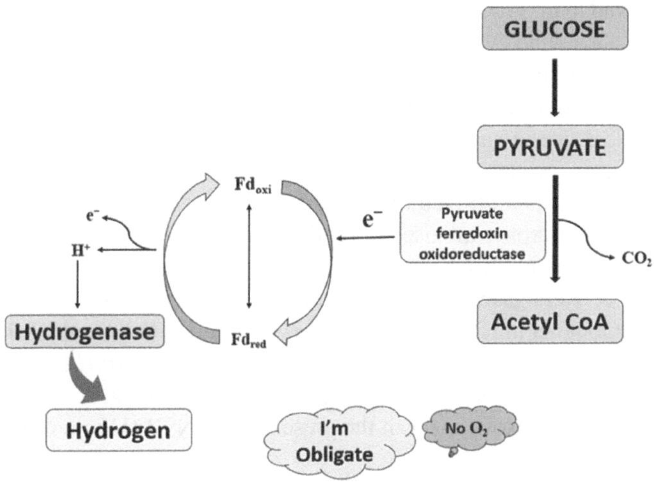

FIGURE 2.4 Production of biohydrogen by obligate anaerobes.

The following is the general reaction of dark fermentation (Equation 2.5):

$$\underset{\text{Organic Substrate}}{C_6H_{12}O_6} + \underset{\text{Water}}{2H_2O} \rightarrow \underset{\text{Carbon dioxide}}{2CO_2} + \underset{\text{Hydrogen}}{4H_2} + \underset{\text{Acetic acid}}{2CH_3COOH} \quad (2.5)$$

2.3.3 Photo-Fermentation

Gest and Kamen in 1949 identified the photo-fermentation process for hydrogen production. This process produces high-purity hydrogen without generating oxygen (Gest & Kamen, 1949). In the photo-fermentation method, biohydrogen is produced by photosynthetic bacteria without O_2 by using rays and biomass through a reaction catalyzed by nitrogenase (Akroum-Amrouche et al., 2013).

$$\underset{\text{Organic Substrate}}{C_6H_{12}O_6} + \underset{\text{Water}}{6\,H_2O} \rightarrow \underset{\text{Carbon dioxide}}{6\,CO_2} + \underset{\text{Hydrogen}}{12\,H_2\left(\textit{Photosynthetic bacteria}\right)} \quad (2.6)$$

In the presence of photosynthetic bacteria, nitrogenase enzymes play a vital role in biohydrogen production, and the synthesis of nitrogenase is highly boosted by light. Photosynthetic bacteria have a higher ability to use their nitrogenase system to create hydrogen (Akroum-Amrouche et al., 2011). In biohydrogen production C/N ratio in substrate plays a vital role (Hay et al., 2013). Factors on which photo-fermentation pathways are dependent include temperature, inoculum types, light intensity, and types of substrates (Basak & Das, 2007). A key factor in changing the metabolic pathways to produce more biohydrogen is temperature. Different organisms have different temperature ranges for optimal biohydrogen production (Akroum-Amrouche et al., 2013). The pigments used in light-harvesting antennas, the properties of light saturation in photosynthesis, and solar energy conversion efficiency vary among organisms (Eroglu & Melis, 2011). Maximum production of biohydrogen is given by an inoculum of photosynthetic bacteria collected from the log phase of the growth curve (Basak & Das, 2007).

Due to their high metabolic elasticity, purple non-sulfur bacteria (PNSB) are utilized in the photo-fermentation procedure and can grow as photoautotrophs and chemoautotrophs. Unlike photosynthetic organisms such as plants, algae, and cyanobacteria, these specialized bacteria can generate hydrogen without needing water or significant energy input, and this process does not create oxygen as a byproduct (Harwood, 2009). Hydrogen, inorganic ions, or carbon compounds act as electron sources to initiate metabolic activities. For metabolic activity, PNSB uses a varied range of carbon-based compounds such as organic acids, amino acids, pyruvate, carbohydrates, acetate, and alcohol. Aromatic organic compounds including benzoate, cinnamate, and chlorobenzoate are also utilized by PNSB to complete their carbon requirements (Gupta et al., 2024). Various substrates and microbes are used to produce biohydrogen as represented in Table 2.4. For biohydrogen production various species of PNSB are used during the photo-fermentation process: *Rhodobacter* sp. (Assawamongkholsiri & Reungsang, 2015), such as *Rhodobacter sphaeroides* (Tao et al., 2008) and *Rhodobacter capsulatus* (Obeid et al., 2010); *Rhodopseudomonas* sp. (Liu et al., 2015), *Rhodopseudomonas palustris* (Bianchi et al., 2010), *Rhodopseudomonas capsulate*; *Rhodospirillum rubrum* (Hosseini et al., 2015); *Rhodovulum* sp., including *Rhodovulum sulfidophilum* (Cai & Wang, 2012). The lignocellulosic matter is pretreated and hydrolyzed into hexose sugar in the presence of PNSB (Hitam & Jalil, 2023).

TABLE 2.4 Production of biohydrogen using photo-fermentation from a variety of substrates and microbes

SUBSTRATE	MICROORGANISM	PH	TEMPERATURE	H_2 YIELD	REFERENCES
Glucose	***Rhodovulum sulfidophilum*** P5	8.0	30°C	7.07 mol H_2/mol glucose	Cai and Wang (2014)
Butyrate assorted with acetate	Mixed culture	7.0	34°C	For acetate at 3.51 mol/kg COD/day For butyrate at 3.33 mol/kg COD/day	Srikanth et al. (2009)
Pulp and paper mill waste mixed with POME	***Rhodobacter sphaeroides*** NCIMB 8253	–	30°C	4.67 mL H_2/mL medium	Budiman et al. (2015)
Organic acids-based lignocellulose	***R. sphaeroides*** ZX-5	7.0	30°C	Acetic acid =1.36 mL H_2/mL media Formic acid=0.67 mL H_2/mL media	Zhu et al. (2010)

2.3.4 Metabolic Pathways of Photo-Fermentative Bacteria

In metabolic pathways, *Rhodopseudomonas*, *Rhodobacter*, *Rhodospirillum*, and *Rhodobium* are anaerobic or photosynthetic bacterial strains that break down hexose sugar to produce biohydrogen using light energy via a mechanism initiated through nitrogenase (Hitam & Jalil, 2023). The bacteria convert the sugars into CO_2 and NADH. Reduction of ferredoxins takes place in the presence of electrons transferred from NADH. The nitrogenase enzyme receives electrons from the reduced ferredoxins via specific complexes to reduce the molecular nitrogen into ammonia, and this results in the generation of the proton gradient. The reversible reduction reaction of the proton is facilitated by the enzyme hydrogenase, resulting in the formation of molecular hydrogen that builds up in the periplasmic region. Nitrogenase works similarly to hydrogenase enzyme in the absence of nitrogen (Gupta et al., 2024). This entire mechanism takes place with the help of hydrolysis of ATP. ATP required in this is generated through photo-phosphorylation (Liu et al., 2021). Metabolic pathways of photo-fermentative bacteria are shown in Figure 2.5, and a detailed mechanism is represented in Figure 2.6.

2.3.5 Direct Biophotolysis

Microalgae harness sunlight to split water molecules during direct biophotolysis, producing hydrogen and oxygen as byproducts. This process occurs through photosynthetic reactions within the microalgal cells:

$$\underset{\text{Water}}{2H_2O} \xrightarrow{\text{solar energy}} \underset{\text{Hydrogen}}{2H_2} + O_2(\text{Oxygen}) \tag{2.7}$$

Photosynthesis in green microalgae is a complex process involving specialized pigments and photosystems. Chlorophyll plays a crucial role in capturing light energy, while Photosystem I (PS I) and Photosystem II (PS II) are key components in light-dependent reactions. PS II is particularly important in oxygen production, accomplishing this through the photolysis of water molecules, which splits water

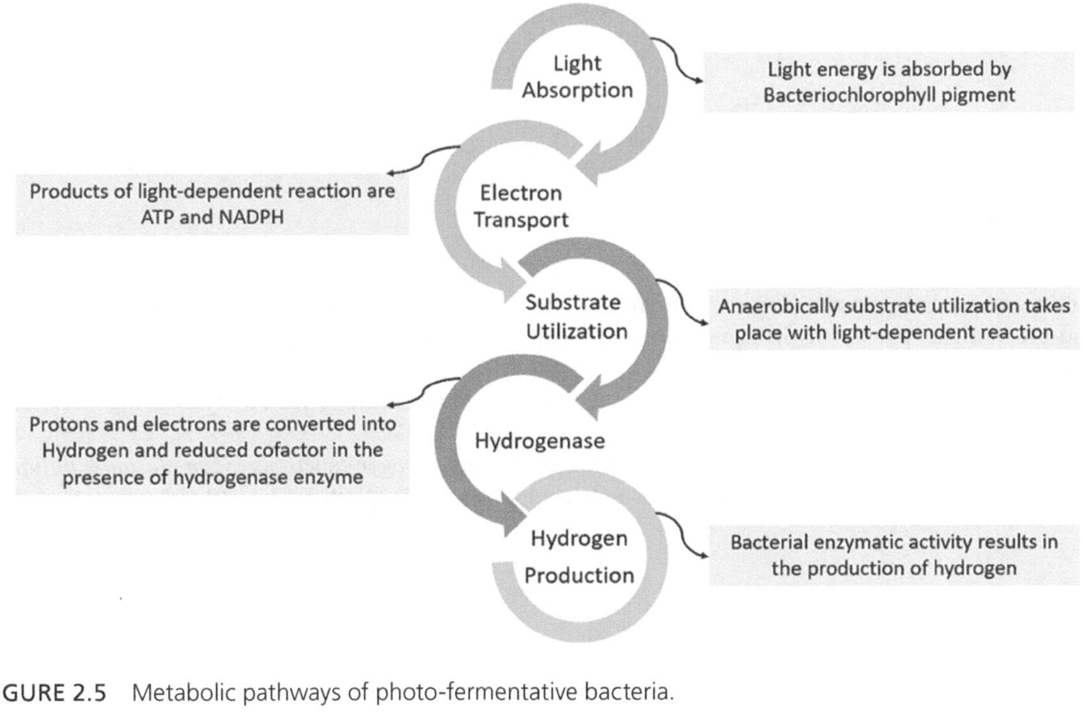

FIGURE 2.5 Metabolic pathways of photo-fermentative bacteria.

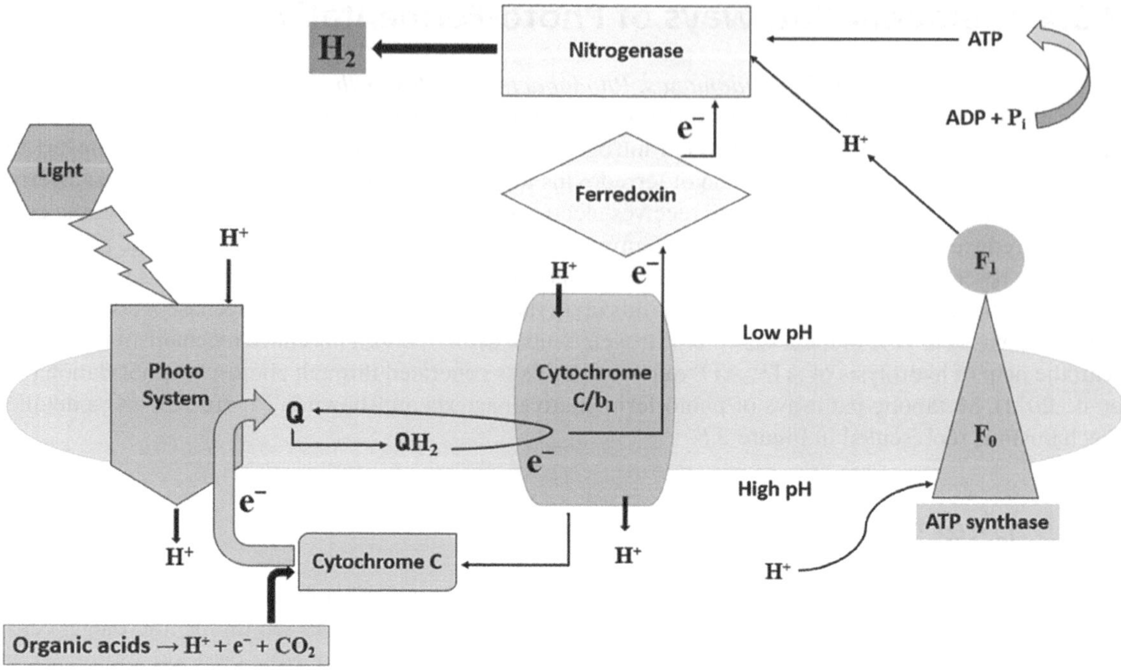

FIGURE 2.6 Detailed mechanism of photo-fermentative bacteria.

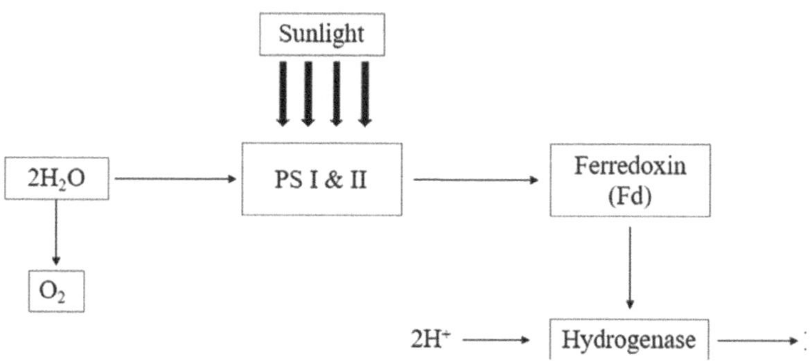

FIGURE 2.7 Direct biophotolysis.

into its constituent elements (Das and Veziroğlu, 2008). Certain microalgal species, including green and blue-green algae, possess hydrogenase enzymes that enable hydrogen production. During this process, when PS II captures sunlight, it generates electrons. These electrons are then shifted to ferredoxin via PS I, initiating a succession of biochemical reactions that can potentially lead to hydrogen generation as shown in Figure 2.7.

Various species are used for the production of biohydrogen such as *Scenedesmus obliquus*, *Chlamydomonas reinhardtii*, *C. moewusii* via direct biophotolysis (Nath & Das, 2004).

2.3.6 Indirect Biophotolysis

This process (indirect biophotolysis) takes place in two steps: (1) using light energy breakdown of water molecules results in the formation of carbohydrates and oxygen; (2) hydrogenase enzyme results in the production of hydrogen gas due to the dissociation of carbohydrates:

$$6CO_2 + 12H_2O + \text{Light energy} \rightarrow 6O_2 + C_6H_{12}O_6 \tag{2.8}$$

$$12H_2O + C_6H_{12}O_6 \rightarrow 6CO_2 + 12H_2 \tag{2.9}$$

During photosynthesis, PS II employs light to drive the extraction of electrons from H_2O. The extracted electrons are then transferred to the complex (Fe-S protein ferredoxin) via PS I. Hydrogen is produced by hydrogenase enzymes which are present in the stroma of the algal chloroplast as shown in Figure 2.8. In this process higher yield of hydrogen takes place and byproducts can be competently converted to hydrogen (Sharma & Arya, 2017).

2.3.7 Microbial Electrolysis Cell

MECs represent an innovative biological approach to hydrogen production. In this process, specialized microorganisms known as exoelectrogens break down organic matter through their metabolic activities. As these microorganisms digest the organic materials, they generate three key components: carbon dioxide, protons (H^+), and electrons. The system harnesses these products through an electrochemical setup. The electrons produced by the microorganism flow to an anode and then travel through an external circuit powered by an electrical source. These electrons ultimately reach a cathode, where they participate in a reduction reaction involving either protons or water molecules. This final step results in the formation of hydrogen gas as the end product.

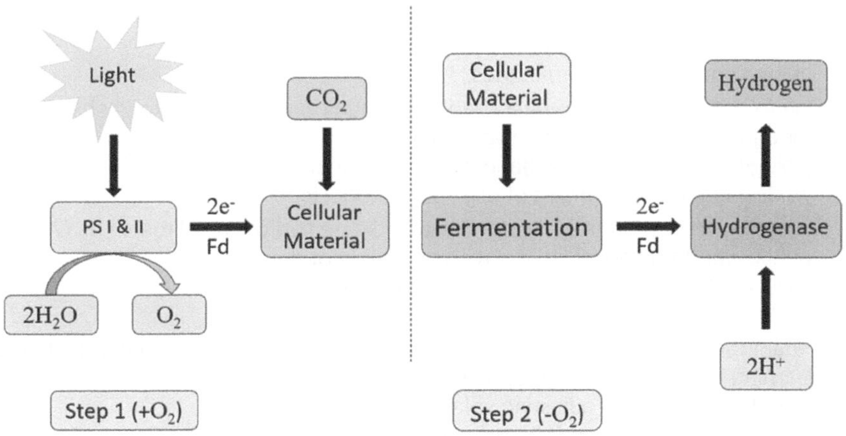

FIGURE 2.8 Indirect biophotolysis.

2.4 STRATEGIES TO INCREASE THE PRODUCTION OF BIOHYDROGEN

The main challenges in the conventional biological generation of hydrogen include poor substrate alteration yield and volatile fatty acid (VFA) gathering. The overall production of hydrogen is very low, to use it on a viable scale (Chandrasekhar et al., 2020). Furthermore, in the dark fermentation process, H_2 production is just 4 mol H_2/mol glucose, whereas theoretical H_2 production could reach 12 mol of H_2/mol glucose, which is a major hindrance in practical applications (Abo-Hashesh & Hallenbeck, 2011). In addition, technological improvements in photobioreactors, proper treatment of effluents, and reuse of remaining carbon fraction for further energy generation would be considered wise for sustainable development (Mohanakrishna et al., 2010).

To devise a path that is better than the conventional processes of biohydrogen generation is the need of the hour. So, various techniques, for example, combined fermentation, sequential fermentation, and cell immobilization, have been used to enhance hydrogen production in biological processes as shown in Table 2.5 and Figure 2.9.

2.4.1 Cell Immobilization Techniques

When compared to other biological processes, the dark fermentation approach produces hydrogen at a higher rate (Singh et al., 2013). Moreover, the fermentation process converts the organic waste into biodegradable materials, thereby making it a more valuable energy resource (Sekoai et al., 2022). The

TABLE 2.5 Key strategies to enhance biohydrogen production

STRATEGY	*DESCRIPTION*	*REFERENCES*
Nanomaterial integration	Adding nanomaterials to biohydrogen manufacturing process to increase output and reactivity. This strategy improves long-term sustainability and cost-effectiveness of biohydrogen generation methods.	Sanghvi et al. (2024)
Dark fermentation optimization	Due to its high production rates, dark fermentation is thought to be the most effective technique for producing biological H_2. On the other hand, issues include decreased yields and the development of organic byproducts.	Teke et al. (2024)
Artificial microbial consortia design	Designing artificial microbial consortia to increase hydrogen yield beyond traditional limits. Precision design of these consortia has achieved yields of 40% higher than the Thauer limit.	Ergal et al. (2020)
Volatile fatty acids optimization	Dynamically optimizing volatile fatty acids using deep learning neural networks to enrich biohydrogen production. This strategy involves adjusting volatile fatty acids concentrations to improve biogas composition and hydrogen yield.	Mahmoodi et al. (2022)
Pretreatment methods	Enhancing H_2 production by using different pretreatment techniques to anaerobic mixed microflora. Pretreatments can increase the efficiency of microbial consortia in producing hydrogen from substrates like dairy wastewater.	Mahmoodi-Eshkaftaki and Mockaitis (2022)
Operational 2023 strategies combination	Combining various operational strategies to enhance biohydrogen production. This includes optimizing conditions such as pH, temperature, and substate concentration to achieve higher yields.	Hidalgo and Martín-Marroquín (2023)

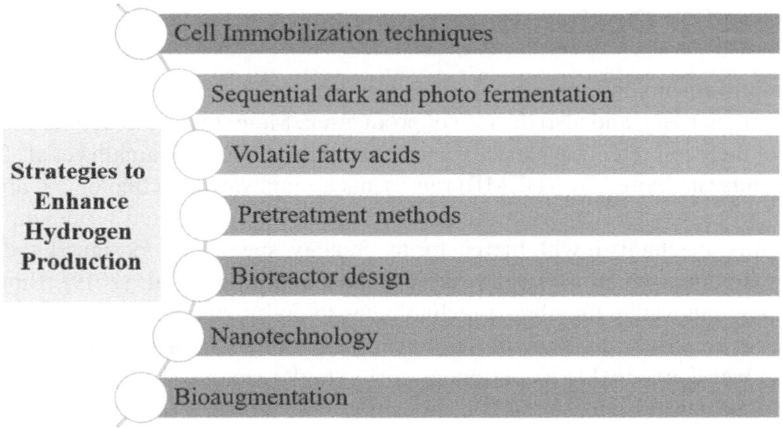

FIGURE 2.9 Different strategies to enhance biohydrogen production.

problem of biomass washout often acts as a problem in continuous or batch operations, so it is necessary to maintain the sufficient density of cells for high production of hydrogen. To raise the biomass content for miscellaneous and for pure cultures this technique can be useful, as found by Singh and Wahid (2015). Therefore, cell immobilization techniques are adopted to create an anaerobic environment that is considered ideal for the generation of oxygen-perceptive fermentation hydrogen. Cell culture purity is critical in research, particularly for hydrogen production processes. When maintaining axenic (sterile) cell cultures proves challenging, this technique becomes valuable in minimizing contamination risks. By implementing specialized methods, researchers can effectively reduce microbial contamination and potentially increase biomass concentration for both mixed and pure culture systems (Singh & Wahid 2015).

2.4.2 Sequential Dark and Light Fermentation Process

Advanced H_2 production techniques are becoming a crucial solution for sustainable energy generation, with sequential dark and photobiological fermentation processes showing particular promise. This innovative approach leverages carbohydrate-rich wastewater and organic waste streams to generate clean energy. However, the method faces significant challenges: accumulation of shorter organic acids during the process, which limits hydrogen yields, and significant expenses involved in processing and handling pure organic acids. The key innovation lies in the collective (dark and photo) fermenting strategy, which offers a more efficient pathway for hydrogen generation compared to traditional methods. By integrating these fermentation techniques, researchers aim to overcome current limitations and develop a more economically viable and environmentally friendly approach to renewable energy production. The primary obstacles include optimizing hydrogen production and reducing the economic barriers related to acid management. Ongoing research focuses on developing more sophisticated fermentation protocols that can maximize hydrogen output while minimizing production costs (Afsar et al., 2011). In innovative biohydrogen production research, a two-stage fermentation process involving both dark and photochemical stages shows promising results for hydrogen generation. This method leverages short-chain organic acids like acetate and lactate, which under specific light conditions can be converted into hydrogen. The biochemical transformation is particularly efficient, generating approximately 6 molecular units of hydrogen for each molecular unit of acetate or lactate processed. This advanced bioenergy technique is being explored as part of a comprehensive research initiative within the European Union's sixth framework program, with initial studies indicating a potential maximum hydrogen yield of 12 molecular units per molecular unit of hexose sugar (Claassen & De Vrije, 2006).

2.4.3 Bioreactor Design Improvements

Technological improvements in bioreactor designs will affect the different parameters such as hydrogen yield, volume of productivity, and also the rate of production. Membrane bioreactors are commonly in use mainly because of their cell retention capacity and separation of VFAs (Wainaina et al., 2019). The hydrogen extractive membrane bioreactor (HEMB) represents an innovative technological approach that combines two critical processes: fermentation and gas separation. This integrated system uniquely combines a membrane separation mechanism with bioreactor technology, specifically designed to enhance hydrogen production and extraction which is developed by Ramírez-Morales et al. (2019). Bioreactor fermentation systems employ innovative membrane technologies to optimize hydrogen production. In a study by Ramírez-Morales et al. (2019), a system utilizing two membranes (M1 and M2) demonstrated promising results. When M1 was connected to a continuous stirred tank reactor (CSTR) and H_2 generation reached a steady state, M2 achieved high hydrogen composition. The reactor's operation led to a decreased partial pressure, resulting in a 16% increase in hydrogen yield. Exploring alternative approaches, Saleem et al. (2018) investigated biohydrogen production using a 1.0L anaerobic dynamic membrane bioreactor (AnDMBR). The reactor featured polyester fabric (monofilament woven meshes) integrated into a crossflow module. Their research revealed that significant hydrogen productionis feasible even at low influent chemical oxygen demand (COD) concentrations ranging from 10 to 39g COD per liter and under low percolation instabilities. Fixed-bed bioreactors (FBRs) have gained prominence in biohydrogen production due to their superior bed porosity and biomass retention capabilities, as noted by Lima et al. (2013). These bioreactors are constructed using various synthetic materials, including polypropylene and polyvinyl chloride (PVC), which can significantly influence productivity. A comparative study by Zhao et al. (2019) investigated three different types of static mattress materials: porous nylon, loofah sponge, and chlorinated polyethylene. The research highlighted that chlorinated polyethylene exhibited superior characteristics, including enhanced bacterial immobilization, improved growth conditions, remarkable buffering capacity, and more efficient hydrogen productivity.

2.4.4 Nanotechnology

The applications of nanotechnology have improved the dark fermentation process in multiple ways by using metal ions, organic metals, and metal oxide nanoparticles. Nanoparticles are considered effective in biohydrogen production, because of their surface characteristics and quantum measurement effects (Pugazhendhi et al., 2019). This technology shortens the lag phase time. The enzymes involved in dark fermentation require metal ions in the form of nanoparticles as cofactors. The metals Co, Si, Ag, Ni, Pd, Ti, Fe, and Cu are among those commonly used (Rao & Basak, 2021).

2.4.5 Additives

Additives are the compounds that work to boost the enzymatic activity and microbial growth. Additionally, these can be utilized to immobilize the bacteria, giving them the ability to withstand environmental stresses, for instance, pH, temperature, and organic loading rates (OLRs; Yang & Wang, 2018). Bao et al. (2013) discovered that adding 0.3g/L of L-cysteine increased the H_2/mol glucose by 1.5mol since L-cysteine lowers the fermentation mechanism's oxidation-reducing efficiency. The inclusion of surfactants during the dark fermentation process has strengthened biohydrogen production in several ways. Surfactants are compounds that spontaneously aggregate into micelles when dissolved in solution above a certain concentration. These molecular assemblies form due to the surfactants' amphiphilic nature, with both hydrophilic and hydrophobic regions. There are mainly two types of surfactants based upon their nature, i.e. iconic or non-iconic. Mostly non-iconic surfactants are used to increase biohydrogen production (Nakama, 2017).

2.4.6 Bioaugmentation

This technique involves the addition of certain external bacteria which are added to the existing microbial consortia, this is mainly done to enhance the efficiency. Additionally, the stability of anaerobic bacteria during the process is another benefit of bioaugmentation. It can stop the fluctuations in the activities of microorganisms that occur because of thermophilic conditions and OLRs (Kotay & Das, 2009). Research conducted by Okonkwo et al. (2020) improved the hydrogen yield by 1.6% and 5% due to fluctuating temperature from 55°C to 35°C when they used *Themoanaerobacter, Thermohydrosulfuricus* DSM-1237, and *Thermotoga neapolitana* DSM-4359 on dark fermentation biohydrogen production culture.

2.4.7 Genetic Engineering

Among the most innovative technologies for improving the dark and photo-fermentation method is genetic engineering. Microorganisms that undergo metabolic manipulation act as the most useful tool in enhancing hydrogen production (Usman et al., 2019). Modifications in genes, removal of hydrogen uptake genes, and development of laboratory reactions are some of the approaches that can be followed to enhance H_2 production (Zhang et al., 2023). By improving the gene expression of Ni-Fe hydrogenase in *Pyrococcus furiosus*, Wu et al. (2018) found the increased hydrogen generation. These authors reduced methyl viologen in the form of an artificial electron donor to create three novel strains that were used to measure the specific hydrogen generation activity, and it was 1.14 times higher in genetically modified strains than in the control.

2.5 FUTURE PROSPECTS

Biohydrogen appears to be one of the best sources of energy for the future. Over the past couple of decades, many research studies on different hydrogen-generating techniques have been carried out. The most sustainable energy source for the manufacture of hydrogen is organic biomass. Biomass is plentiful, environmentally friendly, and simple to use. Green plant's photosynthesis ensures that net CO_2 emissions are almost nil during their life cycle. The classic natural gas reforming method will face competition from the economically feasible thermochemical pyrolysis and gasification methods of producing hydrogen. Another interesting technique for producing hydrogen for future commercial applications is biological dark fermentation. With the development of these technologies, biomass will play a bigger role in building a sustainable hydrogen economy.

REFERENCES

Abo-Hashesh, M., & Hallenbeck, P. C. (2011). Fermentative hydrogen production. In *Microbial Technologies in Advanced Biofuels Production* (pp. 77–92). Springer US.

Acar, C., & Dincer, I. (2019). Review and evaluation of hydrogen production options for a better environment. *Journal of Cleaner Production, 218*, 835–849.

Afsar, N., Özgür, E., Gürgan, M., Akköse, S., Yücel, M., Gündüz, U., & Eroglu, I. (2011). Hydrogen productivity of photosynthetic bacteria on dark fermenter effluent of potato steam peels hydrolysate. *International Journal of Hydrogen Energy, 36*(1), 432–438.

Akroum-Amrouche, D., Abdi, N., Lounici, H., & Mameri, N. (2011). Effect of physico-chemical parameters on biohydrogen production and growth characteristics by batch culture of *Rhodobacter sphaeroides* CIP 60.6. *Applied Energy*, *88*(6), 2130–2135.

Akroum-Amrouche, D., Abdi, N., Lounici, H., & Mameri, N. (2013, March). Biohydrogen production by dark and photo-fermentation processes. In *2013 International Renewable and Sustainable Energy Conference (IRSEC)* (pp. 499–503). IEEE, Ouarzazate, Morocco 7 – 9 March 2013.

Alemahdi, N., Man, H. C., Nasirian, N., & Yang, Y. (2015). Enhanced mesophilic bio-hydrogen production of raw rice straw and activated sewage sludge by co-digestion. *International Journal of Hydrogen Energy*, *40*(46), 16033–16044.

Assawamongkholsiri, T., & Reungsang, A. (2015). Photo-fermentation hydrogen production of Rhodobacter sp. KKU-PS1 isolated from an UASB reactor. *Electronic Journal of Biotechnology*, *18*(3), 221–230.

Baghchehsaraee, B., Nakhla, G., Karamanev, D., & Margaritis, A. (2010). Fermentative hydrogen production by diverse microflora. *International Journal of Hydrogen Energy*, *35*(10), 5021–5027.

Balachandar, G., Khanna, N., & Das, D. (2013). Biohydrogen production from organic wastes by dark fermentation. In Biohydrogen (pp. 103–144). Elsevier.

Barrozo, A., & Orio, M. (2019). Molecular electrocatalysts for the hydrogen evolution reaction: input from quantum chemistry. *Chem Sus Chem*, *12*(22), 4905–4915.

Bao, M. D., Su, H. J., & Tan, T. W. (2013). Dark fermentative bio-hydrogen production: effects of substrate pre-treatment and addition of metal ions or L-cysteine. *Fuel*, *112*, 38–44.

Basak, N., & Das, D. (2007). The prospect of purple non-sulfur (PNS) photosynthetic bacteria for hydrogen production: the present state of the art. *World Journal of Microbiology and Biotechnology*, *23*, 31–42.

Bianchi, L., Mannelli, F., Viti, C., Adessi, A., & De Philippis, R. (2010). Hydrogen-producing purple non-sulfur bacteria isolated from the trophic lake Averno (Naples, Italy). *International Journal of Hydrogen Energy*, *35*(22), 12216–12223.

Budiman, P. M., Wu, T. Y., Ramanan, R. N., & Md. Jahim, J. (2015). Improvement of biohydrogen production through combined reuses of palm oil mill effluent together with pulp and paper mill effluent in photofermentation. *Energy & Fuels*, *29*(9), 5816–5824.

Cai, J., & Wang, G. (2012). Hydrogen production by a marine photosynthetic bacterium, Rhodovulum sulfidophilum P5, isolated from a shrimp pond. *International Journal of Hydrogen Energy*, *37*(20), 15070–15080.

Cai, J., & Wang, G. (2014). Hydrogen production from glucose by a mutant strain of Rhodovulum sulfidophilum P5 in single-stage photofermentation. *International Journal of Hydrogen Energy*, *39*(36), 20979–20986.

Chandrasekhar, K., Kumar, S., Lee, B. D., & Kim, S. H. (2020). Waste-based hydrogen production for circular bioeconomy: Current status and future directions. *Bioresource Technology*, *302*, 122920.

Chandrasekhar, K., Lee, Y. J., & Lee, D. W. (2015). Biohydrogen production: strategies to improve process efficiency through microbial routes. *International Journal of Molecular Sciences*, *16*(4), 8266–8293.

Claassen, P. A., & De Vrije, T. (2006). Non-thermal production of pure hydrogen from biomass: HYVOLUTION. *International Journal of Hydrogen Energy*, *31*(11), 1416–1423.

Das, D., & Veziroğlu, T. N. (2001). Hydrogen production by biological processes: a survey of the literature. *International Journal of Hydrogen Energy*, *26*(1), 13–28.

Das, D., & Veziroglu, T. N. (2008). Advances in biological hydrogen production processes. *International Journal of Hydrogen Energy*, *33*(21), 6046–6057.

Dhanya, B. S., Mishra, A., Chandel, A. K., & Verma, M. L. (2020). Development of sustainable approaches for converting organic waste to bioenergy. *Science of the Total Environment*, *723*, 138109.

Ergal, İ., Gräf, O., Hasibar, B., Steiner, M., Vukotić, S., Bochmann, G., … & Rittmann, S. K. M. (2020). Biohydrogen production beyond the Thauer limit by precision design of artificial microbial consortia. *Communications Biology*, *3*(1), 443.

Eroglu, E., & Melis, A. (2011). Photobiological hydrogen production: recent advances and state of the art. *Bioresource Technology*, *102*(18), 8403–8413.

Fawzy, S., Osman, A. I., Doran, J., & Rooney, D. W. (2020). Strategies for mitigation of climate change: a review. *Environmental Chemistry Letters*, *18*, 2069–2094.

Gest, H., & Kamen, M. D. (1949). Photoproduction of molecular hydrogen by *Rhodospirillum rubrum*. *Science*, *109*(2840), 558–559.

Ghimire, A., Frunzo, L., Pirozzi, F., Trably, E., Escudie, R., Lens, P. N., & Esposito, G. (2015). A review on dark fermentative biohydrogen production from organic biomass: process parameters and use of by-products. *Applied Energy*, *144*, 73–95.

Goswami, R. K., Mehariya, S., Obulisamy, P. K., & Verma, P. (2021). Advanced microalgae-based renewable biohydrogen production systems: a review. *Bioresource Technology*, *320*, 124301.

Gupta, S., Fernandes, A., Lopes, A., Grasa, L., & Salafranca, J. (2024). Photo-fermentative bacteria used for hydrogen production. *Applied Sciences*, *14*(3), 1191.

Hallenbeck, P. C., & Benemann, J. R. (2002). Biological hydrogen production; fundamentals and limiting processes. *International Journal of Hydrogen Energy*, *27*(11–12), 1185–1193.

Han, W., Ye, M., Zhu, A. J., Zhao, H. T., & Li, Y. F. (2015). Batch dark fermentation from enzymatic hydrolyzed food waste for hydrogen production. *Bioresource Technology*, *191*, 24–29.

Harwood, C. S. (2009). Degradation of aromatic compounds by purple nonsulfur bacteria. In The Purple Phototrophic Bacteria (pp. 577–594). Springer Netherlands.

Hay, J. X. W., Wu, T. Y., Juan, J. C., & Md. Jahim, J. (2013). Biohydrogen production through photo fermentation or dark fermentation using waste as a substrate: overview, economics, and future prospects of hydrogen usage. *Biofuels, Bioproducts and Biorefining*, *7*(3), 334–352.

Hidalgo, D., & Martín-Marroquín, J. M. (2023). Enhanced production of biohydrogen through combined operational strategies. *JOM*, *75*(3), 718–726.

Hitam, C. N. C., & Jalil, A. A. (2023). A review on biohydrogen production through photo-fermentation of lignocellulosic biomass. *Biomass Conversion and Biorefinery*, *13*(10), 8465–8483.

Hosseini, S. S., Aghbashlo, M., Tabatabaei, M., Younesi, H., & Najafpour, G. (2015). Exergy analysis of biohydrogen production from various carbon sources via anaerobic photosynthetic bacteria (*Rhodospirillum rubrum*). *Energy*, *93*, 730–739.

Intanoo, P., Chaimongkol, P., & Chavadej, S. (2016). Hydrogen and methane production from cassava wastewater using two-stage up-flow anaerobic sludge blanket reactors (UASB) with an emphasis on maximum hydrogen production. *International Journal of Hydrogen Energy*, *41*(14), 6107–6114.

Kadir, A. A., Azhari, N. W., & Jamaludin, S. N. (2016). An overview of organic waste in composting. In *MATEC Web of Conferences* (Vol. 47, p. 05025). EDP Sciences, Hong Kong, China, April 26–27, 2016.

Kapdan, I. K., & Kargi, F. (2006). Bio-hydrogen production from waste materials. *Enzyme and Microbial Technology*, *38*(5), 569–582.

Khanna, N., & Das, D. (2013). Biohydrogen production by dark fermentation. *Wiley Interdisciplinary Reviews: Energy and Environment*, *2*(4), 401–421.

Kim, S., Choi, K., Kim, J. O., & Chung, J. (2013). Biological hydrogen production by anaerobic digestion of food waste and sewage sludge treated using various pretreatment technologies. *Biodegradation*, *24*, 753–764.

Kotay, S. M., & Das, D. (2009). Novel dark fermentation involving bioaugmentation with the constructed bacterial consortium for enhanced biohydrogen production from pretreated sewage sludge. *International Journal of Hydrogen Energy*, *34*(17), 7489–7496.

Laurinavichene, T., Tekucheva, D., Laurinavichius, K., & Tsygankov, A. (2018). Utilization of distillery wastewater for hydrogen production in one-stage and two-stage processes involving photofermentation. *Enzyme and Microbial Technology*, *110*, 1–7.

Lima, D. M. F., Moreira, W. K., & Zaiat, M. (2013). Comparison of the use of sucrose and glucose as a substrate for hydrogen production in an upflow anaerobic fixed-bed reactor. *International Journal of Hydrogen Energy*, *38*(35), 15074–15083.

Lin, C. Y., & Jo, C. H. (2003). Hydrogen production from sucrose using an anaerobic sequencing batch reactor process. *Journal of Chemical Technology & Biotechnology: International Research in Process, Environmental & Clean Technology*, *78*(6), 678–684.

Lin, C. Y., Nguyen, T. M. L., Chu, C. Y., Leu, H. J., & Lay, C. H. (2018). Fermentative biohydrogen production and its byproducts: a mini-review of current technology developments. *Renewable and Sustainable Energy Reviews*, *82*, 4215–4220.

Liu, X., Huang, L., Rensing, C., Ye, J., Nealson, K. H., & Zhou, S. (2021). Syntrophic interspecies electron transfer drives carbon fixation and growth by Rhodopseudomonas palustris under dark, anoxic conditions. *Science Advances*, *7*(27), eabh1852.

Liu, B. F., Jin, Y. R., Cui, Q. F., Xie, G. J., Wu, Y. N., & Ren, N. Q. (2015). Photo-fermentation hydrogen production by Rhodopseudomonas sp. nov. strain A7 isolated from the sludge in a bioreactor. *International Journal of Hydrogen Energy*, *40*(28), 8661–8668.

Mahmoodi, M., Gustavo, E., Mohammad, M., & Rafiee, R. (2022). Dynamic optimization of volatile fatty acids to enrich biohydrogen production using a deep learning neural network. *Biomass Conversion Biorefinery*, *14*, 8003–8014.

Mahmoodi-Eshkaftaki, M., & Mockaitis, G. (2022). Structural optimization of biohydrogen production: impact of pretreatments on volatile fatty acids and biogas parameters. *International Journal of Hydrogen Energy*, *47*(11), 7072–7081.

Melis, A. (2002). Green alga hydrogen production: progress, challenges, and prospects. *International Journal of Hydrogen Energy*, *27*(11), 1217–1228.

Mishra, P., Krishnan, S., Rana, S., Singh, L., Sakinah, M., & Ab Wahid, Z. (2019). Outlook of fermentative hydrogen production techniques: an overview of dark, photo, and integrated dark-photo fermentative approach to biomass. *Energy Strategy Reviews*, *24*, 27–37.

Mohanakrishna, G., Goud, R. K., Mohan, S. V., & Sarma, P. N. (2010). Enhancing biohydrogen production through sewage supplementation of composite vegetable based market waste. *International Journal of Hydrogen Energy*, *35*(2), 533–541.

Mona, S., Kumar, S. S., Kumar, V., Parveen, K., Saini, N., Deepak, B., & Pugazhendhi, A. (2020). Green technology for sustainable biohydrogen production (waste to energy): a review. *Science of the Total Environment*, *728*, 138481.

Nagarajan, D., Chang, J. S., & Lee, D. J. (2020). Pretreatment of microalgal biomass for efficient biohydrogen production–Recent insights and future perspectives. *Bioresource Technology*, *302*, 122871.

Nakama, Y. (2017). Surfactants. In *Cosmetic Science and Technology* (pp. 231–244). Elsevier.

Nath, K., & Das, D. (2004). Biohydrogen production as a potential energy resource–Present state-of-art. *Journal of Scientific and Industrial Research*, *63*, 729–738.

Nikolaidis, P., & Poullikkas, A. (2017). A comparative overview of hydrogen production processes. *Renewable and Sustainable Energy Reviews*, *67*, 597–611.

Obeid, J., Flaus, J. M., Adrot, O., Magnin, J. P., & Willison, J. C. (2010). State estimation of a batch hydrogen production process using the photosynthetic bacteria Rhodobacter capsulatus. *International Journal of Hydrogen Energy*, *35*(19), 10719–10724.

Okonkwo, O., Papirio, S., Trably, E., Escudie, R., Lakaniemi, A. M., & Esposito, G. (2020). Enhancing thermophilic dark fermentative hydrogen production at high glucose concentrations via bioaugmentation with Thermotoga neapolitana. *International Journal of Hydrogen Energy*, *45*(35), 17241–17249.

Osman, A. I., Abu-Dahrieh, J. K., Cherkasov, N., Fernandez-Garcia, J., Walker, D., Walton, R. I., … & Rebrov, E. (2018). A highly active and synergistic Pt/Mo_2C/Al_2O_3 catalyst for water-gas shift reaction. *Molecular Catalysis*, *455*, 38–47.

Osman, A. I., Deka, T. J., Baruah, D. C., & Rooney, D. W. (2020). Critical challenges in biohydrogen production processes from the organic feedstocks. *Biomass Conversion and Biorefinery*, *13*, 8383–8401.

Pandu, K., & Joseph, S. (2012). Comparisons and limitations of biohydrogen production processes: a review. *International Journal of Advances in Engineering & Technology*, *2*(1), 342.

Paramesh, K., Reddy, N. L., Shankar, M. V., & Chandrasekhar, T. (2018). Enhancement of biological hydrogen production using green alga *Chlorococcum minutum*. *International Journal of Hydrogen Energy*, *43*(8), 3957–3966.

Patel, S. K., Kumar, P., Mehariya, S., Purohit, H. J., Lee, J. K., & Kalia, V. C. (2014). Enhancement in hydrogen production by co-cultures of Bacillus and Enterobacter. *International Journal of Hydrogen Energy*, *39*(27), 14663–14668.

Pugazhendhi, A., Shobana, S., Nguyen, D. D., Banu, J. R., Sivagurunathan, P., Chang, S. W., … & Kumar, G. (2019). Application of nanotechnology (nanoparticles) in dark fermentative hydrogen production. *International Journal of Hydrogen Energy*, *44*(3), 1431–1440.

Rai, P. K., Singh, S. P., Asthana, R. K., & Singh, S. (2014). Biohydrogen production from sugarcane bagasse by integrating dark and photo-fermentation. *Bioresource Technology*, *152*, 140–146.

Ramírez-Morales, J. E., Tapia-Venegas, E., Campos, J. L., & Ruiz-Filippi, G. (2019). Operational behavior of a hydrogen extractive membrane bioreactor (HEMB) during mixed culture acidogenic fermentation. *International Journal of Hydrogen Energy*, *44*(47), 25565–25574.

Rao, R., & Basak, N. (2021). Development of novel strategies for higher fermentative biohydrogen recovery along with novel metabolites from organic wastes: the present state of the art. *Biotechnology and Applied Biochemistry*, *68*(3), 421–444.

Rawat, S., Rautela, A., Yadav, I., Misra, S., & Kumar, S. (2023). A comprehensive review on enhanced biohydrogen production: pretreatment, applied strategies, techno-economic assessment, and future perspective. *Bioenergy Research*, *16*(4), 2131–2154.

Rizwan, M., Shah, S. H., Mujtaba, G., Mahmood, Q., Rashid, N., & Shah, F. A. (2019). Ecofuel feedstocks and their prospect. In Advanced Biofuels (pp. 3–16). Woodhead Publishing.

Rozendal, R. A., Jeremiasse, A. W., Hamelers, H. V., & Buisman, C. J. (2008). Hydrogen production with a microbial biocathode. *Environmental Science & Technology*, *42*(2), 629–634.

Saifuddin, N., & Priatharsini, P. (2016). Developments in bio-hydrogen production from algae: a review. *Research Journal of Applied Sciences, Engineering and Technology*, *12*(9), 968–982.

Saleem, M., Lavagnolo, M. C., & Spagni, A. (2018). Biological hydrogen production via dark fermentation by using a side-stream dynamic membrane bioreactor: effect of substrate concentration. *Chemical Engineering Journal*, *349*, 719–727.

Sampath, P., Brijesh, Reddy, K. R., Reddy, C. V., Shetti, N. P., Kulkarni, R. V., & Raghu, A. V. (2020). Biohydrogen production from organic waste–a review. *Chemical Engineering & Technology*, *43*(7), 1240–1248.

Sanghvi, A. H., Manjoo, A., Rajput, P., Mahajan, N., Rajamohan, N., & Abrar, I. (2024). Advancements in biohydrogen production–a comprehensive review of technologies, lifecycle analysis, and future scope. *RSC Advances*, *14*(49), 36868–36885.

Sarangi, P. K., & Nanda, S. (2020). Biohydrogen production through dark fermentation. *Chemical Engineering & Technology*, *43*(4), 601–612.

Sarma, S., Ortega, D., Minton, N. P., Dubey, V. K., & Moholkar, V. S. (2019). Homologous overexpression of hydrogenase and glycerol dehydrogenase in Clostridium pasteurianum to enhance hydrogen production from crude glycerol. *Bioresource Technology*, *284*, 168–177.

Sekoai, P. T., Chunilall, V., Sithole, B. B., Habimana, O., Ndlovu, S., & Ezeokoli, O. (2022). Elucidating the role of biofilm-forming microbial communities in fermentative biohydrogen process: an overview. *Microorganisms*, *10*, 1924.

Sharma, A., & Arya, S. K. (2017). Hydrogen from algal biomass: a review of the production process. *Biotechnology Reports*, *15*, 63–69.

Silva, F. M. S., Mahler, C. F., Oliveira, L. B., Bassin, J. P. (2018). Hydrogen and methane production in a two-stage anaerobic digestion system by co-digestion of food waste, sewage sludge, and glycerol. *Waste Management*, 76, 339–349.

Singh, L., & Wahid, Z. A. (2015). Methods for enhancing bio-hydrogen production from the biological process: a review. *Journal of Industrial and Engineering Chemistry*, *21*, 70–80.

Singh, L., Wahid, Z. A., Siddiqui, M. F., Ahmad, A., Rahim, M. H. A., & Sakinah, M. (2013). Biohydrogen production from palm oil mill effluent using immobilized *Clostridium butyricum* EB6 in polyethylene glycol. *Process Biochemistry*, *48*(2), 294–298.

Singh, R., Kumar, R., Sarangi, P. K., Kovalev, A. A., & Vivekanand, V. (2023). Effect of physical and thermal pretreatment of lignocellulosic biomass on biohydrogen production by thermochemical route: a critical review. *Bioresource Technology*, *369*, 128458.

Soto, L. R., Byrne, E., van Niel, E. W., Sayed, M., Villanueva, C. C., & Hatti-Kaul, R. (2019). Hydrogen and polyhydroxybutyrate production from wheat straw hydrolysate using Caldicellulosiruptor species and *Ralstonia eutropha* in a coupled process. *Bioresource Technology*, *272*, 259–266.

Srikanth, S., Mohan, S. V., Devi, M. P., Peri, D., & Sarma, P. N. (2009). Acetate and butyrate as substrates for hydrogen production through photo-fermentation: process optimization and combined performance evaluation. *International Journal of Hydrogen Energy*, *34*(17), 7513–7522.

Srivastava, N., Srivastava, M., Kushwaha, D., Gupta, V. K., Manikanta, A., Ramteke, P. W., & Mishra, P. K. (2017). Efficient dark fermentative hydrogen production from enzyme-hydrolyzed rice straw by Clostridium pasteurianum (MTCC116). *Bioresource Technology*, *238*, 552–558.

Tao, Y., He, Y., Wu, Y., Liu, F., Li, X., Zong, W., & Zhou, Z. (2008). Characteristics of a new photosynthetic bacterial strain for hydrogen production and its application in wastewater treatment. *International Journal of Hydrogen Energy*, *33*(3), 963–973.

Teke, G. M., Anye Cho, B., Bosman, C. E., Mapholi, Z., Zhang, D., & Pott, R. W. M. (2024). Towards industrial biological hydrogen production: a review. *World Journal of Microbiology and Biotechnology*, *40*(1), 37.

Tomasik, P., & Horton, D. (2012). Enzymatic conversions of starch. *Advances in Carbohydrate Chemistry and Biochemistry*, *68*, 59–436.

UNEP Emissions gap report (2022) https://www.unep.org/resources/emissions-gap-report-2022. Accessed Mar 2023.

Usman, T. M., Banu, J. R., Gunasekaran, M., & Kumar, G. (2019). Biohydrogen production from industrial wastewater: an overview. *Bioresource Technology Reports*, *7*, 100287.

Vardar-Schara, G., Maeda, T., & Wood, T. K. (2008). Metabolically engineered bacteria for producing hydrogen via fermentation. *Microbial Biotechnology*, *1*(2), 107–125.

Wainaina, S., Parchami, M., Mahboubi, A., Horváth, I. S., & Taherzadeh, M. J. (2019). Food waste-derived volatile fatty acids platform using an immersed membrane bioreactor. *Bioresource Technology*, *274*, 329–334.

Wu, C. H., Ponir, C. A., Haja, D. K., & Adams, M. W. (2018). Improved production of the NiFe-hydrogenase from Pyrococcus furiosus by increased expression of maturation genes. *Protein Engineering, Design and Selection*, *31*(9), 337–344.

Xing, Y., Li, Z., Fan, Y., & Hou, H. (2010). Biohydrogen production from dairy manures with acidification pretreatment by anaerobic fermentation. *Environmental Science and Pollution Research*, *17*, 392–399.

Yagi, T., & Higuchi, Y. (2013). Studies on hydrogenase. *Proceedings of the Japan Academy, Series B*, *89*(1), 16–33.

Yang, G., & Wang, J. (2017). Enhanced hydrogen production from sewage sludge by co-fermentation with forestry wastes. *Energy & Fuels*, *31*(9), 9633–9641.

Yang, G., & Wang, J. (2018). Various additives for improving dark fermentative hydrogen production: a review. *Renewable and Sustainable Energy Reviews*, *95*, 130–146.

Zhang, Q., Dai, W., Wang, X., & Li, J. (2020). Elevated CO_2 concentration affects the defense of tobacco and melon against lepidopteran larvae through the jasmonic acid signaling pathway. *Scientific Reports*, *10*(1), 4060.

Zhang, Y., & Angelidaki, I. (2014). Microbial electrolysis cells turning to be versatile technology: recent advances and future challenges. *Water Research*, *56*, 11–25.

Zhang, J., Xue, D., Wang, C., Fang, D., Cao, L., & Gong, C. (2023). Genetic engineering for biohydrogen production from microalgae. *IScience*, *26*(8).

Zhao, C., Zhang, N., Zheng, H., Zhu, Q., Utsumi, M., & Yang, Y. (2019). Effective and long-term continuous bio-hydrogen production by optimizing fixed-bed material in the bioreactor. *Process Biochemistry*, *83*, 55–63.

Zhu, Z., Shi, J., Zhou, Z., Hu, F., & Bao, J. (2010). Photo-fermentation of Rhodobacter sphaeroides for hydrogen production using lignocellulose-derived organic acids. *Process Biochemistry*, *45*(12), 1894–1898.

3 Cell Immobilisation Technologies for Efficient Biohydrogen Production

Entrapment, Encapsulation, and Adsorption, Types of Carriers, Nano-Based Immobilisation, and Its Prospects

Prasann Kumar

3.1 INTRODUCTION

Biohydrogen has emerged as an alternative, clean, and sustainable energy carrier with high energy content, capable of replacing fossil fuels to address global energy and environmental challenges. Hydrogen combustion also emits only water, unlike hydrocarbons, which produce significant amounts of greenhouse gases and pollutants, making it an environmentally friendly option (Liao et al., 2017; Sanghvi et al., 2024). Biohydrogen production, which typically occurs through biological methods using microorganisms, is a renewable and sustainable alternative that utilises abundant organic materials, including

DOI: 10.1201/9781003568612-3

agricultural waste, wastewater, and biomass. This aligns with international efforts to establish a circular economy, which transforms waste into valuable products while utilising fewer non-renewable resources and generating a smaller carbon footprint. Notably, biohydrogen has gained importance as an energy source and as a means to achieve various international sustainability goals, including reducing greenhouse gas emissions and mitigating climate change, in response to the ever-growing global population and increasing energy demands.

Biohydrogen production using biological systems must be both efficient and stable; however, these factors are among the major bottlenecks for higher-scale biohydrogen production. Microbial hydrogen-producing species of bacteria and algae that produce hydrogen require optimised conditions for high yield, whereas to address the issues of washout and instability due to variations in environmental conditions, typically free-living microorganisms are proven to be candidate organisms, but they are often highly contaminated. Immobilisation traps the microbial cells under a support matrix, allowing them to maintain their metabolic function in a confined area (Qi et al., 2023; Rady et al., 2024). This method enables simplified cell stability, profiling complex metabolic activity, and easy separation and recycling of microbial cells. Immobilising cells, for instance, makes them less sensitive to environmental fluctuations, allowing for the constant generation of biohydrogen.

The biohydrogen production cell immobilisation technologies can be broadly divided into entrapment, encapsulation, and adsorption. Entrapment: You insert the cells into a porous matrix (e.g., alginate or polyacrylamide) so that substrates and products can diffuse, but the cells remain trapped. On the other hand, encapsulation produces a transport-restricted microstructure by interring microbial cells within semi-permeable membranes or capsules, leading to enhanced cellular survival. Adsorption methods are based on physical and/or chemical interactions to bind the cells to solid carriers, such as activated carbon or silica (Mohd Jamaludin et al., 2024; Pachaiappan et al., 2024). All the methods stated above have their significance and disadvantages, and the effectiveness of each technique varies according to the specific biohydrogen production system, including microorganisms used, substrate availability, and operational parameters (Table 3.1).

The carrier's properties, including biocompatibility, porosity, mechanical strength, and chemical and physical properties, can also influence the selection of the immobilisation method, as they are decisive factors in the performance and stability of the immobilised system.

Other recent advances in cell immobilisation technologies are leveraging novel materials and techniques, such as nano-based immobilisation methods, which utilise the unique physical and chemical properties of different nanomaterials. Nano-carriers, such as graphene, carbon nanotubes, and nanofibres, have been reported to significantly enhance biohydrogen production due to their higher surface area for cell adhesion and increased mass transfer rates. Moreover, these materials exhibit robust mechanical strength and chemical stability, making them suitable for long-term industrial applications (Bolonhesi et al., 2023; Pomdaeng et al., 2024). Consolidating nanotechnology and immobilisation strategies set a new milestone in utilising biohydrogen production systems by resolving scaling economics and operational stability issues. These advances demonstrate the potential of immobilisation technologies to facilitate biohydrogen generation as a realistic and sustainable energy carrier.

The growth of biohydrogen devices is also discussed, along with cell immobilisation technologies and their role in enhancing the efficiency and stability of biohydrogen production systems. To this end, the chapter highlights the principles, techniques, and progress made in immobilisation that may ultimately lead to sustainable and scalable biohydrogen production technologies (Figure 3.1). It may, therefore, contribute to achieving a low-carbon, renewable energy future, along with its related economic and environmental benefits (Modzelewska et al., 2024; Wannapokin et al., 2022).

TABLE 3.1 Different types of technology and their application in biohydrogen production

TECHNOLOGY TYPE	*KEY CHARACTERISTICS*	*TYPES OF CARRIERS/ MATERIALS USED*	*ADVANTAGES*	*LIMITATIONS*	*APPLICATIONS IN BIOHYDROGEN PRODUCTION*	*PROSPECTS AND INNOVATIONS*
Entrapment	Cells are physically trapped within a solid matrix, which restricts movement but allows substrate diffusion.	Alginate, agar, polyacrylamide, carrageenan, silica gels.	It is a straightforward process characterised by high cell loading, protection from shear forces, and enhanced stability.	Mass transfer limitations, potential cell leakage, and reduced catalytic efficiency.	Hydrogen production from substrates like glucose and wastewater.	Development of hybrid matrices for improved stability and mass transfer.
Encapsulation	Cells are enclosed within semi-permeable membranes or microcapsules, allowing for the exchange of nutrients.	Chitosan, alginate beads, lipid-based capsules, synthetic polymers.	Controlled microenvironment, enhanced protection, and prolonged operational stability.	Complexity of fabrication, cost, and potential capsule breakage.	Biohydrogen production using immobilised microbial consortia.	Use of innovative materials and stimuli-responsive capsules for optimised performance.
Adsorption	Cells adhere to the surface of solid supports through physical or chemical interactions.	Activated carbon, clay, glass beads, porous ceramics, biochar.	Easy recovery, low cost, reusable supports, and minimal impact on microbial activity.	Weak adhesion under certain conditions, desorption, and limited cell loading.	Adsorbed microbes are used for hydrogen production in bioreactors.	Nano-coatings on carriers to improve adhesion and reusability.
Nano-based immobilisation	Utilises nanoparticles and nanostructures to immobilise cells with enhanced surface area and interaction.	Nanoparticles (e.g., silica, magnetite), nanofibres, nanotubes.	High surface area, enhanced mass transfer, targeted immobilisation, and improved activity.	High cost, potential toxicity of nanomaterials, and scalability issues.	Enhanced hydrogen production through nanoscale precision in microbial immobilisation.	Development of eco-friendly, biodegradable nanomaterials and scalable nano-systems.

Source: Bolonhesi et al. (2023) and Pomdaeng et al. (2024)

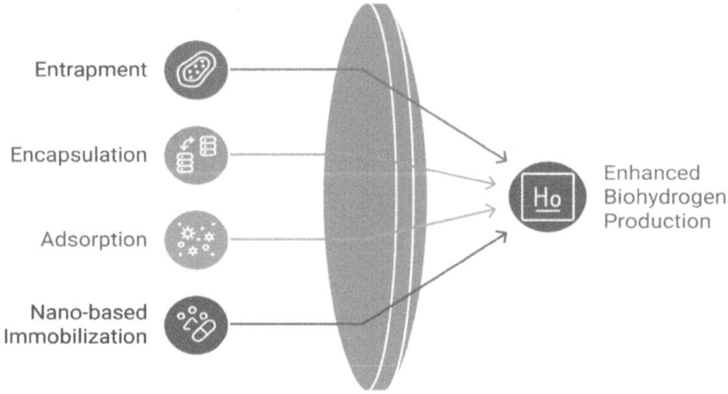

FIGURE 3.1 Strategies for optimal biohydrogen production.

3.2 PRINCIPLES OF CELL IMMOBILISATION

Immobilisation is the physical confinement or anchoring of living cells in a defined space, either within or on a supporting matrix, while retaining their biological activity. The prevalent method preserves high cell densities, shields cells from environmental strain, and enables their reapplication in many biotechnological processes. In industrial biotechnology, immobilisation was first introduced in enzyme technology but later extended to whole cells, providing several benefits, notably in the bioproduction of biohydrogen. Depending on the application's needs, immobilisation can be carried out using various methods, such as entrapment, encapsulation, adsorption, and covalent bonding. The choice of immobilisation depends on the microorganism, substrate, and operational conditions. In this method, cells are entrapped in a porous structure, such as alginate or polyacrylamide, which allows the diffusion of substrates and products while retaining the cells. Encapsulation refers to the process of enclosing cells in semi-permeable membranes or capsules, thereby providing them with a protective microenvironment that further promotes their viability and metabolic activity (Engliman et al., 2020; Modzelewska et al., 2024). Adsorption, a technique based on the surface, involves the physical or chemical binding of cells to carriers, including activated carbon, silica, or natural polymers. The efficiency of these approaches relies on properties such as porosity, biocompatibility, mechanical strength, and the metabolic features of the immobilised cells (Han et al., 2015; Jamali et al., 2019).

These overcome the challenges in traditional microbial hydrogen production systems. One key advantage is increased stability and resilience in response to changing environmental conditions. Backwash also occurs in continuous-flow systems, particularly large-scale systems, due to the potential for washout of free cells. In contrast, immobilised cells are trapped in the matrix or firmly attached to the carrier, which minimises the chances of cell loss and provides reproducibility of results over relatively long durations. This confinement enables the retention of high cell densities, which are essential for achieving high hydrogen yields. Furthermore, bounded systems resist toxic by-products and inhibitory compounds spawned during fermentation, which can adversely affect free cells. The immobilisation matrix usually serves as a barrier that helps to protect cells from harmful agents and improves their activities (Hao et al., 2024; Wannapokin et al., 2022).

The other main benefit of immobilised cells is their simple separation and reuse, making the production of biohydrogen from water more commercially viable. In free-cell systems, separating microbial biomass from the fermentation broth is an energy-intensive and costly process, particularly at the commercial scale. In such cases, immobilised systems would not require these separation processes because the final product would be composed of the product itself, and the matrix or carrier would remain in it. This enables the reuse of the same microbial community multiple times while lowering operational expenses and waste production. In addition, immobilised cells are generally metabolically active for extended periods, as the immobilisation method may allow for a microenvironment that favours cell viability and activity. This extended operational lifespan translates into enhanced productivity and overall cost-effectiveness.

The immobilised arrangement also has side benefits regarding reactor design and scalability. Cells can be immobilised and assimilated into various bioreactor systems (e.g., packed-bed reactors, fluidised-bed reactors (FBRs), and membrane reactors), which offer advantages in terms of mass transfer, substrate utilisation, and product recovery. Packed-bed reactors offer high cell densities and fermentation efficiency, while FBRs offer improved mass transfer and mixing (Moustafa et al., 2023; Pandit et al., 2024). Membrane reactors, for example, enable the selective retention of cells and the separation of hydrogen gas from other metabolites, thereby enhancing the overall process efficiency. With its versatile reactor configurations and the stability and reusability of immobilised cells, immobilisation is a promising strategy for large-scale biohydrogen production.

Regarding biohydrogen production, the differences between immobilised and free cells become apparent in terms of process performance and operational characteristics. While free cells are the most straightforward form to implement and the least expensive to keep, they are fundamentally limited by their sensitivity to environmental variation and operational issues. Free cells are susceptible to washout in both batch and continuous systems, particularly at high flow rates or under high shear forces. Glucose-responsive bacteria manage to counteract the potential acidification of the medium. Still, their effect at the organism or population level is limited and requires continuous glucose feeding to the kinetic apparatus. Cells released free have also been reported to exhibit substrate inhibition and be more affected by toxic by-products, drastically affecting their metabolic activity and hydrogen production rates (Jamali et al., 2019; Min Woon et al., 2024). Immobilised cells, shielded in their matrix or linked to carriers, are much more resilient to such pressures. Unlike free cells, they can perform continuously and sustainably in various working environments (Figure 3.2).

The other significant difference is in the productivity and operational life of the systems. Due to their overwhelming confinement in a physical domain, immobilised cells can have longer-term higher productivity than free cells, which enhances their stabilising capacity. However, a favourable microenvironment is generated during immobilisation to promote prolonged metabolic activity and minimise frequent cell replacement (Table 3.2).

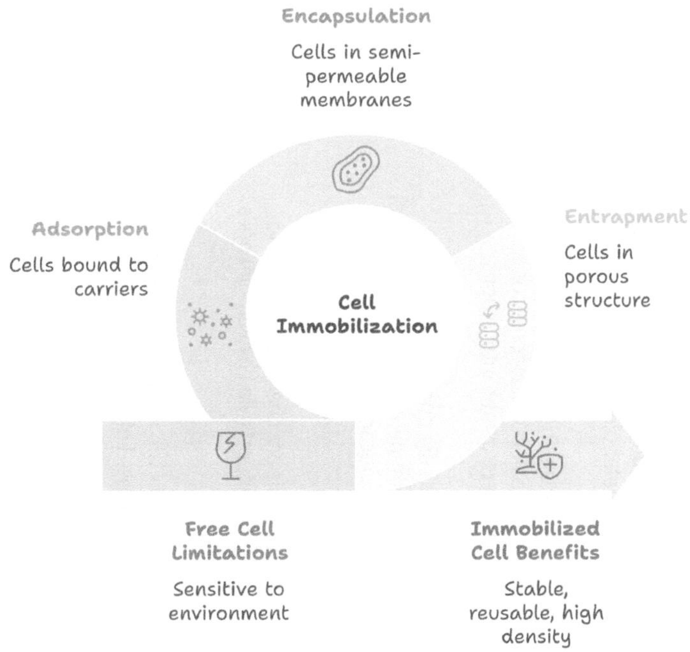

FIGURE 3.2 Exploring cell immobilisation methods and benefits.

TABLE 3.2 Different principles of cell immobilisation

IMMOBILISATION METHOD	*KEY MECHANISM*	*MATERIALS/ SUPPORTS USED*	*ADVANTAGES*	*LIMITATIONS*	*APPLICATIONS*	*FUTURE DIRECTIONS*
Entrapment	Cells are physically confined within a porous matrix, which allows for the diffusion of nutrients and metabolites.	Alginate, polyacrylamide, agar, carrageenan.	A simple method protects cells from environmental stress.	Limited mass transfer, potential cell leakage.	Biohydrogen production, bioremediation.	Development of hybrid materials with enhanced porosity and stability.
Adsorption	Cells adhere to solid supports via weak physical forces or electrostatic interactions.	Activated carbon, glass beads, biochar.	Easy recovery, low cost, minimal impact on cell activity.	Weak binding occurs under certain conditions, resulting in a low loading capacity.	Wastewater treatment and biofuel production.	Surface modifications to carriers for improved binding efficiency.
Encapsulation	Cells are enclosed in semi-permeable membranes or microcapsules.	Chitosan, lipid-based capsules, synthetic polymers.	Controlled microenvironment, prolonged operational stability.	Costly, prone to mechanical failure of capsules.	Enzyme production, biohydrogen generation.	Use of stimuli-responsive or innovative materials for enhanced functionality.
Covalent binding	Cells are chemically bound to supports via covalent linkages.	Functionalised silica, activated polymers.	Strong and stable binding, reusable supports.	Potential damage to cells and high cost of functionalised supports.	Biocatalysis, biosensors.	Biocompatible and non-toxic covalent binding techniques.
Cross-linking	Cells are immobilised by forming cross-linked networks with agents like glutaraldehyde.	Glutaraldehyde, polyacrylamide.	Robust structure, high cell loading.	The toxicity of cross-linking agents and the rigidity they impose limit diffusion.	Industrial enzyme production, bioreactors.	Development of non-toxic cross-linking agents.
Nano-based immobilisation	Nanomaterials offer a high surface area and precise cell binding, resulting in enhanced activity.	Nanoparticles, nanotubes, nanofibres.	High surface area, enhanced catalytic activity.	High cost, potential nanotoxicity, scalability issues.	Biohydrogen production, bioremediation.	Eco-friendly nanomaterials and scalable nano-immobilisation techniques.

Source: Rai et al. (2022), Woo et al. (2022), Ananthi et al. (2024), Han et al. (2015), Pandit et al. (2024)

The overall efficiency of the biohydrogen production process is enhanced and operational costs are reduced, making it a more economically viable option. The cells are immobilised so that they can be used multiple times, eliminating the need for regular inoculation and recovery, thereby streamlining the process and reducing waste generation.

Immobilised systems also provide a much broader range of flexibility and scalability than free-cell systems from an engineering standpoint. The high volumetric productivity provided by immobilised cells facilitates further enhancement of bioreactor performance when incorporated into specific reactor designs, including packed-bed reactors or FBRs, which also optimise mass transfer, substrate utilisation, and product recovery (Han et al., 2015; Jamali et al., 2021). These reactor configurations can be customised according to the specific demands of the biohydrogen production process, thus helping to maximise efficiency and minimise energy consumption. While easier to design and operate, free-cell systems cannot match this flexibility and are often limited by cell density, stability, and product recovery.

Although immobilisation comes with these benefits, it is not without struggles. Generally, although promising, immobilised systems are resource-intensive and time-consuming during preparation compared to their free-cell counterparts, which typically use advanced actives, such as nanocarriers. When selecting the ideal carrier and immobilisation method, it is essential to consider cost, biocompatibility, and scalability. The immobilisation procedure itself can also decrease cell viability or metabolic activity, especially when cells are subjected to intense chemical treatments or physical conditions during the immobilisation process (Rai et al., 2022; Woo et al., 2022). However, the techniques' limitations have been overcome through advances in material science and biotechnology, which have led to new carriers and methods that improve the performance and utility of immobilised systems.

Ultimately, immobilisation is a revolutionary methodology for biohydrogen generation, providing advantages over classical free-cell systems. By capturing microbial cells in or on supportive matrices, immobilisation improves stability, productivity, and reusability, overcoming many of the limitations of free-cell systems. Although the first step of immobilised systems may require more significant resources and technical expertise, the long-term advantages of efficiency, scalability, and cost-effectiveness render immobilisation a highly promising strategy for achieving sustainable biohydrogen production. There is growing evidence for the transformative potential of these systems to enable biohydrogen production as research and implementation in the field advance and novel nanotechnology and carrier materials are identified and introduced into the landscape (Ananthi et al., 2024; Han et al., 2015; Pandit et al., 2024).

3.3 TECHNIQUES FOR CELL IMMOBILISATION

The immobilisation of microbial systems for biohydrogen production has made great strides in developing various techniques, including trapping, encapsulation, and adsorption. However, each strategy has different stability, performance, and efficiency benefits, and together, they are integral parts of modern bioengineering designs. Entrapment keeps microbial cells physically confined within a semi-permeable matrix that allows diffusion of substrates and products while retaining the cells. A few materials frequently used for entrapment are alginate, polyacrylamide, carrageenan, and agar. Alginate exhibits the best characteristics, including biocompatibility, gelation with calcium ions, and cost-effectiveness. While polyacrylamide is highly effective in generating powerful gels, its use is limited due to concerns about potential toxicity. The microenvironment space provided by entrapment matrices promotes cell viability while shielding cells from shear forces, toxic by-products, and environmental fluctuations (Ananthi et al., 2024; Jiao et al., 2024). These matrices' characteristics are their porous structures, which afford rapid mass transfer; thus, substrates and hydrogen gas can diffuse in and out without compromising metabolic activity. This mechanism is especially beneficial in biohydrogen production systems, where the immobilised microbial consortium should optimally convert organic substrates into hydrogen in an anaerobic environment.

The advantages of entrapment in biohydrogen production include improved stability, enhanced yield, and a significantly longer working life compared to free-cell systems. Biohydrogen has been obtained from biochemical substrates such as glucose, wastewater, and lignocellulosic hydrolysates using immobilised *Clostridium* and *Enterobacter* species in alginate beads. They exhibit lower lag phases, better tolerance to environmental stresses, and greater productivity across multiple cycles. Treatments have also enabled high-density bioreactors such as packed-bed reactors and FBRs, where immobilised cells are contained within the reactor with minimal cell washout, resulting in higher volumetric productivity. Although entrapment has advantages, it also has drawbacks. The mechanical integrity of the matrix can deteriorate over time, particularly in continuous systems, leading to cell leakage and a reduction in efficiency. These challenges are gradually being overcome through improvements in materials science, which incorporate nanomaterials or composite hydrogels to enhance the mechanical properties and durability of entrapment matrices (Boshagh et al., 2019; Jamaludin et al., 2024; Woon et al., 2024).

Microencapsulation is another important immobilisation method in which the microbial cells are trapped in semi-permeable membranes or microcapsules. This approach provides a buffered microenvironment that maintains cell viability and activity, allowing for the exchange of substrates and products. Due to their biocompatibility, water-retaining ability, and ability to maintain stable gel systems, hydrogels are beneficial as encapsulating materials.

Microcapsules: Techniques for creating microcapsules, such as extrusion, emulsification, and interfacial polymerisation, allow for control over capsule size, porosity, and mechanical strength (Jamali et al., 2021; Modzelewska et al., 2024; Moustafa et al., 2023). Encapsulation would be a highly advantageous process, especially for applications that expose microbial cells to extreme conditions, such as increased substrate concentrations, toxic by-products, and variations in pH and temperature. The protective covering provides a barrier between the cell and the environment, protecting cells from environmental stressors and improving the stability of their performance.

Encapsulation has been utilised to enhance performance and decrease operational costs in biohydrogen production. By encapsulating cells, they maintain their metabolism and can be reused for several cycles without significantly reducing efficiency. As an illustration, the encapsulation of hydrogen-producing bacteria in polyvinyl alcohol matrices has been utilised to convert wastewater into biohydrogen with sustainable yields. Encapsulation also promotes the co-immobilisation of microbial consortia, allowing for synergistic interactions that improve substrate utilisation and hydrogen synthesis. It was found that encapsulation has some limitations, such as mass transfer limitations, as the membrane is semi-permeable and sets limited costs for the fabrication of capsules. However, the specific requirements for each technology necessitated the design of sophisticated materials, such as nanocomposites and stimuli-responsive hydrogels, that could overcome the aforementioned hurdles to enhance permeability, mechanical strength, and responsiveness to multiple environmental stimuli.

Adsorption is a surface immobilisation process that binds microbial cells to carrier materials through physical or chemical interactions. Cell adhesion and binding between the cells and the carrier surface are defined by three principles of adsorption: electrostatic forces, van der Waals interactions, and hydrophobic effects (Bolonhesi et al., 2023; Han et al., 2015; Pandit et al., 2024). Fully functional absorbents encompass a wide range of potential materials, including common carrier materials such as activated carbon, silica, and diatomaceous earth, as well as natural polymers like chitosan and cellulose. Due to the high surface area and porosity, activated carbon is highly efficient in adsorbing cells and creating a stable microniche. On the other hand, silica-based materials offer chemical stability and can be functionalised to enhance cell anchoring and activity.

Biohydrogen production through adsorption depends on the characteristics of the carrier material and reactor system. Adsorbed cells are commonly incorporated into bioreactors (e.g., fixed-bed and fluidised-bed systems), and the immobilised biomass remains stable in continuous-flow systems. This results in high substrate conversion rates, minimal cell washout, and improved volumetric productivity. For example, fixed-bed reactors utilising hydrogen-producing bacteria carried on activated carbon have been used to produce biohydrogen from organic waste in an efficient manner. If the carrier material allows for a change in surface area, it can promote the diffusion of the substrate and hydrogen gas

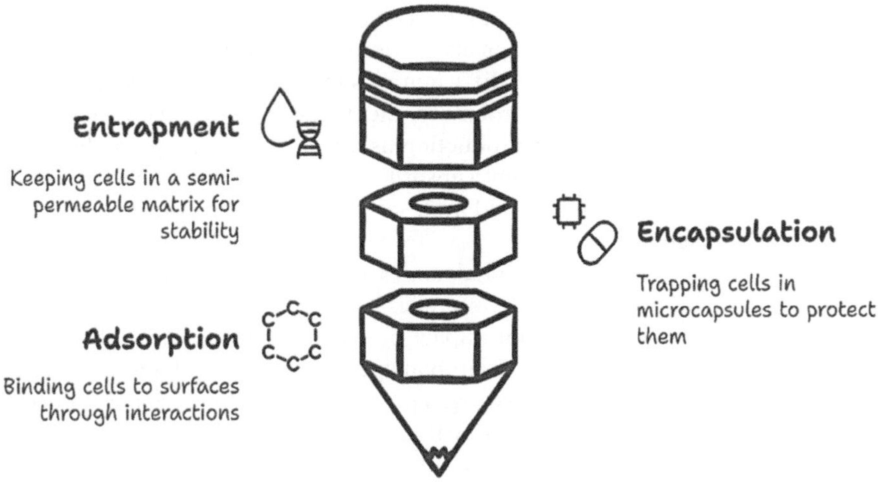

FIGURE 3.3 Strategies for microbial cell immobilisation.

without reducing the activity of the cells (Engliman et al., 2020; Min Woon et al., 2024; Woo et al., 2022). Moreover, adsorption technology enables the use of low-cost and abundantly available materials as carriers, such as agricultural residues, making it an economical option for large-scale applications.

While adsorption has its benefits, it also has downsides. The binding of cells to a carrier surface can be weak enough to result in detachment at high flow rates or shear forces. Furthermore, the acrylic may become saturated over time, which limits its adsorptive capacity and necessitates replacement. The functionalisation of carrier surfaces with chemical groups that promote strong cell binding and the use of composite materials that act synergistically are strategies to improve the stability and/or reusability of the adsorbed systems. Nanotechnology is also a promising means of developing efficient adsorbents due to the high surface area, tailored functionalisation, and improved mass transfer properties of nanostructured materials.

Microbial cell immobilisation via entrapment, encapsulation, and adsorption is essential for advancing biohydrogen production. These single-use bioreactors employ various strategies, materials, and applications to address specific bioprocess challenges, thereby providing enhanced efficiency, stability, and scalability. Using permeable matrices in entrapment shields can support beneficial microbes, such as friendly species, and may help increase productivity. Encapsulation, on the other hand, creates a controlled microenvironment that supports cell activity and allows for their reuse. Carrier–absorption systems are simple and cost-effective, enabling stable and economical production of biohydrogen in various reactor systems (Ananthi et al., 2024; Jamali et al., 2021; Wannapokin et al., 2022). Further developments in materials science, nanotechnology, and bioreactor design continue to improve the possibilities of these immobilisation approaches, leading to eco-friendly and more economical biohydrogen generation systems (Figure 3.3).

3.4 TYPES OF CARRIERS FOR IMMOBILISATION

In various cell immobilisation technologies, natural, synthetic, and hybrid carriers are extensively used, as each category of these carriers offers multiple advantages and is suitable for diverse bioengineering applications, particularly biohydrogen, making them essential. Hydrocolloids are natural carriers in drug delivery applications. Materials such as alginate, agar, chitosan, and other hydrocolloids are frequently utilised as drug delivery materials due to their biocompatibility and widespread availability. Alginate is a natural polysaccharide extracted from brown seaweed that forms a hydrogel in the presence of divalent cations, such as calcium. The gelation is easy and occurs under mild conditions. Hence, it is suitable for immobilising sensitive microbial cells (Han et al., 2015; Jiao et al., 2024; Pandit et al., 2024). Alginate facilitates the diffusion of substrates and products through its polymer network's pores, creating a protective microenvironment that enhances cell viability and activity. Alginate has been utilised in biohydrogen production as a matrix for entrapping hydrogen-producing bacteria, which facilitates steady and efficient substrate conversion. However, alginate has disadvantages, including lower mechanical stability and high sensitivity to degradation over extended periods, which can lead to cell leakage.

Agar is a natural carrier derived from red algae, exhibiting excellent gel-forming properties. It is a rigid and flexible framework for embedding microbes as a biofilm. Additionally, its thermal stability facilitates sterilisation and has no significant impact on its structural integrity. It is also a promising carrier due to its compatibility with various microbial species. Agar-based matrices are typically used in batch and continuous bioreactors for biohydrogen production, yielding stable results even after multiple cycles. Agar is known for its low tolerance to acidic or high-temperature environments, which may limit its industrial applications when grown in such conditions. Chitosan, derived from chitin, possesses many desirable properties, including biodegradability, antimicrobial activity, and the ability to form films, beads, or gels. Its high affinity for hydrogen-bonding interactions facilitates the immobilisation of a portion of microbial cells on the surface (Modzelewska et al., 2024; Moustafa et al., 2023; Woo et al., 2022). The recent modification of hydrogen-producing consortia for enhanced stability and activity, utilising hydrogels and composite materials based on chitosan, has been reported in the scientific literature. Nonetheless, the pH-sensitive behaviours and the higher costs of chitosan relative to other natural carriers might hinder their large-scale applications. Synthetic polysaccharides, such as polyurethane and polyethene glycol (PEG), offer their respective natural counterparts unique advantages, particularly in terms of mechanical strength, chemical stability, and tunability. Polymeric materials, such as polyurethane, are widely used due to their durable and chemically stable nature, which enables the formation of porous structures that facilitate cell adhesion and nutrient exchange. Its mechanical strength provides stability at high flow rates and shear forces; it has been extensively used as a matrix in packed-bed and fluidised-bed bioreactors for biohydrogen production. Polyurethane foams and sponges present a high surface area-to-volume ratio, which can be utilised to enhance the performance of immobilised cell systems. Still, the comparison of polyurethane's origin through synthetic means and the potential for environmental damage must be considered, especially regarding the implications of disposal and recyclability. In contrast, PEG is a hydrophilic polymer that can be cross-linked to yield biocompatible hydrogels with low toxicity (Table 3.3).

The ability of the sol-gel process to control gelation, pore size, and porosity in encapsulation applications is particularly beneficial, given the unique chemical characteristics of sol-gel processing. Similar studies have reported that microencapsulated hydrogen-producing bacteria in PEG-based hydrogels are protected from environmental stresses, enabling continuous hydrogen production over an extended period. Synthetic carriers, such as PEG and polyurethane, provide these advantages but may be cost-prohibitive, and their non-biodegradable nature may limit their use in environmentally sensitive applications (Boshagh et al., 2019; Hao et al., 2024; Rai et al., 2022).

Hybrid carriers combine the best traits of both natural and synthetic materials, offering high performance and versatility. These carriers are composites that inherit the biocompatibility and biodegradability of natural materials, as well as the mechanical strength and stability of synthetic polymers. Other examples reflect combined (alginate-polyacrylamide) composites that exploit the gelling capacity of alginate and the hydrogel properties of polyacrylamide to manufacture materials with enhanced mechanical stability and revised active and degradation resistance matrices. Such hybrids can be highly advantageous in cells stressed by extreme environments, such as continuous-flow reactors, in which native carriers tend to pack in themselves. In contrast, chitosan-polyurethane composites enable the utilisation of chitosan's microbial properties and antimicrobial capabilities as a carrier for rapid and stable biohydrogen production. Additionally, hybrid carriers can incorporate functional additives, such as nanoparticles, to enhance their properties (Engliman et al., 2020; Pandit et al., 2024; Woo et al., 2022).

TABLE 3.3 Different types of carriers for immobilisation

CARRIER TYPE	*EXAMPLES*	*PROPERTIES*	*ADVANTAGES*	*LIMITATIONS*	*APPLICATIONS*	*INNOVATIONS AND TRENDS*
Natural	Alginate, chitosan, agar, cellulose	Biodegradable, biocompatible, porous, non-toxic.	Environmentally friendly, readily available, and cost-effective.	Limited mechanical strength, susceptible to microbial degradation.	Biohydrogen production, enzyme immobilisation, bioremediation.	Hybrid carriers combine natural materials with synthetic reinforcements for improved strength and stability.
Synthetic	Polyacrylamide, polyethene glycol (PEG), polystyrene	Durable, customisable, chemically inert.	High mechanical strength, tuneable properties, and wide availability.	High cost, potential toxicity, and environmental concerns.	Industrial biocatalysis, biosensors, and drug delivery systems.	Development of bioinspired synthetic materials with enhanced biocompatibility.
Inorganic	Silica, glass beads, activated carbon, clay	High surface area, chemical stability, and heat resistance.	Reusable, excellent mass transfer properties, non-biodegradable.	High cost, potential leaching, limited biocompatibility.	Wastewater treatment, hydrogen production, heavy metal removal.	Nanoengineered inorganic materials for improved binding and reusability.
Composite	Alginate-silica, chitosan-carbon nanotubes, polymer-clay	Synergistic properties of combined materials, enhanced porosity.	Improved mechanical strength, tailored properties, and multi-functionality.	Complexity in synthesis leads to higher production costs.	Biofuel production, advanced biosensing, bioreactors.	Integration of nanomaterials and biomolecules for multifunctional applications.
Nanostructured	Nanoparticles (e.g., magnetite, silica), nanofibres, nanotubes	Extremely high surface area, tuneable chemical and physical properties.	Precise immobilisation, enhanced reactivity, and small-scale usage.	High cost, scalability issues, and potential nanotoxicity.	Biohydrogen production, precision medicine, catalysis.	Biodegradable nanostructures and scalable production techniques for broader applications.

Source: Boshagh et al. (2019), Hao et al. (2024), Rai et al. (2022)

Alginate-based composites or other alginate-based nanocomposites supplemented with various fillers, such as graphene oxide or silica nanoparticles, offer superior mechanical strength, thermal stability, and conductivity, making them more suitable for advanced bioreactor systems. Hybrid carriers have advantages but are more complex and expensive than alternatives; therefore, thorough optimisation will be necessary to achieve a given cost-to-performance ratio.

Carrier selection plays a critical role in the performance of immobilised cell systems. Biocompatibility is crucial in this context, and the carrier must not inhibit microbial activity or interfere with metabolic pathways. The inherent biological shortcomings of synthetic carriers often necessitate coating or surface modifications to achieve effective drug delivery; in contrast, 'natural carriers' are typically much more efficient. It is also essential to ensure stability, as the carrier should remain intact during use, considering the effects of shear forces, pH changes, temperature fluctuations, and prolonged use. Instead, hybrid carriers are a Goldilocks solution that improves mechanical and chemical stability compared to the natural side. At the same time, we maintain a closer alignment with the original functionality for delivery that we initiated on the other end (Ananthi et al., 2024; Hao et al., 2024; Jamali et al., 2019). Cost is another significant consideration, particularly for enterprise-level solutions. Compared to their synthetic counterparts, organic carriers such as alginate and agar are generally more economical and environmentally friendly, thereby enhancing the advantages of their use in economically sensitive processes. However, the fundamentally enhanced performances of synthetic and hybrid carriers in a particular application may still be engaging from application prospects, particularly in industrial biohydrogen production.

The selection of a carrier is also related to specific bioprocess requirements, including reactor system type, substrate nature, and target operational parameters. If alginate or other natural carriers are employed, they may suffice when there is some control over time and temperature in the batch reactors. Continuous-flow systems may employ synthetic or hybrid carriers for which long-term stability and shear tolerance are required. The carrier must be compatible with the immobilised microbial consortia, as different strains may have specific requirements for attachment, nutrient diffusion, and metabolic activity. As materials science and biotechnology continue to develop, having a range of carriers at one's disposal makes tailoring solutions to target specific challenges in biohydrogen production more attractive.

Carrier material selection for cell immobilisation is one of the most critical factors strongly influencing the operation and sustainability of biohydrogen production systems. Biomaterials such as alginate, agar, and chitosan exhibit natural biocompatibility and are easily handled, making them suitable for use in various applications. Synthetic carriers, such as polyurethane or PEG, exhibit higher mechanical strength and stability in high-chemical environments, enabling their use in high-stress assemblies. Hybrid carriers are a continuous learning process and can leverage the properties of both natural and synthetic materials, forming flexible and efficient immobilisation systems. By judiciously evaluating individual parameters, such as biocompatibility, stability, cost, etc., carrier selection can be optimised for respective processes, paving the way for more efficient and sustainable biohydrogen production technologies.

3.5 NANO-BASED IMMOBILISATION

Progression in cell immobilisation technologies depends not only on natural carriers but also on synthetic and hybrid carriers, which offer advantages in bioengineering applications, such as biohydrogen production. The most commonly used carriers are commercial polymers, alginate, agar, and chitosan, which are preferred due to their biocompatibility, abundance, and ease of use. Alginate, derived from brown seaweed, is a polysaccharide that undergoes gelation in the presence of divalent cations, such as calcium. The gelation process occurs under mild conditions and is simple, which is advantageous for immobilising sensitive microbial cells (Hao et al., 2024; Jiao et al., 2024; Modzelewska et al., 2024; Woo et al., 2022). The porous structure ensures the high transfer rates associated with biodynamic templates and products. It acts similarly to a microenvironment that promotes the cell's exercise and life. For example, alginate can be used as a preservation matrix for hydrogen producers with stable substrate conversion. Alginate is beneficial, but it also has some limitations, such as low mechanical stability and degradation over long-term use, which can lead to cell leakage.

Another natural carrier is agar, which is isolated from red algae and possesses excellent gelling properties. It creates a firm but pliable network for the entrapment of the aggressive microorganisms. It exhibits exceptional thermal stability, can be sterilised without disrupting its structure, and provides a broad carrier spectrum as it is compatible with many strains of microorganisms. Stable production for multiple batches has been achieved using standard supporting matrices, including agar in batch and continuous bioreactors for biohydrogen generation. Thus far, agar has low-acid or high-temperature tolerance, rendering it unsuitable for specific industrial processes. Chitosan, a biopolymer derived from chitin, is biodegradable and exhibits antimicrobial properties, which can be utilised to form films, beads, or gels. The ability of these molecules to aggregate is particularly effective in cross-linking bacteria, as they have a high affinity for hydrogen-bonding interactions (Min Woon et al., 2024; Pomdaeng et al., 2024; Wannapokin et al., 2022). The great potential of using chitosan in biohydrogen production relies on the elaboration of hydrogels and composites to enhance the stability and activity of hydrogen-producing consortia. Although it provides advantages such as inducing endosomal escape, its pH sensitivity and cost, which are higher than those of other natural carriers, dramatically restrict its application in translation.

Synthetic carriers such as polyurethane and PEG possess distinctive benefits over natural carriers, including mechanical stability, biocompatibility, chemical utilisation, and tuneable structure. Polyurethane is a polymer with good mechanical properties, is resistant to chemical degradation, and has a porous structure that encourages cell adhesion and enables the diffusion of nutrients. Due to its mechanical stability, the polymer is extensively used in packed-bed reactors and FBRs for biohydrogen production, providing reliable performance even at higher flow rates and shear forces. Polyurethane foams and sponges provide surface area-to-volume ratios that can sustain enhanced activity for cell systems when immobilised within the pores of a rigid or flexible base matrix. However, certain aspects must be carefully considered, such as the synthetic origin of polyurethane and its environmental impact during disposal and recycling. On the other hand, PEG is a hydrophilic polymer that has been found to produce hydrogels with high biocompatibility and low toxicity. Due to these varying chemical properties, it is particularly suitable for encapsulation applications where gel, pore size, and permeability can be closely controlled (Jamali et al., 2019; Rai et al., 2022; Wannapokin et al., 2022). PEG-based hydrogels encapsulated hydrogen-producing bacteria in microcapsules, protecting against environmental stresses and supporting sustained hydrogen production over an extended timescale. Nevertheless, synthetic carriers such as PEG and polyurethane may be expensive and are reported to be non-biodegradable materials, which could affect their appeal in environmental applications.

Bottom Line: Hybrid carriers utilise a combination of natural and synthetic materials to produce more adaptable products. These are composites that combine natural materials, which offer biocompatibility and biodegradability, with synthetic polymers, which provide higher mechanical strength and stability. For instance, alginate-polyacrylamide composites utilise alginate's gelation ability and polyacrylamide's strength to create matrices with enhanced mechanical resistance and resistance to degradation. These hybrids are particularly useful in high-stress environments, such as continuous-flow reactors, where natural carriers alone may not be sufficient. Chitosan-polyurethane composites utilise the antimicrobial properties of chitosan, combined with the mechanical stability of polyurethane, to develop carriers for enhanced and stable biohydrogen production. Additionally, functional particles, such as nanoparticles, can be incorporated into hybrid alloys to enhance their functionality further. Alginate-based composite hydrogels containing filler compositions and enhancing their mechanical strength further have improved overall mechanical stability, thermal stability, and ionic and electrical conductivity properties, making these ideal candidates for high-end bioreactor system applications (Moustafa et al., 2023; Pomdaeng et al., 2024; Woo et al., 2022). Hybrid carriers generally provide superior performance but are more complex and expensive to manufacture than traditional carriers and must be optimised to achieve the best performance/cost ratio.

Carrier selection is an essential step in cell immobilisation and significantly impacts the performance of immobilised cell systems. Biocompatibility is necessary as the carrier must not inhibit microbial activity or interfere with metabolic pathways. From this perspective, natural carriers are preferable as compatible building blocks with biological systems, in contrast to synthetic airlines, which require surface modifications or coatings to produce similar outputs. Stability is also critical; during activity, the carrier must maintain structural integrity through phenomena such as exposure to shear forces, pH fluctuations, temperature fluctuations, and duration of use. Hybrid carriers combine the advantages of both types of carriers by providing an intermediate strategy between natural and synthetic carriers, which have improved mechanical and chemical stability compared to their natural counterparts (Engliman et al., 2020; Jamali et al., 2019; Rai et al., 2022). Natural carriers, such as alginate and agar, are typically more affordable and sustainable. As a result, they are more preferred in processes where cost is a key factor. Overall, in specific applications, the improved performance of synthetic and hybrid carriers may justify their price compared to the costs of industrial biohydrogen production processes. Thus, the effects of an industrial biohydrogen production process should be evaluated.

The specific bioprocess requirements (reactor system type, substrate properties, and operational parameters) also dictate the exact carrier used. Natural carriers, such as alginate, may be sufficient in batch reactors, i.e., those operated for a certain period under controlled conditions. Synthetic or hybrid carriers may be more advantageous in continuous-flow systems that require longer-term stability and shear resistance. Given that various strains exhibit unique requirements with regard to attachable, nutrient diffusion, and metabolic activity relative to the carrier that they are engaged with, it is therefore crucial that such a dynamic assessment be updated in the field of multifunctional materials to provide the requisite flexibility and control to interface with immobilised, naturally recruited microbial consortia. As the fields of materials science and biotechnology continue to advance, the scope of carriers is widening, enabling the design of customised solutions to the challenges encountered during biohydrogen production (Engliman et al., 2020; Pandit et al., 2024; Woo et al., 2022).

Cell immobilisation is a crucial step in biohydrogen production systems, and the choice of carrier materials is one of the most critical factors affecting the systems' efficiency and sustainability. To this end, the most common natural carriers are alginate and agar, which are biocompatible and easy to handle (Jiao et al., 2024; Pandit et al., 2024; Pomdaeng et al., 2024). The polymerisation of degradable macromolecular carriers, such as polyurethanes and PEG, yields excellent mechanical strength and chemical stability, encouraging their use in high-stress organs. Hybrid carriers are a new class which combines properties of natural and synthetic materials in an ideal manner to develop flexible and strong immobilisation systems. Researchers and engineers can produce more sustainable and efficient biohydrogen production technologies by selecting the optimal carrier based on biocompatibility, stability, and cost, through a detailed analysis of potential carriers (Figure 3.4).

3.6 CHALLENGES AND LIMITATIONS

Here are the possible reasons that lead to the formidable aspects which may prevent producing H_2 efficiently with immobilised cells: (1) Mass transfer limit (increase in bioreactor volume and wall shear stress): In most cases, using Immobilized Reactor, i.e., a reactor in which cells or enzymes are immobilized on or within a carrier material (IR) is accompanied by the decrease of mass transfer of substrates between the reactor and the cells, especially using the scaffolds or carriers with low porosity or density; (2) Cost and the up-scalability of the nano-based immobilisation: When a massive amount of scaffold with nano-size should be applied, the regeneration of them will be much costly while making it difficult to explain the recyclability for an extended period; (3) Stability and sustain of immobilisation systems: The introduction of porous or semi-permeable carriers may also affect the nature of the life of microorganism. Due to structural and material factors, mass transfer limitations are ubiquitous hurdles in traditional carriers.

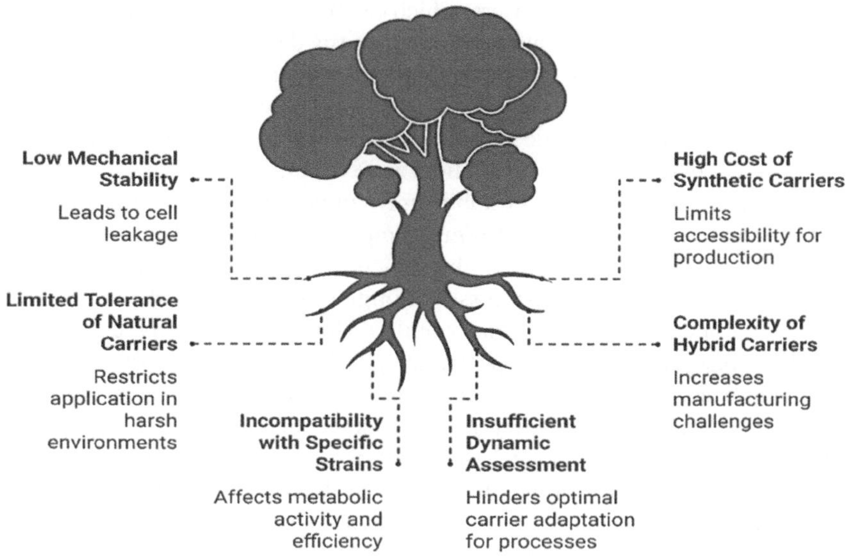

FIGURE 3.4 Ineffective carrier selection for biohydrogen production.

Typically used carriers, such as alginate, agar, and polyacrylamide, often exhibit low porosity or limited pore size, which limits the diffusion of substrates into the immobilised cell matrix and prevents product diffusion from the immobilised microbe into the surrounding medium. Due to the limited spatial distribution between the inflow and outflow of products, low product recovery hampers biohydrogen production by introducing substrate gradients and product accumulation, which can repress microbial metabolism and severely compromise overall productivity. However, the dense packaging of all carriers in packed-bed reactors or biofilm-based systems can increase mass transfer resistance for the active populations residing within, restricting even nutrient diffusion for these populations and creating localised regions where metabolic activity is negligible. To solve this problem, researchers have attempted to find optimised carrier materials to improve porosity and pore size or provide different substrate diffusion routes. Although such adjustments can enhance mass transfer, they can also jeopardise the carrier's mechanical robustness or stability, demanding a proper trade-off between structural integrity and functional performance.

Nano-based immobilisation technologies are a promising alternative to traditional carriers; however, their adoption should consider key parameters such as cost and scalability. Nanofibres, carbon nanotubes, graphene oxide, and magnetic nanoparticles, as nano-based materials, showcase the advantages of the surface area-to-volume ratio, improved mass transfer properties, and highly controllable microenvironments for immobilised cells. These materials can facilitate more effective diffusion of substrate and product than conventional carriers, thereby reducing mass transfer limitations. Nevertheless, the synthesis cost, functionalisation, and large-scale production of nanomaterials significantly deter their potential widespread use. For instance, preparing graphene-based carriers involves complex procedures, such as exfoliation, reduction, and surface modification, which require sophisticated equipment and reagents, resulting in increased costs. Likewise, the preparation of magnetic nanoparticles used in immobilised systems proceeds through thermally initiated processes, such as co-precipitation or heat elaboration, which may not be a realistic proposition for large items. Another limitation of nano-based immobilisation systems is their scalability, as issues related to uniformity and reproducibility during the production process, as well as the strict quality control required to maintain a consistent functional property related to the nanomaterials, can hinder scalability.

Strategies aimed at cost reduction and increased scalability have focused on designing hybrid carriers composed of nano-based materials combined with conventional natural or synthetic matrices. For example, incorporating graphene oxide into an alginate or chitosan hydrogel in a composite carrier can enhance the mechanical properties and mass transfer characteristics of this hydrogel while reducing the overall cost of the system. Using magnetic nanoparticles in combination with polyurethane or PEG leads to the development of magnetically responsive carriers, which can be quickly recovered and reused, making this process more cost-effective for immobilisation. Simultaneously developing nanotechnology and materials science tools, as well as green synthesis methods, and exploring alternative, low-cost precursors for the production of nanomaterials would pave the way for new approaches to address cost and scalability challenges. Nonetheless, ensuring the economic viability of these next-generation nano-enabled strategies will rely on further investment in research and innovation, as well as stronger partnerships between academia, industry, and policymakers, to facilitate the establishment of production pathways and supply chains.

Stable, long-immobilised systems are of utmost importance as they facilitate the operational lifetime of the bioprocess while also ensuring high efficiency. Under conditions such as changes in pH, temperature, and shear forces, multi-step processes of immobilised systems should retain their structural properties, functional properties, and microbial viability. However, traditional carriers such as alginate and agar tend to have restricted durability, particularly in long-term processes or extreme conditions. For example, alginate beads can lose their stability progressively if the exchanged ions dissolve with time in low-calcium environments, triggering cell leakage that reduces the system's performance. Likewise, because agar is hygroscopic, agar-based carriers can experience swelling or loss of structural rigidity when exposed to high temperatures or mechanical forces, impairing the functionality of these biocathodes in continuous-flow reactors. Synthetic carriers, such as polyurethane and PEG, typically offer better stability and long-term reliability due to their superior mechanical properties and resistance to chemical degradation. However, these carriers are non-biodegradable and hurt the environment, so alternative solutions, such as hybrid carriers or biodegradable synthetic polymers, are being explored.

Nano-based immobilisation systems have unique advantages over other support systems because including nanomaterials can provide mechanical robustness, thermal stability, and resistance to the chemical or biological degradation of the carrier. For example, graphene-based carriers possess unique mechanical properties, enabling them to withstand high shear forces and repeated use cycles without significant loss of activity. They offer the advantage of stability over the immobilised systems, as they can be picked out of the system and reused, which limits the necessity of carrier renewal (Boshagh et al., 2019; Engliman et al., 2020; Moustafa et al., 2023; Woo et al., 2022). Additionally, the functionalisation of nanomaterials with appropriate ligands or coatings often improves the compatibility of the immobilised system with microbial cells, ensuring prolonged cell viability and activity. However, it is essential to investigate further and analyse the long-term operational stability and performance of nano-based immobilisation systems to avoid potential challenges related to agglomeration, leaching, or loss of functional properties through operational conditions.

A promising alternative approach for improving stability and longevity in immobilised systems is the use of hybrid carriers that combine the advantages of natural, synthetic, and nano-based materials. Thus, introducing carbon nanotubes or graphene oxide into alginate or chitosan matrices can improve the mechanical strength, inhibit the disintegration of the composite carrier, and ensure the bioreactor's long-term operation. Nanoparticles such as silica or magnetic particles combined with synthetic polymers, i.e., polyurethane or PEG, could also be used to construct carriers with improved stabilisation and reusability (Jamali et al., 2019; Moustafa et al., 2023; Pomdaeng et al., 2024; Rai et al., 2022). The versatility of hybrid carriers also allows for customisation through the alignment of their composition, structure, or functional properties, providing tailored solutions for biohydrogen production and other bioprocesses. Nonetheless, for hybrid carriers to be successful, a good understanding of the interactions between the different components and knowledge of how different fabrication methods of the same materials can lead to other properties make reproducibility and consistency necessary.

Results are discussed in terms of mass transfer limitations, cost and scalability factors, and issues related to the stability and longevity of immobilised systems. These represent critical studies in the development of cell immobilisation technologies for biohydrogen and other processes. The simplicity and affordability of conventional carriers come at the cost of limited performance due to mass transfer resistance or instability. Although nano-based immobilisation systems offer various advantages, including improved mass transfer, stability, and functional characteristics, high costs and scalability issues are significant challenges to the commercialisation of biocatalysts (Ananthi et al., 2024; Jamali et al.,

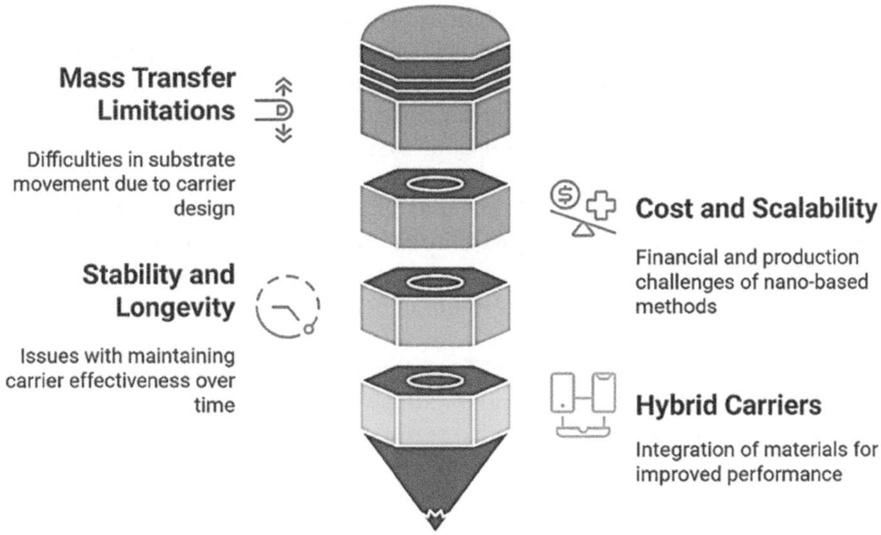

FIGURE 3.5 Overcoming challenges in biohydrogen production.

2019; Wannapokin et al., 2022; Woon et al., 2024). By combining the advantages of natural, synthetic, and nano-based materials, hybrid carriers have the potential to serve as integrated carriers, leading to improved performance and economic viability. They should be directed towards optimising carrier materials and fabrication processes, exploring alternative and sustainable sources of nanomaterials, and developing scalable production routes to realise the widespread use of advanced immobilisation technologies. Overcoming these limitations can significantly accelerate the transition to sustainable energy sources and contribute to the design of efficient and resilient bioprocesses (Figure 3.5).

3.7 FUTURE PROSPECTS

Such solutions can range from the combination of innovative bioreactor designs with nanoscale carriers, utilising genetic engineering to assist in preconditioning cells for acclimation on said carriers, exploring hybrid nano- and micro-immobilisation paradigms to overcome limitations with immobilisation, and ultimately employing artificial intelligence (AI) and machine learning (ML)-driven optimisation of immobilisation paradigms to blanket these (potential) solutions to the limitations plaguing biohydrogen production and other related fields all across biotechnology. Designs of high-performance bioreactors are crucial for maximising the efficiency and productivity of this immobilised system. Adaptations of traditional bioreactors, such as micro and macro mix (homogeneous and heterogeneous flow) type bioreactors, are being developed to address issues encountered in immobilised cell growth in standard bioreactors, including mass transfer limitations, poor mechanical properties, substrate depletion, and other factors that lead to reduced cell viability. Pared-down designs, such as packed-bed reactors, FBRs, and hollow fibre membrane reactors, further ameliorate these problems by improving the spatial arrangement of the carriers and providing superior flow kinetics, along with enhanced mass and heat transfer. Static systems are the most widely used technologies that employ FBRs, which maintain carriers in suspension to allow for uniform contact with substrates while avoiding the clogging usually seen in static systems. Hollow fibre membrane reactors also help construct regulated microenvironments for the immobilisation of cells, with substrates and products segregated by semi-permeable membranes to limit inhibitory effects and enhance productivity. Moreover, the combination of immobilised systems with new and innovative bioreactor design – often enabling not only advances in operational capabilities but also the opportunity for scale-up through controlled operation of key process parameters (e.g., pH temperature, nutrient supply) – also provides a platform to assess further the performance of specific automation strategies in continuous-flow operation.

Another key to this revolution of immobilisation technology is the genetic improvement of compatibility with carriers. Conventional immobilisation systems may be limited by the immobilisation of microbial cells into carrier materials, as cells can leak from the carrier, resulting in low metabolic activity and system instability due to the carrier's incompatibility with microbial cells. Psychoactive agents possess inherent limitations, such as the immobility of most microorganisms during their cultivation. Tethering potential for higher binding surface domain(s) or adhesive protein(s) in the cell can also enhance their binding to the carrier and the stability of cells, thereby reducing cell loss during detachment. Metabolic engineering, for instance, may be used to optimise the pathways, resulting in increased biohydrogen production efficiency and reduced formation of inhibitory end products. For example, one strategy involved using metabolic engineering to redirect substantial carbon flux to hydrogen production by repressing competing pathways or overexpressing rate-limiting enzymes. Furthermore, genetic engineering may also enhance the resistance of immobilised cells to environmental stresses, such as high substrate concentrations, high temperatures, and pH tolerance, thereby increasing the system's operational lifetime. A potential solution to the challenges of conventional immobilisation technologies can be achieved through a synergistic application of advanced bioreactor designs with innovative genetically modified microbial clones.

An innovative strategy is to utilise hybrid nano- and micro-immobilisation systems, which combine the benefits of nanoscale precision with those of micro-scale supports. Nano-based immobilisation is beneficial owing to its high surface area, better mass transfer properties, and ability to manipulate the microenvironments of immobilised cells. However, the scalable and economic quest for purely nano-based systems is beset with staggering complexities. A compromise between performance and practicality can be achieved using hybrid systems that combine nanoscale materials, such as carbon nanotubes, graphene oxide, or magnetic nanoparticles, with microscale carriers, including alginate beads or polymer matrices. For example, microscale carriers functionalised with magnetic nanoparticles enable straightforward recovery and reuse, reducing operating costs and increasing the system's sustainability. Likewise, using porous and composite materials, including graphene oxide in polymeric matrices, enhances mechanical resistance, thermal stability, and mass transfer properties, enabling long-term operation in bioreactors. By modifying both the composition and structure of the carrier material, it is possible to combine the advantages of different strategies; thus, hybrid systems can maximise the improvement of biohydrogen production and other bioprocesses by designing customised systems for the immobilisation and bioconversion of substrates and products. As hybrid nano- and micro-immobilisation systems, the phenomenon of different scales has enormous potential because the two scales can be combined, yielding high efficiencies that will generate a prosperous future for a line of immobilisation technologies.

However, coupling AI and ML with immobilisation processes opens the pathway for emerging micrographs previously unseen in the field. Conventional techniques often fail to consider the interdependencies revealed by science and data in the vast amount of data generated from experimental and operational processes that AI and ML algorithms can uncover. This is particularly crucial in immobilisation systems, as several interacting parameters across carrier material properties, cell sources and compatibility, reactor design, and operational conditions influence system performance. Researchers can harness AI/ML to build predictive models that enable them to optimise multiple parameters, leading to the evolution of high-power-density immobilisation systems. In particular, several ML algorithms can provide predictive insights into the performance of different carrier materials for a specific application based on the physicochemical properties of these materials and also guide the selection of materials for specific applications, significantly reducing the time required for experimental searches. Reactor design, informing system configuration, operating pressure for specific immobilisation systems, and design process optimisation are other areas that can be examined using AI-driven optimisation. AI and ML can also analyse sensor data to calibrate immobilised systems in real time dynamically, maintaining parameters within ideal ranges and enhancing stability and productivity.

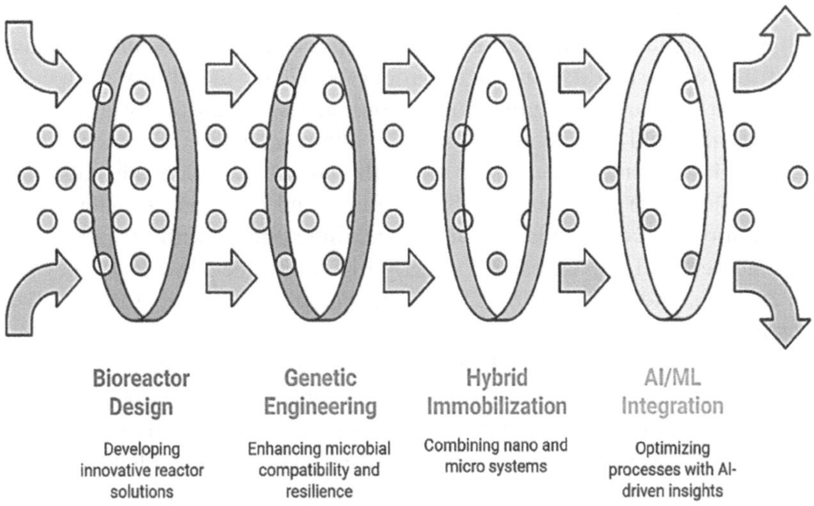

FIGURE 3.6 Optimising biohydrogen production.

AI and ML also facilitate the design of new strategies and materials for immobilisation. For instance, generative algorithms can design new carrier materials with desired characteristics or properties, such as enhanced biocompatibility, mechanical properties, or mass transfer functions. In addition, AI-based simulations can also provide insights into interactions between immobilised cells and carrier materials at the molecular level (i.e., at the interface) to help guide the design of genetically engineered strains or functionalised nanomaterials that could improve compatibility and performance. The use of AI and ML in these fields is not just limited to lab scales. Still, your AI-infused solutions also help scale research to industry levels when combined with AI and ML methods. AI and ML can provide predictive maintenance, process optimisation and quality control capabilities, making them crucial components in solutions that reduce operational costs and increase scalability, while enabling the 24/7 operation of systems commonly found in large-scale bioprocesses.

The advances in modern bioreactor architecture, genetic engineering, and hybrid nano- and micro-immobilisation, combined with AI and ML, represent phenomenal strides in immobilisation technology. They address several critical limitations, including mass transfer limitations, cell compatibility, cost-effectiveness, scalability, and operational feasibility, thereby paving the way for more efficient and sustainable biohydrogen production and other bioprocesses. Advanced bioreactors provide a customisable environment for optimisation technology, while molecular genetics techniques have been utilised to develop microbial strains with enhanced affinity for immobilisation. These systems combine the single-cell precision of nanoscale systems with the structural characteristics of microscale architecture, making them an attractive candidate for next-generation, practical, and efficient integrated devices. AI and ML can provide powerful resources for improving immobilisation methods (e.g., materials and process optimisation) and enabling novel and scalable solutions (Figure 3.6). These innovations represent a holistic approach to overcoming the limitations of immobilisation methods and unlocking the full capabilities of immobilised systems for biotechnological applications.

3.8 CONCLUSION

The critical overview provided here on the extensive recent discoveries and concepts in biohydrogen production technologies reflects the enormous promise of cell immobilisation approaches, innovative bioreactor configurations, and hybrid carriers, along with convergence with other cutting-edge technologies, such as genetic engineering, AI, and ML. Based on a critical assessment of the region, immobilisation technologies such as entrapment, encapsulation, and adsorption offer significant advantages over free-cell systems, including improved stability, enhanced productivity, and the ability to operate under adverse conditions. Natural carriers, such as alginate, agar, and chitosan, can perform reasonably well due to their good biocompatibility and sufficiently low cost. In contrast, synthetic carriers such as polyurethane and PEG were required to enhance mechanical strength and achieve adequate stability. Hybrid systems that leverage the complementary attributes of both natural and engineered materials have successfully overcome these limitations of conventional systems, yielding systems with tuneable trade-offs in performance, scalability, and cost. However, carrier selection is determined by several key criteria, including biocompatibility, stability, and material cost. These criteria are fundamental in optimising biohydrogen production processes, especially on an industrial scale.

The analysis posed several challenges that need to be tackled (e.g., mass transfer limitations, high cost of nano-based immobilisation systems, stability and longevity of immobilised systems). However, the analysis still has limitations. Improvements in mass transfer and controlled microenvironments at the interface around immobilised cells are evident due to advances in bioreactor technology (e.g., packed-bed reactors, FBRs, and hollow fibre membrane reactors) that have addressed many of the problems associated with immobilised cells. Genetic engineering has also enhanced the performance of immobilisation systems by engineering microbial strains with improved adhesion potential, metabolic activity, and tolerance to stress. A comprehensive synthesis and review of this nature were conducted, providing detailed insights into the key processes governing biohydrogen production technologies.

Long-term developments in biohydrogen production technologies will be anything but trivial combining biological, materials science, engineering, and data analytics disciplines. Tools for tuning hybrid (supported) nano- and micro-immobilisation systems will be crucial for scaling and controlling the bioactivity and cost most relevant to nano-convergent technologies. The hybridisation of nanoscale materials delivers the precision and significantly increased surface area benefits of nanoscale materials while incorporating the structural support and robustness of microscale carriers, providing a new approach to maximise productivity without compromising operational feasibility. A future direction would be to explore new combinations of materials and optimise properties to create efficient and friendly carriers for the ecosystem.

Building on the success of several industry stalwarts, the company is looking forward to the idea of genetic engineering to improve microbial strains. Advances in synthetic biology and Clustered Regularly Interspaced Short Palindromic Repeat (CRISPR) genome editing are likely to enable the design of strains with enhanced biohydrogen production capabilities. This implies that metabolic engineering strategies, such as metabolic flux balance, can optimise cellular processes towards hydrogen production and inhibit the production of reactant compounds that may repress the reaction. Additionally, new biohydrogen applications may emerge, as tolerance-resistant traits can be developed to operate in a wide range of conditions. Thus, combining genetic engineering with functionalised carrier materials (e.g., bioactive molecules, adhesive proteins) will promote the interaction between cells and carriers, thereby improving the system's stability.

Finally, another promising research method is the use of AI and ML to optimise biohydrogen production pathways. AI and ML algorithms, which can analyse large datasets and identify relationships not readily apparent to human cognition, present exciting opportunities for optimising immobilisation techniques and bioreactor operations. Predictive models can guide the selection of suitable carrier materials, reactor designs, and conditions, thereby reducing reliance on experimental probe techniques, which are often expensive and time-consuming. Here, AI can mimic the interaction between cells and carriers at the molecular level, allowing for the selective development of new immobilisation systems tailored for a specific application.

Bioreactor design will also continue to evolve, inspired by principles of modular construction, advanced automated sensing, and, yes, AI to make the performance as efficient as practicable. These high-scalability and customisable modular bioreactors will be crucial for biohydrogen production systems adaptable to diverse industrial and ecological conditions. Real-time assessment will be enhanced using advanced sensors, such as biosensors and nanomaterial-based detectors, to ensure that key parameters (pH, temperature, substrate saturation, etc.) remain within an optimal range. These sensors will enable

AI- and ML-based models for adaptive control strategies, maintaining the environment at an optimal level and mitigating system failure.

Sustainability and footprint will also be relevant aspects of future biohydrogen production technologies. The carriers will be developed using renewable and biodegradable raw materials, resulting in a lower environmental impact of the immobilisation systems. Additionally, eco-sustainable materials can be designed as biodegradable carriers, such as bio-based polymers and composites derived from agricultural waste, and used as sustainable alternatives. Additionally, combining biohydrogen production with bioprocesses like wastewater treatment and CO_2 sequestration can facilitate the adoption of circular bioeconomy models, promoting resource efficiency where the outputs of one system become the inputs of another, thereby minimising waste accumulation.

The concept of hybrid systems combining biohydrogen production alongside alternative technologies, such as fuel cells or bioelectrochemical systems, presents an auspicious approach towards integrated energy systems. One potential application of biohydrogen technology is as a feedstock for hydrogen fuel cells, providing a clean and sustainable energy source for mobility and electricity generation. Coupling biohydrogen production with microbially mediated systems, such as microbial electrolysis cells (MECs), can provide an electrochemical potential for the oxidation of organic matter, facilitating substrate hydrogen production and potentially increasing energy recovery from treating organic matter in bioelectrochemical systems. Networked energy systems: This point represents the energy transition for hybrid environmental solutions that generate energy. Science extends to these sources.

REFERENCES

Ananthi, V., Bora, A., Ramesh, U., Yuvakkumar, R., Raja, K., Ponnuchamy, K., Muthusamy, G., & Arun, A. (2024). A review of technologies for sustainable biohydrogen production. *Process Safety and Environmental Protection*, *186*, 944–956. https://doi.org/10.1016/j.psep.2024.04.034

Bolonhesi, I. B. de T. M., Andreani, C. L., de Melo, M. R., Gomes, S. D., & Lopes, D. D. (2023). Can biomass immobilization in hydrolyzed lignocellulosic material enhance biohydrogen production from cassava residues? *Biochemical Engineering Journal*, *190*, 108725. https://doi.org/10.1016/j.bej.2022.108725

Boshagh, F., Rostami, K., & Moazami, N. (2019). Biohydrogen production by immobilized Enterobacter aerogenes on functionalized multi-walled carbon nanotube. *International Journal of Hydrogen Energy*, *44*(28), 14395–14405. https://doi.org/10.1016/j.ijhydene.2018.11.199

Engliman, N. S., Jahim, J. M., Abdul, P. M., Ling, T. P., Tan, J. P., & Ong, C. B. (2020). Effectiveness of fouling mechanism for bacterial immobilization in polyvinylidene fluoride membranes for biohydrogen fermentation. *Food and Bioproducts Processing*, *120*, 48–57. https://doi.org/10.1016/j.fbp.2019.12.004

Han, W., Liu, D. N., Shi, Y. W., Tang, J. H., Li, Y. F., & Ren, N. Q. (2015). Biohydrogen production from food waste hydrolysate using continuous mixed immobilized sludge reactors. *Bioresource Technology*, *180*, 54–58. https://doi.org/10.1016/j.biortech.2014.12.067

Hao, J., Zhou, H., Ma, J., Ren, N., & Xing, D. (2024). Application of recyclable iron-based supplementation for promoting biohydrogen and ethanol co-production. *Fuel*, *371*, 131941. https://doi.org/10.1016/j.fuel.2024.131941

Jamali, N. S., Dzul Rashidi, N. F., Jahim, J. M., O- Thong, S., Jehlee, A., & Engliman, N. S. (2019). Thermophilic biohydrogen production from palm oil mill effluent: effect of immobilized cells on granular activated carbon in fluidized bed reactor. *Food and Bioproducts Processing*, *117*, 231–240. https://doi.org/10.1016/j.fbp.2019.07.012

Jamali, N. S., Md Jahim, J., Mumtaz, T., & Abdul, P. M. (2021). Dark fermentation of palm oil mill effluent by Caldicellulosiruptor saccharolyticus immobilized on activated carbon for thermophilic biohydrogen production. *Environmental Technology & Innovation*, *22*, 101477. https://doi.org/10.1016/j.eti.2021.101477

Jamaludin, N. F. M., Abdullah, L. C., Idrus, S., Engliman, N. S., Tan, J. P., & Jamali, N. S. (2024). Nickel-iron doped on granular activated carbon for efficient immobilization in biohydrogen production. *Bioresource Technology*, *391*, 129933. https://doi.org/10.1016/j.biortech.2023.129933

Jiao, H., Al-Tohamy, R., Li, F., Schagerl, M., Sun, J., & Ali, S. S. (2024). Harnessing wastewater-based microalgae for biohydrogen production. *Process Safety and Environmental Protection*, *190*, 372–385. https://doi.org/10.1016/j.psep.2024.08.019

Liao, Q., Yang, Y.-X., Zhu, X., Chen, R., & Fu, Q. (2017). Pore-scale lattice Boltzmann simulation of flow and mass transfer in bioreactor with an immobilized granule for biohydrogen production. *Science Bulletin, 62*(1), 22–30. https://doi.org/10.1016/j.scib.2016.11.004

Min Woon, J., Shiong Khoo, K., Kazi, M., Nur-e-Alam, M., Tasnim Sahrin, N., Wei Lim, J., Kiatkittipong, W., Ali, S. S., Ho, C.-D., Usman, A., Silas Chidi, B., & Tong, W.-Y. (2024). Hydrogel immobilized microalgae-alginate beads to model the fermentation of phenol-containing wastewater into biohydrogen molecules. *Journal of Molecular Liquids, 407*, 125157. https://doi.org/10.1016/j.molliq.2024.125157

Modzelewska, A., Jackowski, M., Boutikos, P., Lech, M., Grabowski, M., Krochmalny, K., Martínez, M. G., Aragón-Briceño, C., Arora, A., Luo, H., Fiori, L., Xiong, Q., Arshad, M. Y., Trusek, A., Pawlak-Kruczek, H., & Niedzwiecki, L. (2024). Sustainable production of biohydrogen: feedstock, pretreatment methods, production processes, and environmental impact. *Fuel Processing Technology, 266*, 108158. https://doi.org/10.1016/j.fuproc.2024.108158

Mohd Jamaludin, N. F., Engliman, N. S., Abdul Manaf, S. F., Idrus, S., Abdullah, L. C., & Jamali, N. S. (2024). Effect of pH, temperature and agitation on thermophilic biohydrogen production using immobilized cells on carbon composites (GAC-$NiFe_3O_4$). *Chemical Engineering Journal, 485*, 149980. https://doi.org/10.1016/j.cej.2024.149980

Moustafa, H. M. A., Samada, M. K. A., & Khalil, A. I. (2023). Xylulose 5-phosphate production using enzymes immobilized on silica nanoparticles and its application for biohydrogen generation. *Bioresource Technology Reports, 24*, 101665. https://doi.org/10.1016/j.biteb.2023.101665

Pachaiappan, R., Cornejo-Ponce, L., Sagade, A. A., Mani, M., Aroulmoji, V., Femilaa Rajan, V., & Manavalan, K. (2024). A concise review of recent biohydrogen production technologies. *Sustainable Energy Technologies and Assessments, 62*, 103606. https://doi.org/10.1016/j.seta.2024.103606

Pandit, C., Kumar, M., Chander Kuhad, R., Sharma, K., Roy, A., Shukla, R., Pandit, S., Ranjan, N., Mishra, S. K., & Prasad, R. (2024). Exploring the performance of biocatalysts for biohydrogen production. *Biocatalysis and Agricultural Biotechnology, 60*, 103290. https://doi.org/10.1016/j.bcab.2024.103290

Pomdaeng, P., Kongthong, O., Tseng, C.-H., Dokmaingam, P., & Chu, C.-Y. (2024). An immobilized mixed microflora approach to enhancing hydrogen and methane productions from high-strength organic loading food waste hydrolysate in series batch reactors. *International Journal of Hydrogen Energy, 52*, 160–169. https://doi.org/10.1016/j.ijhydene.2023.09.187

Qi, N., Zhao, X., Hu, X., Wang, J., & Yang, C. (2023). Enhancing biohydrogen production by peanut shell carrier assisted fermentation at different hydraulic retention time. *Renewable Energy, 219*, 119492. https://doi.org/10.1016/j.renene.2023.119492

Rady, H. A., Ali, S. S., & El-Sheekh, M. M. (2024). Strategies to enhance biohydrogen production from microalgae: a comprehensive review. *Journal of Environmental Management, 356*, 120611. https://doi.org/10.1016/j.jenvman.2024.120611

Rai, P., Pandey, A., & Pandey, A. (2022). Evaluation of low cost immobilized support matrices in augmentation of biohydrogen potential in dark fermentation process using B. licheniformis AP1. *Fuel, 310*, 122275. https://doi.org/10.1016/j.fuel.2021.122275

Sanghvi, A. H., Manjoo, A., Rajput, P., Mahajan, N., Rajamohan, N., & Abrar, I. (2024). Advancements in biohydrogen production – a comprehensive review of technologies, lifecycle analysis, and future scope. *RSC Advances, 14*(49), 36868–36885. https://doi.org/10.1039/d4ra06214k

Wannapokin, A., Huang, H.-T., Chang, P.-H., Chien, Y.-W., & Hung, C.-H. (2022). Improving production of biohydrogen from COOH-functionalized multiwalled carbon nanotubes through Co-immobilization with Clostridium pasteurianum. *International Journal of Hydrogen Energy, 47*(96), 40704–40713. https://doi.org/10.1016/j.ijhydene.2022.09.095

Woo, W. X., Koh, H. S., Tan, J. P., Yeap, S. K., Abdul, P. M., Indera Luthfi, A. A., & Abdul Manaf, S. F. (2022). An overview on cell and enzyme immobilization for enhanced biohydrogen production from lignocellulosic biomass. *International Journal of Hydrogen Energy, 47*(96), 40714–40730. https://doi.org/10.1016/j.ijhydene.2022.08.164

Woon, J. M., Khoo, K. S., Sahrin, N. T., Ardo, F. M., Lim, J. W., Tong, W.-Y., Chan, Y. J., Lin, C., Goh, P. S., Ansar, S., & Lam, M. K. (2024). Rederiving kinetics to model biohydrogen production from immobilized microalgae alginate beads at various polymerization degrees of alginate under dark fermentative environment. *Algal Research, 82*, 103684. https://doi.org/10.1016/j.algal.2024.103684

4 Recent Membrane-Based Technologies for Biohydrogen Production

Types of Bioreactors, Materials, Techno-Economic Analyses, and Its Future Prospects

Angelaalincy Maria Joseph, Senthilkumar Nangan, and Kanagaraj Kuppusamy

4.1 INTRODUCTION

The dawn of the 21st century has marked a paradigm shift toward the quest for sustainable energy solutions, highlighting the imperative need for alternatives to traditional fossil fuels (Akpan and Olanrewaju, 2023). Among the promising candidates is biohydrogen (BioH), a fuel that burns cleanly, releasing only water as a by-product and offering a renewable and potentially inexhaustible energy source. This chapter introduces the sophisticated membrane-based technologies at the forefront of enhancing BioH production. These technologies not only promise to redefine our energy landscape but also encapsulate the innovative spirit of scientific inquiry and environmental stewardship (Adelekan et al., 2024).

DOI: 10.1201/9781003568612-4

BioH production has captured the attention of researchers and policymakers alike due to its profound implications for energy sustainability and environmental health (Thirumalaivasan et al., 2024b, 2024e). **Nano-phase change material (PCM):** Bridging the gap in energy storage under fluctuating environmental conditions, unlike conventional energy sources that contribute to pollution and greenhouse gas emissions, BioH offers a seamless integration into existing infrastructures with minimal environmental impact. This chapter explores how membrane-based bioreactors (M-BioRs) are pivotal in this green revolution, transforming organic substrates into clean energy through biological processes.

At the heart of advancing BioH production are M-BioRs, which represent a leap in biotechnological engineering (de Araujo et al., 2024). These reactors utilize selective permeable membranes to facilitate the bioconversion processes, ensuring optimal gas transfer and substrate utilization. The chapter details two primary configurations of M-BioRs: the flat-sheet and hollow-fiber systems, each with its unique advantages in scalability and efficiency. Through an in-depth analysis, the text demystifies how these configurations influence the dynamics of microbial communities and hydrogen production rates (Zhou et al., 2015).

In pursuit of higher efficiency and reliability, recent innovations have centered around developing advanced membrane materials (Liu et al., 2018). Hybrid organic-inorganic membranes and nanocomposite materials are at the forefront, enhancing durability and performance while addressing the perennial challenge of biofouling (Ghazi et al., 2023). This section elaborates on how integrating materials like titanium dioxide nanoparticles (NPs) into membranes not only improves their functional properties but also introduces antimicrobial characteristics that extend the operational lifespan of BioRs (Kumar et al., 2022). Understanding the economic implications is crucial for the adoption of new technologies. This chapter discusses comprehensive techno-economic analyses (TEAs) that evaluate the cost-effectiveness of membrane-based BioH (M-BioH) production systems. By comparing these systems with traditional BioH production methods, the text provides a clear economic perspective, highlighting the potential for cost reductions and enhanced market competitiveness through technological innovations (Redwood et al., 2009).

Looking toward the future, the chapter anticipates significant contributions from genetic engineering and nanotechnology in optimizing BioH production (Figure 4.1). It also considers the regulatory and market dynamics that will shape the commercialization of this technology. As governments worldwide pivot toward cleaner energy policies, BioH production is poised to benefit from regulatory support and financial incentives, making it a viable component of the global shift toward renewable energy (Thirumalaivasan et al., 2024a). This introductory chapter sets the stage for a deep dive into the technological, economic,

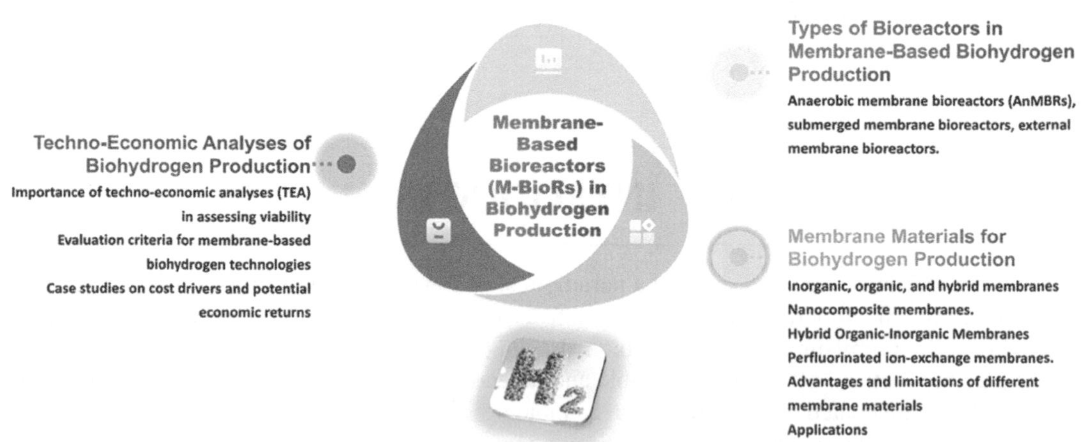

FIGURE 4.1 Schematic representation of basic biohydrogen (renewable hydrogen) production technologies. These technologies are classified into light-dependent and light-independent processes and are essential components of membrane-based advancements for efficient biohydrogen production.

and regulatory aspects of M-BioH production. By bridging interdisciplinary research and technological breakthroughs, it paints a hopeful picture of a sustainable and energy-secure future. The subsequent sections will unfold the complexities and innovations that make BioH a beacon of modern energy solutions, signaling a shift toward a cleaner and more sustainable future in global energy production (Batra, 2023).

4.2 BASIC BioH PRODUCTION AND M-BioRs

Hydrogen is increasingly recognized as a clean energy carrier, essential for transitioning toward sustainable energy systems. Unlike electricity, hydrogen does not exist naturally in its elemental form and is always combined with other substances, making it necessary to extract or produce it (Armaroli and Balzani, 2011). Key sources for hydrogen production include fossil fuels, biomass, and water, with the method of production greatly influencing its environmental footprint and energy efficiency (Shellaiah et al., 2022). Current hydrogen production methods include water electrolysis, thermo-chemical processes such as natural gas reforming, and biological hydrogen production, each offering unique advantages and challenges (Jeje et al., 2024).

Biological hydrogen production methods have attracted significant attention due to their ability to operate under ambient conditions, which minimizes energy input and offers environmental benefits (Qureshi et al., 2022). However, their lower hydrogen yields and slower production rates remain significant challenges that limit commercialization. Biological processes are often more cost-effective than their chemical counterparts, but increasing their efficiency is a critical research focus (Lynd et al., 2022).

The principal methods of biological hydrogen production include biophotolysis, photo-fermentation, and dark fermentation (Melitos et al., 2021). These processes can be categorized into light-dependent and light-independent mechanisms. Both biophotolysis (direct and indirect) and photo-fermentation rely on light to drive the hydrogen production process, while dark fermentation does not (Mishra et al., 2019). The evolution of hydrogen in these biological systems depends on either of two enzymes, hydrogenases or nitrogenases, which catalyze the reactions. Figure 4.2 illustrates the various BioH production technologies and their classification.

Each of these methods holds promise, but they also face unique challenges. For instance, while biophotolysis offers the potential for direct hydrogen production from water using sunlight, the efficiency of these systems is generally low (Nagarajan et al., 2017). Similarly, photo-fermentation can produce hydrogen using organic substrates and light, but it requires complex setups and optimization to scale (Velusamy et al., 2018). Dark fermentation, the primary light-independent process, is relatively simpler but also yields less hydrogen, necessitating further advancements in technology to make it commercially viable (Jain et al., 2024).

BioH production technologies offer a promising, environmentally friendly alternative for hydrogen generation, particularly when integrated with advancements in membrane-based technologies and innovative BioRs (El-Qelish et al., 2022). However, significant improvements in hydrogen yield and process efficiency are required to achieve widespread adoption and commercialization.

M-BioRs have garnered attention for their role in BioH production, a promising source of clean energy. These BioRs combine microbial or enzymatic processes with membrane filtration, enabling efficient hydrogen production by creating a controlled environment for microbial growth, substrate conversion, and selective hydrogen separation (Sołowski et al., 2018). By integrating membranes, M-BioRs enhance microbial contact with substrates while removing inhibitors, which can otherwise limit BioH yield. This separation also allows continuous removal of hydrogen, which can shift the equilibrium and further drive hydrogen production (Ritter and Ebner, 2007).

One of the key advantages of M-BioRs in BioH production is their ability to prevent microbial washout, enabling stable long-term operation and enhancing production efficiency. M-BioRs also facilitate effective substrate utilization, and their modular nature allows for easy scalability, making them adaptable

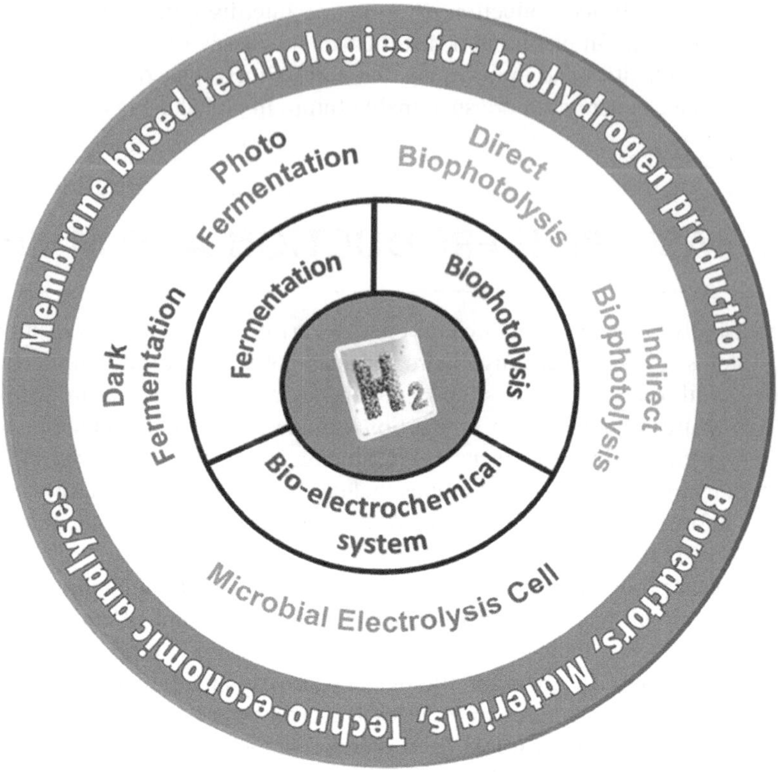

FIGURE 4.2 Overview of the book chapter.

for various production scales. However, challenges such as membrane fouling, high operational costs, and the energy-intensive nature of membrane maintenance remain critical areas for improvement (Wei et al., 2017). Innovations in membrane material, reactor design, and microbial strain development continue to evolve, aiming to overcome these barriers and establish M-BioRs as a viable solution for sustainable BioH production.

4.2.1 Types of M-BioRs

M-BioRs represent a pivotal advancement in the field of BioH production, harnessing the capabilities of specialized membranes to optimize the interactions between gases and microbial substrates. This section delves into the intricacies of M-BioRs, elucidating their operational principles, the distinct configurations available, and their respective efficiencies in the production of BioH (Figure 4.3).

M-BioRs for BioH Production: M-BioRs used for BioH production typically combine biological processes with membrane filtration to enhance hydrogen yield and productivity. These reactors harness microbial metabolism to convert organic substrates into hydrogen gas, while the membrane serves as a selective barrier to separate gases, cells, or liquids (Swaminathan et al., 2024). The integration of membranes improves process efficiency by providing better control over biomass retention and facilitating the separation of hydrogen from the fermentation broth (Thirumalaivasan et al., 2019). There are several configurations of M-BioRs used for BioH production, each offering specific advantages depending on process conditions.

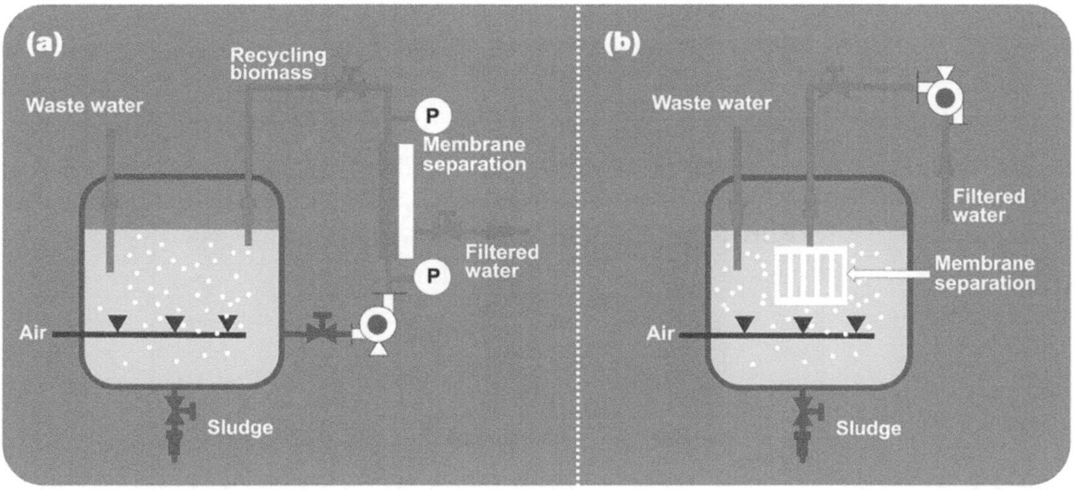

FIGURE 4.3 Comparison of two types of membrane bioreactors. (a) External membrane bioreactor (EM-BioR) with external membrane separation and recycling of biomass. (b) Submerged membrane bioreactor (SM-BioR) with an internal membrane module for solid-liquid separation and in situ gas collection.

4.2.1.1 External Membrane Bioreactor (EM-BioR)

In an external membrane system, the membrane is placed outside the BioR, and a recirculation pump is used to transport the biomass from the reactor to the membrane unit (Sharrer et al., 2007). This configuration allows for efficient separation of BioH gas and biomass, with the retentate being returned to the BioR. The gas is collected from the headspace, and the system allows continuous operation with high control over the retention time of both the microbial culture and the substrates (Li et al., 2019).

4.2.1.2 Submerged Membrane Bioreactor (SM-BioR)

In submerged membrane systems, the membrane is placed directly inside the BioR where hydrogen production occurs. This configuration is more compact and energy-efficient, as it eliminates the need for a recirculation pump (Kaya et al., 2021). The hydrogen gas is collected from the reactor headspace, while the membrane provides solid-liquid separation. Air is often introduced to keep the membrane surface clean and enhance biomass mixing (Jeong et al., 2017).

4.2.1.3 Photoanode-Based Photochemical Cell

A photoanode-based photochemical cell is a promising approach for generating hydrogen through the photolysis of water (Ahmed and Dincer, 2019). In this setup, a semiconductor photoanode absorbs light and initiates charge separation by promoting electrons (e^-) from the valence band to the conduction band, leaving behind positively charged holes (h^+) in the valence band (Abdullah et al., 2024). The excited electrons travel through an external circuit to the cathode, where they reduce protons (H^+) to hydrogen gas (H_2). Meanwhile, the holes in the photoanode participate in the oxidation of water to produce oxygen (O_2) (Ta et al., 2022). This process, as depicted in Figure 4.4, demonstrates the overall mechanism of photoinduced water splitting within the photochemical cell (Podder et al., 2020).

The photoanode material is crucial for the efficiency of the process. It must possess a suitable bandgap to absorb visible light and ensure effective charge separation, while also being stable under illumination in aqueous environments (Moniz et al., 2015). Common materials used include titanium dioxide (TiO_2) and other semiconductors capable of sustaining photochemical reactions under sunlight (Ma et al., 2014).

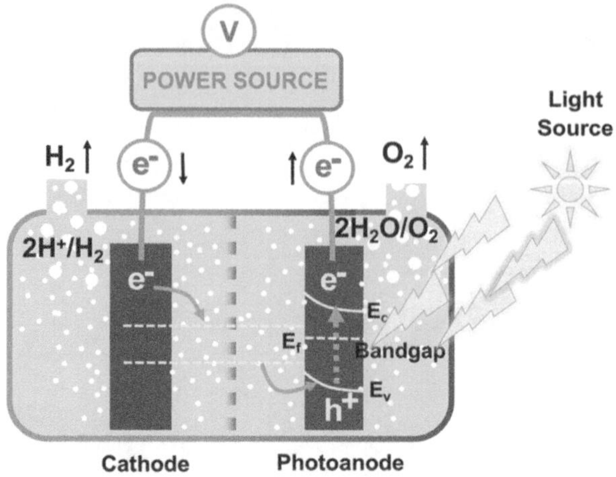

FIGURE 4.4 Schematic diagram of a photoanode-based photochemical cell for biohydrogen production, illustrating the photoinduced charge separation, water oxidation at the photoanode, and hydrogen evolution at the cathode.

The integration of membranes into the system enhances efficiency by separating the produced hydrogen and oxygen gases, thus preventing their recombination and increasing the overall yield (Singla et al., 2022).

In Figure 4.4, the light source excites the photoanode, causing the generation of electron-hole pairs (Braun et al., 2012). The electrons move toward the cathode via an external power source, contributing to the reduction reaction that produces hydrogen, while the holes at the photoanode drive the oxidation of water, releasing oxygen (Bellani et al., 2019). The membrane separates the cathode and photoanode compartments, ensuring the effective capture and separation of the products (Hu, 2019).

In this system, the membrane plays a vital role in facilitating ion transport while keeping the produced gases isolated. This process efficiency largely depends on the materials used for the photoanode and the stability of the membrane, making this an essential area of research for advancing photochemical BioH production technologies.

4.2.2 Types of Membranes in BioRs

The choice of membrane type in BioRs for BioH production is crucial to optimizing the efficiency and sustainability of the process (Jabbari et al., 2019). Flat-sheet membranes offer simplicity, ease of maintenance, and scalability, making them a preferred option for operations prioritizing ease of handling and large-scale deployment (Alkandari and Castro-Dominguez, 2024). However, they occupy more space and may suffer from fouling issues, potentially lowering their overall productivity. In contrast, hollow-fiber membranes, with their high surface-to-volume ratio, provide enhanced gas transfer and occupy less space, making them ideal for more compact systems (Lau and Yong, 2021). However, their complexity, higher susceptibility to clogging, and challenges in maintenance can make them difficult to scale up or operate in large systems (Wang, 2017). Ultimately, the choice between flat-sheet and hollow-fiber membranes depends on specific operational factors, such as reactor size, space constraints, economic considerations, and maintenance capabilities, highlighting the importance of a tailored approach for different BioH production projects (Soto et al., 2022).

4.2.2.1 Flat-Sheet Membranes

Flat-sheet membranes are among the most common configurations used in M-BioRs due to their structural simplicity and operational effectiveness (Chu et al., 2020). These systems comprise multiple layers of membrane materials assembled in a stacked arrangement, resembling sheets laid one atop the other (Hammond, 2004). This design not only facilitates ease of scaleup for industrial applications but also maximizes the membrane area available for microbial activity and gas exchange (Buonomenna, 2013b).

Structural Features and Benefits:

- **Surface Area:** The extensive surface area provided by flat-sheet configurations allows for ample microbial colonization, which is critical for enhancing the bioactivity within the reactor (Huanambal-Sovero et al., 2023).
- **Scalability:** The modular nature of these membranes facilitates straightforward scaling, making them highly adaptable to varying production capacities without significant modifications to the core design (Nunes et al., 2020).
- **Maintenance and Handling:** Flat-sheet membranes are easier to clean and replace due to their accessible arrangement, reducing downtime and maintenance costs (Regula et al., 2014).

Operational Mechanism: In BioH production, these membranes act as critical interfaces where microbial action on organic substrates leads to hydrogen generation (Ayodele et al., 2023). The design of flat-sheet membranes supports a stable environment that maintains adequate humidity and nutrient supply, crucial for microbial health and productivity (He et al., 2020).

4.2.2.2 Hollow-Fiber Membranes

Hollow-fiber membranes provide a different approach to maximizing efficiency in BioH production (Wei et al., 2019). These are constructed from numerous narrow, tube-like fibers bundled together, each acting as an individual BioR (Lau et al., 2022). The hollow structure of these fibers significantly enhances the surface-to-volume ratio, a key factor in improving gas transfer rates and overall reactor productivity (Reichelt et al., 2014).

Structural Features and Benefits:

- **High Surface-to-Volume Ratio:** The compact and dense arrangement of hollow fibers dramatically increases the surface area relative to the volume of the reactor, facilitating efficient gas exchanges and improving the rate of hydrogen production (Zhao et al., 2023).
- **Enhanced Transfer Rates:** The unique configuration of hollow fibers enhances the diffusion of hydrogen and other gases, ensuring that these products are quickly transferred out of the microbial zone to maintain an optimal production environment (Enriquez and Ahring, 2023).
- **Reduced Footprint:** Hollow-fiber systems tend to occupy less space compared to other configurations with equivalent surface area, making them ideal for applications where space is a constraint (Wan et al., 2021).

Operational Mechanism: Hollow-fiber membranes specialize in maintaining a controlled flow of gases and substrates through the reactor (Chang, 2011). The fibers allow for selective permeability oxygen and other inhibitory gases are kept out while allowing hydrogen, the desired product, to be harvested efficiently (Thirumalaivasan et al., 2017). This selective permeation ensures that the internal environment of the BioR remains anaerobic, which is essential for the specific microbial populations that produce hydrogen.

Both flat-sheet and hollow-fiber membranes offer distinct advantages and are tailored to meet specific operational needs in BioH production. The choice between these configurations often depends on factors such as desired production scale, economic considerations, and physical space constraints (Bridge et al., 2013). Understanding these types of M-BioRs allows for a more informed selection of technology based on the specific requirements of a BioH project, leading to more effective and sustainable energy solutions (Venkatesan et al., 2019).

Both flat-sheet and hollow-fiber membranes offer distinct advantages and are tailored to meet specific operational needs in BioH production. The choice between these configurations depends on factors such as desired production scale, economic considerations, and physical space constraints (Hosseini et al., 2023). Understanding the various types of M-BioRs allows for a more informed selection of technology based on the specific requirements of a BioH project, leading to more effective and sustainable energy solutions.

Both flat-sheet and hollow-fiber membranes offer distinct advantages and are tailored to meet specific operational needs in BioH production (Rajapandi et al., 2023). The choice between these configurations depends on factors such as desired production scale, economic considerations, and physical space constraints (Table 4.1). Understanding the various types of M-BioRs allows for a more informed selection of

TABLE 4.1 Comparison of distinct advantages and disadvantages of flat-sheet and hollow-fiber M-BioRs for biohydrogen production

FEATURE	*FLAT-SHEET MEMBRANES*	*HOLLOW-FIBER MEMBRANES*
Advantages		
Surface area	Provides a large, extensive surface area for microbial colonization, enhancing bioactivity.	Offers a high surface-to-volume ratio due to the dense arrangement of fibers, maximizing surface contact with minimal volume.
Scalability	Modular design allows for easy scaling up of production capacity in industrial applications.	Scalable but requires careful integration into reactor design due to the complex structure.
Maintenance and handling	Easy to access, clean, and replace. Maintenance can be performed without fully shutting down the system, reducing downtime.	High packing density, but fibers are more difficult to clean and replace, potentially leading to higher maintenance efforts.
Reactor design flexibility	Relatively simple design, adaptable to various reactor setups and configurations.	Compact design with reduced footprint, ideal for applications with space constraints.
Fouling resistance	Easier to monitor and manage membrane fouling due to the flat, accessible surface.	Generally lower fouling rates due to continuous liquid flow, but internal clogging can be harder to detect and resolve.
Cost-effectiveness	Typically more cost-effective for small and medium-scale systems.	Cost-effective in terms of space-saving designs and high productivity per volume.
Disadvantages		
Space utilization	Requires a larger reactor space to achieve similar surface areas compared to hollow fibers.	Compact design but has a higher capital cost due to its sophisticated structure and manufacturing requirements.
Gas transfer efficiency	Moderate gas transfer efficiency, especially in terms of hydrogen separation.	Superior gas transfer efficiency due to higher surface area and better gas diffusion rates, crucial for hydrogen production.
Energy consumption	Requires higher energy for aeration and membrane cleaning, especially in larger systems.	Typically requires less energy for aeration but may need more energy for efficient substrate flow across the fibers.
Complexity of operation	Simpler operation and control, making it more user-friendly, especially for smaller systems.	More complex in operation due to the need for precise control over flow rates, pressures, and gas transfer.
Durability	More durable in terms of physical wear and tear, especially during cleaning cycles.	The delicate nature of hollow fibers makes them more susceptible to damage, requiring careful handling and operation.
Cost of replacement	Easier and cheaper to replace individual sheets, reducing overall downtime and replacement costs.	Replacing hollow fibers can be more expensive and time-consuming, especially in large bundled configurations.

technology based on the specific requirements of a BioH project, leading to more effective and sustainable energy solutions.

Flat-sheet membranes are advantageous for ease of maintenance, scalability, and cost-effectiveness, particularly in smaller systems. However, they require more space and have moderate gas transfer efficiency (Osman et al., 2024b). Hollow-fiber membranes offer high surface-to-volume ratios, superior gas transfer rates, and compact designs, making them ideal for high-efficiency hydrogen production in space-constrained applications. However, they are more complex to maintain and operate (Arumugam et al., 2023).

4.3 INNOVATIONS IN MEMBRANE MATERIALS

The development of membrane materials for BioH production has evolved significantly in recent years, with a keen focus on overcoming operational challenges such as biofouling, which can drastically impair the efficiency and longevity of BioRs (Thirumalaivasan et al., 2024a). This section delves into two prominent advancements: hybrid organic-inorganic membranes and nanocomposite membranes. These innovations not only enhance the performance of M-BioRs but also ensure their economic viability by reducing maintenance costs and improving durability.

4.3.1 Hybrid Organic-Inorganic Membranes

Hybrid organic-inorganic membranes represent a revolutionary approach in membrane technology, combining the mechanical strength and chemical resistance of inorganic materials with the flexibility and processability of organic polymers (Rahman et al., 2022). This synergy results in membranes that are robust enough to withstand harsh operational conditions while maintaining the necessary flexibility to prevent cracks and structural failures (Qiu et al., 2019).

Hybrid organic-inorganic membranes represent a significant innovation in membrane technology used for BioH production in M-BioRs. These membranes combine the flexibility and processability of organic polymers with the durability, thermal stability, and chemical resistance of inorganic materials (Kudapa et al., 2024). The organic component typically provides mechanical flexibility and enhances the membrane's compatibility with biological systems, while the inorganic component improves the membrane's stability under harsh operating conditions, such as high temperatures and varying pH levels (Laberty-Robert et al., 2011).

The hybrid nature of these membranes allows for greater control over the transport properties of gases and liquids, which is critical for optimizing the separation of BioH from other gases in the BioR (Govindasamy et al., 2022). By adjusting the ratio and type of organic and inorganic components, the permeability and selectivity of the membranes can be fine-tuned, leading to enhanced BioH yield. Compared to conventional membranes, these hybrids offer superior resistance to fouling, extended lifespan, and higher efficiency in separating BioH from complex gas mixtures (Zhen et al., 2019).

Key Features and Benefits:

- **Enhanced Mechanical Properties:** The incorporation of inorganic components like silica or zirconia enhances the structural integrity of the membranes, making them less prone to mechanical deformation under high pressure or temperature variations (Mammeri et al., 2005).
- **Improved Chemical Resistance:** These membranes can resist a wide range of pH levels and aggressive chemicals, thereby ensuring stability and longevity in diverse biochemical environments (Warsinger et al., 2018).

- **Versatility:** The ability to tailor the properties of these membranes by adjusting the organic and inorganic phases allows for customized solutions tailored to specific BioR designs and operational needs (Verma et al., 2024b).

4.3.2 Nanocomposite Membranes

The integration of NPs into the membrane matrix is a cutting-edge innovation aimed at enhancing membrane functionality (Vedarethinam, 2024). Materials such as titanium dioxide or silver are incorporated to leverage their unique properties, such as high surface area and antimicrobial activity, which are critical in enhancing the performance and lifespan of BioRs (Nichols and Davis, 2016).

Nanocomposite membranes incorporate NPs or nanomaterials within a polymer matrix to improve the performance of M-BioRs (Natarajan et al., 2019). These NPs, such as metal oxides, carbon nanotubes (CNTs), and quantum dots, enhance the mechanical strength, thermal stability, and permeability of the membranes (Aryamol et al., 2024; Nair et al., 2023; Ramasubburayan et al., 2024b). The nanocomposite structure increases the surface area available for BioH production, making the process more efficient by allowing faster gas transport through the membrane.

The inclusion of nanomaterials within the membrane can significantly improve its separation capabilities and resistance to fouling (Venkatesan et al., 2023). This leads to a more robust and durable membrane that can sustain longer operational periods without losing efficiency. Nanocomposite membranes offer enhanced permeability without sacrificing selectivity, which is crucial for improving BioH purity (Wu et al., 2020). Additionally, these membranes are more resistant to microbial degradation and physical wear, contributing to the overall sustainability and cost-effectiveness of BioH production (Pachaiappan et al., 2024).

Functional Improvements:

- **Enhanced Gas Separation Efficiency:** NPs can increase the selectivity and permeability of membranes, improving the separation process of hydrogen from other gases and enhancing overall production efficiency (Saini and Awasthi, 2022).
- **Anti-biofouling Properties:** NPs like silver exhibit potent antimicrobial properties, which prevent the growth of biofilms on membrane surfaces, a common issue that leads to decreased membrane efficiency and lifespan (Cruz et al., 2015).
- **Increased Durability:** The presence of NPs contributes to the physical robustness of the membrane, reducing wear and tear and extending the operational life (Si and Guo, 2022).

4.3.3 Additive-Containing Membranes

Additive-containing membranes are a new class of membranes where various functional additives are embedded within the membrane matrix to enhance performance (Nangan et al., 2023a). Carbon-based materials such as carbon dots, carbon quantum dots, and CNTs significantly increase the conductivity, thermal stability, and surface area of the membranes (Nangan et al., 2023a; Ramasubburayan et al., 2023; Verma et al., 2024a). Additionally, heterogeneous materials like clays, zeolites, and silica gel improve the mechanical strength and selective permeability of the membrane. Additives such as 2D porous organic polymers (POPs) and metal-organic frameworks (MOFs) further enhance the separation efficiency by offering molecular sieving properties and increased surface interaction (Alagesan et al., 2016; Arumugam et al., 2023; Kanagaraj and Pitchumani, 2010; Kanagaraj et al., 2023b; Suresh et al., 2014).

The incorporation of additives greatly enhances the functional properties of membranes, making them more effective at selectively filtering BioH from complex gas mixtures. Carbon-based materials, in particular, improve the conductivity and gas permeability of the membrane, leading to faster and more efficient BioH production (Yang et al., 2022). On the other hand, MOFs and porous polymers provide structural frameworks that allow for precise molecular separation, resulting in higher BioH purity and yield compared to traditional BioRs (Aryamol et al., 2024). The increased mechanical and chemical stability from additives also results in longer membrane lifespans, reducing the frequency of replacement and overall operational costs.

4.3.4 Functionalized Membranes

Functionalized membranes are membranes that have been chemically modified to incorporate specific functional groups or molecules, such as supramolecular hosts (e.g., cyclodextrin, synthetic cavitands, cucurbiturils), and nanomaterials like carbon nanomaterials (Chen et al., 2022; Ji et al., 2022; Kanagaraj et al., 2010, 2023a; Rama et al., 2011, 2012b; Tang et al., 2022). These modifications allow for tailored interaction with BioH and other molecules present in the reactor. Supramolecules can form inclusion complexes with certain gases, enhancing the selectivity of BioH separation.

Functionalized membranes play a critical role in improving the selectivity and efficiency of BioH production (Nguyen et al., 2020). By using specific functional groups, these membranes can create highly selective channels that allow only hydrogen molecules to pass through, thus improving BioH purity and reducing contamination from other gases like CO_2 or CH_4. This precise control over molecular transport makes functionalized membranes significantly more efficient than traditional membranes, especially in complex biological systems (MariaJoseph et al., 2024). The use of carbon nanomaterials further enhances the membrane's permeability and strength, contributing to a more stable and efficient production process.

4.3.5 Other Materials

In addition to the advanced membrane categories discussed, other materials, including natural and synthetic polymers, have been utilized in the design of membranes for BioH production (Thirumalaivasan et al., 2024c; Verma et al., 2024a). Materials such as polyethersulfone (PES), cellulose acetate, and polylactic acid (PLA) offer excellent mechanical properties and biocompatibility, making them ideal candidates for BioH production in certain BioR conditions. Functionalized materials, such as surface-modified polymers, provide enhanced interactions with hydrogen and improve the overall separation process (Goddard and Hotchkiss, 2007).

These traditional and functionalized polymeric membranes offer a cost-effective and readily scalable solution for BioH production (Thirumalaivasan et al., 2024d). While they may not offer the same level of permeability or selectivity as some of the more advanced membranes, their ease of fabrication, wide availability, and customizable surface chemistry make them attractive for specific applications (Sadeghi et al., 2018). The use of natural polymers also brings an eco-friendly aspect to BioH production, as these materials are biodegradable and derived from renewable resources. Moreover, the functionalization of these polymers with specific groups can improve their performance, offering a balance between cost and efficiency (Awang et al., 2015).

Membrane innovations in the field of BioH production, including hybrid organic-inorganic, nanocomposite, additive-containing, functionalized, and other material-based membranes, offer significant advancements over conventional membrane technologies (Nair et al., 2023). These innovations provide improved efficiency, selectivity, and stability in M-BioRs, leading to higher BioH yields, lower operational costs, and enhanced sustainability (Bakonyi et al., 2018). By optimizing these membrane materials, researchers and engineers can greatly improve the commercial viability of BioH as a clean energy source (Bakonyi et al., 2013).

4.3.6 Emerging Applications of Membrane Materials

Membrane materials have emerged as versatile and essential components in various industries due to their unique ability to selectively separate, purify, and catalyze reactions (Lee et al., 2016). The diversity in membrane types and materials has allowed for their use in a broad spectrum of applications, including BioH production, where they are critical for enhancing process efficiency (Nangan et al., 2023b). However, beyond BioH, these materials are indispensable in many other fields. Their applications range from separation processes to catalysis, water treatment, and biomedical uses, showcasing their broad utility across multiple industries. Below is a discussion of the various applications of membrane materials (Werber et al., 2016).

Separation of Organic Compounds: Membrane materials are extensively used for separating organic compounds, especially in chemical processing and pharmaceutical industries (Ingole and Ingole, 2014). The selective permeability of these membranes enables the effective separation of components based on molecular size, shape, or chemical properties (Kanagaraj et al., 2022; Peng et al., 2021; Zhang et al., 2021). For example, nanofiltration and reverse osmosis membranes can separate small organic molecules, making them useful in the purification of solvents and recovery of valuable chemicals from waste streams. In BioH production, membrane separation techniques like pervaporation are also applied to enhance the purity of hydrogen gas (Shellaiah et al., 2024).

Separation of Enantiomers: The separation of enantiomers – mirror-image isomers of chiral molecules – is of immense importance in the pharmaceutical industry, as often only one enantiomer of a drug is therapeutically active (Kanagaraj et al., 2017, 2020a; Kanagaraj and Pitchumani, 2013b). Membrane technologies have advanced to a point where chiral-selective membranes are being developed for enantiomeric separation (Thirumalaivasan et al., 2024c). These membranes exhibit chiral recognition properties, making them ideal for separating racemic mixtures into individual enantiomers, an essential process in the production of optically pure drugs.

Catalyst-Loaded Membranes for Organic Reactions: Membrane materials are not just passive barriers but can also be active components in chemical reactions. Catalyst-loaded membranes combine separation with catalysis, allowing for continuous processes where reactants and products are simultaneously separated as reactions occur (Jeganathan et al., 2014; Nangan et al., 2023b; Suresh et al., 2013). These membranes enhance reaction efficiency by localizing the catalyst on the membrane surface or within its pores. This concept is particularly useful in organic synthesis, where catalyst-loaded membranes facilitate reactions like hydrogenation, oxidation, or esterification, and improve product yield (Naha et al., 2020).

Photochemical Reactions: Membrane materials also find application in photochemical reactions (He et al., 2009). In these systems, membranes act as both separation and reaction platforms, where they assist in the controlled exposure of reactants to light, enabling specific photochemical processes (Rao et al., 2018). Such applications are explored in the production of fine chemicals, pharmaceuticals, and even environmental technologies, where membrane photoreactors help in the degradation of pollutants.

Water Treatment (Desalination and Filtration): Membrane-based water treatment technologies have seen widespread use in both industrial and domestic sectors (Yusuf et al., 2020). Desalination, which uses reverse osmosis membranes, is a key process in making seawater drinkable. These membranes remove salts and other impurities, offering a viable solution to the global water scarcity problem (Shemer et al., 2023). Additionally, ultrafiltration and microfiltration membranes are employed for removing suspended particles, bacteria, and viruses in municipal water treatment and industrial wastewater recycling (Hakami et al., 2020).

Food and Beverage Production: In the food and beverage industry, membrane technology plays a crucial role in processes such as concentration, clarification, and sterilization (Charcosset,

2021). Membranes are used for the concentration of fruit juices, dairy products, and alcoholic beverages, as well as for the filtration of beer and wine to improve clarity and shelf life. Membranes offer a non-thermal method for separation, ensuring that heat-sensitive food components are preserved during processing (Picart-Palmade et al., 2019).

Pharmaceutical Manufacturing: Membranes are critical in the pharmaceutical industry for purification, separation, and drug delivery systems (Thirumalaivasan et al., 2024b). In drug production, they help separate active pharmaceutical ingredients (APIs) from by-products and solvents (Buonomenna and Bae, 2015). Additionally, membrane filtration is employed in sterile drug formulation processes, ensuring the removal of microbial contaminants from injectable drugs. In advanced drug delivery systems, membrane-based devices provide controlled release of therapeutic agents, improving drug efficacy and patient compliance (Stamatialis et al., 2008).

Chemical Processing: In chemical industries, membrane materials are used for gas separation, solvent recovery, and the concentration of chemical products (Thirumalaivasan et al., 2024a) For instance, membrane processes such as pervaporation and gas permeation are used to separate hydrogen from gas mixtures or remove organic solvents from reaction mixtures. The use of 2D POPs in membrane-based separations has shown promise in enhancing the efficiency and selectivity of these processes due to their tunable pore structures (Dhakshinamoorthy et al., 2011; Rama et al., 2012a).

Biomedical Applications: In the biomedical field, membranes are indispensable for applications like hemodialysis, drug delivery, and tissue engineering (Rajapandi et al., 2023; Thirumalaivasan and Wu, 2020; Venkatesan et al., 2023). Hemodialysis membranes are used to filter waste products from the blood of patients with kidney failure. Similarly, in tissue engineering, membranes provide scaffolding that supports the growth of new tissues. These materials are also being developed for use in artificial organs and for the controlled release of drugs in the body.

Water Treatment for Domestic and Industrial Supply: Membranes are widely used for water purification in both domestic and industrial settings. Technologies such as ultrafiltration, microfiltration, and reverse osmosis are employed to treat water for drinking and industrial processes (Kanagaraj et al., 2012, 2014; Kanagaraj and Pitchumani, 2014). These membranes remove suspended solids, bacteria, and other contaminants, ensuring safe and clean water supply. Additionally, in industrial water recycling, membrane processes help recover and reuse water, reducing overall water consumption and environmental impact.

Separation in Biotechnological, Chemical, and Metallurgical Industries: In biotechnological and metallurgical industries, membranes are used to separate valuable components from complex mixtures (Thirumalaivasan et al., 2024e). For instance, in biotechnology, membranes are applied in fermentation processes for the continuous removal of by-products, improving process efficiency. In the metallurgical industry, membrane technologies aid in the recovery of metals from leachates and industrial waste, contributing to resource recovery and environmental sustainability (Abraham et al., 2024).

Separation Science (Chiral Recognition for Chromatography and Membranes): Chiral recognition is crucial in fields such as chromatography and membrane-based separations, especially in the pharmaceutical industry (Kanagaraj et al., 2020b; Xiao et al., 2020). Membranes with chiral recognition abilities can separate enantiomers, providing optically pure substances for drug synthesis. These processes are also applied in other fields such as food science and chemical research, where the separation of optical isomers is essential.

2D POPs for Membrane-Based Separations: Recently, the development of 2D POPs has opened new avenues for membrane-based separations (Kanagaraj et al., 2023a). These materials exhibit high surface area, tunable pore sizes, and chemical stability, making them ideal for a wide range of separation processes. They are particularly promising for gas separations, organic solvent nanofiltration, and desalination processes due to their selective permeability and robustness under harsh conditions.

The application of membrane materials spans across various industries, from water treatment and pharmaceutical manufacturing to chemical processing and biomedical applications. Their ability to selectively separate, purify, and catalyze reactions has made them indispensable in modern industrial processes (Kanagaraj and Pitchumani, 2013a). With advancements in membrane technology, such as the development of 2D POPs and chiral-selective membranes, the potential for membrane-based technologies is expanding, offering new opportunities for improving efficiency and sustainability across a wide range of applications.

4.4 EVALUATION CRITERIA AND CASE STUDIES FOR TEAS

Hydrogen is widely regarded as a clean fuel with immense potential to reduce environmental impacts and provide energy security (Prakash et al., 2023). However, the economic feasibility of hydrogen production hinges on the careful selection and optimization of production platforms. TEA and sensitivity analysis are essential tools for assessing the viability of existing hydrogen production technologies, providing insights into cost-effectiveness, energy efficiency, and environmental sustainability (Zulfhazli et al., 2022).

TEAs are essential for evaluating the viability of M-BioH production systems. These analyses combine technical performance metrics with economic factors to assess the overall feasibility and scalability of the technology. Key evaluation criteria include capital and operational costs, energy efficiency, membrane durability, and hydrogen yield (Dincer and Acar, 2015). TEA helps identify cost drivers such as membrane material expenses, maintenance, and energy consumption, while also quantifying potential economic returns, including reduced greenhouse gas emissions and enhanced energy security. Case studies often involve comparing traditional hydrogen production methods with advanced membrane technologies, illustrating cost savings, environmental benefits, and payback periods (Yue et al., 2021). Through these evaluations, TEA guides decision-making for future investments and scaling of BioH production.

As illustrated in Figure 4.5, various feedstocks, such as organic waste, algal biomass, and fossil fuels, can be used across different hydrogen production platforms (Kannah et al., 2021). These platforms include processes like biophotolysis, dark fermentation, gasification, pyrolysis, steam reforming, and water electrolysis, each with distinct cost structures and energy requirements. For example, steam reforming of natural gas, with production costs ranging between 1.25 and 3.50 USD/kg H_2, is currently considered one of the most cost-efficient methods for hydrogen production. Its lower operational costs and established infrastructure make it a preferred choice, though there are ongoing concerns regarding carbon emissions, which necessitate future advancements in carbon capture technologies.

The figure also highlights the potential of renewable processes like water electrolysis and biophotolysis (Sikiru et al., 2024). Although these methods offer environmental benefits by utilizing renewable resources, they face challenges related to higher production costs (e.g., 3.01–4.51 USD/kg H_2 for water electrolysis) (Ramasubburayan et al., 2024a). Thus, research into reducing energy inputs and enhancing efficiency through innovations like membrane reactors is crucial for making hydrogen from these platforms more competitive. TEA also evaluates the sensitivity of hydrogen production costs to fluctuations in feedstock prices, energy consumption, and technological advancements. This holistic approach helps identify key factors influencing the economic feasibility of each process, guiding the development of strategies to optimize hydrogen production.

The integration of TEAs and sensitivity analyses provides valuable insights for optimizing hydrogen production technologies, informing future research directions, and enabling cost reductions. This will play a vital role in transitioning to hydrogen as a mainstream fuel for a sustainable energy future.

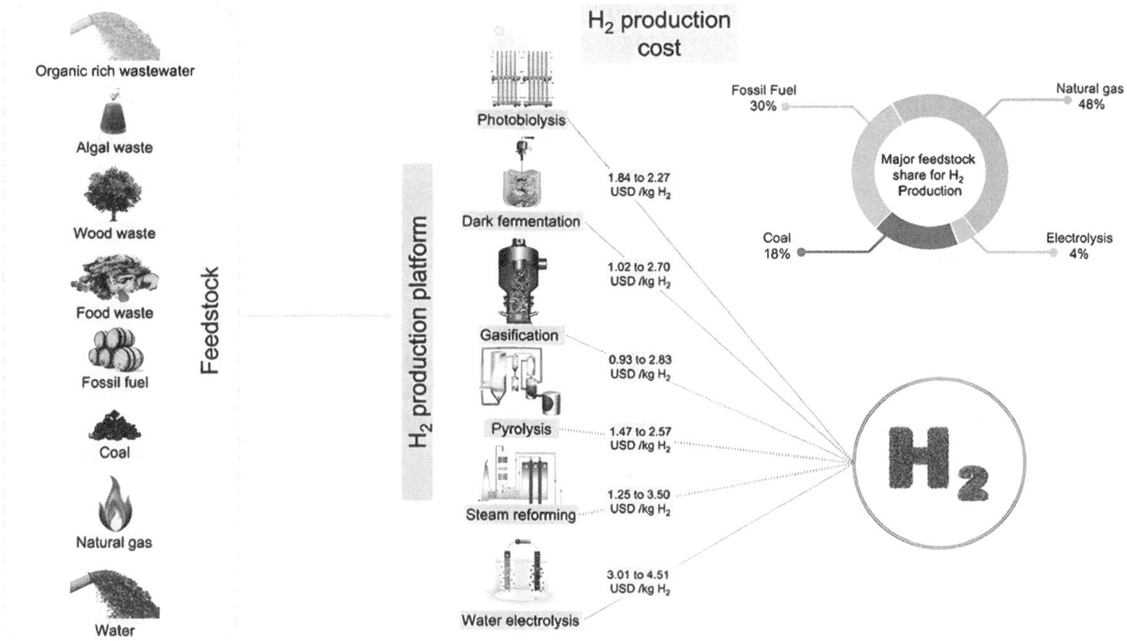

FIGURE 4.5 Overview of hydrogen production platforms, their feedstocks, and associated costs.

4.4.1 Importance of TEAs

TEAs play a critical role in the development and adoption of M-BioH production. As these technologies evolve, TEA helps to assess their economic feasibility and ensure they are viable for large-scale implementation. It provides a comprehensive evaluation by integrating both technical performance, such as hydrogen yield and membrane efficiency, with economic considerations like capital investment (CapEx), operating costs, and potential market value (Ma et al., 2015). One of the primary advantages of TEA is its ability to identify cost drivers, such as membrane material costs, maintenance, and energy consumption, allowing stakeholders to optimize processes and reduce expenses. Additionally, TEA aids in assessing environmental and societal benefits, such as reduced carbon emissions, energy security, and resource efficiency (Munasinghe et al., 2017). By quantifying these aspects, TEA ensures that the development of BioH production technologies is not only technically sound but also economically attractive, thereby increasing the likelihood of their adoption in both industrial and environmental sectors. Furthermore, it enables comparisons between traditional hydrogen production methods and membrane-based systems, highlighting potential economic returns and guiding policy, research, and investment decisions (Liguori et al., 2020). As the demand for sustainable energy solutions increases, TEA becomes indispensable in ensuring that BioH technologies can meet both environmental and economic goals effectively (Debnath et al., 2021).

4.4.2 Evaluation Criteria

- **Cost-Efficiency:** The analysis looks at the capital and operational expenses (OpEx) of membrane-based systems compared to traditional systems. This includes costs related to installation, maintenance, and membrane replacement (Judd, 2017).

- **Return on Investment (ROI):** Evaluating the ROI helps determine the break-even point and long-term financial benefits of investing in advanced membrane technologies (Fernández-Ronco et al., 2013).
- **Market Competitiveness:** By comparing the performance and costs of these systems with existing technologies, stakeholders can assess their competitive edge in the market (Day and Wensley, 1988).

4.4.3 Case Studies

Recent case studies illuminate the economic impact of these innovations (Franceschini, 2015). For instance, a study comparing traditional BioH production setups with those utilizing hybrid membranes found a marked reduction in operational costs due to decreased downtime and lower maintenance requirements (Wheelock Gutierrez, 2023). Another case study highlighted the role of nanocomposite membranes in reducing energy consumption during gas separation, directly correlating to lower production costs (Liang et al., 2019).

Outcomes:

- **Cost Reductions:** Innovations in membrane technology have been shown to significantly reduce both capital and OpEx, making BioH production more financially viable (Osman et al., 2024a).
- **Technological Advancements:** Continuous improvement and innovation in membrane technology can further drive down costs and enhance efficiency, promoting wider adoption (Fane et al., 2011).

The innovations in membrane materials not only address technical challenges like biofouling but also improve the economic attractiveness of M-BioH production systems. Through these advancements, the feasibility and deployment of BioH as a sustainable energy resource are greatly enhanced, offering promising prospects for the energy sector's future (Hosseini and Wahid, 2016).

4.4.4 Applications/Utility/Benefits of TEAs

TEAs offer significant applications, utility, and benefits in the development and optimization of M-BioH production technologies. One key application is in the early-stage assessment of the economic viability of new technologies, providing stakeholders with critical data to evaluate whether a technology is worth pursuing (Farrukh and Holgado, 2020). By combining technical and economic factors, TEA helps identify cost drivers such as membrane materials, operational energy consumption, and maintenance, while also projecting the potential economic returns and payback periods (Mahmud et al., 2021). This aids in resource allocation and investment decisions, ensuring efficient use of capital for research, development, and scaling of BioH production processes.

The utility of TEA extends to guiding process optimization and innovation (Zimmermann et al., 2020). By pinpointing inefficiencies in the system, it informs design improvements, helping to enhance productivity and reduce costs, thus improving the overall cost-benefit ratio (Mantei and Teorey, 1988). Additionally, TEA provides a framework for comparing BioH technologies with conventional hydrogen production methods, allowing for a clear understanding of economic and environmental trade-offs.

The benefits of TEA are multifaceted, including fostering informed decision-making for policymakers, investors, and researchers, while promoting sustainable energy transitions (Brown et al., 2020). It enables stakeholders to quantify not only the economic feasibility but also the long-term societal and environmental benefits, such as lower carbon emissions, increased energy independence, and alignment with renewable energy targets (Parrot and Tierney, 2012). Through these contributions, TEA helps accelerate the commercialization of sustainable hydrogen production technologies (Wang et al., 2024).

4.5 TEAs IN BioH PRODUCTION

TEAs are indispensable tools in the evaluation of technological innovations, especially in the energy sector where new technologies must compete with established systems on both performance and cost (Paul et al., 2021). For M-BioH production, TEA provides a rigorous framework to scrutinize the economic feasibility and sustainability of implementing these technologies at various scales and in different contexts (Rastgar et al., 2023).

4.5.1 Importance of TEAs

The transition toward sustainable energy solutions like BioH necessitates a comprehensive understanding of both the technological and economic aspects of production systems. TEA addresses several critical areas:

Cost Driver Identification: It identifies which components of the production process incur the most significant costs, whether in terms of CapEx (such as BioRs and membrane systems) or OpEx (including energy consumption and maintenance) (Savla et al., 2021).

Assessment of Economic Returns: By projecting the potential financial returns over the lifecycle of the technology, TEA helps stakeholders understand the payback period and the ROI, which are crucial for making informed decisions (Lunelli and Morais, 2024).

Comparative Analysis: Comparing membrane-based systems with traditional BioH production methods illuminates the advantages and potential savings associated with the newer technology, as well as highlighting areas where improvements are necessary (Khoiruddin et al., 2024).

Decision-Making Support: The insights gained from TEA are invaluable for policymakers, investors, and technology developers in making informed decisions about funding, research priorities, and technology deployment (Pólvora et al., 2020).

4.5.2 Framework of TEAs

The TEA of M-BioH production typically follows a structured approach involving several key steps:

- **System Definition and Functional Unit**: Clearly defining the system boundaries and the functional unit (e.g., cost per kilogram of hydrogen produced) to ensure that all relevant costs and benefits are included (Staffell et al., 2019).
- **Capital and Operational Cost Analysis**: Detailed accounting of all CapEx and OpEx associated with the system. This includes costs of materials, construction, labor, energy, and maintenance (Akbar et al., 2021).
- **Performance Metrics Evaluation**: Measurement of system performance in terms of efficiency, hydrogen yield, and reliability, and how these impact the economic model (Acar and Dincer, 2019).
- **Sensitivity and Uncertainty Analysis**: Assessing how changes in key variables (like material costs, energy prices, or production capacity) affect the costs and viability of the system (Awudu and Zhang, 2012).

- **Scenario Analysis**: Evaluating the economics under different scenarios to understand the impacts of policy changes, technological advancements, or market developments (Wiebe et al., 2018).

4.5.3 Case Studies and Economic Viability

Several case studies have underscored the economic viability of M-BioH production, highlighting significant reductions in both CapEx and OpEx due to technological innovations. For instance:

- **Case Study 1**: A pilot project using flat-sheet membrane BioRs demonstrated a 30% reduction in production costs compared to conventional methods due to higher process efficiencies and lower energy requirements (Chang et al., 2020).
- **Case Study 2**: A commercial-scale deployment featuring hollow-fiber membranes showcased a reduction in payback time from 7 to 5 years through improved hydrogen purity and reduced downstream processing costs (Ramdin et al., 2021).

These case studies provide tangible evidence that innovations in membrane technology can dramatically decrease production costs, enhancing the economic competitiveness of BioH against fossil fuels and other renewable energy sources (Manjunathan et al., 2023). TEAs form the backbone of strategic decision-making in the deployment of M-BioH production technologies (Seferlis et al., 2021). By systematically evaluating the economic outcomes and highlighting areas for cost reduction and technological improvement, TEA ensures that investments in BioH technologies are both scientifically sound and economically viable, paving the way for their successful integration into the global energy mix (Glatzel et al., 2024).

4.6 FUTURE PROSPECTS AND INNOVATIONS IN M-BioH PRODUCTION

As the global thrust toward sustainable energy intensifies, the future of BioH production seems particularly promising, driven by groundbreaking advancements in genetic engineering and nanotechnology (Sohail et al., 2019). These technologies are not just supplementary enhancements but are anticipated to be revolutionary, transforming the core of BioH production processes (Zainal et al., 2024).

Figure 4.6 highlights key innovations driving the advancement of M-BioH production. These innovations span across three crucial areas: genetic engineering, nanotechnology, and regulatory frameworks. In the field of genetic engineering, optimizing microbial metabolic pathways and enhancing stress tolerance are essential for increasing BioH production efficiency. By selectively producing hydrogen and minimizing by-products, genetically modified microorganisms can significantly boost BioH yields, making the process more viable on an industrial scale.

Nanotechnology plays a vital role in improving membrane performance, with nano-enhanced materials providing better separation, anti-fouling properties, and increased durability. These advancements allow for greater efficiency in hydrogen capture and separation while reducing operational downtime. Lastly, the development of supportive regulatory frameworks and market dynamics is crucial for the widespread adoption of BioH technologies. Policies promoting clean energy, along with international collaborations and certification standards, will ensure that BioH competes with traditional energy sources and contributes to a sustainable energy future.

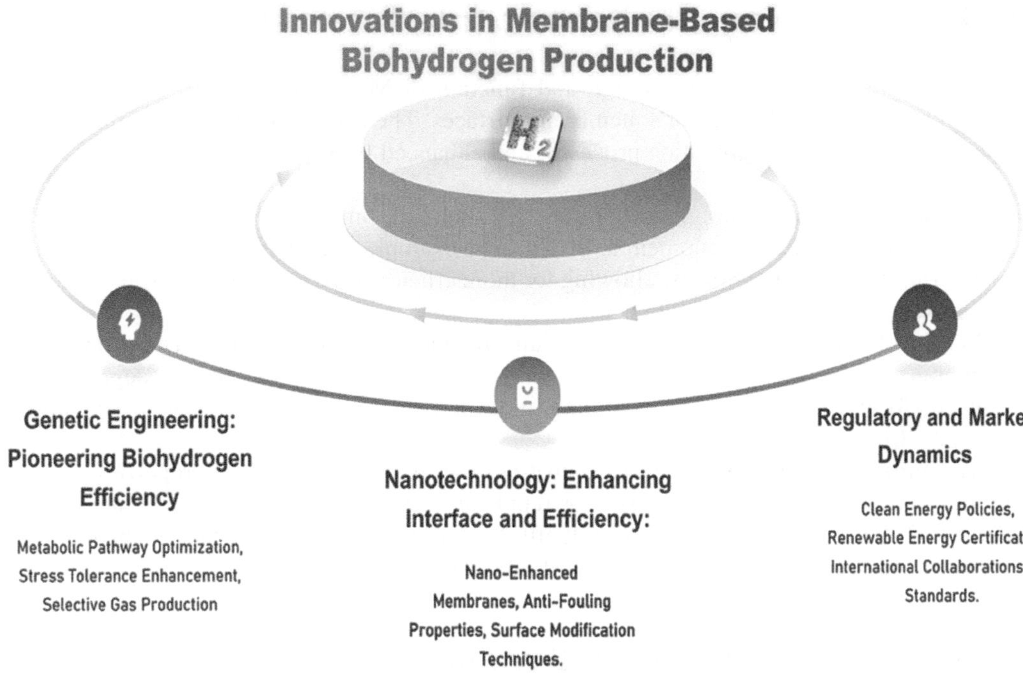

FIGURE 4.6 Future prospects and innovations in membrane-based biohydrogen production – highlighting the role of genetic engineering, nanotechnology, and regulatory frameworks in advancing biohydrogen efficiency, performance, and market integration.

4.6.1 Genetic Engineering: Pioneering BioH Efficiency

Genetic engineering stands poised to redefine the boundaries of BioH production by manipulating the genetic architecture of hydrogen-producing microbes (Pandit et al., 2024). The primary objective here is to engineer microorganisms that can produce hydrogen at increased rates and yields (Garduno Ibarra, 2023). This involves sophisticated genetic modifications to optimize the metabolic pathways that facilitate more efficient bioconversion processes.

Metabolic Pathway Optimization: By editing the genetic sequences that control metabolic pathways, scientists can potentially increase the efficiency of these pathways, leading to increased hydrogen production (Mathews and Wang, 2009). For instance, genes responsible for the metabolic breakdown of substrates can be enhanced, ensuring that more substrate is converted into hydrogen rather than by-products (Rao and Basak, 2021).

Stress Tolerance Enhancement: Genetic modifications can also make microbes more resilient to environmental stresses, such as changes in pH or temperature, which typically inhibit hydrogen production (Lowe et al., 1993). Engineering more robust microbes ensures consistent hydrogen output under variable operational conditions.

Selective Gas Production: Through genetic engineering, it is possible to alter the microbial gas outputs to produce purer hydrogen, reducing the need for costly gas separation processes. This specificity could significantly lower the overall production costs and simplify the technology (Younas et al., 2022).

4.6.2 Nanotechnology: Enhancing Interface and Efficiency

Nanotechnology can revolutionize the design and function of M-BioRs by improving the interaction between microbial cells and the reactor's membrane surfaces. The application of nano-engineered materials can lead to more efficient separation processes and enhanced BioR performance.

Nano-Enhanced Membranes: Incorporating NPs into the membrane matrix can drastically improve its properties (Buonomenna, 2013a). For example, NPs can increase the selectivity and permeability of membranes, allowing for more efficient separation of hydrogen from other gases produced during fermentation.

Anti-Fouling Properties: NPs such as titanium dioxide can confer antibacterial properties to membranes, reducing biofouling. This is crucial as biofouling can significantly degrade membrane performance over time, necessitating frequent replacements and maintenance (Ogunsona et al., 2020).

Surface Modification Techniques: Nano-coatings can modify the surface properties of membranes to enhance microbial adhesion or repulsion, depending on the desired outcome (Yin et al., 2021). This can increase the overall efficiency of the bioconversion process by maximizing the contact between microbes and substrates (Chavan et al., 2022).

4.6.3 Regulatory and Market Dynamics

The scalability and commercial viability of M-BioH production technologies are intricately linked to regulatory and market dynamics (Akram et al., 2024). Government policies play a crucial role in shaping the economic landscape for emerging technologies by offering incentives and removing barriers to adoption.

Clean Energy Policies: Governments worldwide are increasingly prioritizing clean energy technologies. Policies that provide subsidies, tax incentives, or grants for research and development in BioH could accelerate commercialization and adoption (Abdul Malek et al., 2020).

Renewable Energy Certifications: Certification programs that recognize and validate the environmental benefits of renewable technologies can increase market acceptance and consumer trust in BioH technologies (Lewandowski and Faaij, 2006).

International Collaborations and Standards: Harmonizing standards for BioH production and use across borders can facilitate international trade in renewable energy technologies, broadening the market for innovations developed in one country to be used globally (Pelkmans et al., 2014).

As we delve deeper into the possibilities of genetic engineering and nanotechnology in M-BioH production, it becomes evident that these innovations could dramatically shift the efficiency, cost, and scalability of this technology (Saud et al., 2024). Coupled with supportive regulatory frameworks and dynamic market strategies, the future of BioH is not just a vision but an impending reality. These advancements could potentially anchor BioH as a cornerstone of the global renewable energy matrix, making it an indispensable part of our sustainable energy future (Tamilselvan and Shanmugan, 2024).

4.7 CONCLUSION

This chapter has provided a comprehensive exploration of the state-of-the-art in membrane-based technologies for BioH production, highlighting the transformative role of these systems in the renewable energy sector. Through detailed discussions of various BioR configurations, such as flat sheet and hollow fiber, and the integration of innovative membrane materials like hybrid organic-inorganic composites and nanocomposites, we have uncovered the significant strides made toward enhancing the efficiency and economic viability of BioH production. TEAs presented underscore the potential cost-effectiveness and market readiness of these systems, affirming their role in the sustainable energy landscape.

Looking forward, the future of BioH production is imbued with potential, driven by rapid advancements in genetic engineering and nanotechnology. These technological frontiers promise to further optimize production processes and increase yields, setting the stage for BioH to become a cornerstone in the global pursuit of renewable energy solutions. Additionally, the evolving regulatory and market dynamics are expected to play a pivotal role in accelerating the adoption and commercialization of M-BioH technologies. As such, the integration of interdisciplinary research and innovative technological developments ensures that the realm of BioH production remains at the forefront of the energy sector's transformation, heralding a new era of energy sustainability and environmental stewardship.

REFERENCES

Abdul Malek, A., et al., 2020. Prospects, progress, challenges and policies for clean power generation from biomass resources. *Clean Technologies and Environmental Policy*. 22, 1229–1253.

Abdullah, R., et al., 2024. Recent advances in zinc oxide-based photoanodes for photoelectrochemical water splitting. *International Journal of Hydrogen Energy*. 107, 183–207.

Abraham, D. A., et al., 2024. Highly sensitive and selective detection of dopamine using atomic layer deposited HfO_2 ultra-thin films. *Electrochimica Acta*. 497, 144574.

Acar, C., Dincer, I., 2019. Review and evaluation of hydrogen production options for better environment. *Journal of Cleaner Production*. 218, 835–849.

Adelekan, O. A., et al., 2024. Energy transition policies: A global review of shifts towards renewable sources. *Engineering Science & Technology Journal*. 5, 272–287.

Ahmed, M., Dincer, I., 2019. A review on photoelectrochemical hydrogen production systems: Challenges and future directions. *International Journal of Hydrogen Energy*. 44, 2474–2507.

Akbar, A., et al., 2021. Life cycle costing of thermal energy storage system using CAPEX and OPEX. *Journal of Mechanical Science and Technology*. 35, 3223–3229.

Akpan, J., Olanrewaju, O., 2023. Sustainable energy development: History and recent advances. *Energies*. 16, 7049.

Akram, F., et al., 2024. Biohydrogen: Production, promising progressions and challenges of a green carbon-free energy. *Sustainable Energy Technologies and Assessments*. 69, 103893.

Alagesan, M., et al., 2016. Enantiodifferentiating [4+ 4] photocyclodimerization of 2-anthracenecarboxylate mediated by a self-assembled iron tetrahedral coordination cage. *Journal of Photochemistry and Photobiology A: Chemistry*. 331, 95–101.

Alkandari, S. H., Castro-Dominguez, B., 2024. Advanced and sustainable manufacturing methods of polymer-based membranes for gas separation: A review. *Frontiers in Membrane Science and Technology*. 3, 1390599.

Armaroli, N., Balzani, V., 2011. The hydrogen issue. *ChemSusChem*. 4, 21–36.

Arumugam, S., et al., 2023. Heterogenous copper (II) schiff-base complex immobilized mesoporous silica catalyst for multicomponent biginelli reaction. *Journal of Organometallic Chemistry*. 998, 122804.

Aryamol, K., et al., 2024. Recent advances of carbon pathways for sustainable environment development. *Environmental Research*. 250, 118513.

Awang, N., et al., 2015. Functionalization of polymeric materials as a high performance membrane for direct methanol fuel cell: A review. *Reactive and Functional Polymers*. 86, 248–258.

Awudu, I., Zhang, J., 2012. Uncertainties and sustainability concepts in biofuel supply chain management: A review. *Renewable and Sustainable Energy Reviews*. 16, 1359–1368.

Ayodele, D. T., et al., 2023. Factors affecting biohydrogen production: Overview and perspectives. *International Journal of Hydrogen Energy*. 48, 27513–27539.

Bakonyi, P., et al., 2018. A review of the innovative gas separation membrane bioreactor with mechanisms for integrated production and purification of biohydrogen. *Bioresource Technology*. 270, 643–655.

Bakonyi, P., et al., 2013. Biohydrogen purification by membranes: An overview on the operational conditions affecting the performance of non-porous, polymeric and ionic liquid based gas separation membranes. *International Journal of Hydrogen Energy*. 38, 9673–9687.

Batra, G., 2023. Renewable energy economics: Achieving harmony between environmental protection and economic goals. *Social Science Chronicle*. 2, 1–32.

Bellani, S., et al., 2019. Carbon-based photocathode materials for solar hydrogen production. *Advanced Materials*. 31, 1801446.

Braun, A., et al., 2012. Direct observation of two electron holes in a hematite photoanode during photoelectrochemical water splitting. *The Journal of Physical Chemistry C*. 116, 16870–16875.

Bridge, G., et al., 2013. Geographies of energy transition: Space, place and the low-carbon economy. *Energy Policy*. 53, 331–340.

Brown, E., et al., 2020. Low carbon energy and international development: From research impact to policymaking. In Sustaining Natural Resources in a Changing Environment, pp. 112–127. Routledge.

Buonomenna, M., 2013a. Nano-enhanced reverse osmosis membranes. *Desalination*. 314, 73–88.

Buonomenna, M. G., 2013b. Membrane processes for a sustainable industrial growth. *RSC Advances*. 3, 5694–5740.

Buonomenna, M., Bae, J., 2015. Organic solvent nanofiltration in pharmaceutical industry. *Separation & Purification Reviews*. 44, 157–182.

Chang, H., et al., 2020. Membrane applications for microbial energy conversion: A review. *Environmental Chemistry Letters*. 18, 1581–1592.

Chang, S., 2011. Application of submerged hollow fiber membrane in membrane bioreactors: Filtration principles, operation, and membrane fouling. *Desalination*. 283, 31–39.

Charcosset, C., 2021. Classical and recent applications of membrane processes in the food industry. *Food Engineering Reviews*. 13, 322–343.

Chavan, S., et al., 2022. Bioconversion of organic wastes into value-added products: A review. *Bioresource Technology*. 344, 126398.

Chen, Y.-Q., et al., 2022. Metal coordination to a deep cavitand promotes binding selectivities in water. *Chinese Chemical Letters*. 33, 4908–4911.

Chu, C.-J., et al., 2020. A fluorescent turn-on probe for detection of hypochlorus acid and its bioimaging in living cells. *Spectrochimica Acta Part A: Molecular and Biomolecular Spectroscopy*. 233, 118234.

Cruz, M. C., et al., 2015. Plasma deposition of silver nanoparticles on ultrafiltration membranes: Antibacterial and anti-biofouling properties. *Chemical Engineering Research and Design*. 94, 524–537.

Day, G. S., Wensley, R., 1988. Assessing advantage: A framework for diagnosing competitive superiority. *Journal of Marketing*. 52, 1–20.

de Araujo, W. R., et al., 2024. Low-cost biosensor technologies for rapid detection of COVID-19 and future pandemics. *ACS Nano*. 18, 1757–1777.

Debnath, B., et al., 2021. Potential and sustainable utilization of tea waste: A review on present status and future trends. *Journal of Environmental Chemical Engineering*. 9, 106179.

Dhakshinamoorthy, A., et al., 2011. Zn^{2+}-K10-clay (clayzic) as an efficient water-tolerant, solid acid catalyst for the synthesis of benzimidazoles and quinoxalines at room temperature. *Tetrahedron Letters*. 52, 69–73.

Dincer, I., Acar, C., 2015. Review and evaluation of hydrogen production methods for better sustainability. *International Journal of Hydrogen Energy*. 40, 11094–11111.

El-Qelish, M., et al., 2022. Membrane-based technologies for biohydrogen production: A review. *Journal of Environmental Management*. 316, 115239.

Enriquez, F. A., Ahring, B. K., 2023. Strategies to overcome mass transfer limitations of hydrogen during anaerobic gaseous fermentations: A comprehensive review. *Bioresource Technology*. 377, 128948.

Fane, A., et al., 2011. Membrane technology: Past, present and future. In Membrane and Desalination Technologies, pp. 1–45. Humana Press.

Farrukh, C., Holgado, M., 2020. Integrating sustainable value thinking into technology forecasting: A configurable toolset for early stage technology assessment. *Technological Forecasting and Social Change*. 158, 120171.

Fernández-Ronco, M. P., et al., 2013. New considerations in the economic evaluation of supercritical processes: Separation of bioactive compounds from multicomponent mixtures. *The Journal of Supercritical Fluids*. 79, 345–355.

Franceschini, S., 2015. Eco-innovation Dynamics and Sustainability–New Perspectives in Innovation Studies Illuminated through the Case of Lighting and Its Energy Consumption. DTU Management Engineering.

Garduno Ibarra, I. R., 2023. Bioremediation of Brewery Sludge and Hydrogen Production Using Combined Approaches. Université d'Ottawa/University of Ottawa.

Ghazi, R. M., et al., 2023. Health effects of herbicides and its current removal strategies. *Bioengineered*. 14(1), 2259526.

Glatzel, K., Virchow, D., Nakitto, A. M. S., Niyonsenga, S., Babu, S. C., Srivastava, N., Kashandula, P., Ecuru, J. and Osano, P., 2024. Bioeconomy pathways: Experience from Africa, Asia, and Latin America. *IFPRI book chapters*, 116–149.

Goddard, J. M., Hotchkiss, J., 2007. Polymer surface modification for the attachment of bioactive compounds. *Progress in Polymer Science*. 32, 698–725.

Govindasamy, R., et al., 2022. Emerging trends of nanotechnology and genetic engineering in cyanobacteria to optimize production for future applications. *Life*. 12, 2013.

Hakami, M. W., et al., 2020. Ceramic microfiltration membranes in wastewater treatment: Filtration behavior, fouling and prevention. *Membranes*. 10, 248.

Hammond, P. T., 2004. Form and function in multilayer assembly: New applications at the nanoscale. *Advanced Materials*. 16, 1271–1293.

He, D., et al., 2009. Photo-irradiation for preparation, modification and stimulation of polymeric membranes. *Progress in Polymer Science*. 34, 62–98.

He, L., et al., 2020. Enhanced ammonia resource recovery from wastewater using a novel flat sheet gas-permeable membrane. *Chemical Engineering Journal*. 400, 125338.

Hosseini, S. E., Wahid, M. A., 2016. Hydrogen production from renewable and sustainable energy resources: Promising green energy carrier for clean development. *Renewable and Sustainable Energy Reviews*. 57, 850–866.

Hosseini, S. S., et al., 2023. Progress in high performance membrane materials and processes for biogas production, upgrading and conversion. *Separation and Purification Technology*. 310, 123139.

Hu, S., 2019. Membrane-less photoelectrochemical devices for H_2O_2 production: Efficiency limit and operational constraint. *Sustainable Energy & Fuels*. 3, 101–114.

Huanambal-Sovero, V. A., et al., 2023. Permeate microbiome reflects the biofilm microbial community in a gravity-driven woven-fiber microfiltration (WFMF) system for wastewater treatment. *Environmental Science: Water Research & Technology*. 9, 2605–2618.

Ingole, P. G., Ingole, N. P., 2014. Methods for separation of organic and pharmaceutical compounds by different polymer materials. *Korean Journal of Chemical Engineering*. 31, 2109–2123.

Jabbari, B., et al., 2019. Recent progresses in application of membrane bioreactors in production of biohydrogen. *Membranes*. 9, 100.

Jain, R., et al., 2024. Bio-hydrogen production through dark fermentation: An overview. *Biomass Conversion and Biorefinery*. 14, 12699–12724.

Jeganathan, M., et al., 2014. Michael addition of indoles to β-nitrostyrenes catalyzed by HY zeolite under solvent-free conditions. *Tetrahedron Letters*. 55, 2061–2064.

Jeje, S. O., et al., 2024. Advancing the hydrogen production economy: A comprehensive review of technologies, sustainability, and future prospects. *International Journal of Hydrogen Energy*. 78, 642–661.

Jeong, Y., et al., 2017. Treatment of food waste recycling wastewater using anaerobic ceramic membrane bioreactor for biogas production in mainstream treatment process of domestic wastewater. *Water Research*. 123, 86–95.

Ji, J., et al., 2022. The more the slower: Self-inhibition in supramolecular chirality induction, memory, erasure, and reversion. *Journal of the American Chemical Society*. 144, 1455–1463.

Judd, S. J., 2017. Membrane technology costs and me. *Water Research*. 122, 1–9.

Kanagaraj, K., et al., 2010. Per-6-amino-β-cyclodextrin as a reusable promoter and chiral host for enantioselective Henry reaction. *Organic Letters*. 12, 4070–4073.

Kanagaraj, K., et al., 2012. Highly selective fluorescent sensing of fenitrothion using per-6-amino-β-cyclodextrin: Eu (III) complex. *Biosensors and Bioelectronics*. 35, 452–455.

Kanagaraj, K., et al., 2014. Selective "turn-off" fluorescent sensing of mercury ions using aminocyclodextrin: 3-hydroxy-N-phenyl-2-naphthamide complex in aqueous solution. *RSC Advances*. 4, 11714–11722.

Kanagaraj, K., et al., 2017. Chiral buckybowl molecules. *Symmetry*. 9, 174.

Kanagaraj, K., et al., 2020a. pH-controlled chirality inversion in enantiodifferentiating photocyclodimerization of 2-antharacenecarboxylic acid mediated by γ-cyclodextrin derivatives. *Organic Letters*. 22, 5273–5278.

Kanagaraj, K., et al., 2020b. A quinoline-appended cyclodextrin derivative as a highly selective receptor and colorimetric probe for nucleotides. *IScience*. 23, 100927.

Kanagaraj, K., et al., 2022. pH-Controlled enantioselectivity switching of irregular photodimers in photocyclodimerization of 2-anthracenecarboxylic acid mediated with β-cyclodextrin derivatives. *New Journal of Chemistry*. 46, 23066–23076.

Kanagaraj, K., et al., 2023a. Control of reactivity and selectivity in isomerization and rearrangement reactions inside confined spaces. *Chemical Communications*. 59, 9056–9065.

Kanagaraj, K., et al., 2023b. Selective binding and isomerization of oximes in a self-assembled capsule. *Journal of the American Chemical Society*. 145, 5816–5823.

Kanagaraj, K., Pitchumani, K., 2010. Solvent-free multicomponent synthesis of pyranopyrazoles: Per-6-amino-β-cyclodextrin as a remarkable catalyst and host. *Tetrahedron Letters*. 51, 3312–3316.

Kanagaraj, K., Pitchumani, K., 2013a. The aminocyclodextrin/Pd (OAc) 2 complex as an efficient catalyst for the Mizoroki–Heck cross-coupling reaction. *Chemistry–A European Journal*. 19, 14425–14431.

Kanagaraj, K., Pitchumani, K., 2013b. Per-6-amino-β-cyclodextrin as a chiral base catalyst promoting one-pot asymmetric synthesis of 2-aryl-2, 3-dihydro-4-quinolones. *The Journal of Organic Chemistry*. 78, 744–751.

Kanagaraj, K., Pitchumani, K., 2014. Highly selective "turn-on" fluorescent and colorimetric sensing of fluoride ion using 2-(2-hydroxyphenyl)-2, 3-dihydroquinolin-4 (1 H)-one based on excited-state proton transfer. *Chemistry–An Asian Journal*. 9, 146–152.

Kannah, R. Y., et al., 2021. Techno-economic assessment of various hydrogen production methods–A review. *Bioresource Technology*. 319, 124175.

Kaya, D., et al., 2021. Energy efficiency in pumps. Energy Management and Energy Efficiency in Industry: Practical Examples, pp. 329–374. Springer.

Khoiruddin, K., et al., 2024. Enhanced hydrogen production in membrane reactors: A novel approach. *International Journal of Hydrogen Energy*. 83, 946–966.

Kudapa, V. K., et al., 2024. A critical review of fabrication strategies, separation techniques, challenges, and future prospects for the hydrogen separation membrane. *Fusion Science and Technology*. 80(7), 803–825.

Kumar, A., et al., 2022. Recent insights into metallic nanoparticles in shelf-life extension of agrifoods: Properties, green synthesis, and major applications. *Frontiers in Sustainable Food Systems*. 6, 1025342.

Laberty-Robert, C., et al., 2011. Design and properties of functional hybrid organic–inorganic membranes for fuel cells. *Chemical Society Reviews*. 40, 961–1005.

Lau, H. S., et al., 2022. State-of-the-art organic-and inorganic-based hollow fiber membranes in liquid and gas applications: Looking back and beyond. *Membranes*. 12, 539.

Lau, H. S., Yong, W. F., 2021. Recent progress and prospects of polymeric hollow fiber membranes for gas application, water vapor separation and particulate matter removal. *Journal of Materials Chemistry A*. 9, 26454–26497.

Lee, A., et al., 2016. Membrane materials for water purification: Design, development, and application. *Environmental Science: Water Research & Technology*. 2, 17–42.

Lewandowski, I., Faaij, A. P., 2006. Steps towards the development of a certification system for sustainable bio-energy trade. *Biomass and Bioenergy*. 30, 83–104.

Li, L., et al., 2019. Effects of substrate concentration, hydraulic retention time and headspace pressure on acid production of protein by anaerobic fermentation. *Bioresource Technology*. 283, 106–111.

Liang, C. Z., et al., 2019. A review of polymeric composite membranes for gas separation and energy production. *Progress in Polymer Science*. 97, 101141.

Liguori, S., et al., 2020. Opportunities and challenges of low-carbon hydrogen via metallic membranes. *Progress in Energy and Combustion Science*. 80, 100851.

Liu, Y.-Y., et al., 2018. A novel nanomolar highly selective fluorescent probe for imaging mercury (II) in living cells and zebrafish. *Sensors and Actuators B: Chemical*. 277, 673–678.

Lowe, S. E., et al., 1993. Biology, ecology, and biotechnological applications of anaerobic bacteria adapted to environmental stresses in temperature, pH, salinity, or substrates. *Microbiological Reviews*. 57, 451–509.

Lunelli, B. H., Morais, E. R., 2024. Effective early-stage assessment of biorefineries: A comprehensive analysis of how it can guide decision-making in the development of new technologies. In Biorefinery and Industry 4.0: Empowering Sustainability, pp. 279–291. Springer.

Lynd, L. R., et al., 2022. Toward low-cost biological and hybrid biological/catalytic conversion of cellulosic biomass to fuels. *Energy & Environmental Science*. 15, 938–990.

Ma, L.-C., et al., 2015. Integration of membrane technology into hydrogen production plants with CO_2 capture: An economic performance assessment study. *International Journal of Greenhouse Gas Control*. 42, 424–438.

Ma, Y., et al., 2014. Titanium dioxide-based nanomaterials for photocatalytic fuel generations. *Chemical Reviews*. 114, 9987–10043.

Mahmud, R., et al., 2021. Integration of techno-economic analysis and life cycle assessment for sustainable process design–A review. *Journal of Cleaner Production*. 317, 128247.

Mammeri, F., et al., 2005. Mechanical properties of hybrid organic–inorganic materials. *Journal of Materials Chemistry*. 15, 3787–3811.

Manjunathan, J., et al., 2023. Recent advancements in nanotechnological approaches for pollution monitoring and environmental sustainability. *Clean Technologies and Environmental Policy*. 1–17.

Mantei, M. M., Teorey, T. J., 1988. Cost/benefit analysis for incorporating human factors in the software lifecycle. *Communications of the ACM*. 31, 428–439.

MariaJoseph, A., et al., 2024. Rational coupling of selective electrochemical oxidation and reduction reactions for in-situ value-added chemical generation. *Fuel*. 367, 131408.

Mathews, J., Wang, G., 2009. Metabolic pathway engineering for enhanced biohydrogen production. *International Journal of Hydrogen Energy*. 34, 7404–7416.

Melitos, G., et al., 2021. Waste to sustainable biohydrogen production via photo-fermentation and biophotolysis – A systematic review. *Renewable Energy and Environmental Sustainability*. 6, 45.

Mishra, P., et al., 2019. Outlook of fermentative hydrogen production techniques: An overview of dark, photo and integrated dark-photo fermentative approach to biomass. *Energy Strategy Reviews*. 24, 27–37.

Moniz, S. J., et al., 2015. Visible-light driven heterojunction photocatalysts for water splitting – A critical review. *Energy & Environmental Science*. 8, 731–759.

Munasinghe, M., et al., 2017. Economic, social and environmental impacts and overall sustainability of the tea sector in Sri Lanka. *Sustainable Production and Consumption*. 12, 155–169.

Nagarajan, D., et al., 2017. Recent insights into biohydrogen production by microalgae – From biophotolysis to dark fermentation. *Bioresource Technology*. 227, 373–387.

Naha, S., et al., 2020. Nanomolar detection of h2s in an aqueous medium: Application in endogenous and exogenous imaging of hela cells and zebrafish. *ACS Omega*. 5, 19896–19904.

Nair, A., et al., 2023. Multifunctional natural derived carbon quantum dots from Withania somnifera (L.)–Antiviral activities against SARS-CoV-2 pseudoviron. *Environmental Research*. 239, 117366.

Nangan, S., et al., 2023a. Sustainable preparation of luminescent carbon dots from syringe waste and hyaluronic acid for cellular imaging and antimicrobial applications. *Environmental Research*. 237, 116990.

Nangan, S., et al., 2023b. Waste plastics derived nickel-palladium alloy filled carbon nanotubes for hydrogen evolution reaction. *Chemosphere*. 341, 139982.

Natarajan, V., et al., 2019. A far-red to NIR emitting ultra-sensitive probe for the detection of endogenous HOCl in zebrafish and the RAW 264.7 cell line. *Organic & Biomolecular Chemistry*. 17, 3538–3544.

Nguyen, M.-H., et al., 2020. A highly selective fluorescent anthracene-based chemosensor for imaging Zn^{2+} in living cells and zebrafish. *Inorganic Chemistry Communications*. 115, 107882.

Nichols, G., Davis, C. J., 2016. State of the Art Report (SOAR). *Contract*.

Nunes, S. P., et al., 2020. Thinking the future of membranes: Perspectives for advanced and new membrane materials and manufacturing processes. *Journal of Membrane Science*. 598, 117761.

Ogunsona, E. O., et al., 2020. Engineered nanomaterials for antimicrobial applications: A review. *Applied Materials Today*. 18, 100473.

Osman, A. I., et al., 2024a. Membrane technology for energy saving: Principles, techniques, applications, challenges, and prospects. *Advanced Energy and Sustainability Research*. 5, 2400011.

Osman, A. I., et al., 2024b. Machine learning for membrane design in energy production, gas separation, and water treatment: A review. *Environmental Chemistry Letters*. 22, 505–560.

Pachaiappan, R., et al., 2024. A concise review of recent biohydrogen production technologies. *Sustainable Energy Technologies and Assessments*. 62, 103606.

Pandit, C., et al., 2024. Exploring the performance of biocatalysts for biohydrogen production. *Biocatalysis and Agricultural Biotechnology*. 103290.

Parrot, K. W., Tierney, B. X., 2012. Integrated reporting, stakeholder engagement, and balanced investing at American Electric Power. *Journal of Applied Corporate Finance*. 24, 27–37.

Paul, T., et al., 2021. The impact of blockchain technology on the tea supply chain and its sustainable performance. *Technological Forecasting and Social Change*. 173, 121163.

Pelkmans, L., et al., 2014. The role of sustainability requirements in international bioenergy markets. In International Bioenergy Trade: History, Status & Outlook on Securing Sustainable Bioenergy Supply, Demand and Markets, 125–149. Springer.

Peng, C., et al., 2021. Pyrene-tiaraed pillar [5] arene: Strong intramolecular excimer emission applicable for photo-writing. *Chinese Chemical Letters*. 32, 345–348.

Picart-Palmade, L., et al., 2019. Potentialities and limits of some non-thermal technologies to improve sustainability of food processing. *Frontiers in Nutrition*. 5, 130.

Podder, A., et al., 2020. Two-photon active fluorescent indicator for detecting NADH dynamics in live cells and tumor tissue. *Sensors and Actuators B: Chemical*. 324, 128637.

Pólvora, A., et al., 2020. Blockchain for industrial transformations: A forward-looking approach with multi-stakeholder engagement for policy advice. *Technological Forecasting and Social Change*. 157, 120091.

Prakash, P., et al., 2023. Exploring the potential of targeting insulin-like growth factor-1 through network pharmacology, molecular docking, molecular dynamics, and experimental validation of antioxidant and anti-inflammatory activities. *South African Journal of Botany*. 162, 707–718.

Qiu, D., et al., 2019. Mechanical failure and mitigation strategies for the membrane in a proton exchange membrane fuel cell. *Renewable and Sustainable Energy Reviews*. 113, 109289.

Qureshi, F., et al., 2022. Latest eco-friendly avenues on hydrogen production towards a circular bioeconomy: Currents challenges, innovative insights, and future perspectives. *Renewable and Sustainable Energy Reviews*. 168, 112916.

Rahman, M. A., et al., 2022. Organic–inorganic polymer hybrids for water and wastewater treatment. In Inorganic-Organic Composites for Water and Wastewater Treatment, Vol. 1, pp. 29–54. Springer.

Rajapandi, S., et al., 2023. Ziziphus mauritiana-derived nitrogen-doped biogenic carbon dots: Eco-friendly catalysts for dye degradation and antibacterial applications. *Chemosphere*. 338, 139584.

Rama, V., et al., 2011. Syntheses of 5-substituted 1 H-tetrazoles catalyzed by reusable CoY zeolite. *The Journal of Organic Chemistry*. 76, 9090–9095.

Rama, V., et al., 2012a. A multicomponent, solvent-free, one-pot synthesis of benzoxanthenones catalyzed by HY zeolite: Their anti-microbial and cell imaging studies. *Tetrahedron Letters*. 53, 1018–1024.

Rama, V., et al., 2012b. Pyridinium ylide-assisted KY zeolite catalyzed tandem synthesis of polysubstituted cyclopropanes. *Catalysis Communications*. 26, 39–43.

Ramasubburayan, R., et al., 2024a. Correction: Greener approach supported nitrogen-infused carbon dots for biocompatible cellular markers and fluorescent ink based spray-assisted fingerprint analysis. *Waste and Biomass Valorization*. 15, 7139.

Ramasubburayan, R., et al., 2024b. Greener approach supported nitrogen-infused carbon dots for biocompatible cellular markers and fluorescent ink based spray-assisted fingerprint analysis. *Waste and Biomass Valorization*. 15, 6431–6440.

Ramasubburayan, R., et al., 2023. Environmentally benign, bright luminescent carbon dots from IV bag waste and chitosan for antimicrobial and bioimaging applications. *Environmental Research*. 238, 117182.

Ramdin, M., et al., 2021. Electroreduction of CO_2/CO to C_2 products: Process modeling, downstream separation, system integration, and economic analysis. *Industrial & Engineering Chemistry Research*. 60, 17862–17880.

Rao, M., et al., 2018. Photocatalytic supramolecular enantiodifferentiating dimerization of 2-anthracenecarboxylic acid through triplet–triplet annihilation. *Organic Letters*. 20, 1680–1683.

Rao, R., Basak, N., 2021. Development of novel strategies for higher fermentative biohydrogen recovery along with novel metabolites from organic wastes: The present state of the art. *Biotechnology and Applied Biochemistry*. 68, 421–444.

Rastgar, M., et al., 2023. Harvesting blue energy based on salinity and temperature gradient: Challenges, solutions, and opportunities. *Chemical Reviews*. 123, 10156–10205.

Redwood, M. D., et al., 2009. Integrating dark and light bio-hydrogen production strategies: Towards the hydrogen economy. *Reviews in Environmental Science and Bio/Technology*. 8, 149–185.

Regula, C., et al., 2014. Chemical cleaning/disinfection and ageing of organic UF membranes: A review. *Water Research*. 56, 325–365.

Reichelt, E., et al., 2014. Fiber based structured materials for catalytic applications. *Applied Catalysis A: General*. 476, 78–90.

Ritter, J. A., Ebner, A. D., 2007. State-of-the-art adsorption and membrane separation processes for hydrogen production in the chemical and petrochemical industries. *Separation Science and Technology*. 42, 1123–1193.

Sadeghi, I., et al., 2018. Controlling and expanding the selectivity of filtration membranes. *Chemistry of Materials*. 30, 7328–7354.

Saini, N., Awasthi, K., 2022. Insights into the progress of polymeric nano-composite membranes for hydrogen separation and purification in the direction of sustainable energy resources. *Separation and Purification Technology*. 282, 120029.

Saud, A., et al., 2024. Progress in the sustainable development of biobased (nano) materials for application in water treatment technologies. *ACS Omega*. 9, 29088–29113.

Savla, N., et al., 2021. Techno-economical evaluation and life cycle assessment of microbial electrochemical systems: A review. *Current Research in Green and Sustainable Chemistry*. 4, 100111.

Seferlis, P., et al., 2021. Sustainable design, integration, and operation for energy high-performance process systems. *Energy*. 224, 120158.

Sharrer, M. J., et al., 2007. Membrane biological reactor treatment of a saline backwash flow from a recirculating aquaculture system. *Aquacultural Engineering*. 36, 159–176.

Shellaiah, M., et al., 2022. An AIEE active anthracene-based nanoprobe for Zn^{2+} and tyrosine detection validated by bioimaging studies. *Chemosensors*. 10, 381.

Shellaiah, M., et al., 2024. Sensing utilities of cesium lead halide perovskites and composites: A comprehensive review. *Sensors*. 24, 2504.

Shemer, H., et al., 2023. Challenges and solutions for global water scarcity. *Membranes*. 13, 612.

Si, W., Guo, Z., 2022. Enhancing the lifespan and durability of superamphiphobic surfaces for potential industrial applications: A review. *Advances in Colloid and Interface Science*. 310, 102797.

Sikiru, S., et al., 2024. Hydrogen-powered horizons: Transformative technologies in clean energy generation, distribution, and storage for sustainable innovation. *International Journal of Hydrogen Energy*. 56, 1152–1182.

Singla, S., et al., 2022. Hydrogen production technologies-Membrane based separation, storage and challenges. *Journal of Environmental Management*. 302, 113963.

Sohail, M. I., et al., 2019. Environmental application of nanomaterials: A promise to sustainable future. *Comprehensive Analytical Chemistry*. 87, 1–54.

Sołowski, G., et al., 2018. Production of hydrogen from biomass and its separation using membrane technology. *Renewable and Sustainable Energy Reviews*. 82, 3152–3167.

Soto, C., et al., 2022. Recent advances in membrane-based biogas and biohydrogen upgrading. *Processes*. 10, 1918.

Staffell, I., et al., 2019. The role of hydrogen and fuel cells in the global energy system. *Energy & Environmental Science*. 12, 463–491.

Stamatialis, D. F., et al., 2008. Medical applications of membranes: Drug delivery, artificial organs and tissue engineering. *Journal of Membrane Science*. 308, 1–34.

Suresh, D., et al., 2013. Synthesis of 2-substituted 3-ethyl-3H-imidazo [4, 5-b] pyridines catalyzed by Al3+-exchanged K10 clay as solid acids. *Tetrahedron Letters*. 54, 6479–6484.

Suresh, D., et al., 2014. Microwave promoted one-pot synthesis of 2-aryl substituted 1, 3, 4–oxadiazoles and 1, 2, 4–oxadiazole derivatives using Al3+-K10 clay as a heterogeneous catalyst. *Tetrahedron Letters*. 55, 3678–3682.

Swaminathan, P., et al., 2024. A comprehensive review of microbial electrolysis cells integrated with wastewater treatment for hydrogen generation. *Process Safety and Environmental Protection*. 190, 458–474.

Ta, X. M. C., et al., 2022. Alternatives to water photooxidation for photoelectrochemical solar energy conversion and green H_2 production. *Advanced Energy Materials*. 12, 2201358.

Tamilselvan, S. N., Shanmugan, S., 2024. Towards sustainable solar cells: Unveiling the latest developments in bio-nano materials for enhanced DSSC efficiency. *Clean Energy*. 8, 238–257.

Tang, M. M., et al., 2022. Role of rim functions in recognition and selectivity of small-molecule guests in water-soluble cavitand hosts. *Chemistry–An Asian Journal*. 17, e202200466.

Thirumalaivasan, N., et al., 2017. Highly selective turn-on probe for H_2S with imaging applications in vitro and in vivo. *New Journal of Chemistry*. 41, 13510–13515.

Thirumalaivasan, N., et al., 2019. In vitro and in vivo approach of hydrogen-sulfide-responsive drug release driven by azide-functionalized mesoporous silica nanoparticles. *ACS Applied Bio Materials*. 2, 3886–3896.

Thirumalaivasan, N., et al., 2024a. Utilization of fungal and bacterial bioremediation techniques for the treatment of toxic waste and biowaste. *Frontiers in Materials*. 11, 1416445.

Thirumalaivasan, N., et al., 2024b. Nano-PCM materials: Bridging the gap in energy storage under fluctuating environmental conditions. *Process Safety and Environmental Protection*. 189, 1003–1021.

Thirumalaivasan, N., et al., 2024c. Exploring luminescent carbon dots derived from syrup bottle waste and curcumin for potential antimicrobial and bioimaging applications. *Chemosphere*. 354, 141592.

Thirumalaivasan, N., et al., 2024d. Exploring antimicrobial and biocompatible applications of eco-friendly fluorescent carbon dots derived from fast-food packaging waste transformation. *Environmental Research*. 244, 117888.

Thirumalaivasan, N., et al., 2024e. Assessment of sustainability and environmental impacts of renewable energies: Focusing on biogas and biohydrogen (biofuels) production. *Process Safety and Environmental Protection*. 189, 467–485.

Thirumalaivasan, N., Wu, S.-P., 2020. Bright luminescent carbon dots for multifunctional selective sensing and imaging applications in living cells. *ACS Applied Bio Materials*. 3, 6439–6446.

Vedarethinam, V., 2024. Cutting-edge innovations in nanomaterial-enhanced membranes for industrial effluent remediation. In Bioremediation of Emerging Contaminants in Water, Vol. 2, pp. 189–225. ACS Publications.

Velusamy, N., et al., 2018. A hydrogen sulfide triggered self-immolative fluorescent probe for lysosome labeling in live cells. *New Journal of Chemistry*. 42, 1590–1594.

Venkatesan, P., et al., 2023. Eco-friendly, bright luminescent carbon dots and their potential applications for detecting hypochlorous acid in water and live cell imaging. *Journal of Materials Research and Technology*. 24, 6522–6532.

Venkatesan, P., et al., 2019. Redox stimuli delivery vehicle based on transferrin-capped MSNPs for targeted drug delivery in cancer therapy. *ACS Applied Bio Materials*. 2, 1623–1633.

Verma, D., et al., 2024a. A sustainable and green approach towards the utilization of biopolymers for effective wound dressing applications: A detailed review. *Nano-Structures & Nano-Objects*. 37, 101086.

Verma, G., et al., 2024b. Recent advances in advanced micro and nanomanufacturing for wastewater purification. *ACS Applied Engineering Materials*. 2, 262–285.

Wan, C. F., et al., 2021. Design and fabrication of hollow fiber membrane modules. In Hollow Fiber Membranes, pp. 225–252. Elsevier.

Wang, J., 2017. System integration, durability and reliability of fuel cells: Challenges and solutions. *Applied Energy*. 189, 460–479.

Wang, Z., et al., 2024. Recent advances review in tea waste: High-value applications, processing technology, and value-added products. *Science of the Total Environment*. 946, 174225.

Warsinger, D. M., et al., 2018. A review of polymeric membranes and processes for potable water reuse. *Progress in Polymer Science*. 81, 209–237.

Wei, L.-F., et al., 2019. A nano-molar fluorescent turn-on probe for copper (II) detection in living cells. *Methods*. 168, 18–23.

Wei, L.-F., et al., 2017. Fluorescent coumarin-based probe for cysteine and homocysteine with live cell application. *Spectrochimica Acta Part A: Molecular and Biomolecular Spectroscopy*. 183, 204–208.

Werber, J. R., et al., 2016. Materials for next-generation desalination and water purification membranes. *Nature Reviews Materials*. 1, 1–15.

Wheelock Gutierrez, F. E., 2023. A comparative study of low-emissions hydrogen production processes: Technical limitations and future trends.

Wiebe, K., et al., 2018. Scenario development and foresight analysis: Exploring options to inform choices. *Annual Review of Environment and Resources*. 43, 545–570.

Wu, G.-S., et al., 2020. Ultrasensitive and specific two-photon fluorescence detection of hypochlorous acid by a lysosome-targeting fluorescent probe for cell imaging. *Journal of Pharmaceutical and Biomedical Analysis*. 190, 113545.

Xiao, C., et al., 2020. Redox-triggered chirality switching and guest-capture/release with a pillar [6] arene-based molecular universal joint. *Angewandte Chemie International Edition*. 59, 8094–8098.

Yang, W.-Y., et al., 2022. A two photon fluorescent probe for highly selective detection and endogenous imaging of hydrogen sulfide. *Spectrochimica Acta Part A: Molecular and Biomolecular Spectroscopy*. 273, 121043.

Yin, B., et al., 2021. Research and application progress of nano-modified coating in improving the durability of cement-based materials. *Progress in Organic Coatings*. 161, 106529.

Younas, M., et al., 2022. An overview of hydrogen production: Current status, potential, and challenges. *Fuel*. 316, 123317.

Yue, M., et al., 2021. Hydrogen energy systems: A critical review of technologies, applications, trends and challenges. *Renewable and Sustainable Energy Reviews*. 146, 111180.

Yusuf, A., et al., 2020. A review of emerging trends in membrane science and technology for sustainable water treatment. *Journal of Cleaner Production*. 266, 121867.

Zainal, B. S., et al., 2024. Recent advancement and assessment of green hydrogen production technologies. *Renewable and Sustainable Energy Reviews*. 189, 113941.

Zhang, H.-B., et al., 2021. Hydrophobic and metal-coordinated confinement effects trigger recognition and selectivity. *The Journal of Organic Chemistry*. 86, 8873–8881.

Zhao, G., et al., 2023. Monolithic fiber/foam-structured catalysts: Beyond honeycombs and micro-channels. *Catalysis Reviews*. 1–81.

Zhen, G., et al., 2019. Anaerobic membrane bioreactor towards biowaste biorefinery and chemical energy harvest: Recent progress, membrane fouling and future perspectives. *Renewable and Sustainable Energy Reviews*. 115, 109392.

Zhou, J., et al., 2015. High-throughput metagenomic technologies for complex microbial community analysis: Open and closed formats. *MBio*. 6, e02288-14.

Zimmermann, A. W., et al., 2020. Techno-economic assessment guidelines for CO_2 utilization. *Frontiers in Energy Research*. 8, 5.

Zulfhazli, et al., 2022. A systematic review of the techno-economic assessment of various hydrogen production methods of power generation. *Frontiers in Sustainability*. 3, 943145.

5 Biohydrogen Production via Valorization of Microalgal Biomass

Microalgal Strains, Cultivation and Harvesting, Biomass Pretreatment, and Genetic Engineering-Based Strategies for Enhanced Production

Shilpi Tyagi, Deepak Choudhary, Sheetal Soni, and Shweta Jha

5.1 INTRODUCTION

A practical, sustainable solution to the world's energy requirements is the production of biohydrogen from microalgal biomass that can be utilized as an effective strategy to lessen the environmental issues caused by the usage of fossil fuels. Microalgae are attractive systems due to their quick growth rates, ability to fix CO_2, and potential for biological hydrogen production. In the environment, microalgae are important primary producers. Microalgal cells are more efficient than other photosynthetic organisms in performing photosynthesis. Additionally, their shortened life cycle ensures a steady supply of fuel throughout the

DOI: 10.1201/9781003568612-5

year. This chapter reviews some of the key factors influencing the production of biohydrogen, including microalgal strains, growth and harvesting techniques, biomass pretreatment, and genetic engineering (GE)-based production enhancement strategies.

It appears that some microalgal species could be used as an unconventional source of food and medicine. Bio-diesel, bio-gas, bio-oil, and biohydrogen are among the renewable biofuels that can be produced from some members of cyanobacteria and green algae (Maneeruttanarungroj et al., 2010; Khetkorn et al., 2013; Skjånes et al., 2013). Although many algal species, such as *Ankistrodesmus angustus*, *Chlamydobotrys stellata*, *Chlamydomonas acidophila*, and *Chlorella ellipsoidea* have demonstrated the capacity to produce hydrogen gas, most of the studies on hydrogen production from green algae have used *Chlamydomonas reinhardtii* as the model organism (Skjånes et al., 2013).

The two most popular techniques for growing microalgae are controlled closed cultivation systems that use various kinds of bioreactors, and open cultivation systems including raceway ponds, tanks, and open ponds. Open raceway ponds were used in one of the earliest attempts to grow and scale up microalgae (Johnson et al., 1988). Numerous studies have taken place since then to cultivate microalgae through open cultivation methods. An inexpensive startup and cost-efficacy, as well as reduced energy needs for culture mixing, are some of the major benefits of open cultivation system. The drawbacks of these systems include the need for huge spaces to scale up, their vulnerability to contamination (such as that caused by birds), and unfavorable weather. Closed cultivation systems—here referred to as closed photobioreactors, or PBRs—are more effective in terms of quality and can thus overcome the drawbacks of open culture systems, because they can be run under extremely regulated conditions. The type of strain can guide the design and optimization of PBRs. The closed system significantly reduces contamination problems and increases light availability while taking up very little space. Closed PBRs do have certain drawbacks, though, including bio-fouling, overheating, the growth of benthic algae, cleaning problems, a high dissolved oxygen buildup that limits growth, and—most significantly—extremely expensive design and operating capital costs (Chisti, 2008).

Sedimentation, filtration, or centrifugation are methods for collecting microalgae. The size and characteristics of the algal strain determine which harvesting method is best for it. Filtration and centrifugation are applicable to PBRs, whereas sedimentation and flotation harvesting methods are mostly used in open pond cultivation systems. A significant amount of algal culture broth must be manageable for the chosen harvest technique.

Microalgal biomass must be pretreated by chemical, biological, or physical methods, in order to improve the accessibility of substrates needed for the synthesis of biohydrogen. Chemical methods include pretreatment with acids and alkali, ozonolysis, chloroform, iodopropane, and acetylene, biological pretreatment includes cellulolytic and hemicellulolytic microbes, and physical methods include thermal, freezing and thawing, microwave radiation, ultrasonication, and milling (Nagarajan et al., 2020).

In an attempt to reduce the cost of producing hydrogen, microalgal GE approaches have attracted more attention. Many species have undergone GE techniques, such as foreign gene expression, RNA interference (RNAi), gene silencing, and gene disruption, to boost the synthesis and generation of biohydrogen (Chaturvedi et al., 2021). These strategies aim to either increase the electron supply to enzymes or modify the main enzymes that are either directly or indirectly engaged in algae's hydrogen generation.

The synthesis of biohydrogen from microalgae has great potential as a sustainable energy source. The productivity and economic viability of the biohydrogen production process can be greatly increased by optimizing microalgal strains, culture techniques, biomass pretreatment, and GE approaches. Microalgal biohydrogen has the potential to be a sustainable energy source, but further research and development in these areas is essential.

5.2 CULTIVATION OF MICROALGAE

5.2.1 Microalgal Strains for Biohydrogen Production

Microalgae like *C. reinhardtii* and *Scenedesmus obliquus* and other strains show promise for biohydrogen production due to their hydrogenase activity under anaerobic conditions (Table 5.1). Strategies such as nutrient limitation, tailored light, and co-culturing have been shown to improve yields, but challenges like limited efficiency and high energy costs still prevail. Advancement in GE and bioreactor design can overcome these barriers for scalable production.

TABLE 5.1 Various microalgal species, their hydrogen yield, and production strategies

MICROALGAL SPECIES	*HYDROGEN YIELD (ML H_2/g VS)*	*PRODUCTION STRATEGIES*	*REFERENCES*
Chlorella vulgaris	1.75–19	Hydrogen fermentation and effects of different enzymatic pretreatment on hydrogen yield.	Wieczorek et al. (2014)
Chlorella sp.	0.37–7.13	Influences of inoculum–substrate ratio on anaerobic dark fermentation.	Sun et al. (2011)
Nannochloropsis oceanica	2	Dark fermentation, photo-fermentation.	Xia et al. (2013)
Chlamydomonas reinhardtii	16.6[a]	Anaerobic hydrogen production followed by photo-fermentation.	Kim et al. (2006)
Dunaliella tertiolecta	12.6	The high salinity of the slurry was prohibitive to methanogens, resulting in low methane production and high hydrogen.	Lakaniemi et al. (2011)
Scenedesmus obliquus	5.4–34.8	Hydrogen production by mixed culture showed lower H_2/CO_2 ratio than pure culture.	Ortigueira et al. (2015)
Chlorella sorokiniana	63	H_2O_2 showed better effect in treating microalgae biomass.	Roy et al. (2014)
Arthrospira platensis	2.24–8.06[a]	Heating temperature range of 100–121°C, with and without acid addition, was applied in treating microalgae biomass.	Efremenko et al. (2012)
Dunaliella tertiolecta	13	Hydrogen yields increased at least 10% after biological treatment process.	Carver et al. (2011)
Spirulina platensis	0.38[b]	A wide variety of substrates and carbohydrate-rich waste products were used for dark fermentative hydrogen production followed by photo-fermentative hydrogen production.	Sinha and Pandey (2014)
Chlorella pyrenoidosa	81.2	Heat and acid treatment for hydrogen production. Energy was further removed through following photohydrogen production	Xia et al. (2014)

[a] mL H_2/g TS.

[b] mol H_2/mol sugar.

H_2/g VS—hydrogen production per gram of volatile solid.

5.2.2 Cultivation Methods

Microalgae are cultivated in both open and closed systems. While closed systems use a variety of bioreactor designs, including tubular, flat-plate, vertical column, and bag bioreactors, open systems use techniques like open ponds and raceways. The low cost of construction, simplicity of use, ease of maintenance, and scalability of open systems make them advantageous. This method has a number of disadvantages, though, such as a high risk of contamination and little control over culture conditions, including temperature, light intensity, water loss, pH, salinity, and CO_2 distribution. Furthermore, drainage, precipitation, germs, and protozoa can all contaminate open systems. Closed systems, on the other hand, are made to provide greater control over growing conditions while lowering the possibility of contamination. Compared to open systems, closed bioreactors are more space-efficient due to their compact design (Duran et al., 2018; Tan et al., 2020). PBRs are a desirable alternative for growing photosynthetic organisms since they are efficient and scalable. The use of PBRs in bioprocesses to generate important chemical compounds is growing (Chanquia et al., 2022). Generally speaking, PBRs are sealed, light-filled containers that keep the culture away from the outside world. Sterilized filters are used for gas exchange with the atmosphere, greatly lowering the chances of contamination. Compared to open systems, this configuration significantly increases productivity by reducing CO_2 and water loss during the growing phase (Singh and Sharma, 2012). As seen in Figures 5.1 and 5.2, PBRs fall into two general categories: conventional and non-conventional. These days, floating bioreactors (FBRs) are into fashion because they are inexpensive to manufacture, run, and maintain. The usage of FBRs is a unique method of cultivating microalgae and other phototrophic organisms in aquatic environments. The advantage of these systems, which typically float over bodies of water, is that they require less land and can produce using sunlight and natural water sources. FBRs are especially well-suited for the large-scale production of algae because they provide consistent light exposure and efficient gas exchange. Additionally, floating units are easily expandable by varying in size and quantity. However, problems including system stability, nutrient distribution, and potential bio-fouling need to be fixed for the optimum outcomes (Gao et al., 2016; Zeng et al., 2020).

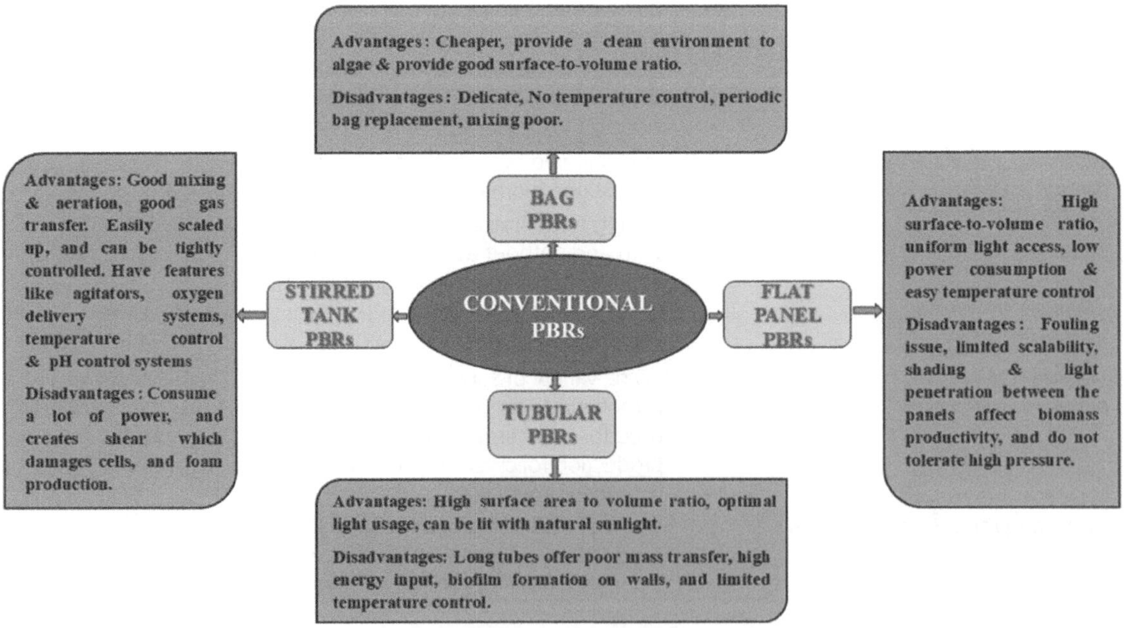

FIGURE 5.1 Conventional photobioreactors (PBRs) including bag PBR, tubular PBR, flat panel PBR, and stirred tank PBR, with their advantages and disadvantages.

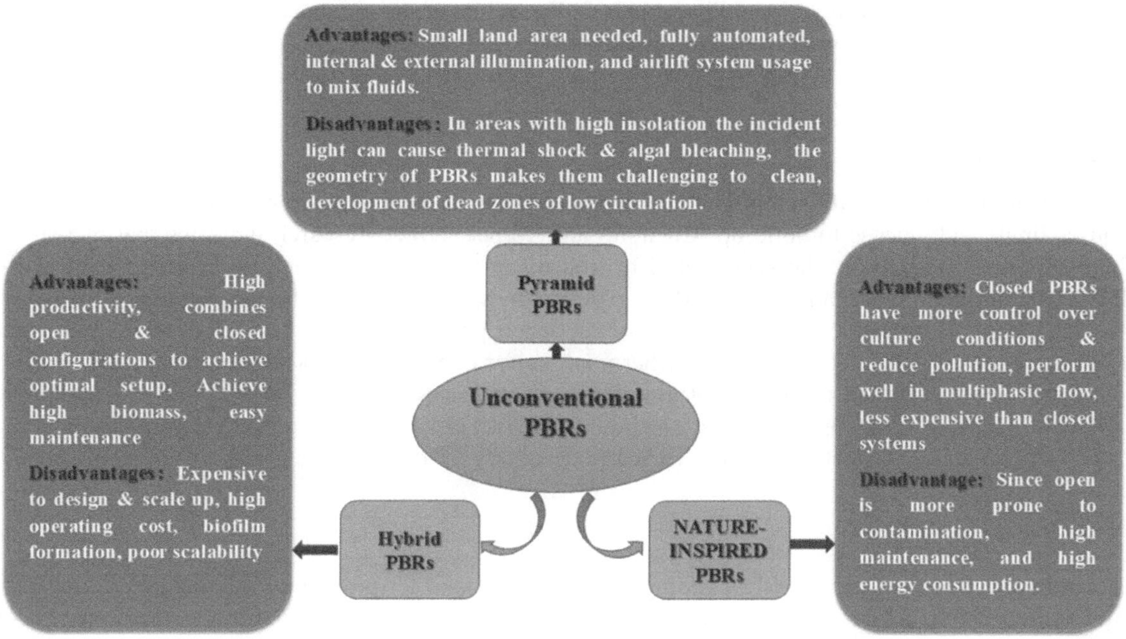

FIGURE 5.2 Unconventional photobioreactors (PBRs) including pyramid PBR, hybrid PBR, and nature-inspired PBR, with their advantages and disadvantages.

5.2.3 Factors Influencing Biohydrogen Production in Microalgae

It has been discovered that a variety of microalgae, including *Euastrum*, *Chlorella*, *Chlamydomonas*, and others, may create hydrogen. However, they require specific conditions, such as the ideal pH, light, temperature, and nutrients, in order to accomplish this successfully (Wang et al., 2020; Ardo et al., 2022), as described below.

5.2.3.1 Temperature

Temperature plays a crucial role in controlling the microalgal growth and the generation of hydrogen. Each enzyme has an ideal temperature range, and it influences the activity of the enzymes involved in biohydrogen generation (Ardo et al., 2022; Woon et al., 2023). Microalgae typically flourish in temperatures between 15°C and 35°C, while some species, such as *Chaetoceros* and *Anacystis nidulans*, may withstand temperatures as high as 40°C. On the other hand, low temperatures and carbon scarcity interfere with photosynthesis, while excessive heat can denature vital enzymes and impair cellular respiration. Additionally, high temperatures deactivate photosynthetic proteins, which hinders the action of the Rubisco enzyme and slows down the synthesis of hydrogen (Ayub et al., 2023).

5.2.3.2 pH

pH of the culture medium determines the microalgal growth and biohydrogen production. The ideal pH ranges for different species vary; marine microalgae need a lower pH because of their low nitrate requirement (Sathinathan et al., 2023). According to Woon et al. (2023), biohydrogen generation is generally most effective at pH values between 4.5 and 8.0. For many strains, the ideal pH range is between 6.5 and 8.5 (Arimbrathodi et al., 2023). *Tetraspora* sp. CU2551, for instance, produces the most biohydrogen at a pH of 7.5 (Maneeruttanarungroj and Phunpruch, 2017). While acidic circumstances (below pH 4.5) may cause metabolism to shift away from hydrogen synthesis, lowering yields, higher pH levels might reduce the lag period and improve hydrogen production rates (Chowdury et al., 2020).

5.2.3.3 Light

Microalgal growth and hydrogen production depend heavily on light. It promotes photosynthesis, which causes lipids and glucose to build up (Cheirsilp and Torpee, 2012). Research indicates that increasing light exposure can improve the synthesis of hydrogen, especially in the anaerobic phase (Anwar et al., 2019). According to Oncel and Sukan (2011), exposure to light increases hydrogen output and decreases the lag phase. However, *C. reinhardtii* may experience photoinhibition when exposed to high light (200 μmol/m²/s) (Dalena et al., 2017). Furthermore, sulfur consumption and light intensity are related; peak hydrogen generation occurs at 200 μmol/m²/s (Kim and Kim, 2012). In sulfur-deficient settings, alternating light-dark photoperiods, like 2–4 hours, also increase hydrogen yields (Vijayaraghavan et al., 2009). Optimization of light condition is therefore necessary for the effective generation of hydrogen.

5.2.3.4 Nutrients

According to Khetkorn et al. (2017), adding nutrients is crucial for promoting microalgal development and hydrogen production. Depending on whether the microalgae grow by autotrophy, heterotrophy, or mixotrophy, different nutrients are needed. For instance, *Chlorella* sp. uses both endogenous and external substrates in its heterotrophic metabolism to produce hydrogen (Morales-Sánchez et al., 2017). Hydrogen production is improved by carbon sources such as glucose, sucrose, and acetate; lag time is decreased by glucose concentrations of about 30 mM (Ahmed et al., 2022). Using maize stalk as a substrate, *Chlorella vulgaris* generated 220 mL/L of hydrogen (Amutha and Murugesan, 2011). For optimum development and hydrogen production, trace elements (such as Mg^{2+}, Zn^{2+}, and Fe^{3+}) and nutrients like nitrogen and phosphorus are also essential (Goswami et al., 2021; Su, 2021). Enzymatic activity and hydrogenase function are improved by proper nutritional balance, which includes trace elements, increasing hydrogen yields (Hawkes et al., 2002). Additionally, hydrogenase inhibition by oxygen can be avoided by reducing sulfur compounds (Veeravalli et al., 2019).

5.3 HARVESTING MICROALGAL BIOMASS

Centrifugation has been recommended by many researchers for the reliable recovery of microalgae (Mohn, 1980; Benemann and Oswald, 1996; Grima et al., 2003). This method uses centrifugal acceleration instead of gravitational force to increase the amount of algal biomass. The primary factors for determining centrifugal separation are particle size and density. Various kinds of centrifuges have been employed, and multiple studies have documented each one's efficiency and dependability. Centrifugation can reach 90–100% harvesting efficiency (Heasman et al., 2000). Centrifugation is quite successful, but the cost is excessive and energy-intensive to be used commercially.

Filtration outperforms other harvesting methods. This method is best suited for algae larger than 70 μm, including filamentous species and agglomerates. Diatomaceous earth or cellulose can improve filtration efficiency (Brennan and Owende, 2010). Conventional pressure or suction filtration is ineffective for tiny algae like *Chlorella*, *Dunaliella*, and *Scenedesmus*. Membrane microfiltration and ultrafiltration are different filtration processes. The downside of these procedures is the high expense of membrane replacement and pumping. Alternative filtering techniques include tangential flow filtration (TFF), pressure filtration, vacuum filtration, microfiltration, ultrafiltration, and dead-end filtration.

In light of the output and initial feedstock amounts, recent research indicates that TFF and pressure filtration can be regarded as energy-efficient dewatering procedures due to their low energy consumption (Danquah et al., 2009). Data regarding the energy consumption and concentration factors of each filtering unit have been provided by Mohn (1980) and Danquah et al. (2009). To optimize algal dewatering, Mohn (1980) investigated several vacuum and pressure filtration systems. To create a slurry with 27% solids, he showed that filtration procedures can reach a concentration factor of 245 times the initial concentration for *Closterium proboscideum*. It is not advised to use this method for large-scale microalgae recovery because of ongoing fouling and the ensuing membrane replacement. Polymer membranes for continuous recovery have not been extensively studied. The hydrodynamic condition, concentration, and microalgal characteristics are some of the variables that affect how well these membranes work. It is impossible to ignore the substantial running costs associated with this dewatering technique, despite its apparent appeal.

Flotation is a method of separation that depends on air bubbles sticking to solid materials. After rising to the liquid's surface, the resultant flocs are collected by filtering and skimming. The type of suspended particles (microalgal cells in the harvesting process) determines the flotation's success. Air bubbles more readily float up the tiny particles, less than 500 μm (Matis et al., 1993). There is comparatively more air-particle contact when the instability of the suspended particles is reduced. The contact angle between the air, solid, and aqueous phases also affects how air bubbles adhere. According to Shelef et al. (1984), the air's propensity to adhere increases with the contact angle. The process used to create bubbles indicates that the most widely utilized flotation techniques are dispersed air flotation, dissolved air flotation (DAF), and electrolytic flotation.

According to reports, up to 6% (w/v) of the algal biomass slurries from algal culture can be recovered using the DAF process by chemical flocculation (Bare et al., 1975). Few studies have demonstrated the technical and financial feasibility of flotation, despite the fact that it has been considered by a number of researchers as a possible harvesting technique. By adding one or more chemicals to the broth, flocculation is utilized to extract microalgal cells. Due to their negative charge, microalgal cell walls stop themselves from aggregating in the suspension. Flocculants, which are polyvalent ions, are added to counteract this negative charge. Two categories of flocculants have been identified: (1) inorganic agents, which include polyvalent metal ions like Al^{3+} and Fe^{+3} that, at the right pH, form polyhydroxy complexes; and (2) polymeric flocculants, which include ionic, nonionic, natural, and synthetic polymers. The former group includes the widely used multivalent flocculants ferrous sulfate, ferric chloride, and aluminum sulfate, whose effectiveness is directly correlated with the ionic charge. According to reports, Fe^{3+} is 80% effective at harvesting several kinds of algae (Knuckey et al., 2006). Algal cells and polyelectrolyte interact for polymer flocculation, which causes algae to bridge and flocs to form. The concentration, molecular weight, and charge of polymers determine how much aggregation occurs.

It has been noted that as the polymer's molecular weight and charge increase, so does its binding capabilities. When choosing a polymer, algal characteristics such the broth's pH, biomass content, and charge are equally crucial to take into account. Using a cationic polyelectrolyte, Tenney et al. (1969) observed successful flocculation in *Chlorella*, but not for an anionic polyelectrolyte. Chitosan was effectively employed by Divakaran and Pillai (2002) as a bioflocculant for *Spirulina*, *Oscillatoria*, *Chlorella*, and *Synechocystis* sp. Because the pH of the media affects the method's effectiveness, freshwater species showed the best results at a pH of 7.0, whereas marine species performed well at lower pH values. Because of their broader variety of applications, less requirements, and decreased sensitivity to media pH, organic flocculants are said to be advantageous. For *Tetraselmis chui*, *Isochrysis* sp., and *Thalassiosira pseudonana*, Heasman et al. (2000) discovered that complete aggregation required just 40 mg/L of chitosan, but *Chaetoceros muelleri* required 150 mg/L. In order to collect *C. vulgaris*, Oh et al. (2001) also investigated microbial flocculants (AM49). It was discovered that this flocculant was superior to other widely used flocculants. When working at pH range of 5–11, a recovery of over 83% was seen, which is greater than that achieved with the cationic polymer polyacrylamide (78%) or aluminum sulfate (72%). When the extra supply of CO_2 is cut off then algae also have the ability to flocculate on their own. Algal cells and calcium, phosphate, magnesium, and carbonate salts precipitate when the CO_2 supply is disrupted during photosynthesis, raising the pH. Auto-flocculation is the outcome of the positively charged ions' interaction and binding with the negatively charged algal surfaces. In a study on auto-flocculation in ponds and laboratory-scale trials, Sukenik and Shelef (1984) found some encouraging outcomes. However, this process may be constrained by the absence of conducive circumstances, particularly light and CO_2.

5.4 BIOMASS PRETREATMENT FOR BIOHYDROGEN PRODUCTION

5.4.1 Physical Pretreatment Methods

5.4.1.1 Thermal

Heat pretreatment inhibits methanogenic activity while promoting the growth of hydrogen-producing bacteria (HPB). While non-spore-forming bacteria, such as methanogens, are eliminated during the process, the development of spores signifies the attendance of HPB (Zhu and Béland, 2006). In mixed anaerobic cultures, heat-shock therapy has shown eradication of methanogens and promotion of hydrogen production (Chaganti et al., 2012). Research has shown that a variety of heat pretreatment methods, especially when employing soluble starch as a substrate, enhance the anaerobic seed sludge's capacity to produce hydrogen. Furthermore, during the synthesis of biohydrogen, pH plays a crucial role in determining methanogenic activity (Zhu and Béland, 2006). Since too much pretreatment might reduce hydrogen output, ideal pretreatment parameters, such as temperature and time, are crucial for optimizing hydrogen production (Sivagurunathan et al., 2017). Using different temperatures and pretreatment times, heat-shock pretreatment of the inoculum for dark fermentation hydrogen generation has shown impressive outcomes (Zhu and Béland, 2006).

5.4.1.2 Freezing and Thawing

A new technique for breaking up microalgal cells and releasing internal components is physical pretreatment that involves freezing and thawing. Large ice crystals created by freeze-drying using a lyophilizer shatter the cells at room temperature (Dixon and Wilken, 2018). A variety of freezing procedures are used, including high-pressure freezing, dehydro-freezing, and the use of anti-freeze proteins; thawing techniques include ohmic thawing and high-pressure microwave. Phase shifts and volume expansion during freezing produce ice that is less dense than liquid water (Kalichevsky et al., 1995). Without using heat, thawing can be utilized to sterilize cyanobacteria and liberate soluble intracellular components (Jan et al., 2008).

5.4.1.3 Microwave Radiation

With benefits including quick heating, low energy consumption, minimum inhibitor production, and efficient cellulose degradation, microwave radiation is a useful pretreatment for lignocellulosic feedstocks (Hu and Wen, 2008; Hong et al., 2004). For instance, switchgrass exposed to microwave radiation and cellulase generated 34.5 g of sugar for every 100 g of biomass. This approach is economical, safe for the environment, and successful at getting rid of microbes. According to one study, sludge's chemical oxygen demand (COD) rose after being heated by a microwave at 850 W for 3 minutes (Thungklin et al., 2011).

5.4.1.4 Ultrasonication

The physical, chemical, and biological qualities of activated sludge are improved by ultrasonication, an effective and environmentally friendly process that shortens treatment times and eliminates the need for chemicals or enzymes (Chu et al., 2001). It works especially well as an inoculum pretreatment for dark fermentation, which produces hydrogen. According to studies, ultrasound treatment increases the amount of hydrogen produced from complicated food waste by up to 75%, increasing the production rate by 104%, and increasing the concentration of total solids (Gadhe et al., 2014).

5.4.1.5 Milling

The initial pretreatment process, milling, is to increase the substrate molecules' surface area. Two-roll, bead, hammer, disc, and colloid milling are common milling methods used to produce biofuel (Kucharska et al., 2018). Wet milling can help lower the high energy requirements of dry milling, which reduces feedstock particle size to 0.2–10 mm (Chang et al., 1997). Specifically, wet disc milling works well to enhance cellulose hydrolysis. Planetary milling is more effective than other techniques at reducing the size of biomass and yielding more glucose while avoiding the production of hazardous byproducts like linolenic acid. Usually, enzymatic hydrolysis or other chemical pretreatments are performed after milling. The kind of biomass determines whether to use dry or wet milling; dry biomass, such as Napier grass, is typically processed using hammer or cryogenic mills, whereas wet biomass, like wheat straw, is best suited for colloid milling or fibrillation (Zeng et al., 2007).

5.4.2 Chemical Pretreatment Methods

5.4.2.1 Acid and Alkali Pretreatment

Because hydrogen-consuming bacteria (HCB) are sensitive to pH variations, pretreatments with acids and alkali are frequently employed to enrich or suppress them, hence increasing hydrogen production (Rafieenia et al., 2017). While alkali pretreatments employ NaOH, KOH, or $Ca(OH)_2$ at concentrations of 1.0–8.0 M and the pH values of 10–12, acid pretreatments usually use HCl, H_2SO_4, or HNO_3 at concentrations of 0.1–6.0 M and pH values of 2.0–4.0 (Mu et al., 2007). Alkali pretreatment improves hydrolysis by breaking down lignocellulosic structures, whereas acid pretreatment aids in the solubilization of hemicellulose but has no discernible effect on lignin (Hendriks and Zeeman, 2009; Mood et al., 2013). Several studies have depicted that pretreatment with $Ca(OH)_2$ improves the digestibility of cellulose and the synthesis of biohydrogen, giving higher yields from wheat straw (Mosier et al., 2005; Alvira et al., 2010). Both acid and alkali pretreatments are successful, but they can cause reactor corrosion and inhibitory compound production. Additionally, the requirement to modify pH raises expenses (Rafieenia et al., 2017).

5.4.2.2 Ozonolysis

For lignocellulosic substrates such as wheat straw, bagasse, and green hay, ozonolysis is a successful oxidative pretreatment that lowers the quantity of lignin and promotes enzymatic hydrolysis to elevate the quantity of fermentable sugars (Bule et al., 2013). Because it doesn't create any hazardous inhibitors, this process is sustainable when used at the ideal temperature and pressure (Quesada et al., 1999). The effectiveness of ozonolysis is influenced by the moisture content of biomass; a higher moisture content results in less lignin oxidation (Rafieenia et al., 2017). For instance, substantial amounts of oxalic and acetic acids were formed by wheat straw treated under particular conditions (40% moisture, 60 L/h ozone flow, 2.7% ozone concentration) (Bellido et al., 2011). For commercial feasibility, important process parameters such as ozone intake, biomass composition, and inhibitor production need to be tuned (Sun and Cheng, 2002). While maintaining the quantity of cellulose and hemicellulose, ozonolysis efficiently breaks down lignin (Khanal, 2008).

5.4.2.3 Chloroform

By inhibiting methyl-CoM reductase (MCR), which stops methanogens from consuming hydrogen, chloroform ($CHCl_3$) is used to suppress methanogenic activity and increase the production of biohydrogen during dark fermentation (Hu and Chen, 2007; Dixon and Wilken, 2018). At room temperature, chloroform concentrations usually vary from 0.0005% to 5%, and treatment durations are 17–24 hours (Hu and Chen, 2007; Dixon and Wilken, 2018). High chloroform concentrations can harm HPB and lower hydrogen yield, even while they successfully suppress HCB (Chidthaisong and Conrad, 2000; Wang and Wan, 2008). However, according to some research, pretreatment with chloroform has no discernible effect on the creation of hydrogen (Luo et al., 2010).

5.4.2.4 Iodopropane

By functioning as a corrinoid antagonist, denaturing the activity of the B12 enzyme, and preventing the synthesis of methyl-CoM, iodopropane inhibits methanogenic activity during dark fermentation and stops the production of methane (Zhu and Béland, 2006; Pita and Perez, 2011). By stopping hydrogen consumption, this suppression of hydrogenotrophic methanogenesis enables greater hydrogen yields (Khanal, 2008). Iodopropane suppresses both methanogens and non-methanogens without influencing the synthesis of hydrogen (Zhu and Béland, 2006).

5.4.2.5 Acetylene

By interfering with methanogenic processes like methanogenesis and ATP production, acetylene, a non-specific methanogenesis inhibitor, increases hydrogen yield (Valdez-Vazquez et al., 2006; Rafieenia et al., 2017). It has beneficial effects on methanogen inhibition and is an inexpensive pretreatment for inoculum. For instance, methanogenic suppression was also observed when the inoculum was treated with 1% acetylene (v/v) using milled paper as a substrate (Valdez-Vazquez et al., 2006).

5.4.3 Biological Pretreatment Method

Because biological pretreatment uses natural cellulolytic and hemicellulolytic microorganisms to break down lignocellulose, it is more energy-efficient and environmentally friendly than physical or chemical approaches (Vats et al., 2013). While enzymatic hydrolysis uses enzymes directly to hydrolyze materials, fungal treatments employing strains such as white-rot fungus break down lignin and some cellulose (Sánchez, 2009). For instance, compared to acid/alkali pretreatment, poplar leaves treated with enzymes such as Viscozyme L produced 44.92 mL/g of biohydrogen (Cui et al., 2010). Furthermore, the hydrogen output of cassava wastewater was greatly increased by sonication and enzyme treatments. However, lignocellulose's slow rate of hydrolysis continues to be a major barrier to its commercial usage (Sun et al., 2002).

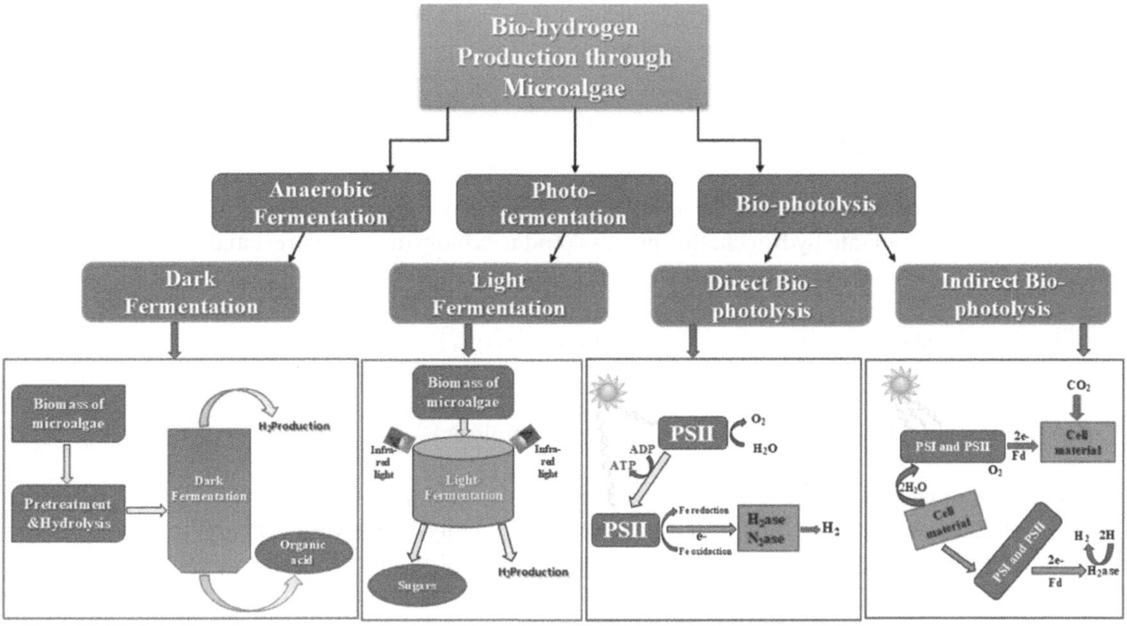

FIGURE 5.3 Various mechanisms of biohydrogen production from microalgae biomass; dark fermentation, light fermentation, direct bio-photolysis, and indirect bio-photolysis.

5.5 BIOHYDROGEN PRODUCTION PROCESSES

Biohydrogen production from microalgae biomass includes dark fermentation, light/photo-fermentation, direct bio-photolysis, and indirect bio-photolysis. Their types and mechanism have been depicted in Figure 5.3.

5.5.1 Photo-Fermentation

The synthesis of biohydrogen using photo-fermentation presents a viable, sustainable, and eco-friendly substitute for traditional techniques. However, obstacles like low production efficiency and high energy requirements for nitrogenase-catalyzed processes prevent its widespread use. Usage of chemical enhancers improves the production of biohydrogen and also the efficiency of light conversion. Iron, molybdenum, ethylenediaminetetraacetic acid (EDTA), and vitamins have shown a great increase in production rates, while copper ions, nickel ions, and diphenyleneiodonium have been shown to have adverse effects. The bacteria, substrates, and quantities utilized all affect how efficient these additions are, highlighting the importance of meticulous photo-fermentation optimization (Budiman and Wu, 2018).

In this novel process, a carbon source is transformed into hydrogen by photosynthetic bacteria, mainly purple non-sulfur (PNS) bacteria, using light energy. Enzymes such as nitrogenase, which is essential for the anaerobic synthesis of hydrogen, employ the electrons produced when the light oxidizes the carbon source. Temperatures between 30°C and 35°C, light intensity between 6 and 6,000 lux, and a pH between 6.8 and 7.5 are ideal for this process.

Hydrogen production rates are still below theoretical expectations, despite the process's potential, and there are still a number of obstacles to overcome before it can be used on a wide scale in industry. To increase sustainability and productivity, continuous efforts are being made to optimize variables includ ing pH, temperature, and light intensity (Melitos et al., 2021).

5.5.2 Bio-Photolysis

Using microorganisms like cyanobacteria or microalgae, which use sunlight to turn water or organic materials (like glucose) into hydrogen, this process produces biohydrogen. Direct and indirect bio-photolysis are the two categories of the process.

$$2H_2O + \text{solar energy} \rightarrow 2H_2 + O_2 \text{ (Direct bio-photolysis)}$$

$$C_6H_{12}O_{16} + 6H_2O \rightarrow 6CO_2 + 12H_2 \text{ (Indirect bio-photolysis)}$$

The reduction of water to hydrogen is catalyzed by hydrogenase, which transfers electrons to ferredoxin after the photosystem absorbs light energy in direct bio-photolysis. However, the oxygen produced during this step inhibits hydrogenase activity. The second stage of indirect bio-photolysis uses the biomass produced by the photosynthetic system, which is primarily composed of carbohydrates, as a source of carbon to make hydrogen. By isolating the synthesis of oxygen from the evolution of hydrogen, indirect bio-photolysis has an advantage over the direct procedure in that it eliminates the inhibition caused by oxygen (Melitos et al., 2021).

5.5.3 Dark Fermentation

Both facultative and obligate anaerobic bacteria can produce hydrogen through a biological process called "dark fermentation," which takes place in the absence of light and oxygen. In order to produce hydrogen, cyanobacteria break down a substrate throughout this process. Lignocellulosic biomass, carbohydrate-rich materials including industrial effluent, crop residues that contain sugar, and municipal solid waste are among the substrates utilized in dark fermentation. The biomass's pretreatment has a big impact on how well dark fermentation works. The kind of microorganisms used and the substrate's sugar level are other variables that impact efficiency (Kamran and Fazal, 2021). Dark fermentation operates under zero light and at high rate.

5.6 GENETIC ENGINEERING TO ELEVATE THE PRODUCTION OF BIOHYDROGEN

To maximize hydrogen generation as a clean and eco-friendly energy source, GE for increased biohydrogen production focuses on altering the metabolic pathways and enzymes of microorganisms. The implementation of GE to enhance the generation of biohydrogen has been made possible by advancements in techniques of molecular biology such as selectable markers (Garcia-Echauri and Cardineau, 2015; Barahimipour et al., 2016), genetic transformation (Yamano et al., 2013; Pratheesh et al., 2014), genomic sequencing (Merchant et al., 2007; Toepel et al., 2013), and use of various promoters (Jinkerson and Jonikas, 2015) in microalgae. The most popular technique for creating nuclear gene mutants for genetic studies in blue-green algae (BGA) is random insertional mutation; still, this approach necessitates the production of numerous mutants and laborious mutant screening.

TABLE 5.2 Genetic engineering techniques to increase the production of biohydrogen from microalgae

ALGAL SPECIES	*TARGET GENE*	*TARGET PROTEIN*	*GENETIC ENGINEERING TECHNIQUE*	*PHENOTYPE*	*REFERENCES*
Anabaena variabilis strain AVM13	*hupSL*	Hydrogen uptake (Hup)-hydrogenase (H_2ase)	Homologous recombination	The slight decrease in acetylene reduction increased the hydrogen (H_2) production rate.	Happe et al. (2000)
Anabaena sp. PCC 7120	*hupL/hoxH*	Hup-H_2ase	Homologous recombination	Increased H_2 production.	Kirst et al. (2012)
Anabaena siamensis TISTR 8012	*hupS*	Hup-H_2ase	Homologous recombination	Higher nitrogenase (N_2ase) activity.	Khetkorn et al. (2016)
Chlamydomonas reinhardtii	L1591-N230Y	D_1 protein	Site-directed mutagenesis	Increased photosynthetic capacity, respiration rate, and relative photosynthesis quantum yield.	Torzillo et al. (2009)
Chlorella sp. DT	*siRNA-PsbO*	Oxygen-evolving complex (OEC)	RNA interference	Leads to the depletion of oxygen (O_2), the induction of HydA, and the sustained production of H_2.	Lin et al. (2013)
Chlamydomonas reinhardtii	*cy6Nac2.49*	Pigment system-II (PSII)	Riboswitch and homologous recombination	Knockdown of the *psbD* gene and less O_2 production.	Surzycki et al. (2007)
Chlamydomonas reinhardtii	*aprl*	Photosynthesis/respiration ratio	DNA insertional mutagenesis; Glycol-aldehyde addition	Respiration rate greater than photosynthesis rate, cells are anaerobic.	Rühle et al. (2008)
Chlamydomonas reinhardtii	CC-849 codon optimized *hemH-lbA*	Leg hemoglobin	Heterologous expression from *Bradyrhizobium* and *Glycine max* in *Chlamydomonas*	Slightly inhibited growth, a lower O_2 content, and H_2ase expressed.	Wu et al. (2010, 2011)
Chlamydomonas reinhardtii	ccPHC	Pyruvate oxidase	Heterologous expression from *E. coli* and *Synechococcus elongatus* PCC 7942 in *Chlamydomonas*	Consumed more O_2 than the wild type.	Xu et al. (2011)

(*Continued*)

TABLE 5.2 (*Continued*) Genetic engineering techniques to increase the production of biohydrogen from microalgae

ALGAL SPECIES	*TARGET GENE*	*TARGET PROTEIN*	*GENETIC ENGINEERING TECHNIQUE*	*PHENOTYPE*	*REFERENCES*
Chlamydomonas reinhardtii	Antisense-*SULP*	Sulfate permease	RNA antisense	Limiting the sulfate supply to the chloroplast, down-regulation of O_2 evolution, and constitutive expression of H_2ase.	Chen et al. (2005)
Chlamydomonas reinhardtii	*pgrl1*	Cyclic electron flow (PGRL1 protein)	DNA insertional mutagenesis	Strongly decreased nonphotochemical quenching (NPQ) of fluorescence.	Tolleter et al. (2011)
Chlamydomonas reinhardtii	*pgr5*	Cyclic electron flow (CEF)	DNA insertional mutagenesis	Completely inhibited CEF, stable PSII, and increased activity of ROS (reactive oxygen species) scavenging enzymes.	Chen et al. (2016)
Chlamydomonas reinhardtii	*stm6*	State transition	DNA insertional mutagenesis	Plenty of starch accumulated, and a low dissolved O_2 concentration.	Volgusheva et al. (2013)
Chlamydomonas reinhardtii	*stmlGLC4lo1*	Light harvesting complex (LHCBM 1,2,3)	RNA interference	Reduced photoinhibition, increased O_2 consumption, and less heat losses.	Oey et al. (2013)
Chlamydomonas reinhardtii	*std3 sda6*	Starch enzyme	DNA insertional mutagenesis	Small amounts of phyto-glycogen are produced.	Jinkerson and Jonikas (2015) and Barahimipour et al. (2016)
Chlamydomonas reinhardtii	CC-2803	RuBisCO large subunit	DNA insertional mutagenesis	Suppressed Calvin-Benson cycles, and decreased O_2 evolution.	Hemschemeier et al. (2008)
Chlamydomonas reinhardtii	Y67A	RuBisCO small subunit	Site-directed mutagenesis	Suppressed the Calvin-Benson cycle, decreased O_2 evolution, and lowered photosynthesis efficiency.	Pinto et al. (2013)
Chlorella sp. DT	*C.s* DT *HydA*	H_2ase	Homologous overexpression in *Chlorella*	*HydAc* expression under oxic condition.	Chien et al. (2012)

(*Continued*)

TABLE 5.2 (*Continued*) Genetic engineering techniques to increase the production of biohydrogen from microalgae

ALGAL SPECIES	*TARGET GENE*	*TARGET PROTEIN*	*GENETIC ENGINEERING TECHNIQUE*	*PHENOTYPE*	*REFERENCES*
Nostoc punctiforme strain NHM5	*hupL*	Hup-H_2ase	Homologous recombination	Increased heterocyst frequency, and increased H_2 production.	Lindberg et al. (2002)
Nostoc sp. PCC 7120	*hupW*	Hup-H_2ase	Homologous recombination	Slightly increased H_2 production	Lindberg et al. (2012) and Yoshino et al. (2007)
Synechocystis strain M55	*ndhB*	The respiratory complex I, NADPH dehydrogenase	Homologous recombination	Producing low amounts of O_2 in the light, and no H_2 uptake during dark-to-light transitions.	Cournac et al. (2004)
Synechocystis sp. PCC 6803	*ctaI/ctaII/cyd*	Respiratory terminal oxidases	Homologous recombination	Respiratory capacity and the amount of photohydrogen produced were found to be inversely correlated.	Gutthann et al. (2007)
Synechocystis sp. PCC 6803	*NarB/nirA*	Nitrate assimilation pathway	Homologous recombination	Higher H_2 production and Hox-H_2ase activities.	Baebprasert et al. (2011)
Synechococcus sp. PCC 7002	*ldhA*	NADH-dependent reduction of pyruvate to D-lactate	Homologous recombination	No D-lactate production and higher concentrations of excreted acetate, alanine, succinate, and H_2 (up to fivefold).	McNeely et al. (2010)

Homologous recombination is used to precisely knock out genes in BGA and green algae chloroplasts (Gonzalez-Ballester et al., 2011). Several techniques and approaches are being developed to increase hydrogen production capacity, and GE was used to boost hydrogen photoproduction efficiency in green microalgae in conjunction with biochemical process strategies (Table 5.2). Using promoters that could operate in an aerobic environment, the modified hydrogenase gene (*hydAc*) was inserted into plasmids to overexpress hydrogenase (hydA).

Given similar conditions, the hydrogen produced from transgenic *Chlorella* sp. DT employing overexpressed homologous HydA was 7–10 times more as compared to the original strain (Chien et al., 2012). A further investigation using the antisense RNA construct in *Chlamydomonas reinhardtii* revealed that hydrogen accumulation and a decrease in photosystem II (PSII) activity were due to the attenuation of sulfate transport gene (*sulP*) expression (Chen et al., 2005). It appears that anti-SulP strains can decrease the quantity of sulfate that reaches the chloroplast, suppressing the oxygen (O_2) evolution and H_2O-oxidation. Mutant *Chlamydomonas* strains with decreased ferredoxin-$NADP^+$ reductase (FNR) activity was developed by employing an RNAi technique. The results also demonstrated that, in response to sulfur deprivation, the FNR-RNAi strains produced 2.5 times as much hydrogen photoproduction as the wild-type strains (Sun et al., 2013). To deal with the O_2 problems, O_2-tolerant hydrogenase has been produced by genetically modified organisms (Wu, 2018). These genetic alterations resulted in a reduced enzyme activity. Inhibiting O_2 synthesis is another method of downregulating PSII (Dubini and Ghirardi, 2015), as described in the following section.

5.6.1 Increasing the Efficacy of Light Capture

There is potential for boosting hydrogen production by decreasing the size of light harvesting complex (LHC) antennas to enhance the energy usage efficiency of light (Kumar et al., 2021). Furthermore, by modifying LHC translation inhibitors, a *C. reinhardtii* mutant with a reduced light-capturing antenna network was created; in this case, heterotrophic growth with adding exogenous glucose enhanced the mutant's hydrogen output by 50%. Chloroplast shrinkage can also boost the photosynthesis effectiveness. A *C. reinhardtii* mutant's chloroplast had a lower chlorophyll concentration per unit cell volume, which improved light transmission and increased efficiency for converting solar energy. In pure culture and co-culture with *Pseudomonas*, the mutant's maximum hydrogen buildup was 1.8–5.2 fold and 2.7–3.1 fold that of the wild type, respectively (Ban et al., 2019).

5.6.2 Modifying the Activities of Hydrogenase and Nitrogenase

Two important enzymes that produce hydrogen in microscopic algae are hydrogenase (H_2ase) and nitrogenase (N_2ase). Because O_2 readily inhibits these enzymes, efforts have been made to enhance their functions through GE. These efforts have included modifying the enzyme structure, expressing foreign genes, modifying the absorption of O_2, and lowering hydrogen consumption (Li et al., 2019, 2022). The metals found in the active site of H_2ase enzymes allow them to be classified as "Fe-Fe, Ni-Fe, and Fe" types (Pandey et al., 2021). Compared to other H_2ases, "Fe-Fe" H_2ase has been shown to have 100 times greater activity (Razu et al., 2019). Additionally, O_2 regulates the transcription, translation, and enzyme activity of microscopic algal H_2ases (Khan and Fu, 2020). The most popular technique for identifying enhanced enzymes in green algae is screening for O_2-tolerant H_2ases produced by random mutation; yet, this process is laborious and high-throughput (Lu and Koo, 2019; El-Mohsnawy et al., 2022; Li et al., 2022). Different microorganism's H_2ases have varying degrees of activity, and overexpression of H_2ase can also boost hydrogen generation. For instance, although *Clostridium acetobutylicum* can overexpress various Fe-Fe H_2ases, it produces very little protein (Sivaramakrishnan et al., 2021). Successful expression of recombinant enzyme from *C. reinhardtii* in *C. acetobutylicum* led to a high H_2ase content (Von Abendroth et al., 2008). Hydrogen is often constantly produced during lack of sulfur. Because sulfur inhibits the production of cysteine (Cys) and methionine (Met) amino acids, PSII is rendered inactive, causing photosynthetic O_2 release to occur more slowly than respiration (Sivaramakrishnan and Incharoensakdi, 2022).

A class of short noncoding RNAs known as microRNAs (miRNAs) is widely distributed in eukaryotic gene regulatory circuits that control "post-transcriptional gene expression" in eukaryotes (Xu et al., 2019). Changes in expression of miRNA may impact numerous metabolic pathways (Lu and Koo, 2019). *C. reinhardtii* "miRNA expression" was elevated in sulfur-deficient growth conditions, and variations in expression were directly correlated with the degree of sulfur shortage (Xu et al., 2019). In order to target a PSII reaction-center protein in *C. reinhardtii*, artificial miRNA gene codon sequences were also built into an extremely expressed blue light-inducible apparatus; following blue light treatment, the PSII responding enter protein was decreased (Wang et al., 2017). To promote hydrogen production in *C. reinhardtii*, exogenous miRNA may also be utilized to regulate the expression of the oxygen-evolving enhancer (OEE_2) gene in PSII. A heat-shock inductive expression vector that contained miRNA targeting OEE_2 was created and effectively transformed into *C. reinhardtii* (Anwar et al., 2019; Banerjee et al., 2020). OEE_2 is a strong PSII-related protein associated with photosynthetic O_2 precipitation, and prior research has demonstrated that OEE_2 expression dramatically decreased following sulfur deprivation. Compared to the control, hydrogen buildup doubled following repeated heat instigation of identical microalgae culture. Genetic manipulation is better than sulfur-deficient circumstances for culture because it eliminates the need to modify the culture media, which makes it more practical for big-scale production (Wang et al., 2017).

BGA have N_2ase to fix nitrogen (N_2) along with H_2ases (Giri et al., 2022). N_2ase catalyzes the "unidirectional" creation of hydrogen, as opposed to the bidirectional proton-based hydrogen generation process (Masukawa et al., 2012). It was once believed that N_2ases could generate hydrogen more readily than H_2ases due to the unidirectional form of the catalytic process; yet, it has now been demonstrated that N_2ase's catalytic effectiveness is significantly lower (Sivaramakrishnan et al., 2021). Since N_2ase needs many electrons, a reducing agent, and ATP to generate significant amounts of hydrogen, genetic modification is essential for using this enzyme to make hydrogen (Mishra et al., 2019).

N_2ase must be manifested in BGA that do not have the *nif* gene family for this gene to be extensively expressed (Thiel, 2019). The non-nitrogen-fixing *Synechococcus elongatus* PCC 7942's genome was modified to express N_2ase by integrating the minimal *nif* gene family from Bacteroides. The issue of lower gene instigation levels brought on by a lengthy *nif* gene cluster is resolved by the division of this minimal *nif* gene family into six parts, every one of which has a separate promoter and terminator. By combining the recombinant strain's biological and electrochemical systems, a bio-electrochemical nitrogen fixation (e-BNF) system was created. Methyl viologen (MV) supplied N_2ase with reducing power, and PSII inhibitor 3-(3,4-dichlorophenyl)-1,1-dimethylurea (DCMU) was employed to inhibit the synthesis of photosynthetic oxygen. Nitrogen fixation and hydrogen production could be carried out continuously by the recombined *S. elongatus* PCC 7942; after 48 hours, the N_2 fixation rate was 21 times higher than that of systems that were just photosynthetic (Dong et al., 2021). Numerous additional genome editing tools are currently being developed by researchers, including the Clustered Regularly Interspaced Short Palindromic Repeat (CRISPR)-Cas9 or Cpf1 system (Baek et al., 2018; Xue et al., 2019), Transcriptional Activator-Like Effector Nucleases (Greiner et al., 2017), "Targeting Induced Local Lesions in Genomes" (Gilchrist and Haughn, 2010; Gonzalez-Ballester et al., 2011), mediated by chemical mutagenesis, Zinc Finger Nuclease (Sizova et al., 2013), etc. The genomes of the mitochondria and chloroplasts have also undergone extensive sequencing alongside the nucleus genome (Hackett et al., 2005). Since *Chlamydomonas*' chloroplast and nuclear DNA can be readily altered, genetic approaches have been used widely to engineer the microalgae's capacity to produce hydrogen (Economou et al., 2014). In particular, the large-scale mutant library (CLiP) and polymerase chain reaction (PCR)-based random mutagenesis mutant collections have been developed to provide the best genetic materials for improving biohydrogen production. Forward and reverse genetics were used to boost the synthesis of biohydrogen (Gonzalez-Ballester et al., 2011).

5.7 CONCLUSIONS AND FUTURE PERSPECTIVES

Biohydrogen production using microalgae is a sustainable source of energy. Various aspects like choice of microalgal strain, type of PBR, biomass pretreatment method, harvesting and cultivation techniques, and GE approaches altogether shape the biohydrogen production. Although some strains of microalgae, like *C. reinhardtii* and *Scenedesmus*, have been found to produce biohydrogen, these strains' productivity and efficiency might differ greatly. It is still difficult to find and screen microalgae strains with a high potential for producing hydrogen under different growth circumstances. It will need more investigation to find and genetically alter strains with increased hydrogenase activity because many strains produce very little hydrogen under normal circumstances.

Growing microalgae on a large scale to produce biohydrogen is frequently expensive. High energy needs, the availability of nutrients, light intensity, and CO_2 supply are a few examples of factors that can dramatically increase operating costs. One of the biggest challenges is still achieving high microalgal biomass yields under controlled settings. Seasonal variations, pollution, and inefficient light use can all impact biomass buildup even with optimization attempts. It is necessary to create economical and effective techniques for harvesting microalgal biomass, such as flocculation, centrifugation, and filtration. Traditional harvesting methods use a lot of energy, which raises the cost of manufacturing as a whole. In order to unleash intracellular hydrogenase enzymes and improve fermentation processes, microalgal biomass must be pretreated before being used to produce biohydrogen. Pretreatment procedures like heat or chemical treatments, however, have the potential to break down important components in the biomass, lowering the final yield and quality. The lack of standardized and financially feasible pretreatment techniques further complicates the process of increasing the production of biohydrogen. Many species of microalgae still have low natural hydrogen generation efficiency through dark fermentation or photo-fermentation. Environmental elements like temperature, pH, and light have a significant impact on the hydrogenase enzyme activity needed for the process. One of the primary bottlenecks is maintaining high hydrogen production naturally or with little assistance. Additionally, several strains show a sharp drop in hydrogen output following extended exposure to either light or shade. The field of GE of microalgal strains is still in its infancy, despite having enormous potential. Although a number of mechanisms for hydrogen synthesis in algae have been discovered, there are currently few genetic tools available for modifying the genomes of microalgae. Complex goals requiring sophisticated bioengineering techniques include introducing foreign genes, enhancing tolerance to oxidative stress brought on by high light or oxygen conditions, and optimizing endogenous mechanisms for hydrogen synthesis.

Microalgae need hydrogenase enzymes to produce biohydrogen; however, these enzymes are very sensitive to oxygen. The efficiency of hydrogen production is limited by the difficulty of maintaining anaerobic or low-oxygen conditions in large-scale cultivation systems. The creation of high-yield, hydrogen-producing microalgae strains with enhanced stress tolerance and optimized metabolic pathways will be made possible by developments in GE, such as CRISPR/Cas9. Energy-efficient, scalable PBRs and systems that use wastewater or CO_2 waste will lower cultivation expenses and increase the efficiency of hydrogen production. The creation of energy-efficient, non-toxic pretreatment techniques will optimize the breakdown of biomass and the production of hydrogen while reducing waste. The production of biohydrogen in integrated systems with other valuable biofuels and chemicals will improve sustainability and economic viability. Production efficiency and scalability will be enhanced by sophisticated bioreactor designs that are monitored in real time for ideal conditions. Future studies will concentrate on integrating microalgal biohydrogen production with carbon capture technology and circular economy principles to ensure its economic and environmental sustainability.

ACKNOWLEDGMENTS

The authors thankfully acknowledge DST (Govt. of Rajasthan) for Biotechnology research grant [F17(2) DST/BT/2024-EAC-00375-6548063/604] and RHTQIS grant [F.25(1)EDU.-4/2021]; Govt. of India for MERU research grant of PM-USHA scheme, CSIR-ASPIRE, DST-SERB Young Scientist grant (SB/YS/LS-39/2014); UGC-BSR Start-up research grant (F.30-50/2014/BSR), UGC-JRF to DC, UGC-CAS and DST-FIST programs in Department of Botany, Jai Narain Vyas University, Jodhpur.

REFERENCES

Ahmed, S. F., Mofijur, M., Nahrin, M., Chowdhury, S. N., Nuzhat, S., Alherek, M., … Mahlia, T. M. I. (2022). Biohydrogen production from wastewater-based microalgae: Progresses and challenges. *International Journal of Hydrogen Energy*, *47*(88), 37321–37342.

Alvira, P., Tomás-Pejó, E., Ballesteros, M., & Negro, M. J. (2010). Pretreatment technologies for an efficient bioethanol production process based on enzymatic hydrolysis: A review. *Bioresource Technology*, *101*(13), 4851–4861.

Amutha, K. B., & Murugesan, A. G. (2011). Biological hydrogen production by the algal biomass *Chlorella vulgaris* MSU 01 strain isolated from pond sediment. *Bioresource Technology*, *102*(1), 194–199.

Anwar, M., Lou, S., Chen, L., Li, H., & Hu, Z. (2019). Recent advancement and strategy on bio-hydrogen production from photosynthetic microalgae. *Bioresource Technology*, *292*, 121972.

Ardo, F. M., Lim, J. W., Ramli, A., Lam, M. K., Kiatkittipong, W., Abdelfattah, E. A., Shahid, M. K., Usman, A., Wongsakulphasatch, S., & Sahrin, N. T. (2022). A review in redressing challenges to produce sustainable hydrogen from microalgae for aviation industry. *Fuel*, *330*, 125646.

Arimbrathodi, S. P., Javed, M. A., Hamouda, M. A., Aly Hassan, A., & Ahmed, M. E. (2023). BioH_2 production using microalgae: Highlights on recent advancements from a bibliometric analysis. *Water*, *15*(1), 185.

Ayub, H. M. U., Nizami, M., Qyyum, M. A., Iqbal, N., & Hasan, M. (2023). Sustainable hydrogen production via microalgae: Technological advancements, economic indicators, environmental aspects, challenges, and policy implications. *Environmental Research*, *244*, 117815.

Baebprasert, W., Jantaro, S., Khetkorn, W., Lindblad, P., & Incharoensakdi, A. (2011). Increased H_2 production in the cyanobacterium *Synechocystis* sp. strain PCC 6803 by redirecting the electron supply via genetic engineering of the nitrate assimilation pathway. *Metabolic Engineering*, *13*(5), 610–616.

Baek, K., Yu, J., Jeong, J., Sim, S. J., Bae, S., & Jin, E. (2018). Photoautotrophic production of macular pigment in a *Chlamydomonas reinhardtii* strain generated by using DNA-free CRISPR-Cas9 RNP-mediated mutagenesis. *Biotechnology and Bioengineering*, *115*(3), 719–728.

Ban, S., Lin, W., Luo, Z., & Luo, J. (2019). Improving hydrogen production of *Chlamydomonas reinhardtii* by reducing chlorophyll content via atmospheric and room temperature plasma. *Bioresource Technology*, *275*, 425–429.

Banerjee, S., Banerjee, S., Ghosh, A. K., & Das, D. (2020). Maneuvering the genetic and metabolic pathway for improving biofuel production in algae: Present status and future prospective. *Renewable and Sustainable Energy Reviews*, *133*, 110155.

Barahimipour, R., Neupert, J., & Bock, R. (2016). Efficient expression of nuclear transgenes in the green alga Chlamydomonas: Synthesis of an HIV antigen and development of a new selectable marker. *Plant Molecular Biology*, *90*, 403–418.

Bare, W. F., Jones, N. B., & Middlebrooks, E. J. (1975). Algae removal using dissolved air floatation. *Journal of Water Pollution Control Federation*, *47*, 153–169.

Bellido, C., Bolado, S., Coca, M., Lucas, S., González-Benito, G., & García-Cubero, M. T. (2011). Effect of inhibitors formed during wheat straw pretreatment on ethanol fermentation by *Pichia stipitis*. *Bioresource Technology*, *102*(23), 10868–10874.

Benemann, J. R., & Oswald, W. J. (1996). Systems and economic analysis of microalgae ponds for conversion of CO {sub 2} to biomass. Final report.

Brennan, L., & Owende, P. (2010). Biofuels from microalgae—A review of technologies for production, processing, and extractions of biofuels and co-products. *Renewable and Sustainable Energy Reviews*, *14*(2), 557–577.

Budiman, P. M., & Wu, T. Y. (2018). Role of chemicals addition in affecting biohydrogen production through photo-fermentation. *Energy Conversion and Management*, *165*, 509–527.

Bule, M. V., Gao, A. H., Hiscox, B., & Chen, S. (2013). Structural modification of lignin and characterization of pre-treated wheat straw by ozonation. *Journal of Agricultural and Food Chemistry*, *61*(16), 3916–3925.

Carver, S. M., Hulatt, C. J., Thomas, D. N., & Tuovinen, O. H. (2011). Thermophilic, anaerobic co-digestion of micro-algal biomass and cellulose for H_2 production. *Biodegradation*, *22*, 805–814.

Chaganti, S. R., Kim, D. H., & Lalman, J. A. (2012). Dark fermentative hydrogen production by mixed anaerobic cultures: Effect of inoculum treatment methods on hydrogen yield. *Renewable Energy*, *48*, 117–121.

Chang, V. S., Burr, B., & Holtzapple, M. T. (1997). Lime pretreatment of switchgrass. In *Biotechnology for Fuels and Chemicals: Proceedings of the Eighteenth Symposium on Biotechnology for Fuels and Chemicals Held May 5–9, 1996, at Gatlinburg, Tennessee* (pp. 3–19). Humana Press.

Chanquia, S. N., Vernet, G., & Kara, S. (2022). Photobioreactors for cultivation and synthesis: Specifications, challenges, and perspectives. *Engineering in Life Sciences*, *22*(12), 712–724.

Chaturvedi, V., Goswami, R. K., & Verma, P. (2021). Genetic engineering for enhancement of biofuel production in microalgae. In *Biorefineries: a step towards renewable and clean energy* (pp. 539–559). Springer Singapore. Cheirsilp, B., & Torpee, S. (2012). Enhanced growth and lipid production of microalgae under mixotrophic culture condition: Effect of light intensity, glucose concentration and fed-batch cultivation. *Bioresource Technology*, *110*, 510–516.

Chen, H. C., Newton, A. J., & Melis, A. (2005). Role of SulP, a nuclear-encoded chloroplast sulfate permease, in sulfate transport and H_2 evolution in *Chlamydomonas reinhardtii*. *Photosynthesis Research*, *84*, 289–296.

Chen, M., Zhang, J., Zhao, L., Xing, J., Peng, L., Kuang, T., … Huang, F. (2016). Loss of algal proton gradient regulation 5 increases reactive oxygen species scavenging and H_2 evolution. *Journal of Integrative Plant Biology*, *58*(12), 943–946.

Chidthaisong, A., & Conrad, R. (2000). Specificity of chloroform, 2-bromoethanesulfonate and fluoroacetate to inhibit methanogenesis and other anaerobic processes in anoxic rice field soil. *Soil Biology and Biochemistry*, *32*(7), 977–988.

Chien, L. F., Kuo, T. T., Liu, B. H., Lin, H. D., Feng, T. Y., & Huang, C. C. (2012). Solar-to-bioH_2 production enhanced by homologous overexpression of hydrogenase in green alga *Chlorella* sp. DT. *International Journal of Hydrogen Energy*, *37*(23), 17738–17748.

Chisti, Y. (2008). Biodiesel from microalgae beats bioethanol. *Trends in Biotechnology*, *26*(3), 126–131.

Chowdury, K. H., Nahar, N., & Deb, U. K. (2020). The growth factors involved in microalgae cultivation for biofuel production: A review. *Computational Water, Energy, and Environmental Engineering*, *9*(4), 185–215.

Chu, C. P., Chang, B. V., Liao, G. S., Jean, D. S., & Lee, D. J. (2001). Observations on changes in ultrasonically treated waste-activated sludge. *Water Research*, *35*(4), 1038–1046.

Cournac, L., Guedeney, G., Peltier, G., & Vignais, P. M. (2004). Sustained photoevolution of molecular hydrogen in a mutant of *Synechocystis* sp. strain PCC 6803 deficient in the type I NADPH-dehydrogenase complex. *Journal of Bacteriology*, *186*(6), 1737–1746.

Cui, M., Yuan, Z., Zhi, X., Wei, L., & Shen, J. (2010). Biohydrogen production from poplar leaves pretreated by different methods using anaerobic mixed bacteria. *International Journal of Hydrogen Energy*, *35*(9), 4041–4047.

Dalena, F., Senatore, A., Tursi, A., & Basile, A. (2017). Bioenergy production from second-and third-generation feedstocks. In *Bioenergy Systems for the Future* (pp. 559–599). Woodhead Publishing.

Danquah, M. K., Ang, L., Uduman, N., Moheimani, N., & Forde, G. M. (2009). Dewatering of microalgal culture for biodiesel production: Exploring polymer flocculation and tangential flow filtration. *Journal of Chemical Technology & Biotechnology: International Research in Process, Environmental & Clean Technology*, *84*(7), 1078–1083.

Divakaran, R., & Sivasankara Pillai, V. N. (2002). Flocculation of algae using chitosan. *Journal of Applied Phycology*, *14*, 419–422.

Dixon, C., & Wilken, L. R. (2018). Green microalgae biomolecule separations and recovery. *Bioresources and Bioprocessing*, *5*(1), 1–24.

Dong, F., Lee, Y. S., Gaffney, E. M., Grattieri, M., Haddadin, H., Minteer, S. D., & Chen, H. (2021). An engineered, non-diazotrophic cyanobacterium and its application in bioelectrochemical nitrogen fixation. *Cell Reports Physical Science 2*, 100444. https://doi.org/10.1016/j.xcrp.2021.100444

Dubini, A., & Ghirardi, M. L. (2015). Engineering photosynthetic organisms for the production of biohydrogen. *Photosynthesis Research*, *123*, 241–253.

Duran, S. K., Kumar, P., & Sandhu, S. S. (2018). A review on microalgae strains, cultivation, harvesting, biodiesel conversion and engine implementation. *Biofuels*. *7269*, 1–12.

Economou, C., Wannathong, T., Szaub, J., & Purton, S. (2014). A simple, low-cost method for chloroplast transformation of the green alga *Chlamydomonas reinhardtii*. *Methods in Molecular Biology*, *1132*, 401–411.

Efremenko, E. N., Nikolskaya, A. B., Lyagin, I. V., Senko, O. V., Makhlis, T. A., Stepanov, N. A., … Varfolomeev, S. D. (2012). Production of biofuels from pretreated microalgae biomass by anaerobic fermentation with immobilized *Clostridium acetobutylicum* cells. *Bioresource Technology*, *114*, 342–348.

El-Mohsnawy, E., Abdelaal, A. S., & El-Sheekh, M. (2022). Direct biohydrogen production from algae. In *Handbook of Algal Biofuels* (pp. 303–330). Elsevier.

Gadhe, A., Sonawane, S. S., & Varma, M. N. (2014). Ultrasonic pretreatment for an enhancement of biohydrogen production from complex food waste. *International Journal of Hydrogen Energy*, *39*(15), 7721–7729.

Gao, F., Liu, Y., & Wei, X. (2016). Development of floating photobioreactors for microalgae cultivation: A review. *Bioresource Technology*, *213*, 204–212.

Garcia-Echauri, S. A., & Cardineau, G. A. (2015). TETX: A novel nuclear selection marker for *Chlamydomonas reinhardtii* transformation. *Plant Methods*, *11*, 1–7.

Gilchrist, E., & Haughn, G. (2010). Reverse genetics techniques: Engineering loss and gain of gene function in plants. *Briefings in Functional Genomics*, *9*(2), 103–110.

Giri, D. D., Dwivedi, H., Alsukaibi, A. K. D., Pal, D. B., Al Otaibi, A., Areeshi, M. Y., … Gupta, V. K. (2022). Sustainable production of algae-bacteria granular consortia based biological hydrogen: New insights. *Bioresource Technology*, *352*, 127036.

Gonzalez-Ballester, D., Pootakham, W., Mus, F., Yang, W., Catalanotti, C., Magneschi, L., … Grossman, A. R. (2011). Reverse genetics in Chlamydomonas: A platform for isolating insertional mutants. *Plant Methods*, *7*, 1–13.

Goswami, R. K., Mehariya, S., Obulisamy, P. K., & Verma, P. (2021). Advanced microalgae-based renewable biohydrogen production systems: A review. *Bioresource Technology*, *320*, 124301.

Greiner, A., Kelterborn, S., Evers, H., Kreimer, G., Sizova, I., & Hegemann, P. (2017). Targeting of photoreceptor genes in *Chlamydomonas reinhardtii* via zinc-finger nucleases and CRISPR/Cas9. *The Plant Cell*, *29*(10), 2498–2518.

Grima, E. M., Belarbi, E. H., Fernández, F. A., Medina, A. R., & Chisti, Y. (2003). Recovery of microalgal biomass and metabolites: Process options and economics. *Biotechnology Advances*, 20(7–8), 491–515.

Gutthann, F., Egert, M., Marques, A., & Appel, J. (2007). Inhibition of respiration and nitrate assimilation enhances photohydrogen evolution under low oxygen concentrations in *Synechocystis* sp. PCC 6803. *Biochimica et Biophysica Acta (BBA)-Bioenergetics*, 1767(2), 161–169.

Hackett, J. D., Scheetz, T. E., Yoon, H. S., Soares, M. B., Bonaldo, M. F., Casavant, T. L., & Bhattacharya, D. (2005). Insights into a dinoflagellate genome through expressed sequence tag analysis. *BMC Genomics*, *6*, 1–13.

Happe, T., Schütz, K., & Böhme, H. (2000). Transcriptional and mutational analysis of the uptake hydrogenase of the filamentous cyanobacterium *Anabaena variabilis* ATCC 29413. *Journal of Bacteriology*, *182*(6), 1624–1631.

Hawkes, F. R., Dinsdale, R., Hawkes, D. L., & Hussy, I. (2002). Sustainable fermentative hydrogen production: Challenges for process optimisation. *International Journal of Hydrogen Energy*, 27(11–12), 1339–1347.

Heasman, M., Diemar, J., O'connor, W., Sushames, T., & Foulkes, L. (2000). Development of extended shelf-life microalgae concentrate diets harvested by centrifugation for bivalve molluscs – A summary. *Aquaculture Research*, 31(8-9), 637–659.

Hemschemeier, A., Fouchard, S., Cournac, L., Peltier, G., & Happe, T. (2008). Hydrogen production by *Chlamydomonas reinhardtii*: An elaborate interplay of electron sources and sinks. *Planta*, *227*, 397–407.

Hendriks, A. T. W. M., & Zeeman, G. (2009). Pretreatments to enhance the digestibility of lignocellulosic biomass. *Bioresource Technology*, *100*(1), 10–18.

Hong, S. M., Park, J. K., & Lee, Y. O. (2004). Mechanisms of microwave irradiation involved in the destruction of fecal coliforms from biosolids. *Water Research*, *38*(6), 1615–1625.

Hu, B., & Chen, S. (2007). Pretreatment of methanogenic granules for immobilized hydrogen fermentation. *International Journal of Hydrogen Energy*, *32*(15), 3266–3273.

Hu, Z., & Wen, Z. (2008). Enhancing enzymatic digestibility of switchgrass by microwave-assisted alkali pretreatment. *Biochemical Engineering Journal*, *38*(3), 369–378.

Jan, T. W., Adav, S. S., Lee, D. J., Wu, R. M., Su, A., & Tay, J. H. (2008). Hydrogen fermentation and methane production from sludge with pretreatments. *Energy & Fuels*, *22*(1), 98–102.

Jinkerson, R. E., & Jonikas, M. C. (2015). Molecular techniques to interrogate and edit the Chlamydomonas nuclear genome. *The Plant Journal*, *82*(3), 393–412.

Johnson, D. A., Weissman, J., & Goebel, R. (1988). An outdoor test facility for the large-scale production of microalgae (No. SERI/TP-231-3325; CONF-880215-4). Solar Energy Research Inst., Golden, CO (USA); Microbial Products, Inc., Fairfield, CA (USA).

Kalichevsky, M. T., Knorr, D., & Lillford, P. J. (1995). Potential food applications of high-pressure effects on ice-water transitions. *Trends in Food Science & Technology*, *6*(8), 253–259.

Kamran, M., & Fazal, M. R. (2021). *Renewable Energy Conversion Systems*. Academic Press.

Khan, S., & Fu, P. (2020). Biotechnological perspectives on algae: A viable option for next generation biofuels. *Current Opinion in Biotechnology*, *62*, 146–152.

Khanal, S. K. (2008). Biohydrogen production: Fundamentals, challenges, and operation strategies for enhanced yield. *Anaerobic Biotechnology for Bioenergy Production: Principles and Applications* (pp. 189–219). Wiley.

Khetkorn, W., Incharoensakdi, A., Lindblad, P., & Jantaro, S. (2016). Enhancement of poly-3-hydroxybutyrate production in *Synechocystis* sp. PCC 6803 by overexpression of its native biosynthetic genes. *Bioresource Technology*, *214*, 761–768.

Khetkorn, W., Khanna, N., Incharoensakdi, A., & Lindblad, P. (2013). Metabolic and genetic engineering of cyanobacteria for enhanced hydrogen production. *Biofuels*, *4*(5), 535–561.

Khetkorn, W., Rastogi, R. P., Incharoensakdi, A., Lindblad, P., Madamwar, D., Pandey, A., & Larroche, C. (2017). Microalgal hydrogen production – A review. *Bioresource Technology*, *243*, 1194–1206.

Kim, D. H., & Kim, M. S. (2012). Semi-continuous photo-fermentative H_2 production by *Rhodobacter sphaeroides*: Effect of decanting volume ratio. *Bioresource Technology*, *103*(1), 481–483.

Kim, M. S., Baek, J. S., Yun, Y. S., Sim, S. J., Park, S., & Kim, S. C. (2006). Hydrogen production from *Chlamydomonas reinhardtii* biomass using a two-step conversion process: Anaerobic conversion and photosynthetic fermentation. *International Journal of Hydrogen Energy*, *31*(6), 812–816.

Kim, M. S., Baek, J. S., Yun, Y. S., Sim, S. J., Park, S., & Kim, S. C. (2006). Hydrogen production from *Chlamydomonas reinhardtii* biomass using a two-step conversion process: Anaerobic conversion and photosynthetic fermentation. *International Journal of Hydrogen Energy*, *31*(6), 812–816.

Kirst, H., Garcia-Cerdan, J. G., Zurbriggen, A., Ruehle, T., & Melis, A. (2012). Truncated photosystem chlorophyll antenna size in the green microalga *Chlamydomonas reinhardtii* upon deletion of the TLA3-CpSRP43 gene. *Plant Physiology*, *160*(4), 2251–2260.

Knuckey, R. M., Brown, M. R., Robert, R., & Frampton, D. M. (2006). Production of microalgal concentrates by flocculation and their assessment as aquaculture feeds. *Aquacultural Engineering*, *35*(3), 300–313.

Kucharska, K., Rybarczyk, P., Hołowacz, I., Łukajtis, R., Glinka, M., & Kamiński, M. (2018). Pretreatment of lignocellulosic materials as substrates for fermentation processes. *Molecules*, *23*(11), 2937.

Kumar, V., Sharma, N., Jaiswal, K. K., Vlaskin, M. S., Nanda, M., Tripathi, M. K., & Kumar, S. (2021). Microalgae with a truncated light-harvesting antenna to maximize photosynthetic efficiency and biomass productivity: Recent advances and current challenges. *Process Biochemistry*, *104*, 83–91.

Lakaniemi, A. M., Hulatt, C. J., Thomas, D. N., Tuovinen, O. H., & Puhakka, J. A. (2011). Biogenic hydrogen and methane production from *Chlorella vulgaris* and *Dunaliella tertiolecta* biomass. *Biotechnology for Biofuels*, *4*, 1–12.

Li, S., Li, F., Zhu, X., Liao, Q., Chang, J. S., & Ho, S. H. (2022). Biohydrogen production from microalgae for environmental sustainability. *Chemosphere*, *291*, 132717.

Li, X., Patena, W., Fauser, F., Jinkerson, R. E., Saroussi, S., Meyer, M. T., … Jonikas, M. C. (2019). A genome-wide algal mutant library and functional screen identifies genes required for eukaryotic photosynthesis. *Nature Genetics*, *51*(4), 627–635.

Lin, H. D., Liu, B. H., Kuo, T. T., Tsai, H. C., Feng, T. Y., Huang, C. C., & Chien, L. F. (2013). Knockdown of PsbO leads to induction of HydA and production of photobiological H_2 in the green alga *Chlorella* sp. DT. *Bioresource Technology*, *143*, 154–162.

Lindberg, P., Devine, E., Stensjö, K., & Lindblad, P. (2012). HupW protease specifically required for processing of the catalytic subunit of the uptake hydrogenase in the cyanobacterium *Nostoc* sp. strain PCC 7120. *Applied and Environmental Microbiology*, *78*(1), 273–276.

Lindberg, P., Schütz, K., Happe, T., & Lindblad, P. (2002). A hydrogen-producing, hydrogenase-free mutant strain of *Nostoc punctiforme* ATCC 29133. *International Journal of Hydrogen Energy*, *27*(11–12), 1291–1296.

Lu, Y., & Koo, J. (2019). O_2 sensitivity and H_2 production activity of hydrogenases—A review. *Biotechnology and Bioengineering*, *116*(11), 3124–3135.

Luo, G., Xie, L., Zou, Z., Wang, W., & Zhou, Q. (2010). Evaluation of pretreatment methods on mixed inoculum for both batch and continuous thermophilic biohydrogen production from cassava stillage. *Bioresource Technology*, *101*(3), 959–964.

Maneeruttanarungroj, C., & Phunpruch, S. (2017). Effect of pH on biohydrogen production in green alga *Tetraspora* sp. CU2551. *Energy Procedia*, *138*, 1085–1092.

Maneeruttanarungroj, C., Lindblad, P., & Incharoensakdi, A. (2010). A newly isolated green alga, Tetraspora sp. CU2551, from Thailand with efficient hydrogen production. *International Journal of Hydrogen Energy*, *35*(24), 13193–13199.

Masukawa, H., Kitashima, M., Inoue, K., Sakurai, H., & Hausinger, R. P. (2012). Genetic engineering of cyanobacteria to enhance biohydrogen production from sunlight and water. *Ambio*, *41*, 169–173.

Matis, K. A., Gallios, G. P., & Kydros, K. A. (1993). Separation of fines by flotation techniques. *Separations Technology*, *3*(2), 76–90.

McNeely, K., Xu, Y., Bennette, N., Bryant, D. A., & Dismukes, G. C. (2010). Redirecting reductant flux into hydrogen production via metabolic engineering of fermentative carbon metabolism in a cyanobacterium. *Applied and Environmental Microbiology*, *76*(15), 5032–5038.

Melitos, G., Voulkopoulos, X., & Zabaniotou, A. (2021). Waste to sustainable biohydrogen production via photo-fermentation and biophotolysis – A systematic review. *Renewable Energy and Environmental Sustainability*, *6*, 45.

Merchant, S. S., Prochnik, S. E., Vallon, O., Harris, E. H., Karpowicz, S. J., Witman, G. B., ... Grossman, A. R. (2007). The Chlamydomonas genome reveals the evolution of key animal and plant functions. *Science*, *318*(5848), 245–250.

Mishra, A. K., Kaushik, M. S., & Tiwari, D. N. (2019). Nitrogenase and hydrogenase: Enzymes for nitrogen fixation and hydrogen production in cyanobacteria. In *Cyanobacteria* (pp. 173–191). Academic Press.

Mohn, F. H. (1980). Experiences and strategies in the recovery of biomass from mass cultures of microalgae. *Algae biomass: Production and use/[sponsored by the National Council for Research and Development, Israel and the Gesellschaft fur Strahlen-und Umweltforschung (GSF), Munich, Germany]; editors, Gedaliah Shelef, Carl J. Soeder.*

Mood, S. H., Golfeshan, A. H., Tabatabaei, M., Jouzani, G. S., Najafi, G. H., Gholami, M., & Ardjmand, M. (2013). Lignocellulosic biomass to bioethanol, a comprehensive review with a focus on pretreatment. *Renewable and Sustainable Energy Reviews*, *27*, 77–93.

Morales-Sánchez, D., Martinez-Rodriguez, O. A., & Martinez, A. (2017). Heterotrophic cultivation of microalgae: Production of metabolites of commercial interest. *Journal of Chemical Technology & Biotechnology*, *92*(5), 925–936.

Mosier, N., Wyman, C., Dale, B., Elander, R., Lee, Y. Y., Holtzapple, M., & Ladisch, M. (2005). Features of promising technologies for pretreatment of lignocellulosic biomass. *Bioresource Technology*, *96*(6), 673–686.

Mu, Y., Yu, H. Q., & Wang, G. (2007). Evaluation of three methods for enriching H_2-producing cultures from anaerobic sludge. *Enzyme and Microbial Technology*, *40*(4), 947–953.

Nagarajan, D., Chang, J. S., & Lee, D. J. (2020). Pretreatment of microalgal biomass for efficient biohydrogen production–Recent insights and future perspectives. *Bioresource Technology*, *302*, 122871.

Oey, M., Ross, I. L., Stephens, E., Steinbeck, J., Wolf, J., Radzun, K. A., ... Hankamer, B. (2013). RNAi knock-down of LHCBM1, 2 and 3 increases photosynthetic H_2 production efficiency of the green alga *Chlamydomonas reinhardtii*. *PLoS One*, *8*(4), e61375.

Oh, H. M., Lee, S. J., Park, M. H., Kim, H. S., Kim, H. C., Yoon, J. H., ... Yoon, B. D. (2001). Harvesting of *Chlorella vulgaris* using a bioflocculant from *Paenibacillus* sp. AM49. *Biotechnology Letters*, *23*, 1229–1234.

Oncel, S. V. S. F., & Sukan, F. V. (2011). Effect of light intensity and the light: Dark cycles on the long term hydrogen production of *Chlamydomonas reinhardtii* by batch cultures. *Biomass and Bioenergy*, *35*(3), 1066–1074.

Pandey, A., Sinha, P., & Pandey, A. (2021). Hydrogen production by sequential dark and photofermentation using wet biomass hydrolysate of *Spirulina platensis*: Response surface methodological approach. *International Journal of Hydrogen Energy*, *46*(10), 7137–7146.

Pinto, T. S., Malcata, F. X., Arrabaça, J. D., Silva, J. M., Spreitzer, R. J., & Esquível, M. G. (2013). Rubisco mutants of *Chlamydomonas reinhardtii* enhance photosynthetic hydrogen production. *Applied Microbiology and Biotechnology*, *97*, 5635–5643.

Pita, F., & Perez, M. (2011). *Pretreatment Application of Mixed Sewage Sludge on Self fermentation for Biohydrogen Production*. University of Cadiz.

Pratheesh, P. T., Vineetha, M., & Kurup, G. M. (2014). An efficient protocol for the Agrobacterium-mediated genetic transformation of microalga *Chlamydomonas reinhardtii*. *Molecular Biotechnology*, *56*, 507–515.

Quesada, J., Rubio, M., & Gómez, D. (1999). Ozonation of lignin rich solid fractions from corn stalks. *Journal of Wood Chemistry and Technology*, *19*(1–2), 115–137.

Rafieenia, R., Girotto, F., Peng, W., Cossu, R., Pivato, A., Raga, R., & Lavagnolo, M. C. (2017). Effect of aerobic pre-treatment on hydrogen and methane production in a two-stage anaerobic digestion process using food waste with different compositions. *Waste Management*, *59*, 194–199.

Razu, M. H., Hossain, F., & Khan, M. (2019). Advancement of bio-hydrogen production from microalgae. In *Microalgae Biotechnology for Development of Biofuel and Wastewater Treatment* (pp. 423–462). Springer.

Roy, S., Kumar, K., Ghosh, S., & Das, D. (2014). Thermophilic biohydrogen production using pre-treated algal biomass as substrate. *Biomass and Bioenergy*, *61*, 157–166.

Rühle, T., Hemschemeier, A., Melis, A., & Happe, T. (2008). A novel screening protocol for the isolation of hydrogen producing *Chlamydomonas reinhardtii* strains. *BMC Plant Biology*, *8*, 1–13.

Sánchez, C. (2009). Lignocellulosic residues: Biodegradation and bioconversion by fungi. *Biotechnology Advances*, *27*(2), 185–194.

Sathinathan, P., Parab, H. M., Yusoff, R., Ibrahim, S., Vello, V., & Ngoh, G. C. (2023). Photobioreactor design and parameters essential for algal cultivation using industrial wastewater: A review. *Renewable and Sustainable Energy Reviews*, *173*, 113096.

Shelef, G., Sukenik, A., & Green, M. (1984). Microalgae harvesting and processing: A literature review.

Singh, R., & Sharma, S. (2012). Development of suitable photobioreactor for algae production – A review. *Renewable Sustainable Energy Review*, *16*, 2347–2353.

Sinha, P., & Pandey, A. (2014). Biohydrogen production from various feedstocks by Bacillus firmus NMBL-03. *International Journal of Hydrogen Energy*, *39*(14), 7518–7525.

Sivagurunathan, P., Anburajan, P., Kumar, G., Arivalagan, P., Bakonyi, P., & Kim, S. H. (2017). Improvement of hydrogen fermentation of galactose by combined inoculation strategy. *Journal of Bioscience and Bioengineering*, *123*(3), 353–357.

Sivaramakrishnan, R., & Incharoensakdi, A. (2022). Cyanobacteria as renewable sources of bioenergy (biohydrogen, bioethanol, and bio-oil production). In *Ecophysiology and Biochemistry of Cyanobacteria* (pp. 431–454). Springer Nature Singapore.

Sivaramakrishnan, R., Shanmugam, S., Sekar, M., Mathimani, T., Incharoensakdi, A., Kim, S. H., … Pugazhendhi, A. (2021). Insights on biological hydrogen production routes and potential microorganisms for high hydrogen yield. *Fuel*, *291*, 120136.

Sizova, I., Greiner, A., Awasthi, M., Kateriya, S., & Hegemann, P. (2013). Nuclear gene targeting in Chlamydomonas using engineered zinc-finger nucleases. *The Plant Journal*, *73*(5), 873–882.

Skjånes, K., Rebours, C., & Lindblad, P. (2013). Potential for green microalgae to produce hydrogen, pharmaceuticals and other high value products in a combined process. *Critical Reviews in Biotechnology*, *33*(2), 172–215.

Su, Y. (2021). Revisiting carbon, nitrogen, and phosphorus metabolisms in microalgae for wastewater treatment. *Science of the Total Environment*, *762*, 144590.

Sukenik, A., & Shelef, G. (1984). Algal autoflocculation—Verification and proposed mechanism. *Biotechnology and Bioengineering*, *26*(2), 142–147.

Sun, J., Yuan, X., Shi, X., Chu, C., Guo, R., & Kong, H. (2011). Fermentation of Chlorella sp. for anaerobic bio-hydrogen production: Influences of inoculum–substrate ratio, volatile fatty acids and NADH. *Bioresource Technology*, *102*(22), 10480–10485.

Sun, Y., & Cheng, J. (2002). Hydrolysis of lignocellulosic materials for ethanol production: A review. *Bioresource Technology*, *83*(1), 1–11.

Sun, Y., Chen, M., Yang, H., Zhang, J., Kuang, T., & Huang, F. (2013). Enhanced H_2 photoproduction by down-regulation of ferredoxin-NADP+ reductase (FNR) in the green alga *Chlamydomonas reinhardtii*. *International Journal of Hydrogen Energy*, *38*(36), 16029–16037.

Surzycki, R., Cournac, L., Peltier, G., & Rochaix, J. D. (2007). Potential for hydrogen production with inducible chloroplast gene expression in *Chlamydomonas*. *Proceedings of the National Academy of Sciences*, *104*(44), 17548–17553.

Tan, J. S., Lee, S. Y., Chew, K. W., Lam, M. K., Lim, J. W., Ho, S. H., & Show, P. L. (2020). A review on microalgae cultivation and harvesting, and their biomass extraction processing using ionic liquids. *Bioengineered*. *11*(1), 116–129.

Tenney, M. W., Echelberger Jr, W. F., Schuessler, R. G., & Pavoni, J. L. (1969). Algal flocculation with synthetic organic polyelectrolytes. *Applied Microbiology*, *18*(6), 965–971.

Thiel, T. (2019). Organization and regulation of cyanobacterial nif gene clusters: Implications for nitrogenase expression in plant cells. *FEMS Microbiology Letters*, 366(7), fnz077.

Thungklin, P., Reungsang, A., & Sittijunda, S. (2011). Hydrogen production from sludge treatment strategy for enhancement of biohydrogen production from complex ultrasonically treated waste-activated sludge. *Water Research*, *35*, 1038–1046.

Toepel, J., Illmer-Kephalides, M., Jaenicke, S., Straube, J., May, P., Goesmann, A., & Kruse, O. (2013). New insights into Chlamydomonas reinhardtii hydrogen production processes by combined microarray/RNA-seq transcriptomics. *Plant Biotechnology Journal*, *11*(6), 717–733.

Tolleter, D., Ghysels, B., Alric, J., Petroutsos, D., Tolstygina, I., Krawietz, D., … Peltier, G. (2011). Control of hydrogen photoproduction by the proton gradient generated by cyclic electron flow in *Chlamydomonas reinhardtii*. *The Plant Cell*, *23*(7), 2619–2630.

Torzillo, G., Scoma, A., Faraloni, C., Ena, A., & Johanningmeier, U. (2009). Increased hydrogen photoproduction by means of a sulfur-deprived *Chlamydomonas reinhardtii* D1 protein mutant. *International Journal of Hydrogen Energy*, *34*(10), 4529–4536.

Valdez-Vazquez, I., Rios-Leal, E., Munoz-Paez, K. M., Carmona-Martinez, A., & Poggi-Varaldo, H. M. (2006) Effect of inhibition treatment, type of Inocula, and incubation temperature on batch H_2 production from organic solid waste. *Biotechnology and Bioengineering* *95*, 342–349.

Vats, S., Maurya, D. P., Shaimoon, M., Agarwal, A., & Negi, S. (2013). Development of a microbial consortium for production of blend of enzymes for hydrolysis of agricultural wastes into sugars. *Journal of Scientific & Industrial Research*, *72*, 585–590.

Veeravalli, S. S., Shanmugam, S. R., Ray, S., Lalman, J. A., & Biswas, N. (2019). Biohydrogen production from renewable resources. In *Advanced Bioprocessing for Alternative Fuels, Biobased Chemicals, and Bioproducts* (pp. 289–312). Woodhead Publishing.

Vijayaraghavan, K., Karthik, R., & Nalini, S. K. (2009). Hydrogen production by *Chlamydomonas reinhardtii* under light driven sulfur deprived condition. *International Journal of Hydrogen Energy*, *34*(19), 7964–7970.

Volgusheva, A., Styring, S., & Mamedov, F. (2013). Increased photosystem II stability promotes H_2 production in sulfur-deprived *Chlamydomonas reinhardtii*. *Proceedings of the National Academy of Sciences*, *110*(18), 7223–7228.

Von Abendroth, G., Stripp, S., Silakov, A., Croux, C., Soucaille, P., Girbal, L., & Happe, T. (2008). Optimized over-expression of [FeFe] hydrogenases with high specific activity in *Clostridium acetobutylicum*. *International Journal of Hydrogen Energy*, *33*(21), 6076–6081.

Wang, A. J., Wang, H. C., Cheng, H. Y., Liang, B., Liu, W. Z., Han, J. L., Zhang, B., & Wang, S. S. (2020). Electrochemistry-stimulated environmental bioremediation: Development of applicable modular electrode and system scale-up. *Environmental Science & Ecotechnology*, *3*, 100050.

Wang, J., & Wan, W. (2008). Comparison of different pretreatment methods for enriching hydrogen-producing bacteria from digested sludge. *International Journal of Hydrogen Energy*, *33*(12), 2934–2941.

Wang, Y., Jiang, X., Hu, C., Sun, T., Zeng, Z., Cai, X., … Hu, Z. (2017). Optogenetic regulation of artificial microRNA improves H_2 production in green alga *Chlamydomonas reinhardtii*. *Biotechnology for Biofuels*, *10*, 1–9.

Wieczorek, N., Kucuker, M. A., & Kuchta, K. (2014). Fermentative hydrogen and methane production from microalgal biomass (*Chlorella vulgaris*) in a two-stage combined process. *Applied Energy*, *132*, 108–117.

Woon, J. M., Khoo, K. S., Akermi, M., Alanazi, M. M., Lim, J. W., Chan, Y. J., … Ardo, F. M. (2023). Reviewing biohydrogen production from microalgal cells through fundamental mechanisms, enzymes and factors that engendering new challenges and prospects. *Fuel*, *346*, 128312.

Wu, C. H. (2018). *Genetic engineering of hydrogenase and related enzymes in the hyperthermophilic archaeon pyrococcus furiosus* (Doctoral dissertation, University of Georgia).

Wu, S., Huang, R., Xu, L., Yan, G., & Wang, Q. (2010). Improved hydrogen production with expression of hemH and lba genes in chloroplast of *Chlamydomonas reinhardtii*. *Journal of Biotechnology*, *146*(3), 120–125.

Wu, S., Xu, L., Huang, R., & Wang, Q. (2011). Improved biohydrogen production with an expression of codon-optimized hemH and lba genes in the chloroplast of *Chlamydomonas reinhardtii*. *Bioresource Technology*, *102*(3), 2610–2616.

Xia, A., Cheng, J., Ding, L., Lin, R., Song, W., Zhou, J., & Cen, K. (2014). Enhancement of energy production efficiency from mixed biomass of *Chlorella pyrenoidosa* and cassava starch through combined hydrogen fermentation and methanogenesis. *Applied Energy*, *120*, 23–30.

Xia, A., Cheng, J., Lin, R., Lu, H., Zhou, J., & Cen, K. (2013). Comparison in dark hydrogen fermentation followed by photo hydrogen fermentation and methanogenesis between protein and carbohydrate compositions in *Nannochloropsis oceanica* biomass. *Bioresource Technology*, *138*, 204–213.

Xu, F. Q., Ma, W. M., & Zhu, X. G. (2011). Introducing pyruvate oxidase into the chloroplast of *Chlamydomonas reinhardtii* increases oxygen consumption and promotes hydrogen production. *International Journal of Hydrogen Energy*, *36*(17), 10648–10654.

Xu, L., Fan, J., & Wang, Q. (2019). Omics application of bio-hydrogen production through green alga *Chlamydomonas reinhardtii*. *Frontiers in Bioengineering and Biotechnology*. *7*, 201. https://doi.org/10.3389/fbioe.2019.00201.

Xue, J. H., Chen, G. D., Hao, F., Chen, H., Fang, Z., Chen, F. F., … Xu, G. L. (2019). A vitamin-C-derived DNA modification catalysed by an algal TET homologue. *Nature*, *569*(7757), 581–585.

Yamano, T., Iguchi, H., & Fukuzawa, H. (2013). Rapid transformation of *Chlamydomonas reinhardtii* without cell-wall removal. *Journal of Bioscience and Bioengineering*, *115*(6), 691–694.

Yoshino, F., Ikeda, H., Masukawa, H., & Sakurai, H. (2007). High photobiological hydrogen production activity of a *Nostoc* sp. PCC 7422 uptake hydrogenase-deficient mutant with high nitrogenase activity. *Marine Biotechnology*, *9*, 101–112.

Zeng, M., Mosier, N. S., Huang, C. P., Sherman, D. M., & Ladisch, M. R. (2007). Microscopic examination of changes of plant cell structure in corn stover due to hot water pretreatment and enzymatic hydrolysis. *Biotechnology and Bioengineering*, *97*(2), 265–278.

Zeng, X., Zhang, L., & Li, Y. (2020). Floating photobioreactors for algae cultivation and carbon capture: Recent advances and challenges. *Algal Research*, *50*, 101999.

Zhu, H., & Béland, M. (2006). Evaluation of alternative methods of preparing hydrogen producing seeds from digested wastewater sludge. *International Journal of Hydrogen Energy*, *31*(14), 1980–1988.

6 Cell Immobilization Strategies for Biohydrogen

Methods, Carriers, and Nano Approaches

Aditi Baldev, Samraggi Singha, Hendrina Marian, Keerthana Karthick, Mohandass Ramya, and Pasupathi Rathinasabapathi

6.1 INTRODUCTION

6.1.1 Overview of Energy Challenges and the Need for Sustainable Alternatives

With the global population exceeding eight billion, the energy demand is rapidly increasing. Fossil fuels, the primary sources of energy, are extracted from the Earth's crust and burned to release energy. However, these non-renewable resources are depleting at an alarming rate and are no longer sufficient to meet growing demands. Moreover, burning fossil fuels emits large quantities of carbon dioxide and other heat-trapping gases into the atmosphere, intensifying climate instability and atmospheric changes. These emissions result in major ecological disruptions, such as more frequent extreme weather events, ocean level rise, and the decline of biodiversity. The excessive exploitation of natural resources to fulfill energy requirements is causing lasting damage to the environment. Furthermore, rising fuel prices are straining households and economies worldwide, making energy less affordable (IEA World Energy Report, 2022). As both the ecological degradation and financial burden caused by conventional power generation become more evident, there is a noticeable transition toward sustainable alternatives. Sustainable

DOI: 10.1201/9781003568612-6

energy refers to sources that can be used without depleting resources for future generations. These green alternatives—including solar, wind, hydropower, geothermal, and hydrogen—emit little to no greenhouse gases during use. Their natural abundance and renewability ensure a stable energy supply, and as technology advances, the declining expenses of clean energy solutions are making them a more viable replacement for fossil fuels.

6.1.2 Biohydrogen: A Promising Renewable Energy Source

Hydrogen, the most plentiful element known in the universe, finds one of its most promising uses as a clean fuel. When used in fuel cells, hydrogen combines with oxygen to generate electricity, and the only emission is water vapor—resulting in a completely clean energy process. Hydrogen is also readily available, constituting around 70% of the Earth's mass, primarily found in the form of water. Through the process of electrolysis—where electrical energy is used to separate hydrogen and oxygen—hydrogen can be extracted, and this method is generally less expensive than refining fossil fuels (Manieniyan et al., 2009). Also, hydrogen has a high calorific value, releasing 122 kJ/g of energy upon combustion (Pandey et al., 2023). In comparison, gasoline provides approximately 42 kJ/g, and coal produces only 32 kJ/g. Hydrogen also boasts a much higher energy density per gram compared to solar and wind sources (Osman et al., 2022; Tripathi & Mohanty, 2024). Its versatility allows it to power vehicles, generate electricity, and provide building heat. However, traditional large-scale methods for producing hydrogen—such as reforming natural gas, converting coal, or using electrolysis—frequently depend on non-renewable energy inputs. These methods consume substantial amounts of energy and release carbon dioxide along with other greenhouse gases, worsening environmental contamination (Demirbas, 2009). This underscores the need for alternative, cleaner hydrogen production methods. Biohydrogen offers a renewable solution. Produced from bio-renewable feedstocks, biohydrogen can be generated through chemical, thermochemical, biological, biochemical, and bio-photolysis methods. Microorganisms like bacteria and algae are employed to either directly produce hydrogen or convert organic materials, such as forestry and agricultural waste, into hydrogen. This makes biohydrogen a more eco-friendly and sustainable option than hydrogen produced via chemical methods. Two main pathways are proposed for microbial hydrogen production. One of the most promising approaches is photo-fermentation production, which utilizes microorganisms capable of using sunlight to break down water molecules into hydrogen and oxygen. Another method produces the hydrogen without the use of light and is known as dark fermentation. Certain microorganisms naturally produce hydrogen, and with the advancement of biotechnologies, these microbial systems could provide clean, renewable hydrogen sources (Demirbas, 2009).

6.1.3 Current Challenges in Industrial Biohydrogen Production

Scientists are increasingly turning their attention to eco-friendly biological techniques like bio-photolysis, dark fermentation, and photo-fermentation for hydrogen energy generation. While bio-photolysis and photo-fermentation methods can result in substantial biohydrogen output, they depend on artificial lighting and involve limited substrate options for photosynthetic microbes, which can significantly drive up expenses during large-scale implementation (Rausch et al., 2014). In contrast, biohydrogen production via dark fermentation can occur without light. However, the overall energy conversion efficiency is low, with dark fermentation yielding only 4.3% and photo-fermentation achieving 5.11% (Zhang et al., 2017). One of the primary challenges is the low biohydrogen production rate and yield (Osman et al., 2023). These low yields are mainly attributed to competing biohydrogen pathways, which reduce the overall efficiency. For instance, biohydrogen production pathways co-produce ethanol. Additionally, certain hydrogen-producing strains like *Clostridium arcticum* and *Clostridium barkeri* utilize hydrogen for their metabolic requirements, resulting in unwanted by-products such as propionic and lactic acids.

The process is further complicated by the presence of biohydrogen-consuming microorganisms, including hydrogenotrophic methanogens, homoacetogens, nitrate-reducing bacteria, and sulfate-reducing bacteria (Chandrasekhar et al., 2015).

Most biohydrogen production processes rely on suspended cell systems. Suspension cell fermentation in biohydrogen production has several disadvantages that hinder its efficiency and scalability. In this method, microbial cells are freely suspended in the fermentation medium, which often leads to cell washout during continuous operations, reducing the overall productivity. Moreover, the suspension system can experience cell damage due to mechanical shear stress, decreasing microbial viability and, consequently, hydrogen yields. The process also demands a high concentration of cells to maintain productivity, leading to increased costs for biomass replenishment. Additionally, suspended cells are not easily reusable, which contributes to resource inefficiency and higher operational costs. To overcome these limitations, cell immobilization techniques have been developed. Immobilizing cells in a fixed matrix or confined space helps maintain cell stability, prevents washout, and reduces shear stress, ultimately improving the longevity and hydrogen production capacity of the microbial cells. This approach also allows for the reuse of cells in repeated cycles, increasing the sustainability and economic viability of biohydrogen production.

6.2 CELL IMMOBILIZATION TECHNOLOGY: AN OVERVIEW

The industrial application of biotechnological strategies is often limited by two key factors: the lack of long-term operational stability and the challenge of recovering and reusing cells. To address these limitations, cell immobilization techniques offer a viable solution (Hartmeier, 2012). Microbial cell immobilization involves enclosing or anchoring microorganisms in a controlled microenvironment, allowing for their economic reuse (Anderson, 1975). This technique can be applied to various biocatalysts, including enzymes, organelles, and both animal and plant cells. Immobilization has found wide applications across biotechnology, pharmaceuticals, environmental science, food industries, and biosensors (Peinado et al., 2005). Initially introduced to replace direct enzyme immobilization methods (Cheetham et al., 1979; Parascandola and Scardi, 1980; Woodward, 1988), cell immobilization offers further advantages by eliminating the need for complex and costly enzyme separation and purification steps. It also enables easy extraction of end products from the reaction environment and effective retrieval of catalytic agents (Junter and Jouene, 2004; Stolarzewicz et al., 2011). Compared to systems using immobilized enzymes, immobilized whole cells present new opportunities, as they inherently contain the desired enzymatic functions without requiring additional supports (Vojtisek and Jirku, 1983). The applications of immobilized microbial cells span from industrial processes to environmental applications. When microorganisms are retained on carriers, they can be used in both batch-linked and continuous production systems, significantly reducing costs by eliminating the need for biocatalyst replenishment (Wada et al., 1979; Park and Chang, 2000; Mrudula and Shyam, 2012). Cell immobilization restricts microbial cells within a localized area, limiting their movement and preserving their biocatalytic potential for multiple or ongoing operations (Dervakos and Webb, 1991; Freeman and Lilly, 1998; Covizzi et al., 2007; Amim et al., 2010). Beginning in the early 1970s, Chibata's team successfully pioneered sustained fermentation techniques for producing L-aspartic acid (Coughlan and Kierstan, 1988), inspiring many subsequent studies into microbial applications involving immobilized cells (Ramakrishna and Prakasham, 1999). The environmental applications of immobilized microbial cells have also been widely reported (Bettmann and Rehm, 1984; Anselmo et al., 1985; Sahasrabudhe et al., 1988). More recently, the use of immobilized fungi has proven beneficial for synthesizing value-added biological products like organic acids, as well as key enzymes such as glucose oxidase and glucoamylase (Abraham, 2003).

6.2.1 Advantages of Cell Immobilization over Unimmobilized Cell System

Ease of Separation of the Biocatalyst: A key benefit of immobilized cell systems is the simplicity in removing the biocatalyst from the surrounding reaction environment. In immobilized systems, cells are physically confined within a matrix or support material, which makes the process of separating them from the fermentation broth or reaction mixture much simpler. This contrasts with free-cell systems where the cells are dispersed throughout the medium, making their separation more labor-intensive and often requiring additional steps such as filtration or centrifugation. While enzymes in cell-free systems are already separated from the cells, they may still require downstream processes to isolate them from the reaction mixture, depending on the specific format of the process. Immobilized systems, especially in continuous reactor setups, allow for easier, more automated, and efficient separation, reducing the need for complex handling and improving the overall operational efficiency.

Reusability of the Biocatalyst, Reducing Operational Costs: A critical advantage of immobilized cells is their ability to be reused multiple times in a production cycle. In biohydrogen production, this translates into significant cost savings, especially at an industrial scale, where the cost of replenishing biocatalysts (e.g., cells or enzymes) can be substantial. The physical confinement of immobilized cells within a matrix helps preserve their structure and functionality over long periods, enabling them to catalyze multiple cycles of the reaction. In contrast, free-cell systems or cell-free enzyme systems often suffer from a faster degradation in stability and activity, leading to higher costs for enzyme production or purification with each use. Immobilization not only increases cell stability but also enhances the cost-effectiveness of bioprocesses by reducing the need for frequent replenishment of biocatalysts.

Compatibility with Continuous Reactors for Biotransformation: Immobilized cell systems are highly compatible with continuous flow reactors, where the reaction mixture continuously passes through the reactor without the need for stopping and restarting the process. This continuous operation increases efficiency and throughput, making immobilized systems particularly advantageous for large-scale industrial applications. In continuous reactors, immobilized cells provide enhanced stability and maintain catalytic activity over long periods, which can be challenging to achieve in cell-free systems. While continuous processes are possible with free enzymes, maintaining enzyme stability and activity in these systems requires careful control and may be more difficult to manage. In contrast, immobilized cells exhibit better performance and resilience in such environments, allowing for sustained biohydrogen production without frequent interruptions.

Higher Biocatalyst Concentration in Smaller Regions: Immobilization allows for the confinement of cells in a smaller, defined space, which results in a higher concentration of biocatalysts within the reaction environment. This confinement increases the surface area-to-volume ratio of the biocatalysts, enhancing the rate of biotransformation reactions and improving productivity. In free-cell systems, achieving similar biocatalyst density can be difficult, as high concentrations of cells or enzymes may lead to limitations such as substrate diffusion issues or increased risks of cell death and enzyme denaturation. In contrast, the immobilized matrix provides a protective environment for the cells, allowing them to operate more efficiently even at higher concentrations. This can lead to a higher reagent flow and faster reaction rates compared to traditional batch reactors, making immobilized cells a more robust option for industrial-scale biohydrogen production.

Reduced Secondary Reactions in Non-Growing Conditions: Immobilized cells, particularly when they are in non-growing or stationary phases, tend to focus solely on the desired biotransformation reactions, minimizing the production of secondary metabolites or by-products that could lower the overall reaction yield. In free-cell systems, particularly in growing conditions, microorganisms may divert some of their metabolic activity toward producing by-products such as organic acids or ethanol, which can reduce the efficiency of biohydrogen production. By maintaining the cells in non-growing conditions through immobilization, these secondary reactions are minimized, allowing for a more efficient and streamlined production process. In cell-free systems, where only specific enzymes are present, there is less concern about unwanted by-products; however, these systems may face limitations in enzyme specificity and stability over time. Immobilization, by keeping the entire cellular machinery intact, ensures that the metabolic environment remains controlled and stable, which is particularly beneficial for more complex biotransformations where multiple metabolic steps are required.

6.3 TYPES OF CELL IMMOBILIZATION TECHNIQUES

Immobilization is essential in enhancing the efficiency of biohydrogen production by improving cell retention, stability, and reuse. There are several types of cell immobilization methods, generally grouped under two main approaches according to their underlying nature: physical and chemical methods.

6.3.1 Physical Immobilization Methods

In physical immobilization, no chemical interactions or covalent bonding are involved. The cells are physically entrapped or confined. The physical immobilization methods are entrapment, encapsulation, and adsorptions methods. Each of these methods has advantages and its limitations (Table 6.1)

6.3.1.1 Entrapment

Entrapment is a widely utilized technique for limiting cell movement, functioning as one of the most straightforward methods of cell immobilization. Classified under active immobilization, this method secures cells through physical or chemical forces. For effective biohydrogen production, the matrix employed must exhibit sufficient permeability to facilitate the diffusion of essential nutrients while simultaneously allowing for the removal of metabolic by-products. A range of supporting materials can serve this function, including polymeric substances with porous structures such as alginate, agar, polyacrylamide,

TABLE 6.1 Different cell immobilization methods

METHOD	*MECHANISM*	*COMMON MATERIALS*	*ADVANTAGES*	*LIMITATIONS*
Adsorption	Cells adhere to the surface	Activated carbon, glass, resins	Simple, reversible	Weak binding, potential cell detachment
Entrapment	Cells entrapped in porous gels	Alginate, agar, silica gel	High stability, protects cells	Diffusion limitations
Encapsulation	Cells enclosed in semi-permeable membranes	Hollow fibers, microcapsules	Controlled environment, high protection	Mass transfer issues, higher cost
Covalent bonding	Covalent attachment of cells to a solid support	Functionalized polymers, cellulose	Strong, stable attachment	May affect cell activity, irreversible
Cross-linking	Cells chemically cross-linked using bifunctional reagents	Glutaraldehyde, toluene diisocyanate	High stability	Toxic cross-linkers, reduced activity

κ-carrageenan, collagen, chitosan, and gelatin, along with other materials like porous metal meshes, silica gel, polyurethane, cellulose triacetate, and polystyrene.

Gel-Based Matrix: The formation of polymer beads involves the assembly of polymer chains that aid in absorbing and retaining substrates. Among the various types of matrices, gel-based matrices are particularly notable due to their biocompatibility, porous nature, and gentle immobilization process. These gels provide a supportive environment that encapsulates cells, enabling them to carry out biological functions while remaining immobilized. Unlike covalent or adsorption-based immobilization techniques, gel-based matrices primarily rely on physical entrapment, which protects the cells without significantly altering their physiological state. To create gelatin and agar beads, a liquid polymer solution is combined with microbial cell mixtures and then molded through a predefined shaping method. Upon lowering the temperature, the polymers solidify, effectively trapping the cell suspension within. Although these gel beads are soft and prone to damage, their durability can be enhanced by incorporating a rigid core made of glass or plastic within a soft gelatin shell. However, this design can limit the diffusion of substrates into the inner core, resulting in reduced activity and productivity of the immobilized cells.

A study by Osamu Hirayama and colleagues focused on producing hydrogen using immobilized cells of the photosynthetic bacterium *Rhodospirillum rubrum* G-9 BM. The researchers aimed to improve the methods of immobilization to enhance the cells' longevity and efficiency in hydrogen production, which could benefit renewable energy and wastewater treatment. In their study, the bacterium was immobilized in either agar or carrageenan gel. They found that carrageenan was more stable and better for sustained hydrogen production. The immobilization method protected the cells from oxygen and contaminants, allowing them to be evenly distributed in the gel. The team also created a continuous hydrogen production system that used a micro-tubing pump to supply substrates and control pH. They monitored hydrogen production every 24 hours over 60 days, finding that the cell activity remained at about 40% of the initial level by the end of the study (Hirayama et al. 1986).

Ion-Exchange Gelation: Gelation through ion exchange is a method used for cell immobilization that involves trapping cells within a gel matrix formed by ionic interactions. This technique is effective in creating a stable, porous gel that allows cells to remain active and viable while being immobilized in a fixed position. The gelation process takes place when a soluble polyelectrolyte in aqueous form is mixed together with an ionic salt solution, leading to gel formation as a result of interactions between the polyelectrolyte and the salt ions. A notable example of this mechanism is calcium alginate gel creation, achieved by combining calcium chloride with a sodium alginate solution. Other polymers produced through ion-exchange gelation include calcium/aluminum carboxymethyl cellulose, aluminum alginate, magnesium pectinate, chitosan polyphosphate, and k-carrageenan. Among these, k-carrageenan and alginate are the most frequently applied in cell immobilization. The stability of ionic gels can be further enhanced through covalent cross-linking (Michael L. Shuler & Kargi, 2002).

Alginate gel entrapment is currently the most widely employed method for biohydrogen production, offering advantages such as cost-effectiveness, mild fermentation conditions, cell reusability, and simplicity in process design. However, alginate gels may lose structural integrity due to elevated levels of certain cations, including calcium (Ca^{2+}), magnesium (Mg^{2+}), and potassium (K^{+}). To overcome this limitation, several strategies have been developed to enhance the mechanical stability of alginate beads. As one instance, Wu et al. demonstrated that incorporating titanium oxide and chitosan into alginate beads led to a threefold rise in biohydrogen output, reaching a yield of 21.3 mmol/L/h. Furthermore, when calcium alginate beads were enhanced with activated carbon, biohydrogen production doubled (Sekoai et al., 2017).

Polymer-Based Matrices: Polymer-based matrices are widely utilized in cell immobilization due to their versatility, biocompatibility, and capacity to form various structures for housing or encapsulating cells. These matrices provide essential physical support to immobilized cells, protecting them from environmental stresses while facilitating the exchange of nutrients, gases, and metabolites. The polymers used for cell immobilization can be natural or synthetic, and they can take the form of gels, beads, films, or other structures based on the intended application. In the formation of polymeric networks, two copolymers are linked together using vinyl group-containing monomers. For instance, when acrylamide is copolymerized with bisacrylamide, it results in the production of polyacrylamide gel beads, which are among the most frequently utilized polymer forms. Other commonly used monomers include methacrylamide, acrylamide, and 2-hydroxyethyl methacrylate, with divinyl compounds like methylene bis-acrylamide often used for cross-linking these monomers.

The process of immobilization through polymerization is relatively straightforward. The polymer precursor mixture is blended with microbial cell culture, allowing the polymer network to form and develop into a solidified matrix. This solid structure is then extruded through a perforated plate to generate consistently sized bead-like particles. Both emulsion and suspension polymerization techniques are also applicable in the fabrication of polymer-based beads for immobilizing cells (Michael L. Shuler & Kargi, 2002). Research has demonstrated that the size of polymer beads significantly influences hydrogen productivity. For instance, *Clostridium butyricum* EB6 cells were immobilized in polyethylene glycol (PEG) beads of varying sizes, revealing that the smallest beads exhibited the highest hydrogen productivity (Singh et al., 2013). Another study investigated how different bead diameters impact hydrogen yield in *Escherichia coli* MC13-4 cells, showing that 6 mm beads provided optimal hydrogen output (Ishikawa et al., 2008). This particular size enabled favorable low-oxygen conditions inside the matrix, promoting enhanced hydrogen production efficiency. These findings underscore the critical role of bead size in ensuring robust cell attachment and maximizing productivity (Gopalakrishnan Kumar et al., 2016) (Figure 6.1).

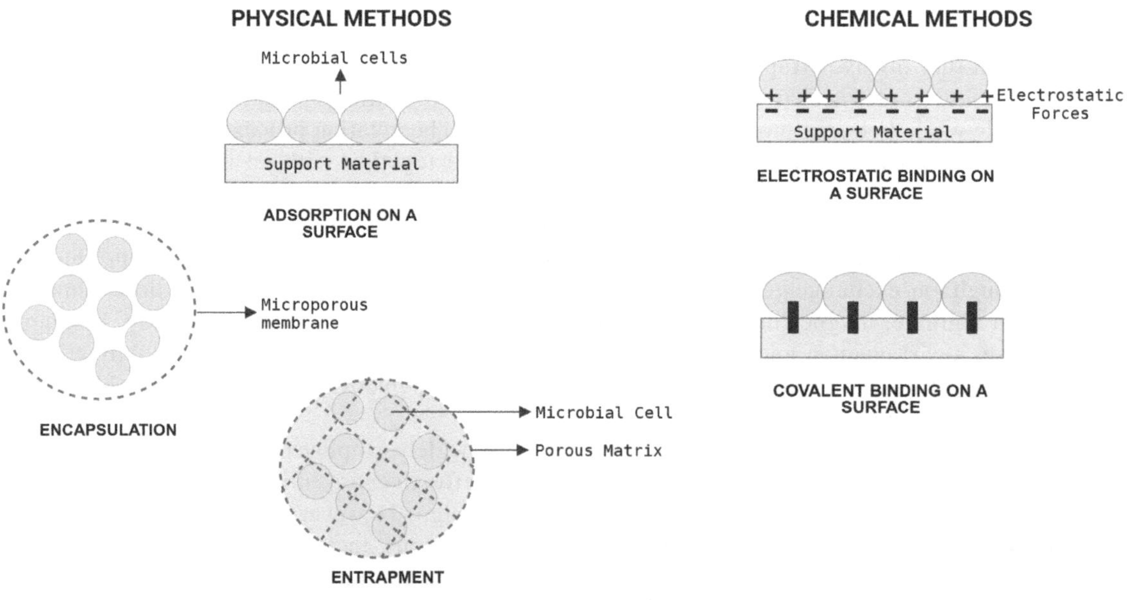

FIGURE 6.1 Overview of cell immobilization methods. This figure illustrates physical and chemical methods for immobilizing microbial cells. Physical methods include Adsorption, where cells adhere to a support material; Encapsulation, trapping cells in a microporous membrane for nutrient exchange; and Entrapment, where cells are immobilized in a porous matrix. Chemical methods involve Electrostatic binding, where cells attach via electrostatic forces, and Covalent binding, forming stable, long-term bonds with the surface. These techniques are widely used in biotechnology to enhance cell stability and reusability in industrial applications.

6.3.1.2 Encapsulation

Encapsulation is a method of cell immobilization that involves enclosing microorganisms within protective materials to enhance their stability, retention, and efficiency. Unlike entrapment, where cells are physically confined within a porous matrix, encapsulation creates a distinct barrier around the cells. This method helps prevent the release of cells into the medium, facilitating more efficient downstream processing of the effluent. In the encapsulation process, cells are suspended in a biocompatible solution, such as alginate, which is then cross-linked to form a gel that encapsulates the cells, creating protective microcapsules. This encapsulation not only improves the stability of the cells against environmental changes but also allows for higher effective cell densities and the potential for reuse of the encapsulated cells, ultimately boosting productivity. Moreover, encapsulation enables controlled mass transfer, optimizing conditions for hydrogen production by regulating the exchange of nutrients, gases, and metabolites (Gopalakrishnan Kumar et al., 2016).

Encapsulation can be effectively applied in various fields, including biohydrogen production. For instance, hydrogen-producing microbes derived from anaerobic sludge have been embedded within polyvinylidene fluoride (PVDF) membranes and utilized to produce hydrogen from challenging fruit-flavored waste substrates (Akinbomi et al., 2015). The microbial cells enclosed in PVDF membranes displayed enhanced hydrogen generation even when exposed to inhibitory agents such as hexanol, myrcene, and octanol. Under both batch and semi-continuous operation modes, the encapsulated cultures showed notably higher biohydrogen yields. The encapsulation process effectively protected the bacteria from inhibitory substances present in the medium, thereby facilitating the valorization of waste nutrients and enhancing the overall hydrogen production process (Gopalakrishnan Kumar et al., 2016). However, encapsulation does present challenges. Diffusion limitations can hinder the flow of nutrients and substrates into the encapsulated cells, potentially impacting metabolic activity. Additionally, the materials used for encapsulation can be costly, and mechanical instability may lead to cell leakage or reduced performance over time. Constraints in oxygen availability and nutrient transport may negatively influence the performance of some encapsulated hydrogen-producing systems. Encapsulation remains a promising technique for enhancing biohydrogen production, particularly when optimized for specific applications.

Microencapsulation: Microcapsules are hollow, spherically shaped structures surrounded by a semi-permeable membrane. In this encapsulation technique, cellular suspensions are placed within the hollow volume of the microcapsules, allowing nutrients to flow in and out through the capsule membrane due to its selective permeability. Microcapsules are often preferred over gel beads because they can support a greater concentration of cells relative to the volume of the carrier matrix. Additionally, limitations in intraparticle diffusion are comparatively reduced in microcapsules, as the inner liquid-filled cavity provides improved nutrient transfer. The composition of the capsule membranes can vary based on the polymers used, including materials like polystyrene, nylon, acrylate, collodion, cellulose acetate–ethyl cellulose, polyester membranes, and polylysine–alginate hydrogel. One significant advantage of this encapsulation method is that it allows low-molecular-weight components to traverse the semi-permeable membrane. Depending on the desired application, different membrane compositions and molecular weight cut-offs may be employed to retain higher molecular weight products inside the capsules while permitting the passage of low-molecular-weight nutrients and by-products. However, the number of components passing through the membrane in microcapsules tends to be lower compared to other immobilization techniques. This limitation can lead to the accumulation of by-products from the metabolic activities of the encapsulated cells, which may hinder hydrogen production. Additionally, the fragility of microcapsules poses a risk of damage during handling and operation.

6.3.1.3 Adsorption

Adsorption is one of the simplest techniques for cell immobilization, as it does not require complex reagents or intricate chemical reactions. This makes it a cost-effective and user-friendly method. The biological activity of immobilized cells is generally retained, and since adsorbed cells can be removed and reused, it is well-suited for large-scale applications. Cell immobilization through adsorption involves anchoring microbial cells onto carrier surfaces through weak intermolecular interactions, such as dispersion forces or Van der Waals attractions. This method generally preserves the catalytic regions of biomolecules, allowing enzymes or microbial cells to retain their functional activity.

Materials selected for supporting cell immobilization may include inorganic options like silica, alumina, ceramics, porous glass, bentonite, clay, and diatomaceous earth, as well as organic alternatives such as starch, cellulose (e.g., diethylaminoethyl (DEAE) cellulose, carboxymethyl cellulose (CMC)), activated carbon, and ion-exchange matrices including Dowex, Amberlite, and Sephadex. Often, these materials need to be chemically or physically pretreated to enhance the immobilization efficiency (Michael L. Shuler & Kargi, 2002). In a reactor system, immobilization via adsorption occurs through direct interactions between the microbial cells and the carrier matrix. This approach offers several advantages, such as improved diffusion of nutrients, maintained biomass integrity, enhanced retention of microbial mass, and better substrate consumption, especially when operating at low hydraulic retention time (HRT), which contributes to increased biohydrogen generation. For example, examined how *Thermoanaerobacterium* bacterial strains interacted with matrices that carried different surface charges. These matrices included two polysaccharide-based materials—negatively charged calcium alginate and positively charged genipin-crosslinked chitosan—and two synthetic acrylic polymers: a hydroxyethyl methacrylate homopolymer (negatively charged) and a positively charged hydroxyethyl methacrylate-based copolymer. The findings revealed that the positively charged support (HM92) promoted stronger microbial attachment, which led to accelerated reaction rates and enhanced hydrogen yield, reaching 3.6 mol of H_2/mol of glucose, surpassing the performance of the remaining matrix materials tested (Gopalakrishnan Kumar et al., 2016). However, compared to other immobilization methods, cell attachment through adsorption is relatively weak and non-specific, which makes it difficult to precisely control the number and orientation of cells on the surface. This can lead to uneven or clustered cell adsorption, potentially affecting the consistency and performance of the immobilized system. Environmental factors, such as pH and temperature, can also affect cell structure and cause desorption. A common challenge with adsorption is cell desorption, especially under strong hydrodynamic conditions, due to the inherently weak binding forces. To address this, stabilization can be achieved by cross-linking the immobilized cells with agents like glutaraldehyde or through covalent binding with the solid platform.

6.3.2 Chemical Immobilization

In chemical immobilization, covalent or ionic bonding is used to secure cells onto a supporting surface called matrix. Covalent bonding and cross-linking are the chemical immobilization methods.

6.3.2.1 Covalent Binding

Covalent binding is a method of cell immobilization in which cells are attached to support surfaces through the formation of strong covalent bonds. Cells are anchored to the support material through reactive sites such as carboxyl, amino, sulfhydryl, and hydroxyl groups, without affecting crucial components of the cells. To prevent any interference with essential functions, protective measures such as competitive inhibition may be employed. The reactive groups present on the matrix are chemically activated using agents such as carbodiimide, glutaraldehyde, or cyanogen bromide. For cell immobilization, covalent binding ensures a robust and sustainable attachment. This method is particularly suited to dynamic environments where immobilized cells need to remain attached, even under mechanical stress, high agitation, or shear forces. Compared to methods like adsorption or entrapment, covalent bonds provide much higher stability and resistance, preventing detachment in high-flow-rate conditions commonly found in bioreactors. Despite its strength, covalent cell immobilization has some limitations. The chemical reactions required for covalent binding may damage cellular structures, such as membranes or proteins, which could impair cell viability and activity. Furthermore, this method is irreversible, making it difficult to remove or replace cells once immobilized. The reagents and procedures involved in covalent immobilization are also more expensive and complex than simpler methods like adsorption.

6.3.3 Nano-Based Immobilization Technologies

Nanomaterials, distinguished by their unique characteristics such as high surface area, tunable porosity, and excellent biocompatibility, are increasingly recognized as promising candidates for cell immobilization in biohydrogen production. These materials create a controlled environment that fosters cell growth and metabolic activity, which can lead to enhanced biohydrogen yields and improved process efficiency. Commonly utilized nanomaterials include carbon-rich materials such as carbon nanotubes (CNTs), graphene, and powdered activated charcoal, known for their high surface area and excellent conductivity. Additionally, nanoparticles composed of metal oxides—like zinc oxide (ZnO), iron oxide (Fe_2O_3), and titanium dioxide (TiO_2)—exhibit catalytic properties that can enhance biohydrogen production by facilitating electron transfer and lowering the activation energy of hydrogen-producing enzymes. Polymer-based nanomaterials, including chitosan, alginate, and polyvinyl alcohol (PVA), can be formed into nanofibrous scaffolds or nanogels, providing a porous and biocompatible environment conducive to cell growth. The use of nanomaterials for cell immobilization offers numerous advantages. They enhance biohydrogen yield due to their favorable environment for metabolic activity and provide improved cell stability by protecting cells from environmental stresses. Furthermore, nanomaterials facilitate mass transfer by enhancing the diffusion of nutrients and oxygen to cells, resulting in increased biohydrogen production rates. They also allow for the creation of highly controlled and reproducible cell immobilization systems, which are essential for process optimization and scalability.

Nanoparticles with pore volumes ranging from 1 to 100 nm exhibit a greater density of functional binding sites, expanded interactive surfaces, and enhanced specificity (Gadhe et al., 2015). The shape of these particles—cubic, spherical, or tetrahedral—can influence their catalytic performance (Singh et al., 2020). Extensive research has been conducted on various organic, inorganic (metallic and metal oxide), and carbon-based nanoparticles for biofuel production and optimization (Ghade et al., 2015). Techniques like magnetic nanoparticles, nanofibers, and nanotubes, along with methods such as Nano-Shear Hybrid Alkaline (NSHA), have demonstrated potential in enhancing biofuel production (Singh et al., 2020). Nickel, as a nanomaterial catalyst, plays a critical role in breaking down lignocellulosic biomass into hemicellulose during biohydrogen production (Chandel et al., 2011). Its characteristics including strong catalytic action, high surface reactivity, and effective adsorption behavior, making it an ideal candidate for this process (Kobayashi et al., 2014). The stability of cellulase, an enzyme vital for cellulose degradation, is influenced by nickel-cobaltite nanoparticles. At a concentration of 1 mM, these nanoparticles maintain the enzyme's stability at high temperatures (80°C) for extended periods (Srivastava et al., 2017). Silica nanoparticles also effectively immobilize cellulase, enhancing its stability and efficiency in biofuel production processes (Datta et al., 2023). Moreover, high-yield results have been achieved when cellulase was covalently attached to mesoporous silica nanoparticles with large internal pore structures (Chang et al., 2011).

Graphite, graphene, and CNTs are prominent in biohydrogen production due to their superior electrical conductivity, mechanical strength, and heat resistance. CNTs, which are cylindrical structures with a nanometer diameter and a hollow interior, are particularly favored for enzyme immobilization because of their increased electrical characteristics and rapid electrode kinetics (Ahmad et al., 2018). However, the high cost of CNTs remains a significant barrier to their widespread adoption (Wang et al., 2006; Shen et al., 2008). Scaling up nano-based immobilization for industrial applications presents complexities due to the intricate nature of nanomaterial production and the need for precise control over production conditions. Additionally, the potential cytotoxic effects of certain nanoparticles raise safety concerns (Zangi et al., 2023). Challenges remain in the application of nanomaterials, particularly regarding the cost of production and the potential toxicity of some materials to cells and the environment. Scaling up nanomaterial-based cell immobilization systems also presents difficulties related to the consistent and reproducible production of nanomaterials for large-scale bioreactor integration.

6.4 FACTORS AFFECTING THE CELL IMMOBILIZATION FOR BIOHYDROGEN PRODUCTION

6.4.1 Carrier Materials

In biohydrogen production, carrier materials, also known as support materials, significantly influence the efficiency and effectiveness of cell immobilization. Carrier materials can be classified as organic or inorganic, each with its unique advantages. Organic materials, such as natural polymers, are valued for their availability, cost-effectiveness, and non-toxic properties, while synthetic organic carriers offer superior mechanical stability, making them suitable for harsh reactor conditions. Inorganic materials, like certain ceramics, provide enhanced durability and resistance to biodegradation, making them ideal for long-term use. These materials provide a stable environment for microbial communities, enhancing hydrogen production by promoting various aspects of microbial metabolism, nutrient retention, and process stability. One of the primary roles of carrier materials is to offer a surface for microbial attachment, which facilitates biofilm formation and improves microbial activity. The ability of a material to enhance microbial attachment directly impacts hydrogen production, as a larger surface area allows for greater biomass retention and higher yields (Han et al., 2011b; Kirli & Kapdan, 2016).

Support materials are instrumental in preserving nutrient availability, making sure that essential nutrients and substrates stay accessible to immobilized microbes, thus promoting efficient metabolic processes. This continuous availability of nutrients maximizes hydrogen production as microbes can consistently access the substrates needed for fermentation and biochemical reactions. Materials with a large surface area, such as activated carbon and biopolymers, further improve contact between microorganisms and substrates, enhancing the rate of biochemical reactions and boosting hydrogen production. The physical stability provided by carrier materials prevents microbial washout in continuous systems, maintaining active populations of key hydrogen-producing strains, which leads to steady output over time (Han et al., 2011b). Additionally, certain support materials encourage biofilm development, offering protection to microbial communities from environmental stress and enhancing their metabolic performance. This increased efficiency contributes to improved hydrogen yields, as biofilms allow microbes to process substrates more effectively. Support materials also enhance mass transfer within reactors by facilitating the movement of gases, such as hydrogen and other substrates. Effective materials with high permeability ensure that hydrogen gas is efficiently released from the liquid phase to the headspace for recovery, preventing gas accumulation that could hinder the process.

For optimal biohydrogen production, support materials should possess specific characteristics, including hydrophilicity, to maintain a moist environment for microbial growth, non-toxicity to avoid inhibiting microbial activity, and non-biodegradability to ensure long-term stability.

Lastly, support materials must be mechanically robust to tolerate acidic environments typical of hydrogen fermentation processes and withstand repeated operational cycles without notable wear. This durability is vital for the long-term economic viability of immobilized biomass systems, as it ensures consistent biohydrogen production over extended periods (Han et al., 2011b; Kirli & Kapdan, 2016). Therefore, selecting the appropriate carrier material is essential to optimize microbial immobilization, improve hydrogen generation, and maximize overall process performance. Additionally, mechanical stability is crucial to withstand physical stresses, while cost-effectiveness is essential for practical, large-scale applications. Other important features include high biomass retention, permeability, flexibility, and porosity, all of which contribute to maximizing microbial efficiency, mass transfer, and hydrogen recovery (Han et al., 2011b; Kirli & Kapdan, 2016).

6.4.2 pH

The acidity or alkalinity of the fermentation medium significantly impacts hydrogenase function, cellular metabolism, and the breakdown of substrates. Maintaining the pH within an optimal range is essential for maximizing hydrogen production while minimizing the activity of microorganisms that consume biohydrogen. Acidic pH levels tend to favor hydrogen-producing species like *Clostridium*, while higher pH levels support the growth of hydrogen-oxidizing methanogens. Variations in pH can alter microbial populations, affect cellular structures and morphologies, and influence biohydrogen production in several ways. Firstly, microbial activity is highly dependent on pH, as different microorganisms thrive in specific pH ranges. For hydrogen-producing bacteria, such as *Clostridium*, a neutral to slightly acidic pH (approximately 6.5–7.5) is ideal for growth and hydrogen production. Significant deviations from this range can inhibit microbial activity and reduce hydrogen yields. Secondly, pH also affects the metabolic pathways microbes use to generate hydrogen. Some pathways are more efficient in acidic conditions, while others perform better in alkaline environments, which directly affects how effectively hydrogen is synthesized.

Additionally, substrate availability is influenced by pH, as it can modify the solubility characteristics of compounds like monosaccharides and carboxylic acids. At certain pH levels, some substrates may become less soluble, limiting their availability for microbial metabolism. Extremely acidic or basic conditions may trigger toxicity by encouraging the formation of harmful by-products or inhibitors, thereby reducing cellular performance and hydrogen output. The formation of biofilms, essential for microbial colonization on support materials, is also pH-dependent. A stable pH can improve biofilm formation and contribute to process stability. Furthermore, pH influences the accessibility of key nutrients—including phosphorus and nitrogen—that are vital for microbial propagation and hydrogen output.

Research highlights the significant impact of pH on hydrogen production. For instance, *C. butyricum* CWBI1009 showed a twofold enhancement in hydrogen generation when maintained at pH 5.2 versus uncontrolled pH conditions. In another study, peak hydrogen productivity (12.27 mmol/L/h) was attained at pH 4.4–4.5 with activated carbon used as the immobilization support. This pH range, combined with a short HRT of 5 hours, effectively inhibited methanogenic archaea, resulting in maximum biohydrogen production. The low pH suppressed competing reactions, while the short HRT limited the growth of hydrogen-consuming microorganisms. Keskin et al. (2012) documented a fivefold rise in hydrogen output when the pH was stabilized between 4.5 and 5.0 during continuous operation using ceramic beads. Similarly, Ren et al. confirmed sustained hydrogen production at pH 6.0 over a 20-day period while employing granular activated carbon. Other studies have identified a broad pH range (4–9) as effective for hydrogen generation, depending on variables like microbial consortia, substrate type, carrier material, and system conditions. For example, Akinbomi et al. (2015) reported enhanced hydrogen fermentation in the pH range of 5.2–5.9, whereas methanogenic activity became dominant at pH 6.8–7.2. Han et al. (2011) found 6.0 to be optimal for biohydrogen production, although the highest yield of 3.21 mol H_2/mol of sucrose was achieved at pH 8.48 using hematite nanoparticles. Zhao et al. (2010) observed that a pH of 10 produced the highest biohydrogen yields due to improved sludge floc solubilization and hydrolysis. Likewise, Mohd Yasin et al. (2011) concluded that the most suitable pH for anaerobic digestion of food waste was between 5.5 and 6.0, which contributed to elevated hydrogen productivity and minimized lag time. These studies highlight the critical importance of pH in optimizing biohydrogen production across processes and microbial systems.

6.4.3 Carbon Source

The carbon source in the biohydrogen production process plays a critical role in influencing cell immobilization by affecting microbial growth, biofilm formation, substrate availability, and overall process efficiency. Different types of carbon sources can lead to variations in how effectively microorganisms attach to support materials, ultimately impacting hydrogen production yields. The nature of the carbon source is particularly important for microbial growth and biofilm development, both of which are essential for effective cell immobilization. Simple carbon sources, such as glucose or fructose, are easily metabolized by hydrogen-producing bacteria, promoting rapid microbial growth and biofilm formation on support materials. This enhances microbial stability and activity, resulting in more efficient hydrogen production. In contrast, complex carbon sources like lignocellulosic biomass (e.g., straw or wood) require enzymatic breakdown, delaying microbial growth and biofilm formation. Such delays can reduce the efficiency of cell immobilization as microbes need more time to colonize the support material (Wu et al., 2002).

The type of carbon source also affects the availability of substrates for microbial metabolism. Simple, readily digestible carbon sources provide immediate energy for microbial growth, accelerating biofilm formation and cell immobilization. However, complex carbon sources, which may contain compounds like lignin, hemicellulose, or cellulose, are harder to degrade and metabolize. Microorganisms involved in biohydrogen production may require pretreatment of these complex substrates to make them more accessible, which can further delay immobilization and reduce overall efficiency. Additionally, different carbon sources can activate specific metabolic pathways in microorganisms, influencing the rate of hydrogen production and immobilization efficiency. Sugars like glucose can be quickly fermented, leading to faster biofilm development. In contrast, complex substrates require initial breakdown by microbes, leading to slower immobilization and less efficient hydrogen production. In some cases, the degradation of complex carbon sources may even produce by-products that inhibit microbial activity or biofilm formation, further impacting the immobilization process (Chu et al., 2011).

Moreover, the carbon source shapes the composition of microbial communities in the biohydrogen production system. Simple carbon sources tend to favor fast-growing hydrogen-producing bacteria like *Clostridium* species, while complex carbon sources may support a more diverse microbial community, including bacteria involved in breaking down complex organic materials. This microbial diversity can affect biofilm stability, as certain species may not directly contribute to hydrogen production but are essential for breaking down the complex substrates. These interactions between microbial species can either enhance or hinder biofilm formation, influencing the overall performance of the immobilization strategy (Rosa et al., 2014). Additionally, the solubility of the carbon source plays a key role in how easily it can be accessed by immobilized cells. Simple carbon sources are typically more soluble, allowing for rapid microbial uptake and metabolic activity, which promotes efficient biofilm formation. Conversely, complex carbon sources may be less soluble, limiting their availability to immobilized cells and slowing down biofilm formation and hydrogen production. This reduced accessibility can further impact the total effectiveness of the immobilization approach, ultimately resulting in decreased hydrogen output (Ishikawa et al., 2008; Seol et al., 2011).

Furthermore, certain carbon sources may produce inhibitory by-products during microbial metabolism, which may result in the buildup of compounds like volatile fatty acids or alcohols that can hinder microbial development or obstruct biofilm formation. These toxic by-products can slow down the immobilization process and reduce hydrogen production efficiency. In conclusion, the choice of carbon source in biohydrogen production has a significant impact on cell immobilization. Simple carbon sources promote rapid microbial growth and biofilm formation, while complex carbon sources may require pretreatment and result in slower immobilization due to their complexity. Understanding the properties of the carbon source and its influence on microbial communities and metabolic pathways is vital for refining biohydrogen generation methods and enhancing immobilization outcomes (Andreani et al., 2015).

6.4.4 Mass Transfer Coefficient

In biohydrogen production using immobilized cells, mass transfer holds a significant influence over the effectiveness of the process. Mass transfer refers to the movement of nutrients, metabolites, and gases through the carrier or support materials that immobilize the microbial cells. The mass transfer coefficient (MTC), which measures how efficiently substances such as gases or solutes are transferred between phases (e.g., liquid to gas or within a liquid phase), significantly impacts cell immobilization and biohydrogen production. Proper mass transfer is crucial for maintaining nutrient availability, preventing the accumulation of inhibitory by-products, and ensuring efficient hydrogen gas release, all of which are essential for optimizing microbial activity in immobilized systems (Beckers et al., 2015).

In biohydrogen production, the MTC affects several key aspects of cell immobilization. First, it plays an integral part in gas-liquid interactions, where hydrogen gas, produced by immobilized microorganisms, needs to be efficiently transferred from the liquid phase to the gas phase. A higher MTC improves gas transfer, preventing the buildup of hydrogen gas around the immobilized cells, which can inhibit their activity and reduce hydrogen production. Second, the MTC influences substrate availability by facilitating the transport of organic matter to the immobilized microbial cells. A higher MTC ensures more effective substrate diffusion through the carrier materials, enabling better microbial metabolism and leading to increased hydrogen production rates (HPRs). Effective mass transfer is particularly important in immobilized systems because poor transfer can lead to nutrient depletion in microenvironments around the cells, limiting their activity. Moreover, the MTC ensures proper nutrient distribution within the immobilized cell matrix. It allows key nutrients—including phosphorus and nitrogen—to reach the immobilized microbes, supporting their growth and biofilm formation. Adequate mass transfer in immobilized systems also reduces the accumulation of toxic by-products, which can occur if inhibitory substances, such as organic acids, are not efficiently removed from the microenvironment around the immobilized cells. A higher MTC ensures that these by-products are effectively transported away from the cell surfaces, promoting healthier microbial communities and more stable hydrogen production. Additionally, the MTC helps regulate important environmental parameters like pH and temperature, which are critical to sustaining optimal biological performance in immobilized systems. Effective mass transfer ensures that these factors remain stable, preventing conditions that could disrupt cell activity and hydrogen production.

In cell immobilization systems, mass transfer becomes even more crucial during scale-up. As the system transitions from laboratory-scale to larger reactors, maintaining effective mass transfer becomes challenging due to changes in flow dynamics and reactor geometry. Inadequate mass transfer in larger systems can lead to uneven distribution of nutrients and gases within the immobilized cell matrix, reducing hydrogen production efficiency. Optimizing the MTC is, therefore, essential for ensuring that immobilized cells continue to function effectively as biohydrogen production processes are scaled up. However, determining and optimizing the MTC in immobilized cell systems present several challenges. Accurately measuring the MTC can be difficult due to the complexity of the biohydrogen production systems, especially when dealing with immobilized cells. The MTC is highly influenced by a range of variables, including temperature, pressure, substrate levels, and flow conditions inside the reactor. Additionally, the geometry of the carrier materials and reactor design can influence mass transfer efficiency. Irregular shapes, internal structures, or dead zones in the immobilization matrix can hinder flow and reduce the effective MTC. In immobilized systems, fouling of the support material can also occur over time, reducing the available surface area for mass transfer and lowering the overall efficiency of the process.

Various factors influence the MTC in immobilized cell systems, including fluid properties such as viscosity and diffusion coefficients, flow conditions (with turbulence enhancing mass transfer), the interfacial surface area between phases, and temperature. Higher temperatures typically increase molecular movement, enhancing mass transfer. Beckers and colleagues (2015) developed a model correlating the HPR with the MTC, demonstrating that HPR increases with the MTC until it reaches a threshold of 0.77/h. They further associated these changes with agitation levels in immobilized reactors, showing that improving mixing can enhance the MTC and consequently boost HPRs. Nath and Chand (1996) proposed another model that estimates MTC based on fluid flow rate and Reynolds number, suggesting that improving mixing in immobilized systems can lead to better mass transfer and, in turn, higher biohydrogen production. The MTC plays a critical role in cell immobilization for biohydrogen production. It directly affects gas-liquid interactions, substrate availability, nutrient distribution, and by-product removal within the immobilized cell matrix. Optimizing the MTC is essential for ensuring efficient hydrogen production, especially in large-scale reactors where maintaining effective mass transfer becomes more challenging.

6.4.5 Microorganisms

Biohydrogen production utilizing immobilized cells has attracted growing interest because of its potential to improve both process efficiency and output. Mixed cultures are frequently employed in dark fermentative hydrogen production, often sourced from natural materials such as anaerobic sludge from wastewater treatment plants, including sewage sludge (Wu et al., 2002; Lin et al., 2009; Barros et al., 2010) and palm oil mill effluent (POME) sludge (Ismail et al., 2011; Mohd Yasin et al., 2011). Several studies have also applied mixed cultures obtained through the spontaneous fermentation of wastewater streams to facilitate biohydrogen production (Fernandes et al., 2013; Nunes Ferraz Júnior et al., 2015). The major benefit of employing a combination of microbial strains lies in their ability to include facultative anaerobic bacteria, which help eliminate residual oxygen and establish favorable oxygen-free conditions for strict anaerobic organisms to carry out hydrogen production. This approach is frequently adopted in microbial enrichment strategies aimed at boosting hydrogen generation. For example, Pachapur et al. (2016) reported that a dual-culture setup comprising the facultative anaerobe *Enterobacter aerogenes* and the obligate anaerobe *C. butyricum* successfully converted glycerol into hydrogen. In this co-culture, *E. aerogenes* effectively removed oxygen, enabling a fully anaerobic environment and reaching a hydrogen output of 19.46±0.95 mmol H_2/L of medium—substantially surpassing the 0.95 mmol H_2/mol of glycerol obtained using single cultures. Typically, sludge intended for hydrogen generation is pretreated to enrich spore-forming *Clostridium* species and suppress unwanted microbial groups like acetogens and methanogens that otherwise utilize hydrogen. Pretreatment strategies commonly include acid exposure (Lee et al., 2004; Ismail et al., 2011), alkali conditioning (Liu et al., 2012), thermal inactivation (Wu et al., 2006; Barros et al., 2011), or hybrid methods combining these approaches (Chu et al., 2011).

Beyond mixed cultures, individual strains of hydrogenogenic bacteria have also been immobilized for targeted biohydrogen production. Members of the *Clostridium* genus, known as Gram-positive, spore-forming facultative anaerobes, are capable of generating as much as 4 mol of H_2/mol of glucose. For instance, *C. butyricum* TISTR1032 immobilized on sugarcane bagasse was used to ferment sugarcane juice, achieving hydrogen productivity of 3.5 L of H_2/L of substrate per day and a yield of 1.52 mol H_2/mol of hexose (Plangklang et al., 2012). Likewise, *C. butyricum* EB6 entrapped in PEG was used to ferment POME, achieving a peak HPR of 22.7 mmol/L/h and a yield of 5.35 L H_2/L of POME (Singh et al., 2013). *Enterobacter* species, classified as Gram-negative facultative anaerobes without spore-forming ability, are also widely applied in hydrogen-producing systems. Han et al. (2011) immobilized *E. aerogenes* ATCC29007 on agar for converting glycerol to hydrogen, achieving a yield of 0.74 mol H_2/mol glycerol. In another experiment, *E. aerogenes* ATCC 13048 was attached to thermally treated USAM (Unsterilized Sludge-Derived Activated Material) granules to generate hydrogen from both clean and waste-derived glycerol. Another hydrogen-yielding bacterium is *Thermotoga neapolitana*, a Gram-negative, thermophilic anaerobe capable of delivering up to 4 mol of H_2/mol of glucose. When anchored to a positively charged hydrogel with built-in pH buffering capability, *T. neapolitana* exhibited an average hydrogen generation rate of 50.6 mL/L/h with a yield of 3.3 mol H_2/mol of glucose (Basile et al., 2012). On porous glass beads and utilizing xylose as a substrate, *T. neapolitana* produced hydrogen at a rate of 5.64 mmol H_2/L/h and a yield of 1.84±0.1 mol H_2/mol xylose. *E. coli* SH5, a facultative and lactate-deficient mutant, was immobilized on agar for hydrogen production from formate, achieving a volumetric maximum production rate of 2.4 L H_2/L/h (Seol et al., 2011).

The immobilization of facultative anaerobes provides several advantages for biohydrogen production, as these organisms can survive in trace amounts of oxygen, exhibiting variable metabolic behaviors depending on oxygen levels. Their rapid growth, efficient substrate utilization, and broad enzyme spectrum make them ideal candidates for co-culturing with strict anaerobes. Additionally, immobilization allows for operation in open systems, reducing the need for strict anaerobic conditions during the start of the bioreactor. Continuous observation and testing of immobilized cultures are critical for tracking their metabolic behavior and enhancing hydrogen production outcomes.

6.4.6 Bead Size

In biohydrogen production, bead size refers to the dimensions of support materials utilized in bioreactors to promote microbial growth. The size of these beads can significantly influence several critical factors, including surface area, mass transfer, and microbial attachment, all of which are essential for optimizing hydrogen production. Typically, bead sizes range from a few hundred micrometers to several millimeters. Smaller beads tend to provide a larger surface area for microbial colonization, thereby enhancing the growth of hydrogen-producing microorganisms, which can lead to increased biomass and higher hydrogen yields. Conversely, larger beads may improve flow characteristics and reduce the risk of clogging within bioreactors.

The selection of bead size is contingent upon various factors, including the specific microbial consortium employed, the type of bioreactor, and the overall design of the biohydrogen production system. It is crucial to balance these factors to enhance both efficiency and yield in biohydrogen production processes. Specifically, bead size can impact biohydrogen production in several ways. First, the surface area available for microbial attachment is often greater in smaller beads, which fosters the proliferation of hydrogen-producing microorganisms. This increase in microbial growth can directly correlate with elevated hydrogen yields. Second, bead size plays a pivotal role in mass transfer dynamics within the reactor. Smaller beads facilitate improved diffusion of substrates and gases, allowing for more efficient substrate access to microorganisms. In contrast, larger beads may create dead zones that hinder substrate exchange, thereby limiting overall production efficiency. Additionally, the flow dynamics within the bioreactor are influenced by bead size. Larger beads can diminish the risk of clogging, ensuring smoother flow and better mixing, both of which are critical for effective gas transfer and microbial activity. Biofilm formation is also affected by bead size. Smaller beads may support denser biofilm development, while larger beads might allow for a more even distribution of microbial communities and better access to nutrients. Furthermore, the choice of bead size can influence HRT, which determines how long substrates remain available for microbial conversion into hydrogen. Lastly, larger beads may be easier to handle and maintain within a bioreactor setup, thereby contributing to operational stability. Consequently, understanding the implications of bead size is essential for optimizing biohydrogen production systems.

6.5 IMMOBILIZED CELL REACTOR CONFIGURATION

Reactor design is critical for optimizing fermentative hydrogen production. In recent decades, a wide variety of reactor systems have been introduced and applied to drive biohydrogen technology closer to real-world, cost-effective deployment. As bioreactors serve as the sites for microbial hydrogen conversion, their characteristics—including type and configuration—substantially influence the reliability of these applications (Figure 6.2). The selection of a bioreactor for immobilized cells is determined by several factors, including the immobilization method employed, the physiological traits of the native cells—whether microbial, plant, or animal, and whether they are viable or nonviable—as well as the hydrodynamic and mass transfer characteristics of the bioreactor. These factors are essential for assessing the system's engineering performance, cell growth, and metabolic behavior.

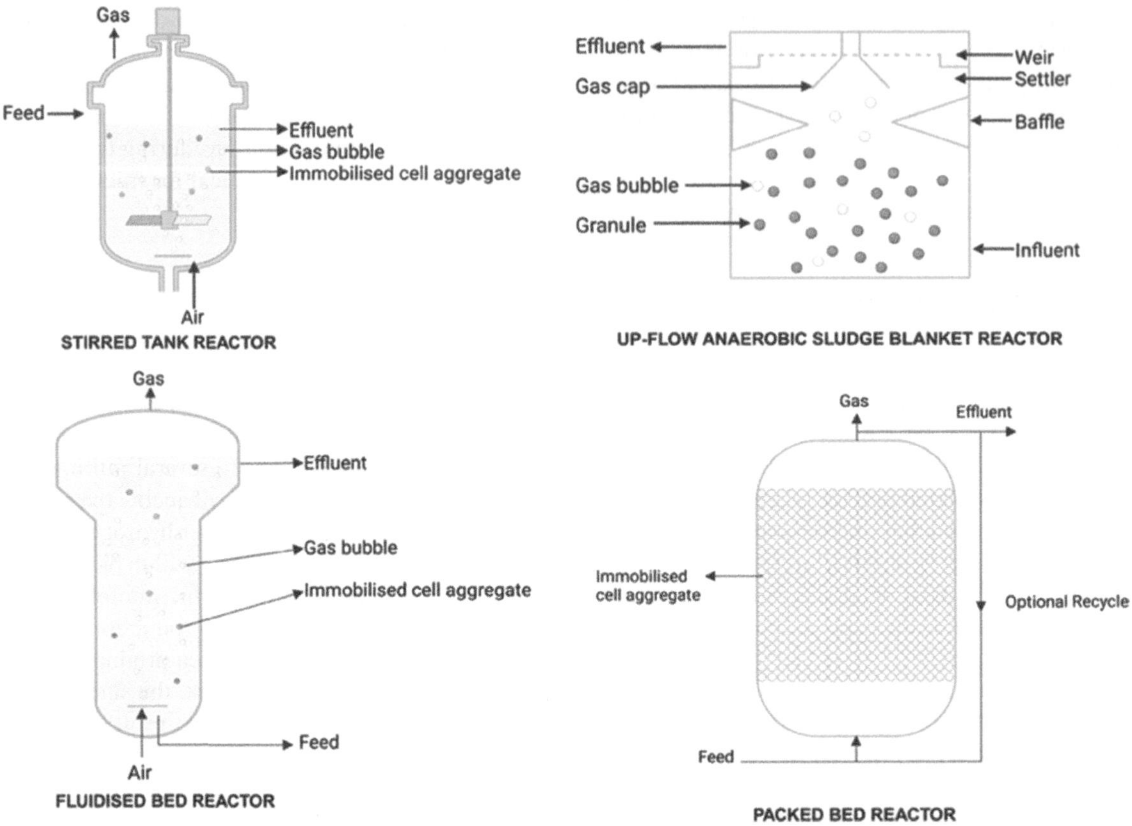

FIGURE 6.2 Bioreactor designs for immobilized cell systems. This figure shows four bioreactor configurations for immobilized cells. The Stirred Tank reactor uses mechanical stirring and air sparging to mix cells, nutrients, and gases. The Up-Flow Anaerobic Sludge Blanket (UASB) Reactor allows the influent to flow upward through a sludge blanket of immobilized cells, separating gas and effluent. The Fluidized bed reactor suspends immobilized cells using upward gas flow, improving mass transfer. The Packed bed reactor has cells immobilized in a packed matrix, with feed passing through for biotransformation and an optional recycle system for better efficiency.

The fermentation process for biohydrogen production can employ various reactor technologies, including packed bed reactors (PBRs), anaerobic fluidized bed reactors (AFBRs), continuously stirred tank reactors (CSTRs), and up-flow anaerobic sludge blanket (UASB) reactors. The design of bioreactor modes is influenced by the specific type and characteristics of the immobilized cells used. Typically, biohydrogen-producing bioreactors with immobilized cells operate in either batch or continuous modes. The efficiency and effectiveness of biohydrogen production are significantly affected by these operational modes, underscoring the importance of careful reactor design in enhancing overall hydrogen production processes.

6.5.1 Continuously Stirred Tank Reactor

CSTR remains one of the most commonly used bioreactors for biohydrogen production under continuous operation, primarily due to its uncomplicated structure, operational ease, efficient mixing, and precise regulation of parameters like temperature and pH. CSTRs support improved retention of microbial biomass, enhanced substrate consumption, and elevated hydrogen yields. Their user-friendly setup, combined with superior mixing and transport efficiency, contributes to their widespread preference among researchers working on continuous hydrogen generation. In these bioreactors, the microbial population responsible for biohydrogen production is fully suspended in the mixed liquor and continuously circulated throughout the reactor by the stirrer.

The performance of a hydrogen-producing reactor in continuous operation is highly dependent on the organic loading rate (OLR) and HRT. While high hydrogen production efficiencies can be achieved, challenges arise at specific lower HRTs due to the washout effect, where biomass is lost from the system. This issue can be mitigated by applying immobilized microbial cultures, which enable greater retention of active biomass within the reactor even under reduced retention durations, thereby enhancing hydrogen production efficiency. For instance, a granular sludge-based CSTR system was developed using anaerobic sludge immobilized in silicone to convert sucrose into hydrogen within a stirred anaerobic setup, achieving peak hydrogen productivity of 15 L/h/L and a yield of 3.5 mol of H_2/mol of sucrose (Wu et al., 2006). A doubling of the HPR per unit biomass was observed following the development of self-aggregated granular sludge, which was linked to shifts in the microbial consortium. More recently, thermally pretreated compost was effectively used as an immobilized inoculum for hydrogen production from soft drink processing wastewater, reaching a maximum hydrogen yield of 55 L/L/day, even under a shortened HRT of just 1.5 hours.

6.5.2 Anaerobic Fluidized Bed Bioreactor

In a fluidized bed bioreactor, immobilized cells are kept in motion by a continuous flow of feed solution. The anaerobic membrane bioreactor (AnMBR) technology combines the advantages of anaerobic digestion with membrane filtration. The membrane effectively retains colloidal and suspended solids, ensuring efficient solid-liquid separation. Within the bioreactor, slow-growing anaerobic microorganisms degrade soluble organic compounds, resulting in biogas production as a by-product. In this system, a flowing medium—whether gaseous or liquid—moves upward through a bed of particulate solid materials, usually serving as biofilm carriers, at sufficient speed to lift the particles and keep them suspended, allowing them to simulate fluid-like properties. This phenomenon, known as fluidization, offers several key advantages in fluidized bed reactors (FBRs). It boosts the efficiency of mass transfer between the immobilized microbial biomass and available nutrients.

Additional operational benefits associated with FBRs include enhanced substrate diffusion due to dynamic mixing, improved hydrogen output under reduced HRT, and elevated retention of microbial populations. The design also features low contamination rates since it does not rely on a shaft seal. Compared to PBRs, fluidized beds better handle the integration of a third, gaseous phase, promoting enhanced multiphase interaction that alleviates diffusion barriers and reduces the risk of inhibitory accumulation. The overall functionality of FBR systems relies on multiple factors, including granular density, settling velocities, fluid drag properties, minimum velocities required to initiate fluidization for different particle diameters, and the overall pattern of bed expansion. Selecting appropriate support material is critical; it should have a high density to withstand high superficial velocity conditions and prevent the washout of immobilized materials. Various support materials, including expanded clay, polyethylene terephthalate (PET), ethylene vinyl acetate copolymer, polystyrene, activated carbon, modified alginate, and ground tire particles, have been investigated for their effectiveness in hydrogen production (Patrick Thabang Sekoai, 2017).

Experiments have shown that an AFBR employing a blended matrix composed of sodium alginate, activated carbon, silicone, and acrylic latex reached an optimal hydrogen yield of 2.67 mol H_2/mol sucrose and a generation rate of 0.93 L/h/L (Wu et al., 2008). Additionally, when pretreated anaerobic sludge was immobilized onto expanded clay, the system achieved hydrogen productivity of 1.21 L/h/L with a yield of 2.59 mol H_2/mol glucose. The study demonstrated that expanded clay is effective in enhancing microbial adhesion, resulting in greater biomass accumulation within the system. However, fluidized bed bioreactors can experience unstable bed expansion due to changes in bead density over time, and axial back-mixing can occur, which may reduce efficiency, particularly in cases of product inhibition (Barros et al., 2010).

6.5.3 Packed Bed Reactor

The packed bed bioreactor is designed as a vertical column structure that contains a bed populated with immobilized microbial cultures, through which the nutrient-rich substrate stream is passed. This system operates using a plug-flow pattern, minimizing axial dispersion and promoting uniform flow. Its columnar design allows for easy scalability. However, PBRs operate under conditions of reduced hydraulic turbulence, leading to mass transfer limitations for microbial cultures. Consequently, this results in decreased substrate conversion rates and hydrogen production. Lower hydrogen output and efficiency observed in packed bed configurations are mainly linked to the emergence of pH stratification and uneven microbial distribution. Despite efforts to control the effluent pH above 6.0, consistently high molar yields of hydrogen (H_2) could not be maintained. This inconsistency was attributed to the development of pH gradients throughout the length of the reactor column, which caused non-uniform microbial performance across different regions. To address issues related to mass transfer resistance and pH heterogeneity, the use of fluidized bed or expanded bed reactor systems with recirculation flow has been recommended as a more effective strategy for enhancing HPRs and yields (Rachman et al., 1998).

One advantage of PBRs is their greater resistance to biomass washout compared to CSTRs. The choice of support materials significantly influences biomass retention and, consequently, hydrogen production in fixed-bed reactors. These support materials are like those used in FBRs and include expanded clay, charcoal, porous ceramic, low-density polyethylene, glass beads, Raschig rings, and various biological materials such as bamboo stems, rice straw, bagasse, and coir. Kumar and Das (2001) investigated how various supporting media affected the immobilization of *Enterobacter cloacae* IIT-BT 08 within packed bed systems. Their results indicated that coir, because of its fibrous and grooved surface, provided increased surface availability and emerged as the most effective immobilization substrate in comparison to bagasse and rice straw. Coir yielded the best results in parameters like biomass retention (0.44 g dry cells/g of dry carrier), volumetric packing (100 g/L of reactor space), biomass loading (44 g dry cells per reactor volume), and hydrogen generation rate (62 mmol/L/h). Further evaluations of fixed-bed designs with rhomboid, tapered, or tubular geometries, running under a HRT of 1.08 hours, showed that the rhomboid bioreactor attained the top hydrogen productivity of 1.60 L/h/L, outperforming the tapered design (1.46 L/h/L) and the tubular configuration (1.40 L/h/L). Biohydrogen output was also enhanced using high-temperature-adapted strains immobilized on Raschig rings within packed bed setups (Pekguzel et al., 2016), producing hydrogen at rates 7–11 times greater than those achieved with freely suspended cells. In another investigation, *Clostridium tyrobutyricum* JM1 was fixed onto polyurethane foam to enable continuous hydrogen generation in a packed bed configuration using glucose as the main organic feedstock. This configuration attained a peak hydrogen output of 7.2 L H_2/L/day at an HRT of 2 hours, along with a glucose-to-product conversion rate of 97.4% (Jo et al., 2008).

6.5.4 Up-flow Anaerobic Sludge Blanket Reactor

Recent studies have indicated that the UASB configuration represents a highly effective bioreactor platform for hydrogen generation. The UASB reactor system has been widely adopted for hydrogen production because of its ability to sustain dense microbial populations and maintain robust treatment performance. The operation of a UASB reactor is based on its granular sludge bed, which becomes expanded as wastewater flows vertically upward through it. In this process, wastewater ascends through a dense biomass layer, where anaerobic microbes break down the organic compounds. The upward movement, coupled with gravitational settling, maintains the structure of the sludge bed, with flocculants aiding in particle aggregation. Various carrier media applied in these reactors include ceramic rings, expanded clay, alginate gels, activated carbon, stones, and ceramic balls. The UASB reactor provides multiple advantages over traditional PBR systems, including its natural sludge aggregation, resilience under challenging operational conditions, increased tolerance to higher organic loading, efficient nutrient diffusion, and improved gas output.

Nonetheless, UASB reactors encounter operational drawbacks such as lengthy startup durations, prolonged periods required for developing a stable microbial bed, and difficulties in sustaining strict anaerobic conditions. In one investigation, *Clostridium* sp. LS2 was embedded in PEG and employed to produce biohydrogen from POME using a UASB configuration (Singh et al., 2013). This setup resulted in a hydrogen generation rate of 336 mL H_2/L/h and a hydrogen yield of 0.35 L H_2/g of COD removed. Another study employed ceramic spheres as carrier media for biohydrogen production in both UASB and CSTR configurations. The UASB system outperformed its counterpart, delivering a hydrogen generation rate of 2.7 L H_2/L/day at a retention period of 3 hours. The immobilized system showed a greater ability to resist biomass washout than the suspended cell approach. As a result, it could function efficiently even at a considerably reduced hydraulic retention period, achieving significantly elevated hydrogen productivity. The influence of substrate loading on hydrogen output was also studied using a UASB bioreactor filled with granular sludge. A high hydrogen yield of 1.64 mol H_2/mol of glucose was achieved at a loading rate of 12.8 g COD/L/day. When anaerobic sludge immobilized on CNTs was used for hydrogen generation in a UASB setup, the system achieved an impressive production rate of 5.55 L/L/day and a yield of 2.45 mol H_2/mol glucose (Liu et al., 2012).

The exploration of biohydrogen production using various types of immobilized cell bioreactors has demonstrated significant advancements in optimizing yield and efficiency. The comparative analysis of different immobilization techniques reveals that each method offers distinct advantages, with some enhancing cell stability and activity while others improving overall reactor performance and HPRs.

6.6 FUTURE PERSPECTIVES OF CELL IMMOBILIZATION FOR HYDROGEN PRODUCTION

The future perspectives in cell immobilization for hydrogen production are promising, driven by ongoing advancements in biotechnology, materials science, and process optimization. Several key areas highlight the potential improvements and innovations that could greatly enhance the effectiveness, scalability, and sustainability of hydrogen production through immobilized cells. One of the most important future directions is the development of advanced immobilization materials. Current immobilization techniques often use natural polymers like alginate or synthetic polymers, but future innovations could involve the use of nanomaterials, smart polymers, or biocompatible hydrogels. These advanced materials could improve the mechanical strength, mass transfer, and durability of immobilized systems. Additionally, novel materials could enhance cell viability and protect the microorganisms from environmental stress, extending their operational lifespan in hydrogen production reactors.

Optimization of immobilization techniques is another critical area for future research. Innovative methods such as microencapsulation, layered coatings, and biofilm-based immobilization may provide greater control over the microenvironment of the cells. These methods could improve substrate diffusion and metabolite removal, leading to higher hydrogen yields. Future work may also focus on creating immobilization systems that can better withstand industrial conditions such as fluctuating pH, temperature, and substrate concentrations, making them more practical for large-scale biohydrogen production. The integration of multi-microbial systems into immobilized matrices is another exciting avenue for future exploration. Different microbial species can have complementary metabolic pathways, and by co-immobilizing them, the entire metabolic process could become more efficient. For example, combining fermentative bacteria with photosynthetic bacteria could optimize hydrogen production under various environmental conditions, including dark and light cycles. This strategy would enhance substrate utilization and increase overall hydrogen yields by leveraging synergies between microbial species.

Scalability and reactor design will be critical in the future of cell immobilization for hydrogen production. As the technology matures, large-scale bioreactors that support immobilized systems must be designed to maximize efficiency. For example, membrane bioreactors, fluidized bed systems, and packed bed configurations are some setups that could be explored. These designs need to optimize factors such as gas-liquid mass transfer, retention time, and system stability while ensuring that immobilized cells maintain high productivity over time. In the future, the development of reactors tailored specifically for immobilized cell systems will be essential for the commercialization of biohydrogen production. A major future direction is the application of genetic engineering to improve the performance of immobilized cells. Microorganisms could be genetically modified to enhance hydrogen production efficiency, increase tolerance to stressors, and improve substrate utilization. These engineered cells, when immobilized, could operate more efficiently and consistently, even under suboptimal conditions. Additionally, synthetic biology approaches could be used to engineer novel microbial consortia that work synergistically to enhance hydrogen production, particularly in immobilized systems where cell-to-cell communication and substrate exchange can be optimized.

The future of cell immobilization for hydrogen production will likely involve greater integration with renewable and waste-based feedstocks. Immobilized cells offer the advantage of being highly adaptable to various substrates, including lignocellulosic biomass and organic waste streams. As interest grows in the circular economy and waste-to-energy processes, the use of immobilized cells to convert agricultural, industrial, or municipal waste into hydrogen will become increasingly viable. Future research may focus on improving the efficiency of immobilized systems to degrade complex feedstocks and convert them into hydrogen. In conclusion, the future of cell immobilization for hydrogen production is bright, with significant potential for innovations in materials, reactor designs, and microbial engineering. These advancements will play a key role in making biohydrogen production more efficient, scalable, and sustainable, ultimately supporting the worldwide shift to clean and renewable energy alternatives.

6.7 CONCLUSION

Biohydrogen, a clean and renewable energy source, faces challenges in large-scale production due to low conversion efficiencies and unwanted side reactions. Cell immobilization technology addresses these limitations by enhancing stability, promoting reusability, and improving overall biohydrogen yields. Techniques such as entrapment, encapsulation, and adsorption improve cell retention, reduce washout, and enhance tolerance to harsh fermentation conditions, ultimately increasing hydrogen production efficiency. Looking ahead, the integration of nanotechnology and research on self-healing and self-regenerating immobilization matrices holds promise for cost-effective and sustainable biohydrogen production processes. Cell immobilization is a transformative technology poised to advance biohydrogen production and unlock its full potential as an eco-friendly and renewable energy solution.

REFERENCES

Abraham, T.E. (2003). Studies on chromium(VI) adsorption-desorption using immobilized fungal biomass. *Bioresource Technology*, *87*(1), 17–26. https://doi.org/10.1016/S0960-8524(02)00222-5

Akinbomi, J., Wikandari, R., & Taherzadeh, M.J. (2015). Enhanced fermentative hydrogen and methane production from an inhibitory fruit-flavored medium with membrane-encapsulated cells. *Membranes*, *5*(4), 616–631. https://doi.org/10.3390/membranes5040616

Andreani, C., et al. (2015). Biohydrogen production: pH influence on fermentative bacteria. *Renewable Energy*, 75, 18–23.

Barros, A.R., Cavalcante de Amorim, E.L., Reis, C.M., Shida, G.M., & Silva, E.L. (2010). Biohydrogen production in anaerobic fluidized bed reactors: Effect of support material and hydraulic retention time. *International Journal of Hydrogen Energy*, *35*(8), 3379–3388. https://doi.org/10.1016/j.ijhydene.2010.01.108

Beckers, F., et al. (2015). Mass transfer coefficient correlations in biohydrogen production reactors. *Chemical Engineering Science*, 137, 226–234.

Chandrasekhar, K., Lee, Y.J., & Lee, D.W. (2015). Biohydrogen production: Strategies to improve process efficiency through microbial routes. *International Journal of Molecular Sciences*, *16*, 8266–8293. https://doi.org/10.3390/ijms16048266

Chang, R.H.Y., Jang, J., & Wu, K.C.W. (2011). Cellulase immobilized mesoporous silica nanocatalysts for efficient cellulose-to-glucose conversion. *Green Chemistry*, *13*(10), 2844–2850. https://doi.org/10.1039/C1GC15563F

Cheetham, P.S.J., Blunt, K.W., & Bucke, C. (1979). Physical studies on cell immobilization using calcium alginate gels. *Biotechnology and Bioengineering*, 21(12), 2155–2168.

Chu, J., Lin, C., & Xu, K. (2011). Influence of pH on biohydrogen production from glucose by anaerobic mixed cultures. *Bioresource Technology*, 102(3), 2203–2209.

Datta, T., Ahmed, A., Jha, S.K., Srivastava, R., Singh, P., & Naik, B. (2023). Enhancement of biohydrogen production by the application of microbial nanotechnology. In: *Bionanotechnology towards Green Energy* (pp. 275–294). CRC Press.

Demirbas, A. (2009). Political, economic and environmental impacts of biofuels: A review. *Applied Energy*, 86(Supplement 1), S108–S117. https://doi.org/10.1016/j.apenergy.2009.04.036

Gadhe, A., Sonawane, S.S., & Varma, M.N. (2015). Enhancement effect of hematite and nickel nanoparticles on biohydrogen production from dairy wastewater. *International Journal of Hydrogen Energy*, *40*(13), 4502–4511. https://doi.org/10.1016/j.ijhydene.2015.01.151

Han, S., Liu, J., & Wang, Y. (2011). Effects of carrier materials on biohydrogen production using anaerobic fermentation. *International Journal of Hydrogen Energy*, 36(7), 4674–4680.

Hirayama, O., Uya, K., Hiramatsu, Y., Yamada, H., & Moriwaki, K. (1986). Photoproduction of hydrogen by immobilized cells of a photosynthetic bacterium, *Rhodospirillum rubrum G-9 BM. Agricultural and Biological Chemistry*, *50*(4), 891–897.

Ishikawa, T., et al. (2008). Effect of pH on hydrogen production by thermophilic bacteria. *Journal of Bioscience and Bioengineering*, 106(3), 306–309.

Ismail, N., et al. (2011). Biohydrogen production by mixed anaerobic cultures from palm oil mill effluent sludge. *Bioresource Technology*, 102(11), 6036–6043.

Jo, J.H., Lee, D.S., Park, D., & Park, J.M. (2008). Biological hydrogen production by immobilized cells of *Clostridium tyrobutyricum* JM1 isolated from a food waste treatment process. *Bioresource Technology*, *99*(14), 6666–6672. https://doi.org/10.1016/j.biortech.2007.11.067

Keskin, T., Giusti, L., & Azbar, N. (2012). Continuous biohydrogen production in immobilized biofilm system versus suspended cell culture. *International Journal of Hydrogen Energy*, *37*(5), 1418–1424. https://doi.org/10.1016/j.ijhydene.2011.10.013

Kirli & Kapdan, 2016: Kirli, S., & Kapdan, I. K. (2016). Carrier materials effect on hydrogen production by immobilized anaerobic bacteria. *Renewable Energy*, 86, 1284–1290.

Kobayashi, H., Hosaka, Y., Hara, K., Feng, B., Hirosaki, Y., & Fukuoka, A. (2014). Control of selectivity, activity, and durability of simple supported nickel catalysts for hydrolytic hydrogenation of cellulose. *Green Chemistry*, *16*(2), 637–644. https://doi.org/10.1039/C3GC41357H

Kumar, G., Mudhoo, A., Sivagurunathan, P., Nagarajan, D., Ghimre, A., Lay, C.H., Lin, C.Y., Lee, D.J., & Chang, J.S. (2016). Recent insights into the cell immobilization technology applied for dark fermentative hydrogen production. *Bioresource Technology*, *219*, 725–737. https://doi.org/10.1016/j.biortech.2016.08.065

Kumar, N., & Das, D. (2001). Continuous hydrogen production by immobilized *Enterobacter cloacae* IIT-BT 08 using lignocellulosic materials as solid matrices. *Enzyme and Microbial Technology*, *29*(*4–5*), 280–287. https://doi.org/10.1016/S0141-0229(01)00394-5

Lin, C. Y., et al. (2009). Biohydrogen production using mixed cultures of anaerobic bacteria. *International Journal of Hydrogen Energy*, 34(4), 1366–1375.

Liu, Z., Lv, F., Zheng, H., Zhang, C., Wei, F., & Xing, X.H. (2012). Enhanced hydrogen production in a UASB reactor by retaining microbial consortium onto carbon nanotubes (CNTs). *International Journal of Hydrogen Energy*, *37*(14), 10619–10626. https://doi.org/10.1016/j.ijhydene.2012.04.057

Manieniyan, V., Thambidurai, M., & Selvakumar, R. (2009). Study on energy crisis and the future of fossil fuels. *Proceedings of SHEE*, *10*, 2234–3689.

Mrudula, S., & Shyam, N. (2012). Immobilization of *Bacillus megaterium* MTCC 2444 by Ca-alginate entrapment method for enhanced alkaline protease production. *Brazilian Archives of Biology and Technology*, 55, 135–144. https://doi.org/10.1590/S1516-89132012000100017

Mohd Yasin, N. H., et al. (2011). Optimization of pH for biohydrogen production from food waste by mixed anaerobic cultures. *International Journal of Hydrogen Energy*, 36(22), 14674–14680.

Nath, K., & Chand, S. (1996). Estimation of mass transfer coefficients in immobilized systems. *Biotechnology and Bioengineering*, 51(6), 579–584.

Osman, A.I., Mehta, N., Elgarahy, A.M., et al. (2022). Hydrogen production, storage, utilization, and environmental impacts: A review. *Environmental Chemistry Letters*, *20*, 153–188. https://doi.org/10.1007/s10311-021-01322-8

Osman, A.I., Deka, T.J., Baruah, D.C., et al. (2023). Critical challenges in biohydrogen production processes from organic feedstocks. *Biomass Conversion and Biorefinery*, *13*, 8383–8401. https://doi.org/10.1007/s13399-020-00965-x

Pandey, A.K., Park, J., & Ko, J., et al. (2023). Machine learning in fermentative biohydrogen production: Advantages, challenges, and applications. *Bioresource Technology*, *370*, 128502. https://doi.org/10.1016/j.biortech.2022.128502

Pachapur, V.V., et al. (2016). Co-culture system improves hydrogen production from glycerol. *International Journal of Hydrogen Energy*, 41(20), 8676–8684.

Pekguzel, E., Gundogdu, T., & Azbar, N. (2016). Enhancement of biohydrogen production via thermophilic cell culture immobilized on glass beads and Raschig rings of different sizes in a packed bed reactor. *Chemical and Biochemical Engineering Quarterly*, *29*(4), 541–547. https://doi.org/10.15255/CABEQ.2014.2082

Plangklang, P., et al. (2012). Biohydrogen production using immobilized Clostridium butyricum on sugarcane bagasse. *Renewable Energy*, 39(1), 232–237.

Rachman, M.A., Nakashimada, Y., Kakizono, T., & Nishio, N. (1998). Hydrogen production with high yield and high evolution rate by self-flocculated cells of *Enterobacter aerogenes* in a packed-bed reactor. *Applied Microbiology and Biotechnology*, *49*(4), 450–454. https://doi.org/10.1007/s002530051197

Rausch, B., Symes, M.D., & Chisholm, G., et al. (2014). Decoupled catalytic hydrogen evolution from a molecular metal oxide redox mediator in water splitting. *Science*, *345*(6202), 1326–1330. https://doi.org/10.1126/science.1257443

Rosa, L., et al. (2014). The impact of pH on hydrogen production by mixed cultures. *International Journal of Hydrogen Energy*, 39(13), 7169–7178.

Sekoai, P.T., Awosusi, A.A., Yoro, K.O., et al. (2017). Microbial cell immobilization in biohydrogen production: A short overview. *Critical Reviews in Biotechnology*. https://doi.org/10.1080/07388551.2017.1312274

Seol, E., et al. (2011). Hydrogen production from formate by immobilized Escherichia coli. *International Journal of Hydrogen Energy*, 36(15), 9212–9219.

Shen, J.W., Wu, T., Wang, Q., & Kang, Y. (2008). Induced stepwise conformational change of human serum albumin on carbon nanotube surfaces. *Biomaterials*, *29*(28), 3847–3855. https://doi.org/10.1016/j.biomaterials.2008.05.024

Sivagurunathan, P., Kumar, G., Bakonyi, P., et al. (2016). A critical review on issues and overcoming strategies for the enhancement of dark fermentative hydrogen production in continuous systems. *International Journal of Hydrogen Energy*, *41*(6), 3820–3836. https://doi.org/10.1016/j.ijhydene.2015.12.081

Shuler, M.L., & Kargi, F. (2002). *Bioprocess Engineering: Basic Concepts* (2nd ed.). Prentice Hall.

Singh, N., Dhanya, B.S., & Verma, M.L. (2020). Nano-immobilized biocatalysts and their potential biotechnological applications in bioenergy production. *Materials Science for Energy Technologies*, *3*, 808–824. https://doi.org/10.1016/j.mset.2020.09.006

Singh, L., Siddiqui, M.F., Ahmad, A., et al. (2013). Application of polyethylene glycol immobilized *Clostridium sp.* LS2 for continuous hydrogen production from palm oil mill effluent in upflow anaerobic sludge blanket reactor. *Biochemical Engineering Journal*, *70*, 158–165. https://doi.org/10.1016/j.bej.2012.10.010

Srivastava, N., Srivastava, M., Kushwaha, D., et al. (2017). Efficient dark fermentative hydrogen production from enzyme hydrolyzed rice straw by *Clostridium pasteurianum* (MTCC116). *Bioresource Technology*, *238*, 552–558. https://doi.org/10.1016/j.biortech.2017.04.035

Tripathi, P., & Mohanty, K. (2024). Fermentative biohydrogen production for sustainable energy. In: Bisaria, V. (eds) *Handbook of Biorefinery Research and Technology: Production of Biofuels and Biochemicals*. Springer.

Wada, M., Kato, J., & Chibata, I. (1979). A new immobilization of microbial cells. *European Journal of Applied Microbiology and Biotechnology*, 8(4), 241–247. https://doi.org/10.1007/BF00508788

Wang, Y., Maspoch, D., Zou, S., et al. (2006). Controlling the shape, orientation, and linkage of carbon nanotube features with nanoaffinity templates. *Proceedings of the National Academy of Sciences*, *103*(7), 2026–2031. https://doi.org/10.1073/pnas.0511022103

Woodward, J.; Zachry, G. S. (1982). Immobilization of cellulase through its carbohydrate side chains - a rationale for its recovery and reuse. *Enzyme and Microbial Technology*, 4(4), 245–248.

Wu, S.Y., Hung, C.H., Lin, C.N., et al. (2006). Fermentative hydrogen production and bacterial community structure in high-rate anaerobic bioreactors containing silicone-immobilized and self-flocculated sludge. *Biotechnology and Bioengineering*, *93*(5), 934–946. https://doi.org/10.1002/bit.20800

Wu, S.Y., Lin, C.N., & Chang, J.S. (2008). Hydrogen production with immobilized sewage sludge in three-phase fluidized-bed bioreactors. *Biotechnology Progress*, *19*, 828–832. https://doi.org/10.1021/bp0201354

Wu, T.-Y., Mohammad, A.W., & Ng, D.K.S. (2002). Effects of pH on hydrogen production from glucose by anaerobic mixed cultures. *Journal of Chemical Technology & Biotechnology*, 77(9), 1098–1105.

Zangi, A.R., Amiri, A., Borzouee, F., et al. (2023). Immobilized nanoparticles-mediated enzyme therapy; Promising way into clinical development. *Discover Nano*, *18*(1), 55. https://doi.org/10.1186/s11671-023-03823-7

Zhang, Z., Li, Y., Zhang, H., et al. (2017). Potential use and energy conversion efficiency analysis of fermentation effluents from photo and dark fermentative bio-hydrogen production. *Bioresource Technology*, *245*(A), 884–889. https://doi.org/10.1016/j.biortech.2017.08.110

Zhao, Y.-F., Wang, D., Song, Y.-J., & Xu, Q. (2010). Effect of pH on biohydrogen production by anaerobic fermentation from food waste. *Bioresource Technology*, 101(2), 444–449.

7 Recent Advances in Photo-Fermentative Biohydrogen Production

Substrate Pretreatment, Strain Improvement, Bioreactor Optimization

Senthilkumar Nangan, Angelaalincy Maria Joseph, and Kanagaraj Kuppusamy

7.1 INTRODUCTION

As the world grapples with the escalating consequences of climate change and the depleting reserves of fossil fuels, the quest for renewable and sustainable energy solutions has never been more urgent (Krane, 2017). Among the plethora of renewable energy technologies, biohydrogen (BioH) production through photo-fermentation represents a particularly promising frontier (Melitos et al., 2021). This biological approach not only promises to harness the power of the sun but also offers a way to convert organic waste into valuable energy, embodying the principles of a circular economy (Unuofin et al., 2023). Photo-fermentation leverages the unique abilities of photosynthetic bacteria (PSB) to convert organic substrates directly into hydrogen using sunlight as the energy source (Melis and Melnicki, 2006). This process offers dual benefits: it helps in waste reduction while producing hydrogen, a clean and high-energy fuel. However, despite its potential, the technology faces several challenges that hinder its application on a commercial scale (Aryamol et al., 2024; Zou et al., 2022). These challenges revolve primarily around the efficiency of substrate conversion, the robustness of microbial strains, and the overall effectiveness of the bioreactor (BioR) employed.

DOI: 10.1201/9781003568612-7

This chapter explores the latest advancements in the field of photo-fermentative BioH production with a particular focus on three critical aspects: substrate pretreatment, microbial strain improvement, and BioR design optimization (Buschke et al., 2013). Each of these areas holds the key to unlocking the technology's potential and is currently the subject of intense research and innovation. The introduction of advanced membrane-based technologies has revolutionized substrate pretreatment processes (Fane et al., 2011). By employing methods such as nanofiltration and reverse osmosis, researchers have been able to concentrate organic substrates and eliminate inhibitors that prevent bacterial activity, thus enhancing the bioavailability of the substrate for microbial action (Nguyen et al., 2020; Olaniran et al., 2013). These advancements are crucial as they directly influence the efficiency and yield of hydrogen production, paving the way for more effective and scalable systems.

Parallel efforts in the field of genetic and metabolic engineering have led to significant improvements in microbial strains (Nielsen, 2001). By manipulating the genetic makeup of PSB, such as *Rhodopseudomonas palustris* and *Rhodobacter sphaeroides*, scientists have enhanced these organisms' natural ability to produce hydrogen (Wang et al., 2016). These engineered strains can thrive under a variety of conditions and show increased resistance to environmental stresses, which are often amplified in membrane-integrated systems. Moreover, the optimization of BioR designs has seen innovative configurations that integrate membranes for *in situ* hydrogen separation and purification (Bakonyi et al., 2018; MariaJoseph et al., 2024; Thirumalaivasan et al., 2024d). These designs are tailored to maximize light exposure, optimize microbial interaction with substrates, and efficiently separate hydrogen as it is produced, thereby enhancing the overall productivity of the system (Nirmala et al., 2023).

This chapter will delve deeply into each of these areas, examining the current state of research, the latest technological innovations, and the potential for future development (Figure 7.1). By addressing the technological and economic challenges and showcasing the latest scientific advancements, this chapter

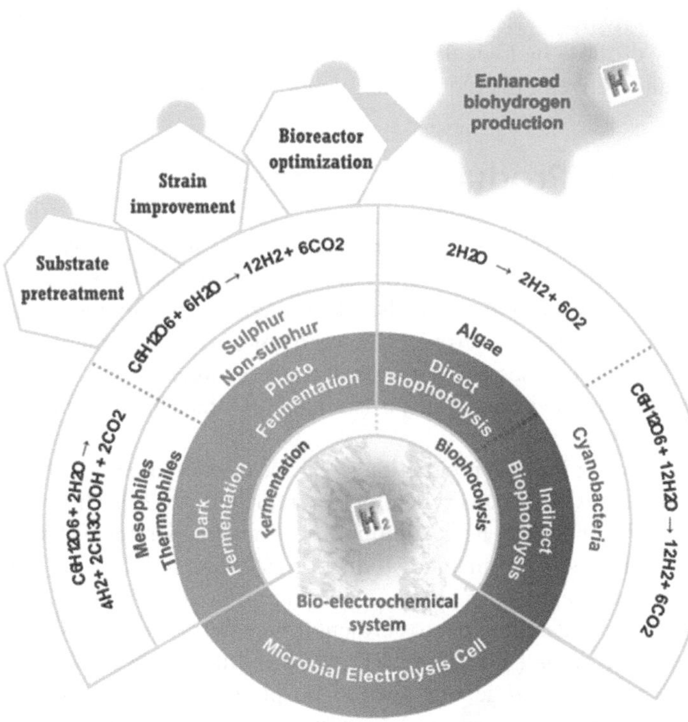

FIGURE 7.1 Overview of the biohydrogen production pathways and optimization strategies covered in this chapter. The figure highlights the roles of different fermentation processes, biophotolysis, and microbial electrolysis in enhancing hydrogen production, supported by strain improvement, substrate pretreatment, and bioreactor optimization techniques.

aims to paint a comprehensive picture of the potential of photo-fermentative BioH production as a key player in the future renewable energy landscape.

7.2 BIOH – FERMENTATION-BASED PRODUCTION TECHNOLOGIES

Hydrogen is increasingly recognized as a clean energy carrier, comparable to electricity in its versatility and environmental benefits. However, due to its natural occurrence in combination with other elements, hydrogen must be produced from various resources, including fossil fuels, biomass, and water (Balat and Kırtay, 2010). The environmental impact and energy efficiency of hydrogen production depend on the methods used (Dincer and Acar, 2015). Among the most advanced technologies are water electrolysis, thermo-chemical processes such as natural gas reforming, and biological hydrogen production (Li et al., 2020; Nair et al., 2023; Nangan et al., 2023b).

In recent years, biological hydrogen production has attracted significant attention due to its potential for cost-effective and environmentally friendly processes. Biological methods can be conducted at ambient temperatures and pressures, which minimizes energy consumption (Gogate and Pandit, 2004). However, the primary challenge lies in overcoming low hydrogen yields (HYs) and production rates, which hinder the commercialization of these technologies (Ishaq et al., 2022).

BioH production can be classified into several fermentation-based methods, including biophotolysis (direct and indirect), photo-fermentation, and dark fermentation. Biophotolysis and photo-fermentation require light, while dark fermentation is independent of light (Nagarajan et al., 2017). These processes rely on specific enzymes – hydrogenases or nitrogenases – for hydrogen evolution. Despite their potential, improving the efficiency of these biological processes remains essential for scaling up BioH production and making it a viable alternative in the global energy market (Srirangan et al., 2012).

7.2.1 Types of BioH Production Process

BioH and biogas (BioG) are two vital biofuels offering cleaner alternatives to hydrogen production derived from fossil fuels, aligning with global sustainability objectives. BioH production plays a crucial role in the bio-economy by utilizing low-cost biomass feedstock through various methods such as dark fermentation, light fermentation, biophotolysis, microbial electrochemical cells, and integrated co-fermentation (as shown in Figure 7.2). These processes emphasize renewable sources, aiming to make BioH production more economically viable and environmentally sustainable (Oumer et al., 2018; Shellaiah et al., 2024; Thirumalaivasan et al., 2024c).

An important advancement in BioH production is the use of microalgae, which leverage stored glycogen and address oxygen sensitivity issues in the process (Jiao et al., 2024). This method has the potential to significantly improve HYs and efficiency. Additionally, promising strategies like microbial electrolysis and biowaste gasification offer ways to enhance production rates and reduce costs (Chaurasia and Mondal, 2021). To establish BioH as a competitive and sustainable energy source, future advancements may involve combining technologies such as dark and photo-fermentation, microbial electrolysis, and thermochemical processes like biowaste gasification (Aziz et al., 2021; Naha et al., 2020; Thirumalaivasan et al., 2024b). The involvement of specific bacterial genera, such as purple non-sulfur (PNS) bacteria, further underscores the potential of BioH biological approaches in efficient BioH production (Thirumalaivasan et al., 2024e). In conclusion, BioH production technologies are essential for a transition to cleaner energy systems. Continued innovation in both biological and thermochemical methods will be key to making BioH a viable and impactful alternative to conventional hydrogen sources (Figure 7.2).

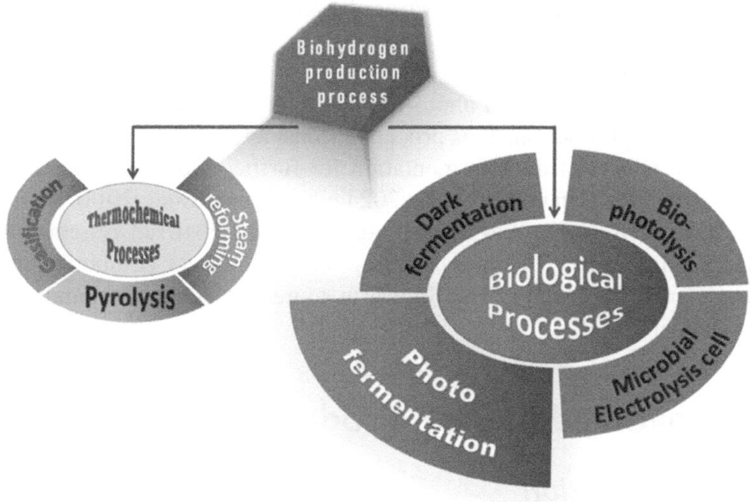

FIGURE 7.2 Overview of the selected BioH production technologies. The figure illustrates two main categories of biohydrogen production processes: thermochemical processes and biological processes.

BioH production using biowaste is a promising area of research due to its potential for sustainable energy generation. Two main processes, thermochemical and biological, are commonly employed for BioH production (Arregi et al., 2018). Thermochemical processes such as pyrolysis, gasification, and steam reforming involve high-temperature treatments, while biological processes utilize microorganisms to convert organic materials into hydrogen (Kumar et al., 2020a). Both dark fermentation and photo-fermentation are widely researched biological methods, where biowaste serves as a feedstock for microbial action to generate hydrogen (Jain et al., 2024).

Dark fermentation, illustrated in Figure 7.3a, is an anaerobic process in which biowaste is broken down by microorganisms in a continuous stirred tank bioreactor (CSTR). The process is efficient but

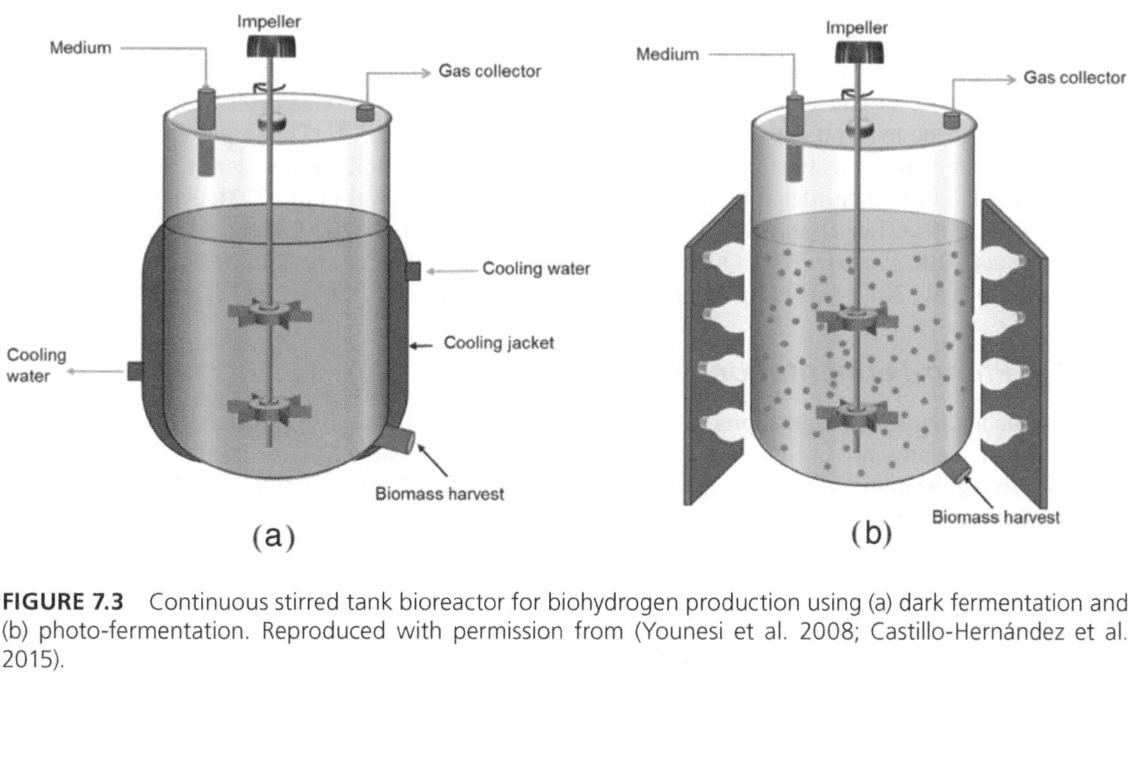

FIGURE 7.3 Continuous stirred tank bioreactor for biohydrogen production using (a) dark fermentation and (b) photo-fermentation. Reproduced with permission from (Younesi et al. 2008; Castillo-Hernández et al. 2015).

yields a relatively low amount of hydrogen. However, the scalability and simplicity of the method make it a focus of industrial research.

Photo-fermentation, shown in Figure 7.3b, involves the use of PSB to convert biowaste into hydrogen under light conditions. The CSTR used in photo-fermentation typically includes light panels around the reactor, enhancing hydrogen production. This process is considered complementary to dark fermentation, as it can utilize the by-products of dark fermentation for further hydrogen generation, thus improving overall efficiency (Thirumalaivasan et al., 2024a, 2024e; Yang and Wang, 2018).

Both processes have been explored extensively using magnetic stirred systems, as mentioned in the work of Castillo-Hernández et al. (2015), highlighting the potential for combining thermochemical and biological processes to increase the HY from biowaste sources (Castillo-Hernández et al., 2015; Ismail et al., 2008; Karmann et al., 2019).

7.2.2 Process Involved in BioH Production

7.2.2.1 Photo-Fermentation

Photo-fermentative BioH production is carried out by anoxygenic bacteria, primarily through the nitrogenase enzyme. This enzyme plays a key role in converting organic carbon substrates into BioH and carbon dioxide, while also fixing molecular nitrogen (LaSarre et al., 2017). The process is mainly facilitated by Gram-negative purple bacteria, particularly from the class Alpha-proteobacteria. Within this group, common BioH-producing species include *R. sphaeroides*, *Rhodobacter capsulatus*, *R. palustris*, and *Rhodospirillum rubrum* (Koku et al., 2003; Larimer et al., 2004b).

In laboratory-scale experiments, a wide range of organic carbon substrates, especially organic acids and specific sugars, have been explored for BioH production by PNS bacteria (Tiang et al., 2020). However, certain species of these bacteria lack the necessary transport proteins for metabolizing sugars such as glucose and fructose (Jeckelmann and Erni, 2020; Manjunathan et al., 2023; Ramasubburayan et al., 2024a). A promising approach is the use of effluents rich in acetic and butyric acids from dark fermentation, which can be further utilized in photo-fermentation to enhance BioH yields (Rai and Singh, 2016; Teke et al., 2024).

Considering the environmental and economic challenges associated with pure substrates, recent research has shifted toward using waste streams, such as food waste, crop residues, or wastewater, for BioH production. Studies have shown that photo-fermentative bacteria can achieve BioH production rates comparable to those of conventional substrates when using waste materials, promoting the concept of a circular bio-economy (Potts et al., 2018; Uys, 2019; Wang et al., 2017). This highlights the potential for utilizing commercial waste streams as a sustainable source for BioH production (Chen et al., 2020; Connolly et al., 2014; Demirdelen et al., 2020; Du et al., 2018; Jaiswal et al., 2021; Klotz et al., 2006; Larimer et al., 2004a; Lee et al., 2011; Li et al., 2018; Marvuglia et al., 2020; Mechold et al., 1996; Patil and Gawande, 2016; Pinjari, 2023; Quehenberger et al., 2017; Rabha et al., 2017; Rai and Singh, 2016; Sabaté Brescó et al., 2017; Saifuddin and Priatharsini, 2016; Shen et al., 2018; Xiao et al., 2017).

7.2.2.2 Photo-Fermentation

Photo-fermentation is a biological process where PSB convert various substrates into hydrogen and carbon dioxide in the presence of light. Sunlight serves as the primary energy source for this light-dependent process, driving the conversion of organic compounds into BioH. The production of molecular hydrogen in photo-fermentation occurs through the nitrogenase enzyme, which catalyzes organic compounds under nitrogen-limiting conditions (Androga et al., 2012; Manjunathan et al., 2023).

In this process, PSB, such as *Rhodobacter* species, utilize organic acids like acetic acid as electron donors. These electrons are transferred to nitrogenase by ferredoxin, with the energy supplied by adenosine triphosphate (ATP). Additional ATP is required to drive the reduction of protons to hydrogen in the absence of nitrogen. This efficient use of organic acids and the nitrogenase enzyme makes photo-fermentation an attractive method for BioH production under suitable environmental conditions. The overall reactions of hydrogen production are reported below:

$$C_6H_{12}O_6 + 6H_2O + \text{light energy} \rightarrow 12H_2 + 6CO_2, \Delta G0 = +3.2\ \text{kJ} \quad (7.1)$$

The conversion of acetic acid into hydrogen and CO_2 through a nitrogen-deficient fermentative process occurs as in Equation 7.2:

$$CH_3COOH + 2H_2O + \text{light} \rightarrow 4H_2 + 2CO_2 \quad (7.2)$$

The organic acid can be oxidized with sunlight as the energy source, utilizing the tricarboxylic acid (TCA) cycle (Manish and Banerjee, 2008; Mudhoo et al., 2011).

Various substrates, including glucose, sucrose, succinate, and volatile fatty acids (VFAs), can be utilized as carbon sources by PNS bacteria to generate hydrogen gas. Common PNS bacteria used in photo-fermentative hydrogen production include *R. sphaeroides* O.U001, *R. capsulatus*, *R. sphaeroides*-RV, *Rhodobacter sulfidophilus*, *R. palustris*, and *R. rubrum* (Shellaiah et al., 2022; Sørensen and Mortensen, 2005).

One promising method involves using PNS PSB in fermentative processes to fully convert substrates into hydrogen and carbon dioxide. These bacteria generate ATP and high-energy electrons in anaerobic conditions, and nitrogenase facilitates hydrogen production by reducing protons with the help of ATP and reduced ferredoxin. Unlike cyanobacteria or green algae, these bacteria cannot split water during photolysis, so they rely on organic substrates (such as organic acids) as electron donors (Levin et al., 2004). Despite the potential benefits of this process, challenges remain, such as the high energy requirements of nitrogenase, low light conversion efficiency, and the costly hydrogen separation processes. Ongoing research aims to overcome these limitations (Hallenbeck and Ghosh, 2009; Rokhina et al., 2009).

7.2.2.3 Dark Fermentation

Dark fermentation is a highly effective process for BioH production through biomass conversion, offering a higher hydrogen production rate (HPR) and simpler operational requirements compared to photo-fermentation. This process involves a catabolic reaction facilitated by anaerobic bacteria, which does not require direct solar energy input (Logan et al., 2008; Łukajtis et al., 2018). In dark fermentation, substrates like carbohydrates and proteins are metabolized by bacteria in the absence of light, resulting in the formation of carboxylic acids, hydrogen gas, carbon dioxide, and organic solvents (Mohanraj et al., 2019). However, the accumulation of organic acids such as acetic and butyric acid presents a challenge, as it shifts metabolic pathways, leading to reduced HYs (Holladay et al., 2004; Velusamy et al., 2018).

To ensure optimal hydrogen production, strict anaerobic conditions must be maintained, as oxygen inhibits the process. Common bacteria used in dark fermentation include *Escherichia coli*, *Clostridium*, and *Enterobacter* species, which facilitate the conversion of organic substrates into hydrogen. Because maintaining strict anaerobic conditions is challenging, facultative anaerobic bacteria such as *E. coli* and *Enterobacter* are often preferred due to their lower sensitivity to dissolved oxygen and their ability to continue producing hydrogen even in the presence of trace amounts of oxygen (Shin et al., 2007; Yoshida et al., 2005). By-products such as acetate, butyrate, carbon dioxide, and organic solvents are also formed during this process. Various organic substrates and wastewater streams can be employed as electron donors in dark fermentation, making this method cost-effective compared to others.

The distribution of products in dark fermentation is influenced by several factors, including the oxidation state of the substrate, microbial species used, and environmental parameters such as pH, temperature, and hydrogen partial pressure (Hallenbeck and Ghosh, 2009; Kapdan and Kargi, 2006). Hydrogen production studies have been conducted across four temperature ranges: ambient (15°C–30°C), mesophilic

(30°C–39°C), thermophilic (50°C–64°C), and hyper-thermophilic (>65°C). For example, research has shown that *R. palustris* P4 and *Citrobacter* sp. Y1 can produce hydrogen from sugars over a broad pH (5–9) and temperature (25°C–40°C) range, with maximum HYs of 2.8 mmol-H_2/mmol-glucose and 2.5 mol-H_2/mol-glucose, respectively (Oh et al., 2004; Oh et al., 2002; Wei et al., 2017).

Moreover, hydrogen production is influenced by the partial pressure of hydrogen in the reactor. Mandal et al. demonstrated that lowering the total pressure in the headspace of the reactor from 760 to 380 mm Hg during batch fermentation with *Enterobacter cloacae* increased the molar yield of hydrogen from 1.9 to 3.9 mol-H_2/mol-glucose (Mandal et al., 2006). Therefore, achieving optimal hydrogen production depends on the specific microbial cultures and operational conditions used in the fermentation process.

In dark fermentation, pyruvate is the key intermediate produced during the catabolism of carbohydrate-rich substrates. Hydrogen is subsequently produced through one of two enzyme systems that catalyze pyruvate (Khan et al., 2016).

Pyruvate:

$$\text{formate lyase}(\text{PFL})\,\text{Pyruvate} + \text{CoA} \rightarrow \text{acetyl-CoA} + \text{formate}$$

Pyruvate:

$$\text{ferredoxin oxido reductase }(\text{PFOR})\ \text{Pyruvate} + \text{CoA} + 2\text{Fd}(\text{ox}) \rightarrow \text{acetyl-CoA} + \text{CO}_2 + 2\text{Fd}(\text{red}).$$

$$\text{acetyl-CoA+formate} \rightarrow \ (\text{FHL as catalyst})\,\text{CO}_2\text{+H}_2$$

The highest reported BioH production yield for strict anaerobic bacteria is 4 moles of hydrogen per mole of glucose, whereas facultative anaerobes typically produce around 2 moles of hydrogen per mole of glucose.

7.2.2.4 Biophotolysis

Biophotolysis is a hydrogen and carbon dioxide production process involving photoautotrophic organisms like microalgae and cyanobacteria under anaerobic conditions (Singer et al., 2018; Thirumalaivasan et al., 2019). These organisms use sunlight as an energy source and carbon dioxide as a carbon source to split water, producing BioH (Sverdrup et al., 2008). A key advantage of biophotolysis is its simplicity, as it doesn't require additional substrates, relying solely on water, sunlight, and carbon dioxide for growth and hydrogen production (Abd Rahman et al., 2018; Serov et al., 2015). Biophotolysis is further divided into two categories: direct biophotolysis and indirect biophotolysis.

7.2.2.5 Direct Biophotolysis

In direct biophotolysis, water is converted into hydrogen and chemical energy under light irradiation, as described in Equation 7.3 (Yu and Takahashi, 2007). This process closely resembles the mechanisms of photosynthesis in plants and algae, where light energy is utilized to drive the splitting of water molecules, producing hydrogen as a by-product (Thirumalaivasan et al., 2019; Türker et al., 2008). Direct biophotolysis is considered an efficient method of BioH production due to its reliance on simple inputs like water and light, mirroring natural processes that have evolved in photoautotrophic organisms for energy conversion.

$$2\text{H}_2\text{O} \rightarrow \ (h\gamma,\ \text{light catalysed reaction})\ 2\text{H}_2 + \text{O}_2 \tag{7.3}$$

In Equation 7.3, $h\gamma$ (light) denotes the energy of a photon (where h is Planck's constant and γ is the frequency of light). Photosystems I (PSI) and II (PSII), along with ferredoxin, participate in converting light into chemical energy as hydrogen (H_2) molecules. The absorbed light energy drives the linear transfer of electrons to ferredoxin. The mechanism of electron transfer through the hydrogenase enzyme is outlined in Equations 7.4 and 7.5, demonstrating how electrons are utilized in the production of hydrogen from water (Ghirardi et al., 2000):

$$2H_2O \rightarrow 4H^+ + 4e^- + O_2 \tag{7.4}$$

$$4H^+ + 4e^- \rightarrow 2H_2 \tag{7.5}$$

Hydrogenase catalyzes the conversion of protons (H_2) into hydrogen (H_2) by accepting electrons from reduced ferredoxin, facilitating BioH production through one-stage direct biophotolysis using water and solar energy in algae photosynthesis systems (Balamurugan et al., 2017; Rahman et al., 2015). A key limitation of this process is the oxygen sensitivity of the hydrogenase enzyme, which inhibits its activity and reduces HYs. To address this issue, a two-stage direct biophotolysis process has been proposed, where oxygen is removed via respiration using exogenous or endogenous substrates, such as sulfur deprivation. This method helps minimize oxygen's inhibitory effects. Despite these efforts, hydrogen production remains low, with reported yields around 0.07 mmol/L h, primarily due to the oxygen sensitivity of the enzymes involved in the process. This demonstrates the challenge of optimizing the biophotolysis process for efficient hydrogen production.

7.2.2.6 Indirect Biophotolysis

Hydrogen production through indirect biophotolysis occurs in two sequential stages. First, biomass is produced via photosynthesis, using carbohydrates as substrates. In the second stage, this carbohydrate-rich biomass undergoes fermentation, leading to hydrogen production (Larminie et al., 2003). This two-stage process effectively combines photosynthesis and fermentation for BioH generation from renewable biomass sources.

$$6H_2O + 6CO_2 + \text{light energy} \rightarrow C_6H_{12}O_6 + 6O_2 \tag{7.6}$$

$$C_6H_{12}O_6 + 2H_2O \rightarrow 4H_2 + 2CH_3COOH + 2CO_2 \tag{7.7}$$

Following these steps results in the temporal and/or spatial separation of oxygen and hydrogen evolutions, preventing enzyme deactivation and reducing the risk of an explosive gas mixture (Bélafi-Bakó et al., 2006; Podder et al., 2020). A simple separation method is essential for hydrogen purification, as CO_2 can be easily removed from the produced H_2/CO_2 mixture. Cyanobacteria species such as *Gloeobacter* sp., *Synechocystis* sp., and *Synechococcus* sp. are commonly used in the indirect biophotolysis process (Chu et al., 2020; Das and Veziroglu, 2008). *Anabaena* species and strains are frequently employed due to their high HPRs. In addition, mutant strains of *Anabaena variabilis* can also participate in the indirect biophotolysis process, achieving HPRs of approximately 0.355 mmol/h/L. These advancements underscore the potential of biophotolysis for sustainable hydrogen production, with cyanobacteria playing a significant role in improving efficiency.

7.3 ENHANCEMENT STRATEGIES FOR PHOTO-FERMENTATIVE BIOH PRODUCTION

Enhancement strategies for photo-fermentative BioH production focus on optimizing both biological and process parameters to increase HY, with a particular emphasis on PSB such as *R. sphaeroides*, *R. palustris*, and *R. capsulatus* (Li et al., 2024). These bacteria convert organic matter into hydrogen under light conditions, utilizing the TCA cycle to generate protons and electron donors while PSI absorbs light to drive the hydrogen production process (Antal et al., 2012; Thirumalaivasan et al., 2017; Wei et al., 2019). Key strategies include improving substrate availability through effective pretreatment methods, such as chemical or enzymatic hydrolysis, which break down complex biomass into fermentable sugars. Genetic engineering of photo-fermentative bacteria has been explored to enhance hydrogen production pathways, increase resistance to environmental stresses, and reduce by-product formation (Gupta et al., 2024; Rajapandi et al., 2023; Venkatesan et al., 2019). Additionally, optimizing environmental factors like light intensity, temperature, and pH significantly impacts microbial efficiency. Co-culturing with other microorganisms or integrating dark fermentation with photo-fermentation systems can further boost HYs by recycling waste products (Laxman Pachapur et al., 2015). Collectively, these strategies aim to overcome current limitations, making BioH production more sustainable and efficient (Figure 7.4).

7.3.1 Substrate Pretreatment

Lignocellulosic biomass and organic waste, often used as substrates in BioH production, have a complex and compact structure that limits the release of fermentable sugars (Arumugam et al., 2023; Bhatia et al., 2021; Natarajan et al., 2019). Pretreatment methods, including physical, chemical, biological, and combined approaches, are essential for breaking down these substrates to enhance sugar availability and improve hydrogen production. Effective substrate pretreatment increases bioavailability, enabling more efficient microbial conversion of organic matter into hydrogen (Shellaiah et al., 2019; Wang et al., 2021).

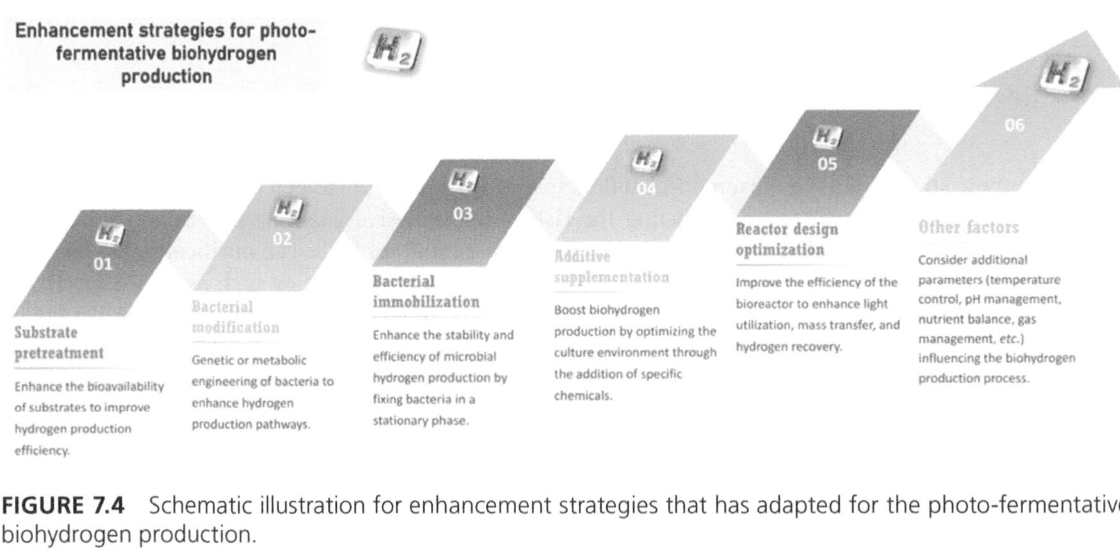

FIGURE 7.4 Schematic illustration for enhancement strategies that has adapted for the photo-fermentative biohydrogen production.

7.3.2 Bacterial Modification

The metabolic efficiency of PSB plays a central role in BioH production. Genetic engineering and strain improvement have emerged as critical strategies for increasing HY (Sivaramakrishnan et al., 2021). By modifying bacterial genomes, researchers aim to enhance resistance to environmental stresses, improve metabolic pathways, and boost HPRs. Bacterial modification has proven effective in overcoming metabolic barriers and optimizing microbial performance in various production conditions (Montaño López et al., 2022; Thirumalaivasan and Wu, 2020).

7.3.3 Bacterial Immobilization

Bacterial immobilization is a technique that involves attaching bacteria to carriers, which separates them from the surrounding environment. This approach retains higher biomass, improves stability, and enhances HPRs and HY (Hitam and Jalil, 2023; Venkatesan et al., 2023; Wu et al., 2020). Immobilization helps overcome the limitations of low biomass and bioactivity, especially in continuous reactors, by creating an optimal environment for bacterial growth and activity, leading to more stable and efficient BioH production (Chandrasekhar et al., 2015; Shellaiah et al., 2021).

7.3.4 Additive Supplementation

Adding supplementary additives to the fermentation process is a relatively simple yet effective method to intensify BioH production. Metal ions essential for bacterial growth, enzyme synthesis, and photocatalytic nanomaterials that enhance light absorption are commonly added to improve bacterial metabolism and HY (Zhao et al., 2022). Additives such as buffer solutions also help maintain optimal conditions for bacterial activity, further boosting hydrogen production (Akhlaghi and Najafpour-Darzi, 2020; Velusamy et al., 2019).

7.3.5 Reactor Design Optimization

Optimizing the design of photobioreactors is critical for large-scale BioH production. Key design factors include uniform light distribution, appropriate mixing modes, and high biomass retention (Nangan et al., 2023a; Sirohi et al., 2022; Yang et al., 2022). Innovative reactor designs, such as baffle and up-flow systems, ensure that the bacterial activity is maintained and that the fermentation medium is effectively mixed. These improvements help sustain continuous hydrogen production, enhancing both the efficiency and scalability of the process (Acar and Dincer, 2019).

7.3.6 Research Needs and Future Directions

As hydrogen demand grows, sustainable BioH production must be improved. While many enhancement strategies have been explored, most studies remain at the batch scale (Sivaranjani et al., 2023). Future research should focus on scaling up BioH production, particularly in continuous reactor systems. Strategies such as improving substrate pretreatment, optimizing bacterial modification and immobilization, using additives, and refining reactor designs will be essential for advancing BioH production technologies (Cheng et al., 2022).

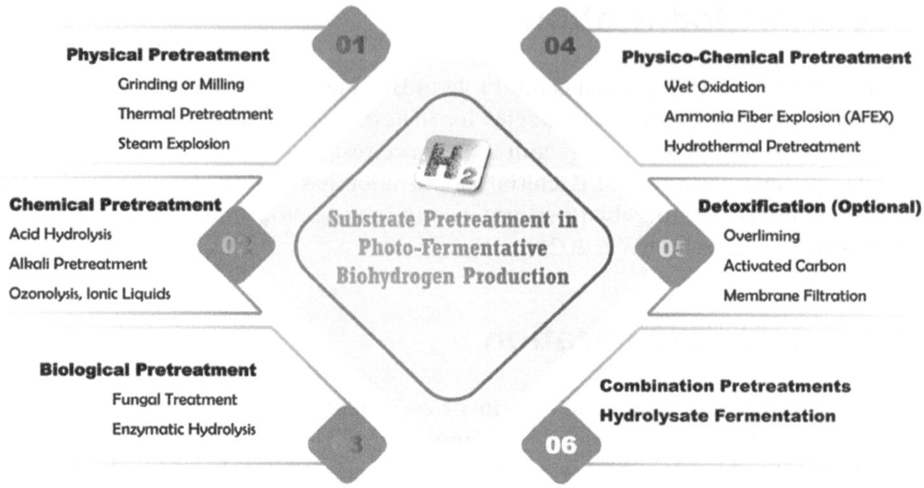

FIGURE 7.5 Schematic illustration of different types of substrate pretreatment in photo-fermentative biohydrogen production.

Enhancement strategies for photo-fermentative BioH production are essential for overcoming current limitations and scaling up the process (Zhang et al., 2021). Substrate pretreatment, bacterial modification, and immobilization have been shown to improve efficiency, while additive supplementation and reactor design optimization further intensify production (Figure 7.5). The integration of these strategies will be critical to advancing the field and meeting future hydrogen demands (Patel et al., 2024) (Table 7.1).

TABLE 7.1 Overview of substrate pretreatment methods, descriptions, examples, and references for photo-fermentative biohydrogen production

STEP/METHOD	*DESCRIPTION*	*EXAMPLE*	*REF*
Physical Pretreatment			
Grinding or milling	Reduces the particle size of the substrate, increasing surface area for enzymatic or microbial action.	Grinding corn stover increases surface area, aiding microbial access	Yogyta et al. (2023)
Thermal pretreatment	Involves heating the substrate to high temperatures (usually 120°C–200°C). Breaks down hemicellulose and improves enzyme accessibility but can also produce inhibitory compounds like furfural and hydroxymethylfurfural.	Thermal pretreatment of agricultural waste at 150°C improves hydrolysis yield	Serrano-Martínez et al. (2024)
Steam explosion	The substrate is treated with steam under high pressure, followed by sudden pressure release, causing the material to explode into finer particles and disrupt the cell wall structure.	Steam explosion of rice straw disrupts the cell wall, making cellulose more accessible	Serrano-Martínez et al. (2024)

(Continued)

TABLE 7.1 (*Continued*) Overview of substrate pretreatment methods, descriptions, examples, and references for photo-fermentative biohydrogen production

STEP/METHOD	*DESCRIPTION*	*EXAMPLE*	*REF*
Chemical Pretreatment			
Acid hydrolysis	Dilute or concentrated acids (e.g., sulfuric acid) break down hemicellulose and partially hydrolyze cellulose into fermentable sugars like glucose and xylose. Care must be taken to control inhibitors.	Acid hydrolysis of sugarcane bagasse releases glucose for fermentation	Aguilar et al. (2002)
Alkali pretreatment	Alkali treatments (e.g., sodium hydroxide, ammonia) are used to break down lignin and disrupt the cellulose structure, making sugars more accessible for fermentation.	Alkali pretreatment of wheat straw enhances its digestibility for biohydrogen production	Wang et al. (2015)
Ozonolysis	Ozone is used to degrade lignin and hemicellulose without producing harmful inhibitors, although it can be expensive.	Ozonolysis of lignocellulosic biomass improves its bioconversion to fermentable sugars	Saini and Kaur (2024)
Ionic liquids	Ionic liquids dissolve lignin and cellulose, facilitating biomass breakdown. Effective but costly and potentially toxic to microorganisms.	Ionic liquid pretreatment of corn stover releases more sugars than traditional methods	Chai et al. (2024)
Biological Pretreatment			
Fungal treatment	Certain fungi, especially white-rot fungi, degrade lignin using enzymes like lignin peroxidase and manganese peroxidase. Slow but environmentally friendly.	White-rot fungi break down lignin in wood chips, improving digestibility	Millati et al. (2011)
Enzymatic hydrolysis	Commercial enzymes like cellulases and hemicellulases are used to convert cellulose and hemicellulose into fermentable sugars. Specific and effective but costly for large-scale use.	Enzymatic hydrolysis of corn stover using cellulase releases sugars for biohydrogen production	Saha et al. (2016)
Physico-chemical Pretreatment			
Wet oxidation	Combines thermal and chemical treatment by heating biomass in the presence of oxygen or air at high temperatures, degrading hemicellulose and lignin but potentially producing inhibitors.	Wet oxidation of wheat straw improves sugar recovery and reduces recalcitrance	Grover and Choudhury (2024)
Ammonia fiber explosion (AFEX)	Biomass is treated with liquid ammonia at high pressure, followed by sudden pressure release. This disrupts lignin and increases cellulose accessibility.	AFEX pretreatment of corn stover enhances microbial digestion and biohydrogen production	Hu et al. (2024)
Hydrothermal pretreatment	Uses hot compressed water (160°C–240°C) to break down hemicellulose and make cellulose more accessible, without added chemicals.	Hydrothermal pretreatment of wheat straw improves biohydrogen yield by making sugars more available for fermentation	Hu et al. (2024)

(*Continued*)

TABLE 7.1 (*Continued*) Overview of substrate pretreatment methods, descriptions, examples, and references for photo-fermentative biohydrogen production

STEP/METHOD	*DESCRIPTION*	*EXAMPLE*	*REF*
Detoxification (Optional)			
Overliming	Calcium hydroxide is added to neutralize acidic hydrolysates and precipitate inhibitors like furfural.	Overliming detoxifies hydrolysate from acid pretreatment of corn stover	Toscan et al. (2024)
Activated carbon	Activated carbon adsorbs inhibitory compounds, like phenolics, from hydrolysates.	Activated carbon treatment removes phenolics from lignocellulosic hydrolysate for improved fermentation	Ananthi et al. (2024)
Membrane filtration	Physically separates inhibitory compounds from fermentable sugars using selective membranes.	Membrane filtration separates inhibitors from sugar-rich hydrolysate for fermentation	Gyan et al. (2024)
Combination pretreatments	Combines physical, chemical, and biological methods to optimize sugar release and minimize inhibitory compounds.	Combining steam explosion with enzymatic hydrolysis improves sugar release from wheat straw	Preechakun et al. (2024)
Hydrolysate fermentation	After pretreatment, the hydrolysate is fermented by bacteria, typically under anaerobic or micro-aerobic conditions, to produce biohydrogen.	Hydrolysate of sugarcane bagasse is fermented by *Rhodobacter* to produce hydrogen. *Rhodopseudomonas* ferments rice straw hydrolysate into biohydrogen	Hassan et al. (2024) and Zhang et al. (2024)

7.4 SUBSTRATE PRETREATMENT IN PHOTO-FERMENTATIVE BIOH PRODUCTION

Physical Pretreatment: Physical pretreatment methods aim to reduce particle size and enhance the surface area available for microbial action, which improves the efficiency of subsequent enzymatic or microbial hydrolysis (Alvira et al., 2010). Techniques like grinding and milling are simple and effective but may require significant energy input, especially for hard lignocellulosic materials. Steam explosion, while more advanced, can lead to partial degradation of hemicellulose and an increase in the availability of cellulose (Yu et al., 2022). However, it may also produce some inhibitory compounds that can affect BioH production. Overall, physical pretreatment alone is not always sufficient, especially for lignin-rich substrates, and often needs to be paired with chemical or biological methods for optimal results (Brémond et al., 2018).

Chemical Pretreatment: Chemical pretreatment methods, such as acid or alkali treatments, play a critical role in breaking down the complex structures of biomass, particularly lignin and hemicellulose (Yoo et al., 2020). Acid hydrolysis is highly effective at liberating fermentable sugars from hemicellulose but can result in the production of toxic by-products like furfural and hydroxymethylfurfural, which can inhibit microbial growth. Alkali pretreatment, on the other hand, is more effective in disrupting lignin, allowing for better cellulose accessibility. However, the cost and environmental concerns of using strong chemicals must be considered (Cano-Ruiz and McRae, 1998). Ozonolysis and the use of ionic liquids offer alternatives, but they are not yet widely adopted due to their cost and potential toxicity (Beil et al., 2021).

In this context, chemical pretreatment holds promise but requires careful optimization to balance sugar yield and the formation of inhibitory compounds, which can be a bottleneck for BioH production (Sharmila et al., 2022).

Biological Pretreatment: Biological pretreatment is an eco-friendly alternative, leveraging the natural ability of certain fungi and enzymes to break down lignin and cellulose. White-rot fungi, for instance, are highly effective at degrading lignin but operate at much slower rates compared to chemical methods (Leonowicz et al., 1999). Enzymatic hydrolysis, which uses cellulases and hemicellulases, is more specific and can effectively release sugars, though its scalability is challenged by the high cost of enzymes (Kuhad et al., 2016). Despite being environmentally sustainable and producing fewer inhibitors, biological methods alone may not achieve the necessary efficiency for large-scale BioH production and often require pretreatment combinations for better results.

Physico-chemical Pretreatment: Physico-chemical methods combine both chemical and physical processes to achieve more effective breakdown of biomass (Bensah and Mensah, 2018). Wet oxidation and ammonia fiber explosion (AFEX) are commonly used to simultaneously degrade hemicellulose and lignin while increasing the digestibility of cellulose (Biswas et al., 2015). These methods, especially AFEX, have been shown to improve the efficiency of downstream hydrolysis and fermentation. However, they also require high energy inputs and sometimes generate inhibitory by-products, which must be managed. Hydrothermal pretreatment is another technique gaining popularity due to its minimal chemical use, but the high operating temperatures and pressures involved can be costly (Sarker et al., 2021). The discussion around physico-chemical methods highlights the need for optimization to minimize energy costs and the generation of inhibitors while maximizing sugar release for effective BioH production (Soares et al., 2020).

Combination Pretreatments: Given the limitations of individual pretreatment methods, combining techniques has become a common approach to maximizing substrate breakdown while minimizing inhibitory by-product formation. For example, steam explosion followed by enzymatic hydrolysis can provide the benefits of both methods, where physical disruption of the biomass structure enhances enzymatic access to cellulose (Yang et al., 2011). Similarly, acid pretreatment followed by a detoxification step like overliming can optimize sugar release while reducing inhibitor concentrations (Wang and Feng, 2010). While combination pretreatments can be highly effective, they also increase process complexity and may require additional energy or resources, raising concerns about economic viability. This combination approach is seen as a promising strategy to overcome the individual limitations of various methods, but it needs to be evaluated carefully for industrial-scale applications (Chaubey et al., 2013).

Detoxification (Optional): Detoxification is often necessary after chemical or thermal pretreatments due to the formation of inhibitory compounds like furfural, phenolics, and acetic acid. These inhibitors can significantly reduce microbial efficiency in the hydrogen production process. Methods like overliming and activated carbon treatment are effective at neutralizing or removing inhibitors but add additional steps and cost to the process (Aryamol et al., 2024; Thirumalaivasan et al., 2024a; Verma et al., 2024). Membrane filtration is another advanced technique that selectively removes inhibitors while retaining fermentable sugars. While detoxification improves the efficiency of fermentation, it may not always be feasible at large scales due to its added cost and operational complexity. Therefore, the detoxification step is critical in scenarios where chemical or thermal pretreatments are used, but it introduces another layer of process optimization, particularly in terms of cost-effectiveness (Falciglia et al., 2018).

Hydrolysate Fermentation: After pretreatment, the resulting hydrolysate is rich in sugars and can be fermented by photo-fermentative bacteria to produce BioH. The success of this fermentation depends largely on the efficiency of the pretreatment process and the removal of any inhibitory compounds. Microorganisms like *Rhodobacter* and *Rhodopseudomonas* thrive in anaerobic conditions, using light as an energy source to convert organic substrates into hydrogen

(Hallenbeck and Liu, 2016). This stage is crucial because any inefficiencies in pretreatment directly affect the HY. If inhibitors remain, they can reduce microbial activity and hydrogen production (Chen et al., 2021). It's important to emphasize that hydrolysate fermentation is the culmination of all previous steps. The effectiveness of the fermentation process is highly dependent on the quality of the pretreatment, making upstream optimization crucial (Iram et al., 2020).

Optimization Considerations: In conclusion, the success of substrate pretreatment in BioH production largely depends on the careful optimization of each method. Factors such as cost, energy input, scalability, and the balance between sugar yield and inhibitor formation must be considered. Physical, chemical, and biological methods each offer unique advantages and drawbacks, and a combination of these methods often provides the best results (Jamkhande et al., 2019). However, the economic feasibility of such combinations remains a challenge. Detoxification, although optional, becomes critical when using harsher chemical pretreatments (Cannella et al., 2014). Optimizing these processes for specific substrates and conditions will be key to making photo-fermentative BioH production more viable on a commercial scale.

7.4.1 Current Challenges with Substrate Utilization

The efficiency of BioH production is critically dependent on the accessibility and quality of organic substrates presented to microbial systems (Bharathiraja et al., 2016). One of the primary challenges in substrate utilization is the inherent complexity of organic compounds, which can vary widely in their chemical compositions and physical properties. These complexities often impede the microbial breakdown processes necessary for efficient hydrogen production (Cabrol et al., 2017). Additionally, many organic substrates contain natural or synthetic inhibitors – such as heavy metals, salts, and non-biodegradable organics – that can severely reduce the activity and health of microbial communities. These inhibitors can stem from agricultural runoffs, industrial waste inputs, or even from the treatment processes themselves, posing significant challenges to maintaining robust and efficient microbial operations (Tabassum et al., 2021).

7.4.2 Advances in Membrane-Based Pretreatment Methods

To address these challenges, recent advancements in membrane technology have revolutionized the pretreatment of substrates for BioH production. Membrane-based methods such as nanofiltration and reverse osmosis are at the forefront of these innovations, offering dual benefits: they concentrate the organic substrates to enhance their bioavailability and simultaneously remove a broad spectrum of inhibitors that hinder microbial efficiency (Chen et al., 2014).

7.4.2.1 Enhancing Substrate Bioavailability Through Nanofiltration

Nanofiltration membranes are engineered to have pore sizes on the order of nanometers, which allows them to selectively separate different molecular sizes. These membranes are particularly effective in removing low-molecular-weight inhibitors while concentrating organic substrates (Kim et al., 2018). By adjusting the molecular weight cut-off of the nanofiltration membranes, operators can tailor the pretreatment process to the specific needs of the microbial consortia, optimizing the feedstock for enhanced hydrogen production (Kumar et al., 2020b).

7.4.2.2 Reverse Osmosis in Removing Inhibitors

Reverse osmosis employs a high-pressure process that forces water and small molecules through a semi-permeable membrane, leaving behind larger molecules and assorted impurities, including many inhibitors problematic in BioH processes. This technique is invaluable not only for concentrating substrates but also for achieving significant reductions in salinity and other inhibitive factors. By creating a purer, more refined substrate solution, reverse osmosis significantly increases the efficiency of the microbial systems involved in hydrogen production (Islam et al., 2021).

7.4.2.3 Impact on Hydrogen Production Efficiency

The integration of these advanced membrane technologies into the substrate pretreatment phase has marked a significant step forward in BioH production. By improving substrate quality and consistency, these technologies reduce the metabolic stress on microbial populations, thereby enhancing their longevity and productivity (Liu et al., 2021). Moreover, the reduced presence of inhibitors allows for a broader operational window and less stringent control measures, reducing operational costs and complexity.

7.4.2.4 Case Studies and Applications

Practical applications of these technologies have demonstrated substantial improvements in HY. For instance, studies involving the pretreatment of agricultural waste streams with nanofiltration have shown an increase in hydrogen production by up to 30%, highlighting the potential of these technologies to turn even low-quality organic wastes into valuable resources for renewable energy production.

The advancements in membrane technologies for substrate pretreatment represent a pivotal development in the photo-fermentative production of BioH. As these technologies continue to evolve and integrate with other aspects of hydrogen production systems, they pave the way for more sustainable, efficient, and economically viable solutions in renewable energy landscapes. Future research will undoubtedly expand upon these foundations, exploring new materials and configurations for membranes and further enhancing the synergy between substrate pretreatment and microbial efficiency (Zhen et al., 2019).

7.5 STRAIN IMPROVEMENT FOR ENHANCED BIOH PRODUCTION

Strain improvement for enhanced BioH production focuses on optimizing the genetic and metabolic capabilities of microorganisms to increase HYs. Traditional methods, such as selective breeding, have been used to identify naturally efficient strains, but recent advancements in genetic engineering have unlocked new possibilities for modifying microbial pathways (Alam et al., 2021). Techniques such as gene editing, metabolic pathway optimization, and the introduction of foreign genes have been applied to enhance key aspects of hydrogen production, such as improving electron flow, reducing by-product formation, and increasing resistance to environmental stresses (Younas et al., 2022). By modifying strains like *R. sphaeroides* and *R. palustris*, researchers aim to overcome natural metabolic limitations and improve hydrogen production efficiency. These genetically enhanced strains are critical for advancing BioH production, making the process more feasible for large-scale industrial applications (Figure 7.6).

7.5.1 Genetic and Metabolic Engineering of PSB

The drive toward a sustainable energy future has intensified efforts in BioH production, with PSB at the forefront of this research. Species such as *R. palustris* and *R. sphaeroides* have been pivotal due to their inherent capabilities to produce hydrogen through photo-fermentation. Recent advances in genetic and metabolic engineering have significantly enhanced these capabilities, opening new avenues for increased HYs.

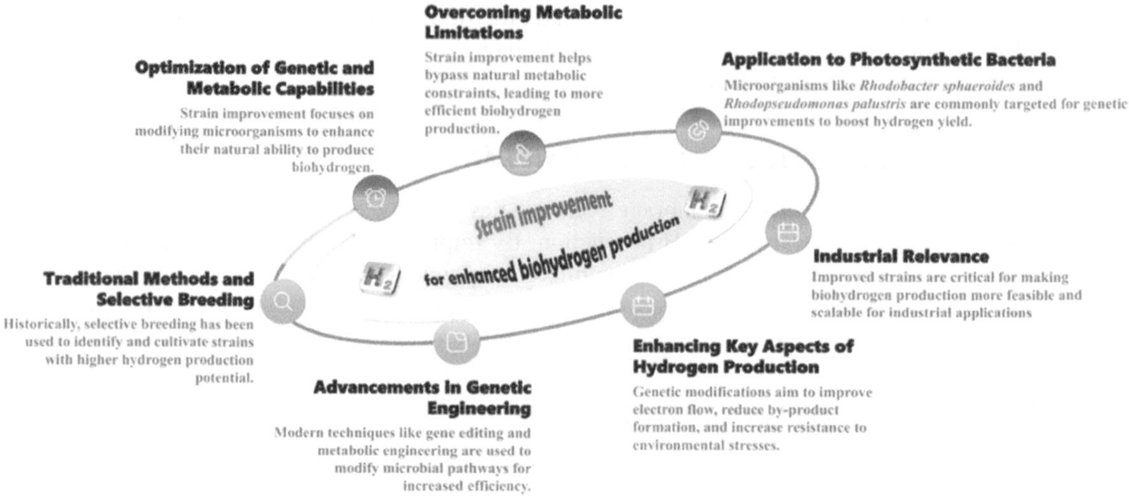

FIGURE 7.6 Schematic illustration of various strategies used for the strain improvement for enhanced biohydrogen production.

The advent of clustered regularly interspaced short palindromic repeats (CRISPR)-Cas systems and other gene-editing tools has revolutionized our approach to microbial engineering (Zhang, 2019). By targeting specific genes involved in the photosynthetic pathway, researchers have been able to increase the rate of hydrogen production. Modifications in the genes responsible for the uptake hydrogenase and nitrogenase enzymes have shown particularly promising results. These enzymes play critical roles in the metabolic pathways for hydrogen production, and enhancing their expression or activity directly boosts HY (Hallenbeck et al., 2012).

Beyond genetic modifications, metabolic engineering has allowed for the optimization of entire biochemical pathways that influence hydrogen production. By rerouting metabolic fluxes, suppressing competitive pathways, and enhancing the pathways leading to hydrogen production, these engineered bacteria can achieve higher efficiency and productivity under photo-fermentative conditions. For instance, the optimization of the Calvin cycle to increase the availability of reducing equivalents necessary for hydrogen production is one such strategy that has yielded fruitful results (Resnick, 2004).

7.5.2 Role of Membrane Integration in Supporting Engineered Strains

The integration of membrane technologies not only aids in substrate pretreatment but also plays a crucial role in creating optimal conditions for genetically engineered microbial strains. This synergy between membrane technology and strain engineering is critical, especially under non-ideal and stress conditions that typically hinder microbial performance.

Membrane systems can be designed to regulate the microenvironment of the BioR, including parameters such as pH, temperature, and the concentration of substrates and by-products. For example, selective permeability of membranes can be utilized to retain certain substrates while removing inhibitory by-products like organic acids, which can significantly impact bacterial growth and hydrogen production efficiency.

Stress conditions such as oxidative stress or nutrient limitations can severely impact the viability and productivity of microbial strains. Membranes can be used to modulate the exposure of these engineered bacteria to stressors by controlling the influx and efflux of molecules within the BioR (Bremer and Krämer, 2019). Furthermore, membranes can facilitate the continuous removal of hydrogen from the reactor, thus preventing product inhibition and promoting a more sustained production rate.

The design of BioRs incorporating these membranes must consider both the physical and biological requirements of the engineered strains. For example, photo-BioRs optimized for light distribution can enhance photosynthetic efficiency, while membrane-integrated systems can improve the separation and purity of hydrogen produced. This holistic approach to BioR design ensures that the engineered capabilities of the microbial strains are fully realized in practical applications (Yadav et al., 2012).

7.6 OPTIMIZATION OF BIOR DESIGNS AND ITS PARAMETERS

The quest for sustainable hydrogen production has led to significant advancements in BioR technology. Central to these developments is the design and optimization of BioRs tailored to maximize the efficiency of hydrogen production through photo-fermentation. Innovations in BioR design primarily focus on integrating membrane technologies that facilitate *in situ* hydrogen separation and purification. This integration not only streamlines the hydrogen production process but also enhances the purity of the produced hydrogen, making it more viable for commercial use.

Novel BioR configurations have explored various geometries and operational strategies to optimize light exposure and nutrient distribution, which are critical for the growth and activity of PSB. For instance, tubular BioRs with semi-transparent, light-diffusing membranes provide uniform light distribution that maximizes the photosynthetic activity while maintaining the desired culture conditions. Additionally, these BioRs often incorporate dynamic membrane systems that can adjust their permeability and selectivity in response to changes in the microbial environment, thereby maintaining optimal conditions for hydrogen production (Figure 7.7).

FIGURE 7.7 Schematic illustration for optimizing the design elements ensures that bioreactors are better suited for large-scale, efficient biohydrogen production.

7.6.1 Operational Parameters Affecting BioH Production

The performance of BioH BioRs is significantly influenced by several operational parameters. Key among these are light intensity, reactor configuration, and the type of membrane used. Each of these parameters must be meticulously optimized to enhance the efficiency and output of hydrogen production. Light intensity directly affects the rate of photosynthesis in bacteria, thus influencing hydrogen production. Optimal light management involves not just the intensity, but also the duration and quality (wavelength) of light. Advanced light-emitting diode (LED) systems allow for precise control over these aspects, enabling tailored light regimes that optimize bacterial photosynthetic efficiency (Serôdio et al., 2018). The configuration of the reactor plays a crucial role in determining the mass transfer efficiency and the distribution of bacteria within the system. Innovative designs such as spiral and vertically stacked BioRs have been developed to maximize the surface area exposed to light, thereby increasing the potential for hydrogen production. These configurations also facilitate easier removal of hydrogen from the culture medium, reducing the risk of hydrogen inhibition.

The choice of membrane is pivotal in the separation and purification stages of hydrogen production. Membranes must be selectively permeable to allow hydrogen to pass through while retaining bacterial cells and substrates. Materials like polyvinylidene fluoride (PVDF) and polytetrafluoroethylene (PTFE) are commonly used due to their high chemical resistance and mechanical stability, which are essential for operating under the harsh conditions sometimes present in BioRs (Table 7.2) (Woźniak-Budych, 2023).

Membrane integration plays a transformative role in supporting engineered bacterial strains, particularly under non-ideal and stress conditions. By creating controlled environments, membranes can help maintain optimal pH, temperature, and osmotic conditions that are crucial for the survival and productivity of genetically modified strains (Abraham et al., 2024; Kanagaraj and Pitchumani, 2010). These engineered strains often possess enhanced hydrogen-producing capabilities but may require specific

TABLE 7.2 Key operational parameters that influence biohydrogen production, along with methods to optimize these conditions and improve overall hydrogen yield

OPERATIONAL PARAMETER	*METHOD OF ENHANCEMENT*	*DESCRIPTION*
Light intensity	Adjusting light intensity and using light-distributing systems	Photosynthetic bacteria require light for hydrogen production. Optimal light intensity increases hydrogen yield.
Temperature	Maintaining optimal temperature (30°C–40°C for most PSB)	Temperature affects microbial metabolism. Controlling it within optimal ranges ensures maximum biohydrogen output.
pH level	Buffering and regular monitoring	Maintaining a stable pH (6.8–7.5) enhances microbial activity and hydrogen production efficiency.
Substrate concentration	Optimizing substrate loading	Excessive substrate can lead to substrate inhibition, while low concentrations limit hydrogen yield.
Nutrient availability	Adding essential nutrients or co-substrates	Adequate nutrients (e.g., nitrogen, phosphorus) support microbial growth and boost hydrogen production.
Mixing/aeration	Optimizing mixing strategies and controlling aeration levels	Proper mixing ensures uniform nutrient distribution and prevents gas accumulation that may inhibit bacterial growth.

(*Continued*)

TABLE 7.2 (*Continued*) Key operational parameters that influence biohydrogen production, along with methods to optimize these conditions and improve overall hydrogen yield

OPERATIONAL PARAMETER	*METHOD OF ENHANCEMENT*	*DESCRIPTION*
Hydraulic retention time (HRT)	Adjusting HRT to optimal values	Longer retention times may enhance hydrogen yield but could reduce system throughput. Balance is crucial.
Partial pressure of hydrogen	Using gas stripping or degassing techniques	Lowering the partial pressure of hydrogen in the reactor prevents inhibition of the hydrogen-producing pathways.
Inhibitor removal	Applying detoxification methods	Inhibitors like furfural and phenolics can accumulate; removing them enhances microbial efficiency.
CO_2 management	Optimizing CO_2 removal strategies	Excess CO_2 can inhibit hydrogen production, so effective removal or utilization improves the overall process.

environmental conditions to function effectively. Membrane systems can isolate and protect these sensitive strains from fluctuations in external conditions, thereby stabilizing the BioH production process.

Recent developments in membrane technologies, such as nanofiltration and reverse osmosis, have proven effective not only in substrate pretreatment by concentrating organic substrates and removing inhibitors but also in enhancing the overall operational stability of BioRs. By improving the bioavailability of substrates, these technologies ensure a steady supply of nutrients to the microbial cells, thus supporting sustained and efficient hydrogen production. Through these advanced configurations and careful management of operational parameters, modern BioRs are set to revolutionize the field of BioH production. By addressing both biological and engineering challenges, these systems pave the way toward establishing photo-fermentative hydrogen production as a key player in the renewable energy sector (Singh and Sarma, 2022).

7.7 OTHER PARAMETERS FOR BIOH PRODUCTION ENHANCEMENT

To improve BioH production efficiency, various parameters, beyond traditional microbial and environmental factors, have been explored. These include the use of advanced materials, additives, and functionalized systems that contribute to optimizing the hydrogen production process (Zhakeyev et al., 2017). Below are some key parameters and techniques that have been applied for enhancing BioH production.

7.7.1 Functionalized Membranes

Functionalized membranes play a significant role in improving BioH production by enhancing separation efficiency and controlling the mass transfer of gases and substrates. These membranes can be made using nanocomposites, supramolecules (e.g., cyclodextrins, metal organic frameworks (MOFs), covalent organic frameworks (COFs), cavitands, and synthetic cavitands) (Alagesan et al., 2016; Ji et al., 2022; Kanagaraj et al., 2012; Kanagaraj et al., 2023a; Kanagaraj et al., 2023b; Tang et al., 2022; Xiao et al., 2020). In BioH production, functionalized membranes can be used to separate hydrogen from other gases, improve the catalytic process, and enhance product purification. Additionally, these membranes are widely used in catalysis, water purification, biomedical applications, and sensors, which demonstrates their versatility in optimizing BioH systems.

7.7.2 Additive-Based Enhancements

Various additives have been applied to BioR to enhance BioH production by influencing microbial activity, improving reaction conditions, or modifying the substrate availability. Some of the most commonly used additives are described in the following sections.

7.7.2.1 Heterogeneous Materials

Clays, zeolites, MOFs, and COFs are examples of porous materials that have been used as catalysts or adsorbents to improve HY (Jeganathan et al., 2014; Rama et al., 2011; Rama et al., 2012a; Rama et al., 2012b; Suresh et al., 2013; Suresh et al., 2014). These materials can help in enhancing the surface area for microbial attachment or act as carriers for immobilized enzymes.

7.7.2.2 Carbon-Based Materials

Carbon nanomaterials, including carbon nano-allotropes, carbon quantum dots derived from plant or waste materials, carbon nanotubes (CNTs), fullerenes (C60), and carbon dots, have been used to improve microbial efficiency in hydrogen production (Rajapandi et al., 2023). These materials enhance electron transfer and provide a conductive surface for bacterial growth, increasing BioH production efficiency.

7.7.3 Nanoparticles (NPs)

The use of NPs, particularly metal NPs, has proven beneficial in improving hydrogen production. Metal NPs such as silver, gold, and platinum enhance the catalytic efficiency of enzymes involved in hydrogen generation (Thirumalaivasan, 2024). They also promote electron transfer during metabolic processes. These particles can be used to functionalize membranes, act as carriers for microbial immobilization, or be directly added to the microbial culture to enhance overall BioH yields.

7.7.4 Other Parameters

Metal Ions: Metal ions, such as nickel, iron, magnesium, molybdenum, and zinc, are essential cofactors for key enzymes involved in hydrogen production, like hydrogenase. By adding these metal ions to the system, microbial metabolism and enzyme activity are enhanced, leading to an increase in HY. These metal ions improve the efficiency of both electron transfer processes and the overall catalytic activity of the hydrogen-producing enzymes.

Buffer Solutions: Buffers, such as phosphate or acetate buffers, are added to the fermentation medium to maintain a stable pH, which is crucial for microbial activity. Stable pH conditions ensure optimal enzyme function and HPRs by preventing harmful pH shifts that could disrupt microbial metabolism.

Nitrogen Sources: Providing additional nitrogen sources such as ammonium sulfate or urea is another approach to enhancing BioH production. Nitrogen is a vital nutrient for microbial growth, and its availability can stimulate microbial metabolism, thereby increasing HYs.

Trace Elements: Trace elements, including cobalt, copper, selenium, and tungsten, are involved in several metabolic pathways and are essential for enzymatic functions related to hydrogen production. Supplementing the BioR with trace elements supports microbial growth and ensures that the necessary cofactors for enzymatic hydrogen production are available.

Vitamins: Vitamins such as biotin, riboflavin, and thiamine act as cofactors in microbial metabolic processes, supporting cell growth and energy production. The addition of vitamins can enhance the growth rate and efficiency of hydrogen-producing microorganisms, improving BioH production.

Sulfate or Sulfide Compounds: Sulfur-containing compounds, such as sulfate or sulfide, serve as electron donors and play a role in stimulating microbial metabolism. These compounds can enhance hydrogen production by improving the redox balance within the microbial community and supporting sulfur metabolism in certain microbial strains.

Photocatalytic Nanomaterials: Photocatalytic nanomaterials, such as titanium dioxide NPs, improve light absorption in photo-BioRs. By increasing the efficiency of light absorption, these materials enhance the photosynthetic activity of microorganisms, leading to higher hydrogen production under light conditions (Kanagaraj et al., 2022; Kanagaraj et al., 2020a; Rao et al., 2018).

Organic Acids: Organic acids, such as acetic acid and butyric acid, can be added as co-substrates in BioH production. These acids serve as carbon sources, improving the availability of electrons for hydrogen production and enhancing the overall efficiency of the process (Ramasubburayan et al., 2024b).

Redox Mediators: Redox mediators, such as methyl viologen and neutral red, are compounds that facilitate electron transfer within microbial cells (Kanagaraj et al., 2020b). By improving electron transport, these mediators enhance the efficiency of hydrogen production, leading to increased HYs.

Carbon Sources: Additional carbon sources, such as glucose, sucrose, and glycerol, are often added to BioRs to support microbial growth and provide more substrate for hydrogen production. The presence of sufficient carbon sources ensures that microorganisms have the necessary fuel to drive the BioH production process (Nair et al., 2023; Thirumalaivasan et al., 2024b, 2024c). These parameters, when optimized and applied correctly, can significantly enhance the efficiency of BioH production. By using functionalized membranes, nanomaterials, and a variety of additives, researchers can create more favorable conditions for microbial activity, leading to greater HYs and more efficient production processes.

7.8 FUTURE DIRECTIONS AND CHALLENGES

As the quest for renewable energy solutions intensifies, photo-fermentative BioH production stands out due to its unique utilization of solar energy and organic substrates. This chapter examines the potential future research directions and challenges that must be navigated to scale these technologies from laboratory settings to industrial applications (Sovacool et al., 2020). It emphasizes how membrane technologies have catalyzed advances in substrate pretreatment, strain improvement, and BioR optimization, setting the stage for significant breakthroughs in the field.

One of the primary future directions for photo-fermentative BioH production involves its integration into existing industrial infrastructures (Putatunda et al., 2023). Current systems are often designed for lab-scale experiments, and a significant gap exists in transitioning these setups to larger scales suitable for commercial production. Research must focus on modular reactor designs that can be seamlessly integrated with existing renewable energy processes, potentially in symbiosis with biomass, wind, or solar power operations to maximize energy efficiency and reduce operational costs (Butturi et al., 2019).

Continued innovation in membrane technologies is critical to scaling up photo-fermentative BioH production. Recent advancements have demonstrated how nanofiltration and reverse osmosis can enhance substrate bioavailability by concentrating organic substrates and removing inhibitory compounds. Future research should explore the development of more robust, energy-efficient, and selective membranes that can operate under a wider range of conditions, further increasing the yield and purity of hydrogen produced (Yin and Yip, 2017).

The economic viability of BioH production is currently hampered by high capital and operational costs, particularly associated with the construction and maintenance of BioRs and the energy-intensive nature of membrane technologies (Ahmed et al., 2021). Research must therefore focus on cost-effective materials and methods, such as the utilization of waste-derived substrates and the development of low-cost, high-efficiency membranes (Samadi et al., 2022). Additionally, improving the longevity and durability of these systems in industrial environments will help reduce maintenance costs and enhance economic feasibility.

Technical challenges in scaling BioH production include maintaining optimal conditions for microbial cultures over extended periods and managing the integration of BioR systems with hydrogen purification and storage facilities (Tapia-Venegas et al., 2015). Future developments should aim at creating more resilient microbial strains through genetic and metabolic engineering that are capable of withstanding variations in operational conditions. Moreover, optimizing operational parameters like light intensity, pH levels, and nutrient supply in scaled-up systems will be essential for maintaining high efficiency and stability (Benner et al., 2022).

As global energy demands continue to rise, the environmental impact of traditional fossil fuels becomes increasingly unsustainable. Photo-fermentative BioH production offers a cleaner alternative, with the potential to significantly reduce greenhouse gas emissions and dependence on non-renewable energy sources (Mishra et al., 2019). Long-term research should focus on lifecycle assessments to quantify the environmental benefits and explore synergies with other renewable energy technologies.

Looking ahead, the field of photo-fermentative BioH production is poised for transformative growth. Key areas for future research include the exploration of hybrid systems that combine photo-fermentation with other renewable energy technologies, the use of artificial intelligence and machine learning for process optimization, and the development of global collaborations to standardize and promote the adoption of BioH technologies. The future of photo-fermentative BioH production is rich with opportunities and challenges. By addressing these challenges head-on and continuing to innovate in the areas of substrate pretreatment, strain improvement, and BioR design, this field can significantly contribute to the global portfolio of sustainable energy solutions. The integration of advanced membrane technologies will be crucial in overcoming existing barriers, paving the way for a more sustainable and economically viable BioH production landscape (Osman et al., 2024).

7.9 CONCLUSION

This chapter has highlighted significant advancements in the field of photo-fermentative BioH production, which holds great promise as a sustainable and environmentally friendly energy source. Through innovative approaches in substrate pretreatment, strain improvement, and BioR optimization, the chapter has demonstrated how membrane-based technologies are fundamentally transforming the efficiency and feasibility of BioH production. These advancements not only enhance the HY from various organic substrates but also streamline the operational processes to support commercial scalability.

Despite these technological strides, several challenges remain that must be addressed to fully realize the potential of photo-fermentative BioH production. The economic and technical hurdles, including the high cost of membrane technologies and the need for robust microbial strains capable of sustained production under variable conditions, are critical areas requiring further innovation and research. Moreover, integrating these systems into existing industrial processes poses a substantial challenge, necessitating a multidisciplinary approach to design modular and efficient BioRs that can be scaled up for industrial applications.

As we move forward, the integration of advanced technologies and continuous research will be crucial in overcoming these barriers, paving the way for BioH to become a cornerstone of the renewable energy landscape. This chapter closes with a hopeful vision where photo-fermentative BioH production not only contributes significantly to reducing our carbon footprint but also exemplifies a successful transition to a circular economy. The path is set for future innovations to enhance the scalability and economic viability of this promising technology, making it an integral part of global energy solutions.

REFERENCES

Abd Rahman, S. N., et al., 2018. Simulation of PEMFC stack for portable power generator application. *Jurnal Kejuruteraan*. 1, 1.

Abraham, D. A., et al., 2024. Highly sensitive and selective detection of dopamine using atomic layer deposited HfO_2 ultra-thin films. *Electrochimica Acta*. 497, 144574.

Acar, C., Dincer, I., 2019. Review and evaluation of hydrogen production options for better environment. *Journal of Cleaner Production*. 218, 835–849.

Aguilar, R., et al., 2002. Kinetic study of the acid hydrolysis of sugar cane bagasse. *Journal of Food Engineering*. 55, 309–318.

Ahmed, S. F., et al., 2021. Biohydrogen production from biomass sources: Metabolic pathways and economic analysis. *Frontiers in Energy Research*. 9, 753878.

Akhlaghi, N., Najafpour-Darzi, G., 2020. A comprehensive review on biological hydrogen production. *International Journal of Hydrogen Energy*. 45, 22492–22512.

Alagesan, M., et al., 2016. Photocyclodimerization of 2-anthracenecarboxylate mediated by a self-assembled iron tetrahedral coordination cage. *Journal of Photochemistry and Photobiology A*. 331, 95.

Alam, K., et al., 2021. Synthetic biology-inspired strategies and tools for engineering of microbial natural product biosynthetic pathways. *Biotechnology Advances*. 49, 107759.

Alvira, P., et al., 2010. Pretreatment technologies for an efficient bioethanol production process based on enzymatic hydrolysis: A review. *Bioresource Technology*. 101, 4851–4861.

Ananthi, V., et al., 2024. A review on the impact of various factors on biohydrogen production. *International Journal of Hydrogen Energy*. 52, 33–45.

Androga, D. D., et al., 2012. Photofermentative hydrogen production in outdoor conditions. In *Hydrogen Energy—Challenges and Perspectives*, pp. 77–120. In Tech.

Antal, T., et al., 2012. Photosynthetic hydrogen production: Mechanisms and approaches. In *State of the Art and Progress in Production of Biohydrogen*, pp. 25–54. Bentham Science.

Arregi, A., et al., 2018. Evaluation of thermochemical routes for hydrogen production from biomass: A review. *Energy Conversion and Management*. 165, 696–719.

Arumugam, S., et al., 2023. Heterogenous copper (II) schiff-base complex immobilized mesoporous silica catalyst for multicomponent biginelli reaction. *Journal of Organometallic Chemistry*. 998, 122804.

Aryamol, K., et al., 2024. Recent advances of carbon pathways for sustainable environment development. *Environmental Research*. 250, 118513.

Aziz, M., et al., 2021. Hydrogen production from biomasses and wastes: A technological review. *International Journal of Hydrogen Energy*. 46, 33756–33781.

Bakonyi, P., et al., 2018. A review of the innovative gas separation membrane bioreactor with mechanisms for integrated production and purification of biohydrogen. *Bioresource Technology*. 270, 643–655.

Balamurugan, G., et al., 2017. New phenazine based AIE probes for selective detection of aluminium (III) ions in presence of other trivalent metal ions in living cells. *Analyst*. 142, 4721–4726.

Balat, H., Kırtay, E., 2010. Hydrogen from biomass–present scenario and future prospects. *International Journal of Hydrogen Energy*. 35, 7416–7426.

Beil, S., et al., 2021. Toward the proactive design of sustainable chemicals: Ionic liquids as a prime example. *Chemical Reviews*. 121, 13132–13173.

Bélafi-Bakó, K., et al., 2006. Integration of biohydrogen fermentation and gas separation processes to recover and enrich hydrogen. *International Journal of Hydrogen Energy*. 31, 1490–1495.

Benner, P., et al., 2022. Lab-scale photobioreactor systems: Principles, applications, and scalability. *Bioprocess and Biosystems Engineering*. 45, 791–813.

Bensah, E. C., Mensah, M. Y., 2018. Emerging physico-chemical methods for biomass pretreatment. In *Fuel Ethanol Production from Sugarcane*, pp. 41–62. In Tech.

Bharathiraja, B., et al., 2016. Biohydrogen and Biogas–An overview on feedstocks and enhancement process. *Fuel*. 185, 810–828.

Bhatia, S. K., et al., 2021. Renewable biohydrogen production from lignocellulosic biomass using fermentation and integration of systems with other energy generation technologies. *Science of the Total Environment*. 765, 144429.

Biswas, R., et al., 2015. Wet explosion: A universal and efficient pretreatment process for lignocellulosic biorefineries. *BioEnergy Research*. 8, 1101–1116.

Bremer, E., Krämer, R., 2019. Responses of microorganisms to osmotic stress. *Annual Review of Microbiology*. 73, 313–334.

Brémond, U., et al., 2018. Biological pretreatments of biomass for improving biogas production: An overview from lab scale to full-scale. *Renewable and Sustainable Energy Reviews*. 90, 583–604.

Buschke, N., et al., 2013. Metabolic engineering of industrial platform microorganisms for biorefinery applications–optimization of substrate spectrum and process robustness by rational and evolutive strategies. *Bioresource Technology*. 135, 544–554.

Butturi, M., et al., 2019. Renewable energy in eco-industrial parks and urban-industrial symbiosis: A literature review and a conceptual synthesis. *Applied Energy*. 255, 113825.

Cabrol, L., et al., 2017. Microbial ecology of fermentative hydrogen producing bioprocesses: Useful insights for driving the ecosystem function. *FEMS Microbiology Reviews*. 41, 158–181.

Cannella, D., et al., 2014. PEI detoxification of pretreated spruce for high solids ethanol fermentation. *Applied Energy*. 132, 394–403.

Cano-Ruiz, J., McRae, G., 1998. Environmentally conscious chemical process design. *Annual Review of Energy and the Environment*. 23, 499–536.

Castillo-Hernández, A., et al., 2015. Start-up and operation of continuous stirred-tank reactor for biohydrogen production from restaurant organic solid waste. *International Journal of Hydrogen Energy*. 40, 17239–17245.

Chai, Y., et al., 2024. Upgrading lignocellulosic biomass to high-value products through the pretreatment driven by bio-based ethylene glycol solvent. *Chemical Engineering Journal*. 496, 153797.

Chandrasekhar, K., et al., 2015. Biohydrogen production: Strategies to improve process efficiency through microbial routes. *International Journal of Molecular Sciences*. 16, 8266–8293.

Chaubey, R., et al., 2013. A review on development of industrial processes and emerging techniques for production of hydrogen from renewable and sustainable sources. *Renewable and Sustainable Energy Reviews*. 23, 443–462.

Chaurasia, A. K., Mondal, P., 2021. Hydrogen production from waste and renewable resources. In *Hydrogen Fuel Cell Technology for Stationary Applications*, pp. 22–46. IGI Global.

Chen, J. L., et al., 2014. Toxicants inhibiting anaerobic digestion: A review. *Biotechnology Advances*. 32, 1523–1534.

Chen, Y., et al., 2020. Mortality and recurrent vascular events after first incident stroke: A 9-year community-based study of 0·5 million Chinese adults. *The Lancet Global Health*. 8, e580–e590.

Chen, Y., et al., 2021. Recent advance in inhibition of dark fermentative hydrogen production. *International Journal of Hydrogen Energy*. 46, 5053–5073.

Cheng, D., et al., 2022. Advanced strategies for enhancing dark fermentative biohydrogen production from biowaste towards sustainable environment. *Bioresource Technology*. 351, 127045.

Chu, C.-J., et al., 2020. A fluorescent turn-on probe for detection of hypochlorus acid and its bioimaging in living cells. *Spectrochimica Acta Part A: Molecular and Biomolecular Spectroscopy*. 233, 118234.

Connolly, D., et al., 2014. A comparison between renewable transport fuels that can supplement or replace biofuels in a 100% renewable energy system. *Energy*. 73, 110–125.

Das, D., Veziroglu, T. N., 2008. Advances in biological hydrogen production processes. *International Journal of Hydrogen Energy*. 33, 6046–6057.

Demirdelen, T., et al., 2020. Green touch for hydrogen production via alkaline electrolysis: The semi-flexible PV panels mounted wind turbine design, production and performance analysis. *International Journal of Hydrogen Energy*. 45, 10680–10695.

Dincer, I., Acar, C., 2015. Review and evaluation of hydrogen production methods for better sustainability. *International Journal of Hydrogen Energy*. 40, 11094–11111.

Du, X., et al., 2018. Parameter estimation of activated sludge process based on an improved cuckoo search algorithm. *Bioresource Technology*. 249, 447–456.

Falciglia, P. P., et al., 2018. A review on the microwave heating as a sustainable technique for environmental remediation/detoxification applications. *Renewable and Sustainable Energy Reviews*. 95, 147–170.

Fane, A., et al., 2011. Membrane technology: Past, present and future. In *Membrane and Desalination Technologies*, pp. 1–45. Humana Press.

Ghirardi, M. L., et al., 2000. Microalgae: A green source of renewable H_2. *Trends in Biotechnology*. 18, 506–511.

Gogate, P. R., Pandit, A. B., 2004. A review of imperative technologies for wastewater treatment I: Oxidation technologies at ambient conditions. *Advances in Environmental Research*. 8, 501–551.

Grover, C., Choudhury, B., 2024. Alkaline wet oxidative pretreatment and acid hydrolysis of wheat straw for squalene and monomethyl branched chain fatty acids rich lipid production in Lentibacillus salarius BPIITR. *Biocatalysis and Agricultural Biotechnology*. 57, 103062.

Gupta, S., et al., 2024. Photo-fermentative bacteria used for hydrogen production. *Applied Sciences*. 14, 1191.

Gyan, K., et al., 2024. A review of the advances in detoxification strategies of lignocellulosic hydrolysate for bio-based succinic acid production. *Biocatalysis and Agricultural Biotechnology*. 103323.

Hallenbeck, P. C., et al., 2012. Strategies for improving biological hydrogen production. *Bioresource Technology*. 110, 1–9.

Hallenbeck, P. C., Ghosh, D., 2009. Advances in fermentative biohydrogen production: The way forward? *Trends in Biotechnology*. 27, 287–297.

Hallenbeck, P. C., Liu, Y., 2016. Recent advances in hydrogen production by photosynthetic bacteria. *International Journal of Hydrogen Energy*. 41, 4446–4454.

Hassan, Q., et al., 2024. A review of hydrogen production from bio-energy, technologies and assessments. *Energy Harvesting and Systems*. 11, 20220117.

Hitam, C., Jalil, A. A., 2023. A review on biohydrogen production through photo-fermentation of lignocellulosic biomass. *Biomass Conversion and Biorefinery*. 13, 8465–8483.

Holladay, J. D., et al., 2004. Review of developments in portable hydrogen production using microreactor technology. *Chemical Reviews*. 104, 4767–4790.

Hu, M., et al., 2024. Ammonia fiber expansion (AFEX) combined with white-rot fungi pretreatment to improve enzymatic hydrolysis of lignocellulose. *Biomass Conversion and Biorefinery*. 14, 10085–10099.

Iram, A., et al., 2020. Ideal feedstock and fermentation process improvements for the production of lignocellulolytic enzymes. *Processes*. 9, 38.

Ishaq, H., et al., 2022. A review on hydrogen production and utilization: Challenges and opportunities. *International Journal of Hydrogen Energy*. 47, 26238–26264.

Islam, A. K., et al., 2021. Bio-hydrogen production from wastewater: A comparative study of low energy intensive production processes. *Clean Technologies*. 3, 156–182.

Ismail, K. S. K., et al., 2008. Biological hydrogen production from CO: Bioreactor performance. *Biochemical Engineering Journal*. 39, 468–477.

Jain, R., et al., 2024. Bio-hydrogen production through dark fermentation: An overview. *Biomass Conversion and Biorefinery*. 14, 12699–12724.

Jaiswal, K. K., et al., 2021. Recent trends in the development and diversification of sericulture natural products for innovative and sustainable applications. *Bioresource Technology Reports*. 13, 100614.

Jamkhande, P. G., et al., 2019. Metal nanoparticles synthesis: An overview on methods of preparation, advantages and disadvantages, and applications. *Journal of Drug Delivery Science and Technology*. 53, 101174.

Jeckelmann, J.-M., Erni, B., 2020. Transporters of glucose and other carbohydrates in bacteria. *Pflügers Archiv-European Journal of Physiology*. 472, 1129–1153.

Jeganathan, M., et al., 2014. Michael addition of indoles to β-nitrostyrenes catalyzed by HY zeolite under solvent-free conditions. *Tetrahedron Letters*. 55, 2061–2064.

Ji, J., et al., 2022. The more the slower: Self-inhibition in supramolecular chirality induction, memory, erasure, and reversion. *Journal of the American Chemical Society*. 144, 1455–1463.

Jiao, H., et al., 2024. Recent advances in sustainable hydrogen production from microalgae: Mechanisms, challenges, and future perspectives. *Ecotoxicology and Environmental Safety*. 270, 115908.

Kanagaraj, K., et al., 2012. Highly selective fluorescent sensing of fenitrothion using per-6-amino-β-cyclodextrin: Eu (III) complex. *Biosensors and Bioelectronics*. 35, 452–455.

Kanagaraj, K., et al., 2022. pH-Controlled enantioselectivity switching of irregular photodimers in photocyclodimerization of 2-anthracenecarboxylic acid mediated with β-cyclodextrin derivatives. *New Journal of Chemistry*. 46, 23066–23076.

Kanagaraj, K., et al., 2020a. pH-controlled chirality inversion in enantiodifferentiating photocyclodimerization of 2-antharacenecarboxylic acid mediated by γ-cyclodextrin derivatives. *Organic Letters*. 22, 5273–5278.

Kanagaraj, K., Pitchumani, K., 2010. Solvent-free multicomponent synthesis of pyranopyrazoles: Per-6-amino-β-cyclodextrin as a remarkable catalyst and host. *Tetrahedron Letters*. 51, 3312–3316.

Kanagaraj, K., et al., 2023a. Control of reactivity and selectivity in isomerization and rearrangement reactions inside confined spaces. *Chemical Communications*. 59, 9056–9065.

Kanagaraj, K., et al., 2023b. Selective binding and isomerization of oximes in a self-assembled capsule. *Journal of the American Chemical Society*. 145, 5816–5823.

Kanagaraj, K., et al., 2020b. A quinoline-appended cyclodextrin derivative as a highly selective receptor and colorimetric probe for nucleotides. *IScience*. 23, 100927.

Kapdan, I. K., Kargi, F., 2006. Bio-hydrogen production from waste materials. *Enzyme and Microbial Technology*. 38, 569–582.

Karmann, S., et al., 2019. Fed-batch cultivations of Rhodospirillum rubrum under multiple nutrient-limited growth conditions on syngas as a novel option to produce poly (3-hydroxybutyrate)(PHB). *Frontiers in Bioengineering and Biotechnology*. 7, 59.

Khan, M., et al., 2016. Comparing the value of bioproducts from different stages of anaerobic membrane bioreactors. *Bioresource Technology*. 214, 816–825.

Kim, S., et al., 2018. Removal of contaminants of emerging concern by membranes in water and wastewater: A review. *Chemical Engineering Journal*. 335, 896–914.

Klotz, M. G., et al., 2006. Complete genome sequence of the marine, chemolithoautotrophic, ammonia-oxidizing bacterium *Nitrosococcus oceani* ATCC 19707. *Applied and Environmental Microbiology*. 72, 6299–6315.

Koku, H., et al., 2003. Kinetics of biological hydrogen production by the photosynthetic bacterium *Rhodobacter sphaeroides* OU 001. *International Journal of Hydrogen Energy*. 28, 381–388.

Krane, J., 2017. Climate change and fossil fuel: An examination of risks for the energy industry and producer states. *MRS Energy & Sustainability*. 4, E2.

Kuhad, R. C., et al., 2016. Revisiting cellulase production and redefining current strategies based on major challenges. *Renewable and Sustainable Energy Reviews*. 55, 249–272.

Kumar, G., et al., 2020a. Thermochemical conversion routes of hydrogen production from organic biomass: Processes, challenges and limitations. *Biomass Conversion and Biorefinery*. 1–26.

Kumar, R., et al., 2020b. Sustainable production and purification of succinic acid: A review of membrane-integrated green approach. *Journal of Cleaner Production*. 277, 123954.

Larimer, F. W., et al., 2004a. Complete genome sequence of the metabolically versatile photosynthetic bacterium Rhodopseudomonas palustris. *Nature Biotechnology*. 22, 55–61.

Larimer, M. E., et al., 2004b. Predicting drinking behavior and alcohol-related problems among fraternity and sorority members: Examining the role of descriptive and injunctive norms. *Psychology of Addictive Behaviors*. 18, 203.

Larminie, J., et al., 2003. *Fuel Cell Systems Explained*. Wiley.

LaSarre, B., et al., 2017. Microbial mutualism dynamics governed by dose-dependent toxicity of cross-fed nutrients. *The ISME Journal*. 11, 337–348.

Laxman Pachapur, V., et al., 2015. Co-culture strategies for increased biohydrogen production. *International Journal of Energy Research*. 39, 1479–1504.

Lee, J. W., et al., 2011. Microbial production of building block chemicals and polymers. *Current Opinion in Biotechnology*. 22, 758–767.

Leonowicz, A., et al., 1999. Biodegradation of lignin by white rot fungi. *Fungal Genetics and Biology*. 27, 175–185.

Levin, D. B., et al., 2004. Biohydrogen production: Prospects and limitations to practical application. *International Journal of Hydrogen Energy*. 29, 173–185.

Li, Q., et al., 2018. Production of succinate from simply purified crude glycerol by engineered Escherichia coli using two-stage fermentation. *Bioresources and Bioprocessing*. 5, 1–10.

Li, S., et al., 2020. Hydrogen production: State of technology. In *IOP Conference Series: Earth and Environmental Science*, Vol. 544, p. 012011. IOP Publishing.

Li, S., et al., 2024. A review of green biohydrogen production using anoxygenic photosynthetic bacteria for hydrogen economy: Challenges and opportunities. *International Journal of Hydrogen Energy*. 54, 218–238.

Liu, H., et al., 2021. Microbial physiological engineering increases the efficiency of microbial cell factories. *Critical Reviews in Biotechnology*. 41, 339–354.

Logan, B. E., et al., 2008. Microbial electrolysis cells for high yield hydrogen gas production from organic matter. *Environmental Science & Technology*. 42, 8630–8640.

Łukajtis, R., et al., 2018. Hydrogen production from biomass using dark fermentation. *Renewable and Sustainable Energy Reviews*. 91, 665–694.

Mandal, B., et al., 2006. Improvement of biohydrogen production under decreased partial pressure of H_2 by Enterobacter cloacae. *Biotechnology Letters*. 28, 831–835.

Manish, S., Banerjee, R., 2008. Comparison of biohydrogen production processes. *International Journal of Hydrogen Energy*. 33, 279–286.

Manjunathan, J., et al., 2023. Recent advancements in nanotechnological approaches for pollution monitoring and environmental sustainability. *Clean Technologies and Environmental Policy*. 1–17.

MariaJoseph, A., et al., 2024. Rational coupling of selective electrochemical oxidation and reduction reactions for in-situ value-added chemical generation. *Fuel*. 367, 131408.

Marvuglia, A., et al., 2020. Advances and challenges in assessing urban sustainability: An advanced bibliometric review. *Renewable and Sustainable Energy Reviews*. 124, 109788.

Mechold, U., et al., 1996. Functional analysis of a relA/spoT gene homolog from Streptococcus equisimilis. *Journal of Bacteriology*. 178, 1401–1411.

Melis, A., Melnicki, M. R., 2006. Integrated biological hydrogen production. *International Journal of Hydrogen Energy*. 31, 1563–1573.

Melitos, G., et al., 2021. Waste to sustainable biohydrogen production via photo-fermentation and biophotolysis – A systematic review. *Renewable Energy and Environmental Sustainability*. 6, 45.

Millati, R., et al., 2011. Biological pretreatment of lignocelluloses with white-rot fungi and its applications: A review. *BioResources*. 6, 5224–5259.

Mishra, P., et al., 2019. Outlook of fermentative hydrogen production techniques: An overview of dark, photo and integrated dark-photo fermentative approach to biomass. *Energy Strategy Reviews*. 24, 27–37.

Mohanraj, S., et al., 2019. Metabolic engineering and molecular biotechnology of biohydrogen production. *Biohydrogen*. Elsevier. 413–434.

Montaño López, J., et al., 2022. Physiological limitations and opportunities in microbial metabolic engineering. *Nature Reviews Microbiology*. 20, 35–48.

Mudhoo, A., et al., 2011. Biohydrogen production and bioprocess enhancement: A review. *Critical Reviews in Biotechnology*. 31, 250–263.

Nagarajan, D., et al., 2017. Recent insights into biohydrogen production by microalgae – From biophotolysis to dark fermentation. *Bioresource Technology*. 227, 373–387.

Naha, S., et al., 2020. Nanomolar detection of H_2S in an aqueous medium: Application in endogenous and exogenous imaging of hela cells and zebrafish. *ACS Omega*. 5, 19896–19904.

Nair, A., et al., 2023. Multifunctional natural derived carbon quantum dots from *Withania somnifera* (L.)–Antiviral activities against SARS-CoV-2 pseudoviron. *Environmental Research*. 239, 117366.

Nangan, S., et al., 2023a. Sustainable preparation of luminescent carbon dots from syringe waste and hyaluronic acid for cellular imaging and antimicrobial applications. *Environmental Research*. 237, 116990.

Nangan, S., et al., 2023b. Waste plastics derived nickel-palladium alloy filled carbon nanotubes for hydrogen evolution reaction. *Chemosphere*. 341, 139982.

Natarajan, V., et al., 2019. A far-red to NIR emitting ultra-sensitive probe for the detection of endogenous HOCl in zebrafish and the RAW 264.7 cell line. *Organic & Biomolecular Chemistry*. 17, 3538–3544.

Nguyen, M.-H., et al., 2020. A highly selective fluorescent anthracene-based chemosensor for imaging Zn^{2+} in living cells and zebrafish. *Inorganic Chemistry Communications*. 115, 107882.

Nielsen, J., 2001. Metabolic engineering. *Applied Microbiology and Biotechnology*. 55, 263–283.

Nirmala, N., et al., 2023. A review on biological biohydrogen production: Outlook on genetic strain enhancements, reactor model and techno-economics analysis. *Science of the Total Environment*. 896, 165143.

Oh, Y.-K., et al., 2004. Photoproduction of hydrogen from acetate by a chemoheterotrophic bacterium Rhodopseudomonas palustris P4. *International Journal of Hydrogen Energy*. 29, 1115–1121.

Oh, Y.-K., et al., 2002. Fermentative hydrogen production by a new chemoheterotrophic bacterium Rhodopseudomonas palustris P4. *International Journal of Hydrogen Energy*. 27, 1373–1379.

Olaniran, A. O., et al., 2013. Bioavailability of heavy metals in soil: Impact on microbial biodegradation of organic compounds and possible improvement strategies. *International Journal of Molecular Sciences*. 14, 10197–10228.

Osman, A. I., et al., 2024. Membrane technology for energy saving: Principles, techniques, applications, challenges, and prospects. *Advanced Energy and Sustainability Research*. 5, 2400011.

Oumer, A., et al., 2018. Bio-based liquid fuels as a source of renewable energy: A review. *Renewable and Sustainable Energy Reviews*. 88, 82–98.

Patel, S. K. S., et al., 2024. Recent developments in hydrogen production, storage, and transportation: challenges, opportunities, and perspectives. *Fire*. 7, 233.

Patil, R., Gawande, R., 2016. A review on solar tunnel greenhouse drying system. *Renewable and Sustainable Energy Reviews*. 56, 196–214.

Pinjari, S., 2023. Development of physisorption materials for hydrogen storage properties at room temperature.

Podder, A., et al., 2020. Two-photon active fluorescent indicator for detecting NADH dynamics in live cells and tumor tissue. *Sensors and Actuators B: Chemical*. 324, 128637.

Potts, R., et al., 2018. Environmental dynamics during the onset of the Middle Stone Age in eastern Africa. *Science*. 360, 86–90.

Preechakun, T., et al., 2024. Detoxification of hemicellulose-enriched hydrolysate from sugarcane bagasse by activated carbon and macroporous adsorption resin. *Biomass Conversion and Biorefinery*. 14, 14559–14574.

Putatunda, C., et al., 2023. Current challenges and future technology in photofermentation-driven biohydrogen production by utilizing algae and bacteria. *International Journal of Hydrogen Energy*. 48, 21088–21109.

Quehenberger, J., et al., 2017. Sulfolobus–a potential key organism in future biotechnology. *Frontiers in microbiology*. 8, 2474.

Rabha, D., et al., 2017. Experimental investigation of thin layer drying kinetics of ghost chilli pepper (Capsicum Chinense Jacq.) dried in a forced convection solar tunnel dryer. *Renewable Energy*. 105, 583–589.

Rahman, S., et al., 2015. Overview of biohydrogen production technologies and application in fuel cell. *American Journal of Chemistry*. 5, 13–23.

Rai, P. K., Singh, S., 2016. Integrated dark-and photo-fermentation: Recent advances and provisions for improvement. *International Journal of Hydrogen Energy*. 41, 19957–19971.

Rajapandi, S., et al., 2023. Ziziphus mauritiana-derived nitrogen-doped biogenic carbon dots: Eco-friendly catalysts for dye degradation and antibacterial applications. *Chemosphere*. 338, 139584.

Rama, V., et al., 2011. Syntheses of 5-substituted 1 H-tetrazoles catalyzed by reusable CoY zeolite. *The Journal of Organic Chemistry*. 76, 9090–9095.

Rama, V., et al., 2012a. A multicomponent, solvent-free, one-pot synthesis of benzoxanthenones catalyzed by HY zeolite: Their anti-microbial and cell imaging studies. *Tetrahedron Letters*. 53, 1018–1024.

Rama, V., et al., 2012b. Pyridinium ylide-assisted KY zeolite catalyzed tandem synthesis of polysubstituted cyclopropanes. *Catalysis Communications*. 26, 39–43.

Ramasubburayan, R., et al., 2024a. Correction: Greener approach supported nitrogen-infused carbon dots for biocompatible cellular markers and fluorescent ink based spray-assisted fingerprint analysis. *Waste and Biomass Valorization*. 15, 7139.

Ramasubburayan, R., et al., 2024b. Greener approach supported nitrogen-infused carbon dots for biocompatible cellular markers and fluorescent ink based spray-assisted fingerprint analysis. *Waste and Biomass Valorization*. 15, 6431–6440.

Rao, M., et al., 2018. Photocatalytic supramolecular enantiodifferentiating dimerization of 2-anthracenecarboxylic acid through triplet–triplet annihilation. *Organic Letters*. 20, 1680–1683.

Resnick, R. J., 2004. *The economics of Biological Methods of Hydrogen Production*. Massachusetts Institute of Technology.

Rokhina, E. V., et al., 2009. Low-frequency ultrasound in biotechnology: State of the art. *Trends in Biotechnology*. 27, 298–306.

Sabaté Brescó, M., et al., 2017. Pathogenic mechanisms and host interactions in Staphylococcus epidermidis device-related infection. *Frontiers in Microbiology*. 8, 1401.

Saha, B. C., et al., 2016. Biological pretreatment of corn stover with white-rot fungus for improved enzymatic hydrolysis. *International Biodeterioration & Biodegradation*. 109, 29–35.

Saifuddin, N., Priatharsini, P., 2016. Developments in bio-hydrogen production from algae: A review. *Research Journal of Applied Sciences, Engineering and Technology*. 12, 968–982.

Saini, A., Kaur, S. J., 2024. Lignocellulosic biomass pretreatment for enhanced bioenergy recovery. *Biomass Energy for Sustainable Development*. CRC Press, pp. 205–241.

Samadi, A., et al., 2022. Waste-derived low-cost ceramic membranes for water treatment: Opportunities, challenges and future directions. *Resources, Conservation and Recycling*. 185, 106497.

Sarker, T. R., et al., 2021. Hydrothermal pretreatment technologies for lignocellulosic biomass: A review of steam explosion and subcritical water hydrolysis. *Chemosphere*. 284, 131372.

Serôdio, J., et al., 2018. An LED-based multi-actinic illumination system for the high throughput study of photosynthetic light responses. *PeerJ*. 6, e5589.

Serov, A., et al., 2015. Nano-structured non-platinum catalysts for automotive fuel cell application. *Nano Energy*. 16, 293–300.

Serrano-Martínez, V. M., et al., 2024. Steam Explosion-Based Method for the Extraction of Cellulose and Lignin from Rice Straw Waste. *Applied Sciences*. 14, 2059.

Sharmila, V. G., et al., 2022. Trends in dark biohydrogen production strategy and linkages with transition towards low carbon economy: An outlook, cost-effectiveness, bottlenecks and future scope. *International Journal of Hydrogen Energy*. 47, 15309–15332.

Shellaiah, M., et al., 2019. Cysteamine-capped gold-copper nanoclusters for fluorometric determination and imaging of chromium (VI) and dopamine. *Microchimica Acta*. 186, 1–8.

Shellaiah, M., et al., 2024. Sensing Utilities of Cesium Lead Halide Perovskites and Composites: A Comprehensive Review. *Sensors*. 24, 2504.

Shellaiah, M., et al., 2022. An AIEE active anthracene-based nanoprobe for Zn^{2+} and tyrosine detection validated by bioimaging studies. *Chemosensors*. 10, 381.

Shellaiah, M., et al., 2021. A pH cooperative strategy for enhanced colorimetric sensing of Cr (III) ions using biocompatible L-glutamic acid stabilized gold nanoparticles. *Microchemical Journal*. 160, 105754.

Shen, Z., et al., 2018. Synergy of lignocelluloses pretreatment by sodium carbonate and bacterium to enhance enzymatic hydrolysis of rice straw. *Bioresource Technology*. 249, 154–160.

Shin, J.-H., et al., 2007. Fermentative hydrogen production by the newly isolated Enterobacter asburiae SNU-1. *International Journal of Hydrogen Energy*. 32, 192–199.

Singer, S., et al., 2018. Anaerobic membrane gas extraction facilitates thermophilic hydrogen production from Clostridium thermocellum. *Environmental Science: Water Research & Technology*. 4, 1771–1782.

Singh, N., Sarma, S., 2022. Biological routes of hydrogen production: A critical assessment. In *Handbook of Biofuels*, pp. 419–434. Elsevier.

Sirohi, R., et al., 2022. Design and applications of photobioreactors-a review. *Bioresource Technology*. 349, 126858.

Sivaramakrishnan, R., et al., 2021. Insights on biological hydrogen production routes and potential microorganisms for high hydrogen yield. *Fuel*. 291, 120136.

Sivaranjani, R., et al., 2023. A comprehensive review on biohydrogen production pilot scale reactor technologies: Sustainable development and future prospects. *International Journal of Hydrogen Energy*. 48, 23785–23820.

Soares, J. F., et al., 2020. Dark fermentative biohydrogen production from lignocellulosic biomass: Technological challenges and future prospects. *Renewable and Sustainable Energy Reviews*. 117, 109484.

Sørensen, H. P., Mortensen, K. K., 2005. Soluble expression of recombinant proteins in the cytoplasm of Escherichia coli. *Microbial Cell Factories*. 4, 1–8.

Sovacool, B. K., et al., 2020. Sociotechnical agendas: Reviewing future directions for energy and climate research. *Energy Research & Social Science*. 70, 101617.

Srirangan, K., et al., 2012. Towards sustainable production of clean energy carriers from biomass resources. *Applied Energy*. 100, 172–186.

Suresh, D., et al., 2013. Synthesis of 2-substituted 3-ethyl-3H-imidazo [4, 5-b] pyridines catalyzed by Al^{3+}-exchanged K10 clay as solid acids. *Tetrahedron Letters*. 54, 6479–6484.

Suresh, D., et al., 2014. Microwave promoted one-pot synthesis of 2-aryl substituted 1, 3, 4–oxadiazoles and 1, 2, 4–oxadiazole derivatives using Al^{3+}-K10 clay as a heterogeneous catalyst. *Tetrahedron Letters*. 55, 3678–3682.

Sverdrup, G., et al., 2008. *Status of Renewable Hydrogen Production in the United States*. National Renewable Energy Lab (NREL), Golden, CO (United States).

Tabassum, N., et al., 2021. Progress in microbial fuel cells for sustainable management of industrial effluents. *Process Biochemistry*. 106, 20–41.

Tang, M. M., et al., 2022. Role of Rim Functions in Recognition and Selectivity of Small-Molecule Guests in Water-Soluble Cavitand Hosts. *Chemistry–An Asian Journal*. 17, e202200466.

Tapia-Venegas, E., et al., 2015. Biohydrogen production by dark fermentation: Scaling-up and technologies integration for a sustainable system. *Reviews in Environmental Science and Bio/Technology*. 14, 761–785.

Teke, G., et al., 2024. Towards industrial biological hydrogen production: A review. *World Journal of Microbiology and Biotechnology*. 40, 37.

Thirumalaivasan, N., 2024. Letter to the Editor "assessing the performance of virus-mimic nanoparticles against other nano-drug delivery systems in efficiency, biocompatibility, and patient results". *Oral Oncology Reports*. 10, 100436.

Thirumalaivasan, N., et al., 2017. Highly selective turn-on probe for H_2S with imaging applications in vitro and in vivo. *New Journal of Chemistry*. 41, 13510–13515.

Thirumalaivasan, N., et al., 2019. In vitro and in vivo approach of hydrogen-sulfide-responsive drug release driven by azide-functionalized mesoporous silica nanoparticles. *ACS Applied Bio Materials*. 2, 3886–3896.

Thirumalaivasan, N., et al., 2024a. Utilization of fungal and bacterial bioremediation techniques for the treatment of toxic waste and biowaste. *Frontiers in Materials*. 11, 1416445.

Thirumalaivasan, N., et al., 2024b. Nano-PCM materials: Bridging the gap in energy storage under fluctuating environmental conditions. *Process Safety and Environmental Protection*. 189, 1003–1021.

Thirumalaivasan, N., et al., 2024c. Exploring luminescent carbon dots derived from syrup bottle waste and curcumin for potential antimicrobial and bioimaging applications. *Chemosphere*. 354, 141592.

Thirumalaivasan, N., et al., 2024d. Exploring antimicrobial and biocompatible applications of eco-friendly fluorescent carbon dots derived from fast-food packaging waste transformation. *Environmental Research*. 244, 117888.

Thirumalaivasan, N., et al., 2024e. *Assessment of Sustainability and Environmental Impacts of Renewable Energies: Focusing on Biogas and Biohydrogen (Biofuels) Production*. Process Safety and Environmental Protection.

Thirumalaivasan, N., Wu, S.-P., 2020. Bright luminescent carbon dots for multifunctional selective sensing and imaging applications in living cells. *ACS Applied Bio Materials*. 3, 6439–6446.

Tiang, M. F., et al., 2020. Recent advanced biotechnological strategies to enhance photo-fermentative biohydrogen production by purple non-sulphur bacteria: An overview. *International Journal of Hydrogen Energy*. 45, 13211–13230.

Toscan, A., et al., 2024. Comparison of liquid hot water and saturated steam pretreatments to evaluate the enzymatic hydrolysis yield of elephant grass. *Biomass Conversion and Biorefinery*. 14, 8057–8070.

Turker, L., Gumus, S., & Tapan, A. (2008). Biohydrogen production: molecular aspects. *Journal of Scientific and Industrial Research*. 67(11), 994.

Unuofin, J. O., et al., 2023. Embracing the future of circular bio-enabled economy: Unveiling the prospects of microbial fuel cells in achieving true sustainable energy. *Environmental Science and Pollution Research*. 30, 90547–90573.

Uys, A., 2019. *Age Estimation of Living South African Individuals: A Multifactorial Model*. University of Pretoria.

Velusamy, N., et al., 2019. FRET-based dual channel fluorescent probe for detecting endogenous/exogenous H_2O_2/H_2S formation through multicolor images. *Journal of Photochemistry and Photobiology B: Biology*. 191, 99–106.

Velusamy, N., et al., 2018. A hydrogen sulfide triggered self-immolative fluorescent probe for lysosome labeling in live cells. *New Journal of Chemistry*. 42, 1590–1594.

Venkatesan, P., et al., 2023. Eco-friendly, bright luminescent carbon dots and their potential applications for detecting hypochlorous acid in water and live cell imaging. *Journal of Materials Research and Technology*. 24, 6522–6532.

Venkatesan, P., et al., 2019. Redox stimuli delivery vehicle based on transferrin-capped MSNPs for targeted drug delivery in cancer therapy. *ACS Applied Bio Materials*. 2, 1623–1633.

Verma, D., et al., 2024. A sustainable and green approach towards the utilization of biopolymers for effective wound dressing applications: A detailed review. *Nano-Structures & Nano-Objects*. 37, 101086.

Wang, B., Feng, H., 2010. *Detoxification of Lignocellulosic Hydrolysates*, p. 233. Wiley Online Library.

Wang, D., et al., 2015. Comparing the hydrolysis and biogas production performance of alkali and acid pretreatments of rice straw using two-stage anaerobic fermentation. *Biosystems Engineering*. 132, 47–55.

Wang, H., et al., 2021. Alkaline thermal pretreatment of waste activated sludge for enhanced hydrogen production in microbial electrolysis cells. *Journal of Environmental Management*. 294, 113000.

Wang, Y., et al., 2017. Pubchem bioassay: 2017 update. *Nucleic Acids Research*. 45, D955–D963.

Wang, Y., et al., 2016. Advances in the genetic modification of Rhodobacter sphaeroides to improve hydrogen production. *Renewable and Sustainable Energy Reviews*. 60, 1312–1318.

Wei, L.-F., et al., 2019. A nano-molar fluorescent turn-on probe for copper (II) detection in living cells. *Methods*. 168, 18–23.

Wei, L.-F., et al., 2017. Fluorescent coumarin-based probe for cysteine and homocysteine with live cell application. *Spectrochimica Acta Part A: Molecular and Biomolecular Spectroscopy*. 183, 204–208.

Woźniak-Budych, M. J., 2023. Polymeric membranes for biomedical applications. *Physical Sciences Reviews*. 8, 1181–1211.

Wu, G.-S., et al., 2020. Ultrasensitive and specific two-photon fluorescence detection of hypochlorous acid by a lysosome-targeting fluorescent probe for cell imaging. *Journal of Pharmaceutical and Biomedical Analysis*. 190, 113545.

Xiao, C., et al., 2020. Redox-triggered chirality switching and guest-capture/release with a pillar [6] arene-based molecular universal joint. *Angewandte Chemie International Edition*. 59, 8094–8098.

Xiao, Y., et al., 2017. Mapping the ecological networks of microbial communities. *Nature Communications*. 8, 2042.

Yadav, V. G., et al., 2012. The future of metabolic engineering and synthetic biology: Towards a systematic practice. *Metabolic Engineering*. 14, 233–241.

Yang, B., et al., 2011. Enzymatic hydrolysis of cellulosic biomass. *Biofuels*. 2, 421–450.

Yang, G., Wang, J., 2018. Various additives for improving dark fermentative hydrogen production: A review. *Renewable and Sustainable Energy Reviews*. 95, 130–146.

Yang, W.-Y., et al., 2022. A two photon fluorescent probe for highly selective detection and endogenous imaging of hydrogen sulfide. *Spectrochimica Acta Part A: Molecular and Biomolecular Spectroscopy*. 273, 121043.

Yin, H., Yip, A. C., 2017. A review on the production and purification of biomass-derived hydrogen using emerging membrane technologies. *Catalysts*. 7, 297.

Yogyta, K., et al., 2023. Comparative analysis of various methods to enhance the bioethanol production. *International Journal of Renewable Energy Technology*. 14, 192–216.

Yoo, C. G., et al., 2020. The critical role of lignin in lignocellulosic biomass conversion and recent pretreatment strategies: A comprehensive review. *Bioresource Technology*. 301, 122784.

Yoshida, A., et al., 2005. Enhanced hydrogen production from formic acid by formate hydrogen lyase-overexpressing Escherichia coli strains. *Applied and Environmental Microbiology*. 71, 6762–6768.

Younas, M., et al., 2022. An overview of hydrogen production: Current status, potential, and challenges. *Fuel*. 316, 123317.

Yu, J., Takahashi, P., 2007. Biophotolysis-based hydrogen production by cyanobacteria and green microalgae. *Communicating Current Research and Educational Topics and Trends in Applied Microbiology*. 1, 79–89.

Yu, Y., et al., 2022. Steam explosion of lignocellulosic biomass for multiple advanced bioenergy processes: A review. *Renewable and Sustainable Energy Reviews*. 154, 111871.

Zhakeyev, A., et al., 2017. Additive manufacturing: Unlocking the evolution of energy materials. *Advanced Science*. 4, 1700187.

Zhang, C., et al., 2024. Improving photofermentative hydrogen productivity of photosynthetic bacteria using a formulated Fe and Mo metal supplemented lignocellulosic substrate. *International Journal of Hydrogen Energy*. 49, 516–531.

Zhang, F., 2019. Development of CRISPR-Cas systems for genome editing and beyond. *Quarterly Reviews of Biophysics*. 52, e6.

Zhang, Q., et al., 2021. Enhancement strategies for photo-fermentative biohydrogen production: A review. *Bioresource Technology*. 340, 125601.

Zhao, Q., et al., 2022. Using nanomaterials to increase the efficiency of chemical production in microbial cell factories: A comprehensive review. *Biotechnology Advances*. 59, 107982.

Zhen, G., et al., 2019. Anaerobic membrane bioreactor towards biowaste biorefinery and chemical energy harvest: Recent progress, membrane fouling and future perspectives. *Renewable and Sustainable Energy Reviews*. 115, 109392.

Zou, C., et al., 2022. Industrial status, technological progress, challenges, and prospects of hydrogen energy. *Natural Gas Industry B*. 9, 427–447.

8 Advancements in Nanotechnology and the Use of Nanomaterials in Valorizing Biohydrogen Production by Biomass Conversion

Manoj Kumar, Aditi Sarker, Asim Kumar Jana, and Nikhil Gauravarapu Navlur

8.1 INTRODUCTION

The global energy demand continues to escalate due to rapid population growth and industrialization (Sinharoy et al., 2020). The finite nature of fossil fuels, coupled with their price instability and adverse environmental effects, has intensified the global pursuit of alternative energy solutions (Chen et al., 2023; Epelle et al., 2022; Kumar et al., 2024). Among these, renewable energy is increasingly seen as a sustainable option for mitigating the challenges of fossil fuels (Shelare et al., 2023; Singh et al., 2020). Biofuels, in particular, have gained recognition as eco-friendly, efficient, readily producible, and cost-effective renewable energy alternatives (Bosu and Rajamohan, 2022). Liquid and gaseous biofuels exemplify this category, such as biogas, biomethane, bioethane, bio-butane, biohydrogen, bio-butanol, biodiesel, bio-oil, and bioethanol. Of these, biohydrogen has garnered significant interest due to its unique advantages, including the generation of water vapor as a clean by-product, its high energy density of 121 kJ/g, and

 DOI: 10.1201/9781003568612-8

the ability to be generated from diverse feedstocks and naturally occurring microorganisms. Moreover, biohydrogen production under ambient temperature and pressure conditions enhances its feasibility for large-scale applications (Kumar et al., 2019; Sinharoy et al., 2020). The research on biohydrogen production experienced a surge in interest and research between 2016 and 2024 (Figure 8.1).

Biohydrogen production can be achieved by several biological techniques, including direct and indirect bio-photolysis and fermentation methods, particularly dark and photo-fermentation (Akhlaghi and Najafpour-Darzi, 2020; Bhatia et al., 2021). These approaches employ diverse organic substrates and microorganisms to produce hydrogen (Song et al., 2022). Among various substrates, lignocellulosic biomass is particularly notable because of its high cellulose content, renewable properties, and widespread availability, making it a promising candidate for large-scale biohydrogen production (Megía et al., 2021). Although biomass conversion to biohydrogen is regarded as one of the most sustainable strategies in biofuel development, numerous barriers must be overcome for its industrial-scale application (Singh et al., 2020). Key obstacles include the complex structure of lignin in cellulosic biomass, inefficient cellulose-to-sugar conversion processes, the high cost of cellulolytic enzymes, and low yields in substrate-to-product conversion, all of which hinder the advancement of viable cellulosic biohydrogen technologies (Yue et al., 2021). Cellulolytic enzymes are essential in facilitating the conversion of biomass into biohydrogen. Challenges such as the high cost of enzymes, low production efficiency, and limited enzyme activity pose significant barriers to achieving higher sugar yields, directly influencing biohydrogen production (Kumar et al., 2019). Overcoming these obstacles requires strategies to enhance cellulose accessibility for bioconversion and improve enzyme production and performance. Advancing these aspects is critical for the development of sustainable and economically feasible cellulosic biohydrogen production technologies.

Various strategies have been developed to enhance enzyme functionality, boost yield, and increase the efficiency of biohydrogen generation. Despite these efforts, achieving implementation that is both sustainable and cost-effective remains a significant hurdle (Bolivar et al., 2022). Nanotechnology has emerged as a promising tool in various biochemical applications (Priya et al., 2021). Owing to their unique properties, nanomaterials present great potential for enhancing the conversion of biomass into biohydrogen (Singh et al., 2020; Tang et al., 2023). Attributes such as superior electrical conductivity and a high surface-to-volume ratio enable catalytic improvements across different stages of cellulosic biohydrogen production (Epelle et al., 2022). Through biological processes such as photo-fermentation, bio-photolysis, and dark fermentation, a variety of nanomaterials, such as Cu, Ag, Fe, Ni, Au, and Ti,

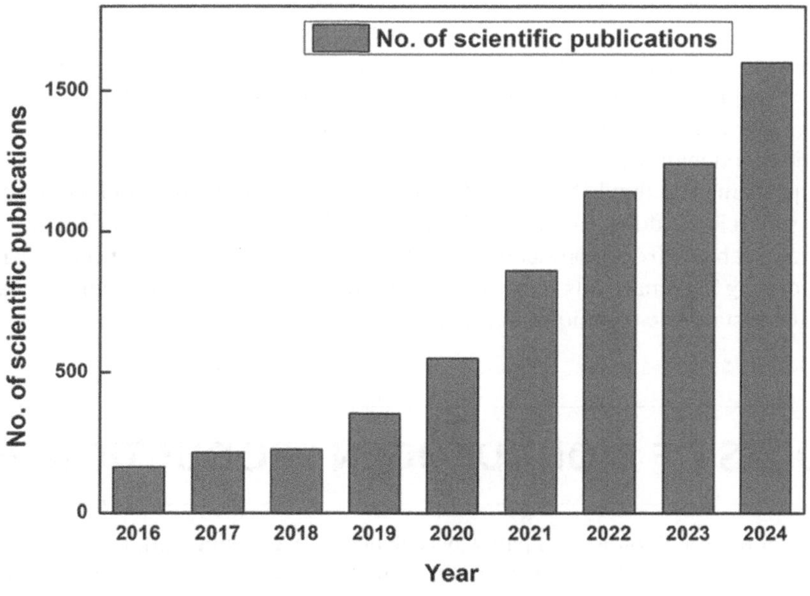

FIGURE 8.1 Annual scientific publications in supplementation of nanoparticles in biohydrogen production from 2016 to 2024.

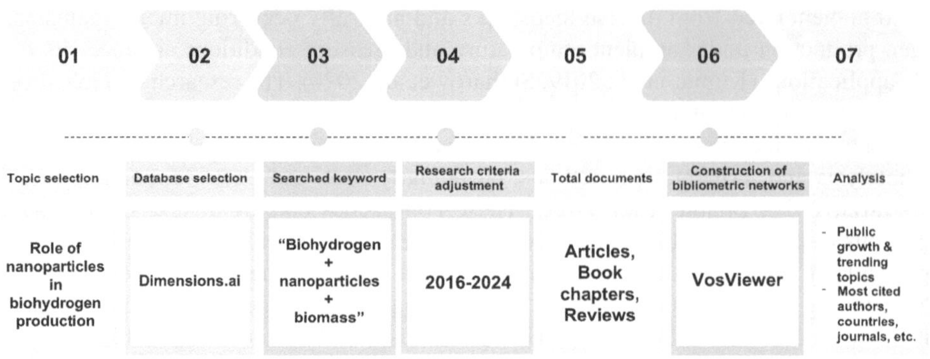

FIGURE 8.2 Schematic flow representation of data collection and visualization for bibliometric analysis.

have been effectively used to improve the production of biohydrogen (Deivayanai et al., 2022; Velvizhi et al., 2020; Younas et al., 2022). By eliminating any remaining oxygen during the fermentation process and lowering the oxidation-reduction potential, these substances also function as oxygen scavengers. This modification increases the production of biohydrogen by creating an anaerobic environment that is favorable for hydrogenase activity. Moreover, the use of nanomaterials in the pre-treatment of lignocellulosic biomass aids in lignin degradation, which improves sugar yields and accelerates the conversion process (Deivayanai et al., 2022). They also enhance the efficiency of production by stabilizing cellulase enzymes against variations in temperature and pH, promoting the effective hydrolysis of cellulose into fermentable sugars (Akhlaghi & Najafpour-Darzi, 2020; Bosu and Rajamohan, 2022; Epelle et al., 2022). The data collection and visualization methodology for the bibliometric study are represented in Figure 8.2. The VOSViewer software has been utilized to analyze the intricate literature and identify the author's keyword clusters that correspond to different research themes. The keyword used for the bibliometric analysis is "Biohydrogen + nanoparticles + biomass." The clustering of the most prominent sources concerning citation and countries concerning bibliographic coupling in applying nanoparticles (NPs) in biohydrogen production is noticed in Figure 8.3. Citation represents the frequency with which other documents in the dataset have cited a particular source. In bibliographic coupling, countries are analyzed based on authors/countries. It identifies which countries are leading contributors to a specific area of research and reveals its pattern of international collaboration and partnerships (Jannat et al., 2024). This analysis revealed *Bioresource Technology* and the *Journal of Environmental Management* as the two most significant journals in the sector, and India and China showed the most prominent bibliographic coupling.

This chapter highlights the diverse approaches inspired by remarkable physicochemical characteristics and catalytic functionalities of nanomaterials, focusing on their role in enhancing biohydrogen production. Key aspects include the different modes of biohydrogen production and factors affecting NPs to boost biohydrogen yields. Additionally, this chapter provides an in-depth examination of how different nanomaterials impact biohydrogen production and its mechanism. Recent advancements in this field suggest that incorporating nanomaterials into biological routes for biohydrogen production can pave the way for innovative and sustainable techniques for biohydrogen generation.

8.2 MODES OF BIOHYDROGEN PRODUCTION PROCESS

Hydrogen production through biological pathways can be achieved using dark fermentation, direct photolysis, indirect bio-photolysis, and photo-fermentation methods. Additionally, hybrid systems that combine dark fermentation processes with photosynthetic approaches represent an innovative strategy for enhancing biohydrogen yield and overall process efficiency (Mehra et al., 2025). The biohydrogen production from various feedstocks via different modes is summarized in Table 8.1.

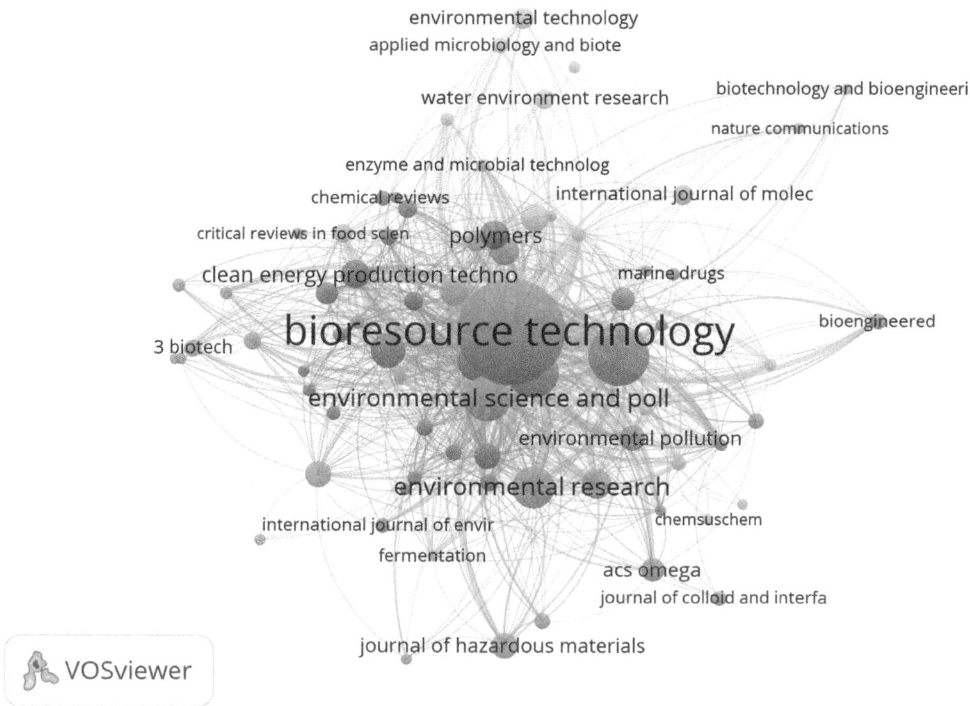

(a)

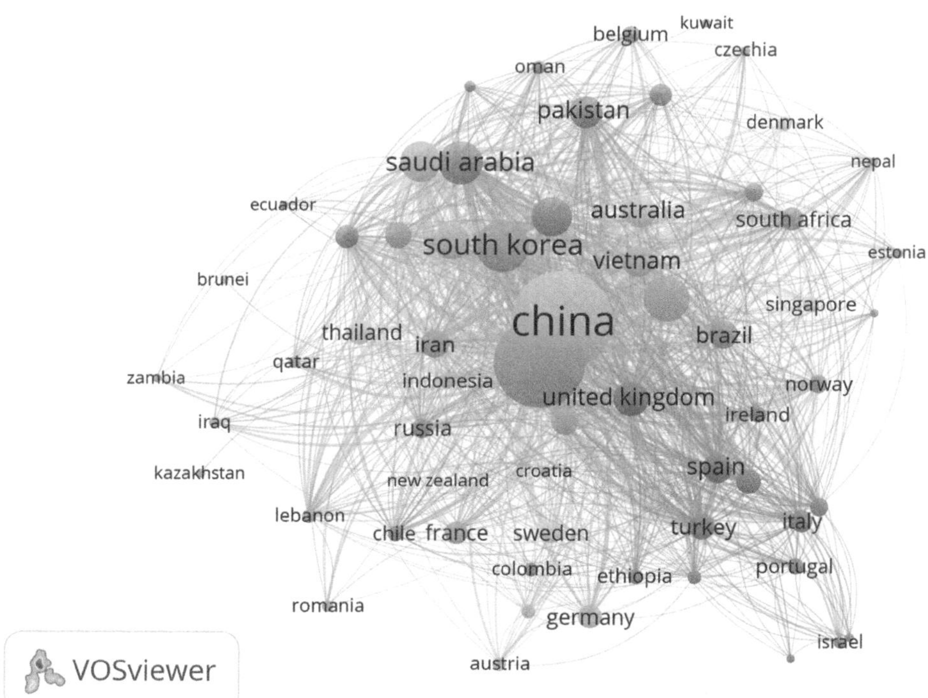

(b)

FIGURE 8.3 (a) Clustering of the most prominent sources concerning citation in the application of nanoparticles in biohydrogen production, (b) clustering of the countries concerning bibliographic coupling.

TABLE 8.1 Biohydrogen yield from various feedstock via different modes of production

SL. NO.	*FEEDSTOCK*	*INOCULUM*	*BIOHYDROGEN YIELD*	*MODE OF PRODUCTION*	*REFERENCE*
1.	Xylose (10 g/L)	***Thermoanaerobacterium thermosaccharolyticum*** KKU19	2.09 ± 0.002 mol H_2/mol xylose	Dark fermentation	Khamtib and Reungsang (2012)
2.	Corn stover hydrolysate	***Thermoanaerobacterium thermosaccharolyticum*** W16	108.5 mmol/L	Dark fermentation	Ren et al. (2010)
3.	Glucose Xylose	***Thermoanaerobacterium thermosaccharolyticum*** W16	2.42 mol H_2/mol glucose 2.19 mol H_2/mol xylose	Dark fermentation	Ren et al. (2008)
4.	Glucose Fructose Sucrose Acetate	***Rhodopseudomonas*** strain A7	33 mL H_2/60 mL media 30 mL H_2/60 mL media 43 mL H_2/60 mL media 160 mL H_2/60 mL media	Photo-fermentation	Liu et al. (2015)
5.	Succinate (50 mM) Butyrate (50 mM)	Photosynthetic bacterial strain ZX-5	223 ± 10 mL/35 mL succinate 324 ± 15 mL/35 mL butyrate	Photo-fermentation	Tao (2008)
6.	Glucose (10 g/L)	***Platymonas subcordiformis***	138.45 ± 3.39 mL	Bio-photolysis	Dudek et al. (2018)
7.	Benzoate (600 mg/L)	***Lyngbya*** species of cyanobacteria	17.05 μmol H_2/g Chlorophyll a	Bio-photolysis	Shi and Yu (2016)
8.	Bagasse	adhE inactivated ***Klebsiella oxytoca*** HP1	162.1 ± 7.5 mL H_2/g bagasse	Hybrid system	Wu et al. (2010)
9.	Beet molasses	***Caldicellulosiruptor saccharolyticus, Rhodobacter capsulatus*** hupmutant	13.7 mol H_2/mol hexose	Hybrid system	Özgür et al. (2010)
10.	Starch wastewater	Free mixed culture bacteria, ***Pseudomonas*** and ***Rhodopseudomonas palustris*** sp.	1.43 ± 0.24 mol H_2/mol glucose	Hybrid system	Nasr et al. (2015)
11.	Cheese whey wastewater	***Enterobacter aerogenes*** MTCC 2822, ***Rhodopseudomonas*** BHU 01	3.40 mol H_2/mol lactose	Hybrid system	Rai et al. (2012)
12.	Fruit and vegetable waste and cheese whey powder	***Clostridium butyricum, Rhodopseudomonas palustris***	793.7 mL H_2/g COD	Hybrid system	Niño-Navarro et al. (2020)

8.2.1 Dark Fermentation

It is widely accepted that dark fermentation can produce biohydrogen with various substrates and operate under ambient conditions at lower energy levels than in conventional methods. Despite these advantages, this mode is constrained by low hydrogen yields and incomplete substrate utilization.

It can largely be attributed to the accumulation of acidic metabolites produced due to microbial activity (Alvarado-Cuevas et al., 2013; Nikolaidis and Poullikkas, 2017). Additionally, the practical hydrogen yield often falls short of the theoretical maximum. Acetic acid (CH_3COOH) is the main by-product, and the maximum yield that can be achieved is 4 mol of H_2 per mol of glucose ($C_6H_{12}O_6$). Process efficiency is significantly decreased by forming various by-products, such as lactic acid and ethanol (Chaganti et al., 2012). Using fermentative bacteria like *Clostridium*, *Citrobacter*, *Rhodopseudomonas*, and *Enterobacter* generates biohydrogen on a large scale. Numerous tests have utilized these bacteria to produce biohydrogen and break down organic waste effectively (Corneli et al., 2016; Ramprakash and Muthukumar, 2018). The enzyme hydrogenase aids the dark fermentation process for creating biohydrogen. It can be categorized in various ways due to its metal binding capacity. Improving hydrogenase's bioactivity has an enormous possibility of raising the total production yield. Pyruvate is then converted into acetyl-CoA, while ferredoxin undergoes reduction. This process results in the generation of acetate and ATP. The reduced ferredoxin is subsequently re-oxidized by the hydrogenase enzyme, facilitating the synthesis of biohydrogen (Das and Veziroglu, 2008). The type of substrate and the number of microorganisms used have a significant impact on efficiency. Moreover, it operates independently of light and is unaffected by the oxygen inhibition of the hydrogenase enzyme. Additionally, it generates a variety of metabolic intermediates that hold added value and can be utilized for synthesizing other commercially relevant products (da Silva Veras et al., 2017). Despite its benefits, the anaerobic microorganisms cannot metabolize these by-products, leading to their accumulation and a subsequent sharp decrease in pH. As a result, it limits the biohydrogen yield and necessitates additional treatment strategies to optimize production (Khamtib and Reungsang, 2012; Nikhil et al., 2018).

8.2.2 Photo-Fermentation

The organic and volatile fatty acids (VFAs) produced during dark fermentation can be substrates for photo-fermentative bacteria. This enables biohydrogen production through the action of nitrogenase enzymes under light-driven, nitrogen-deficient conditions (Chandra et al., 2015). Common photo-fermentative bacteria include species of *Rhodopseudomonas*, *Rhodobacter*, *Rhodovulum*, and *Anabaena* (Corneli et al., 2016). These microorganisms metabolize organic wastes, converting them into carbon dioxide and hydrogen through nitrogenase enzymes. The conversion occurs in an anaerobic environment with ATP as the energy source. Photo-fermentation is a promising method for producing hydrogen in large quantities, with peaks of up to 8 mol per mole of acetic acid (Das and Veziroglu, 2008). The hydrogen production method has certain benefits, including increased H_2 yields without feedback of nitrogenase in the presence of oxygen. Additionally, the absence of photosystem (PS) II in the photo-fermentative bacteria reduces their sensitivity toward oxygen and enhances their capabilities of biohydrogen production. The functioning of such bacteria is versatile, and they can switch to photosynthesis in carbon-rich environments and nitrogen-adequate conditions. In contrast, they can engage in biohydrogen synthesis in nitrogen-deprived environments. Despite these advantages, photo-fermentation faces challenges, including relatively low productivity of biohydrogen and prolonged reaction time (Liu et al., 2015; Sambusiti et al., 2015). This limits its commercial feasibility. Addressing these constraints through advanced biotechnological interventions and process optimization remains a critical area of research for realizing its full potential.

8.2.3 Bio-Photolysis

Bio-photolysis is a photo-driven process primarily observed in green algae and cyanobacteria. While it shares its fundamental principles with plant photosynthesis, the key distinction lies in its optimization for hydrogen generation rather than carbon-based biomass synthesis. Notable algal species in this mode include *Chlamydomonas reinhardtii*, *Scenedesmus*, and *Chlorella* (Giannelli and Torzillo, 2012; Oey et al., 2016; Rashid et al., 2011). Bio-photolysis can be categorized into two pathways: direct and indirect bio-photolysis. Direct bio-photolysis involves cyanobacteria and microalgae, utilizing nitrogenase and bi-directional hydrogenase enzymes. Both PS I and PS II are involved in this process, triggering the photolysis of water, producing hydrogen ions and oxygen. Photon absorption excites electrons in PS II. These are then transported to ferredoxin via PS I. Ferredoxin subsequently donates these electrons to the hydrogenase enzyme that catalyzes the reduction of hydrogen ions to molecular hydrogen (H_2) (da Silva Veras et al., 2017; Dincer and Acar, 2015). Indirect bio-photolysis is a multistep process facilitated by pigments such as carotenoids, chlorophyll, and phycobiliproteins. These pigments enable cyanobacteria to capture sunlight effectively. The captured light can be transformed into chemical energy via photosynthesis. It results in the production of biomass, including glucose and other monomeric sugars. When water and light are present, this biomass undergoes hydrolysis, producing H_2. This hydrolysis is catalyzed by enzymes nitrogenase and hydrogenase (da Silva Veras et al., 2017; Nikolaidis and Poullikkas, 2017). Cyanobacteria capable of biohydrogen production can be categorized into nitrogen-fixing strains (such as *Anabaena* sp., *Oscillatoria*) and non-nitrogen-fixing strains (such as *Synechococcus*, *Lyngbya* sp.). Nitrogen-fixing cyanobacteria utilize two distinct pathways for biohydrogen production: heterocystous nitrogen fixation and non-heterocystous nitrogen fixation (Shi and Yu, 2016). This approach to biohydrogen production is energetically demanding, requiring a substantial amount of ATP along with traces of CO_2 being produced as a by-product of hydrogen (Manish & Banerjee, 2008).

8.2.4 Hybrid System

The hybrid system for biohydrogen production represents a significant advancement over the conventional techniques. It is aimed at enhancing both the yield and overall productivity of biohydrogen. This approach to biohydrogen production leverages the synergistic interaction between non-photosynthetic and photosynthetic bacteria. By integrating these two biological pathways, the hybrid system achieves higher production rates, improved biohydrogen yield, and greater process efficiency (Liu et al., 2010; Zagrodnik and Łaniecki, 2017). During dark fermentation, non-photosynthetic bacteria generate short-chain fatty acids such as acetic, lactic, butyric, etc. These cause a sharp decline in pH, negatively impacting the bacterial activity and reducing the hydrogen yield. However, in hybrid systems, these organic acids serve as substrates for the photo-autotrophic microorganisms. They use acids as carbon sources, producing additional hydrogen and carbon dioxide during photosynthesis. This reaction maximizes the overall efficiency of biohydrogen production (da Silva Veras et al., 2017; Sambusiti et al., 2015; Zagrodnik and Łaniecki, 2017).

8.3 NANOTECHNOLOGY IN BIOHYDROGEN PRODUCTION: ROLE OF NANOMATERIALS IN CATALYSIS

Hydrogenase and nitrogenase are two essential enzymes in biohydrogen generation that mediate hydrogen production by diverse biological systems. The reversible oxidation of hydrogen into protons and electrons is mediated by the metalloenzyme hydrogenase, which has a bimetallic catalytic site. The redox potential of the molecule that binds to the enzyme affects the reaction's outcome (Das, 2001; Kim and Kim, 2011). The nitrogenase enzyme catalyzes hydrogen evolution through nitrogen fixation in anoxic environments. They are primarily discovered in bacteria and archaea, and their efficiency in producing hydrogen is markedly lower than that of hydrogenase. Targeting those enzymes has, therefore, been the primary recognition of techniques to boost the output of biohydrogen, which has stronger efficiency of electron shipping in microbial structures (Nagarajan et al., 2017). Biohydrogen technology calls for pyruvate ferredoxin oxidoreductase, similar to hydrogenase and nitrogenase. This enzyme lowers ferredoxin and aids in the breakdown of pyruvate into acetyl-CoA. The reduced ferredoxin ultimately transfers the electrons to hydrogenase, appreciably affecting the hydrogen yield (Wu et al., 2017). The primary techniques for generating biohydrogen at some point of dark fermentation are formate and NADH-dependent pathways. The formate hydrogen manufacturing pathway includes breaking formate into hydrogen, which the enzyme formate hydrogen lyase enables. This technique has a theoretical yield of 2 mol of hydrogen in step with 1 mol of glucose (Nikhil et al., 2014). As an opportunity, the hydrogenase enzyme promotes the NADH-based pathway of hydrogen generation. This can produce up to 4 mol of hydrogen for every mole of glucose (Hallenbeck, 2009). Some results demonstrate how hydrogen generation can be enhanced by employing NPs to target and optimize specific enzyme pathways. Due to their huge surface areas, NPs have exceptional electron absorption capabilities, and their quantum size directly affects the electron transfer rate. This special characteristic greatly increases the catalytic efficiency of enzymes and offers better selectivity for synthesizing biohydrogen via several biological pathways (Macrelli et al., 2014). On supplementation into the fermentation medium, these NPs interact with the microbial cells and penetrate the cells through the cell membranes. Once internalized, they improve the enzymes' activities, with metallic clusters as their catalytic locales. Initially, they encourage the electron exchange between ferredoxin and hydrogenase. Further, it channels the generated electron toward hydrogenase to promote the evaluation of hydrogen (Han et al., 2011; Mohanraj et al., 2014a). Notably, metals or metallic NPs such as Fe^{2+} and Ni^{2+} also promote hydrogenase activity. Research revealed how Fe^0 NPs specifically contribute to the synthesis of iron-containing ferredoxin proteins. This significantly accelerates the electron transfer between ferredoxin and hydrogenase (Yang and Wang, 2018). Supplementation of nano-TiO_2 also improved the activity of the nitrogenase enzyme, which occurs due to suppressing hydrogenase uptake. As a result, increased hydrogen evolution by photosynthetic bacteria was observed. This was followed by improved hydrolysis of polysaccharides and increased bacterial growth (Zhao and Chen, 2011). Besides titanium dioxide, iron, nickel, and molybdenum NPs are also known to affect nitrogenase activity. They regulate the gene expression of nitrogenase enzymes, leading to a notable increment in hydrogen production (Özgür et al., 2010). The strategic supplementation of NPs at optimal concentrations could improve production efficiency and support the metabolic activity and growth of the fermenting microbial communities.

8.4 TYPES OF NANOMATERIALS FOR BIOHYDROGEN PRODUCTION

The application of NPs has recently shown significant potential in enhancing biohydrogen production. This can be attributed to their high surface area-to-volume ratio and strong electron exchange potential. The NPs collectively facilitate smoother metabolic processes in microorganisms (Patel et al., 2018). Commonly used NPs in biohydrogen production include metal and non-metal-based such as Gold (Au), Copper (Cu), Palladium (Pd), Iron (Fe), Nickel (Ni), SiO_2, Fe_2O_3, and TiO_2. Carbon-based NPs like carbon nanotubes (CNTs), graphene, and various nanocomposites have also been widely employed in this context. Table 8.2 summarizes the effect of different NPs on the yield of biohydrogen.

8.4.1 Gold NPs

The utilization of gold nanoparticles (Au NPs) has been shown to have significant advantages in enhancing enzyme bioactivity, immobilizing capabilities, and catalytic activity across different chemical reactions (Zhang and Shen, 2007). A study examined the catalytic activity of Au NPs in sucrose degradation by anaerobic cultures of *Clostridium butyricum* under unheated and pre-heating conditions. Using Au NPs with different sizes (ranging from 5 to 10 and up to 20 nm) has demonstrated that biohydrogen production is more productive in these conditions than in control groups. Pre-heated 5 nm Au NPs demonstrated superior performance, producing a maximum cumulative biohydrogen output of 4.48 mol and

TABLE 8.2 Effect of different nanoparticles on the yield of biohydrogen

SL NO.	*NANOPARTICLES*	*SUBSTRATE*	*INOCULUM*	*OPERATIONAL CONDITIONS*	*BIO-H_2 YIELD INCREASE (%)*	*REFERENCES*
1.	Zerovalent iron (Fe^0)	Grass	***Clostridium*** sp.	pH 7.0, 37°C	15.9	Yang and Wang (2018)
2.	Fe_2O_3	Glucose	Thermophilic mixed bacteria	pH 5.5, 60°C	34.38	Engliman et al. (2017)
3.	Fe_3O_4	Glucose Sugarcane bagasse	Anaerobic sludge Anaerobic sludge	pH 7.0, 35°C pH 5.0, 30°C	26.4 69.6	Zhao et al. (2011) Reddy et al. (2017)
4.	Nickel oxide	Palm oil mill effluent	***Bacillus anthracis*** PUNAJAN 1	pH 7.0, 37°C	151	Mishra et al. (2018)
5.	TiO_2	Acetate Glucose	***Rhodopseudomonas*** sp. ***Rhodobacter sphaeroides***	pH 7.0, 37°C pH 8.0, 34°C	8 63.27	Liu et al. (2017) Pandey et al. (2015)
6.	Silver	Glucose	***Clostridium butyricum***	pH 8.0–9.4, 35.5°C	67.5	Zhao et al. (2013)
7.	SiO_2	Air:CO_2 (97:3)	***Chlamydomonas reinhardtii*** CC124	pH 4.2, 28°C	45.2	Giannelli and Torzillo (2012)
8.	Palladium	Glucose	***Enterobacter cloacae*** + mixed culture	pH 7.0, 37°C	6.4	Mohanraj, Anbalagan, et al. (2014b)
9.	Gold	Acetate Artificial wastewater	Electrochemically active bacteria	- -	56 67.5	Khan et al. (2013)
10.	Maghemite	Starch wastewater	***Clostridium*** sp. + ***Rhodopseudomonas palustris***	pH 6.7, 30°C	57.8	Nasr et al. (2015)
11.	Powdered activated carbon	Glucose Sucrose	Anaerobic sludge	pH 5.5, 37°C	62.5 73	Wimonsong and Nitisoravut (2015) Wimonsong and Nitisoravut (2014)
12.	Granulated activated carbon	Starch wastewater	Acidogenic mixed culture	pH 5.5, 28°C	94.5	Mohan et al. (2008)

achieving a 56% yield. This was particularly impressive. The results emphasize the significance of Au NP size in determining biohydrogen production efficiency (Zhang and Shen, 2007). When the reaction was facilitated in a single-chamber reactor with sodium acetate, these NPs provided recombinant charges that allowed protons to reduce on the surface (Khan et al., 2013). Additionally, biohydrogen production has been improved by photoinduced fermentation employing Au NPs. For instance, introducing silica-core gold-shell NPs resulted in the fermentation of acetate and malate in the *Rhodopseudomonas palustris* CGA009 strain. Without preventing bacterial growth, the procedure produced 4.38 mmol of hydrogen with malate and 2.31 mol of hydrogen with acetate, a 115% increase in hydrogen generation. These NPs' improved effectiveness is explained by their increased surface area and smaller size, which efficiently aid in microbial adhesion to the active sites, a crucial step in the synthesis of biohydrogen (Ji et al., 2021). Despite their encouraging results, their expensive cost hampers the practical use of Au NPs in producing biohydrogen. One significant barrier to these NPs' widespread use is their cost (Nasr et al., 2015).

8.4.2 Palladium NPs

During fermentative biohydrogen production, palladium nanoparticles (Pd NPs) have significantly affected substrates and microbial activity. Their catalytic function in raising hydrogenase activity and hydrogen yields is principally to blame for this (Deplanche et al., 2010). Investigating the impact of Pd NPs made with leaf extract from *Coriandrum sativum* on the synthesis of biohydrogen revealed a 0.6% and 6.4% increase in yield, yielding 1.48 and 2.48 mol of hydrogen. A concentration of 5.0 mg/L was added during the investigation. The experiment was performed in a medium dominated by mixed cultures based on glucose and *Enterobacter cloacae* 811 and 101. Furthermore, although Pd^{2+} ions prolonged the lag phase and decreased yields, these NPs considerably decreased the lag phase linked to hydrogen generation. *E. cloacae* 811, 101 showed a lower hydrogen output than the mixed culture. Surprisingly, Pd NPs did not impair the bioactivity of the individual or mixed cultures, even at higher concentrations (20 mg/L) (Mohanraj et al., 2014a). Additionally, Pd NPs have demonstrated encouraging outcomes in the catalytic synthesis of hydrogen from formic acid. A carbon catalyst was created using nitrogen-doped activated carbon (AC) and Pd NPs, which have an average size of about 2.8 nm. This facilitates hydrogen generation from formic acid (Yao et al., 2020). Furthermore, Pd NPs that have reduced particle sizes to less than 2 nm and are attached to porous carbon through hydrolysis with dimethylamine-borane (DMAB) contribute to enhanced hydrogen generation (Wen et al., 2020). These advancements indicate the adaptability of Pd NPs in speeding up biohydrogen production.

8.4.3 Silver NPs

Silver nanoparticles (Ag NPs) are becoming more widespread in various fields, including biological research, protein engineering, and waste management. Their use in synthesizing biohydrogen and as a substrate for fermentation has been widely discussed (Strapasson et al., 2021). It is also important that these NPs promote acidogenesis, a crucial process in biohydrogen production (P. Sharma et al., 2022), to aid bacteria and algae growth. There is evidence of a significant improvement in fermentative biohydrogen and total acidogenesis, with the ability to generate 2.48 mol of hydrogen for every mole of glucose. Glucose and *C. butyricum* were introduced into the anaerobic batch reactor through immunization. Production was 67.6% higher than in the control group. It was also found that using Ag NPs increased hydrogen production and decreased the time required for hydrogen synthesis (Zhao et al., 2013). Researchers used *Clostridium beijerinckii* to create these NPs from two different sources, and their relative effects were compared. Heena and DU-C2 actinomycete were used in the green synthesis process to create Ag NPs. Ag NPs obtained from Heena significantly increased the overall biohydrogen production to 1.71 mol H_2 per glucose molecule after 72 hours of cultivation. Less quantity was produced by the one synthesized from the DU-C2 actinomycetes compound (I. Khan et al., 2021). Furthermore, the investigation examined how the silver oxide nanoparticles (AgO NPs) affect biohydrogen yield. The homogeneous structure of the chlorella species (85 nm particle size) was used. It acted as a reducing agent in the production of AgO NPs (Yildirim et al., 2022). A study investigated the influence of AgO (100–600 mg/L) on biohydrogen production by *Clostridium* species during dark fermentation. It was observed that the production of H_2 increased by 17% when AgO NPs were present at 400 mg/L.

8.4.4 Copper NPs

Copper NPs, also known as Cu NPs in the fermentation process, have been shown to enhance hydrogen generation. Cu NPs affect the *C. butyricum* CWBI1009 strain's ability to ferment glucose efficiently. At a concentration of 10^{-6} mol/L, Cu NPs with a 2–3 nm particle size encased in porous SiO_2 greatly enhanced hydrogen yield (Beckers et al., 2013). However, Cu NPs have also exhibited inhibitory effects on anaerobic biohydrogen production processes. For instance, when glucose was used as a substrate, *E. cloacae* 811, 101, and *C. acetobutyricum* NCIM 2337 experienced reduced hydrogen production with increasing

Cu NP concentrations. At 2.5 mg/L, H_2 yields decreased by 2.9% and 3.5%, respectively, compared to control values of 1.44 and 1.74 mol. At 12.5 mg/L, the reductions were even more pronounced, with yields dropping by 72.2% and 56.9%. Higher concentrations of Cu NP were associated with a decrease in the acetate-to-butyrate ratio, which negatively affected hydrogen production. On the other hand, larger acetate-to-butyrate ratios were linked to better H_2 yields, while higher propionate and alcohol levels produced less hydrogen (H. Li et al., 2018; Muri et al., 2018). Intense antibacterial activity is the main reason for Cu NPs' negative effects on biohydrogen generation. These results emphasize the importance of keeping Cu NP concentrations below a critical threshold (less than 2.5 mg/L) to maximize hydrogen production during anaerobic processes and avoid adverse effects on microbial activity (Mohanraj et al., 2016).

8.4.5 Nickel NPs

It has been shown that nickel ions (Ni^{2+}) and nickel nanoparticles (Ni NPs) can enhance hydrogen generation by promoting fermentative electron transfer and raising hydrogenase enzyme activity (Taherdanak et al., 2015). Several studies have demonstrated that they can increase hydrogen production. Adding Ni^{2+} and Ni NPs at concentrations below 10 mg/L in acid hydrolysis of maize stover led to a significant increase in hydrogen production (Sun et al., 2020). NiO NPs synthesized from *Eichhornia crassipes* extract further demonstrated their utility in regulating hydrogen production. With a particle size of 9.1 nm and 20 mg/L (concentration), NiO NPs achieved a 37.78% increase in hydrogen yield compared to control (Zhang et al., 2021). The combination of NiO and Fe_2O_3 NPs has shown synergistic effects. Research reported that co-adding NiO and Fe_2O_3 NPs during the fermentative process of dairy wastewater significantly enhanced biohydrogen yields (Gadhe et al., 2015). This combined NP approach increased hydrogen production by 1.2–1.5 times, achieving a maximum hydrogen yield improvement of 27%. The synergistic effect likely results from enhanced hydrogenase and ferredoxin oxidoreductase activities. Despite these benefits, excessive concentrations of Ni NPs can adversely affect the overall production. A study revealed a reduction in the production when these NPs were used at a concentration of 5 mg/L, with the control producing 60% more hydrogen (Taherdanak et al., 2015). Conversely, adding 25 mg/L Ni^{2+} yielded 55% hydrogen yield (Mishra et al., 2018). High concentrations of NiO NPs (above 3 mg/L) were found to be toxic to bacterial cells, reducing cell counts by up to 62.24%. In summary, Ni and NiO NPs enhance biohydrogen production substantially, but their performance is highly dose-sensitive. Optimized concentrations are crucial to maximize hydrogen yields without compromising microbial viability and activity.

8.4.6 Titanium Dioxide NPs

Titanium dioxide nanoparticles (TiO_2 NPs) have demonstrated significant potential in enhancing biohydrogen production efficiency during various processes. The impact of TiO_2 NPs during anaerobic digestion and utilizing UV-treated sugarcane bagasse and acid hydrolysis resulted in a biohydrogen of 101.6 mL/g volatile solids (VS) (Jafari and Zilouei, 2016). Similarly, investigating the effect of these NPs utilizing *Rhodobacter sphaeroides* NMBL-02 under visible light showed an increase in biohydrogen yield by 154% compared to the control (Pandey et al., 2015). When *C. reinhardtii* cells were encapsulated under sulfur-depleted conditions within the TiO_2 cells, they exhibited double the hydrogen production efficiency compared to the non-encapsulated counterparts (Stojkovic et al., 2015). It was also found that these NPs enhanced hydrogen production from waste-activated sludge through photo-fermentation. The hydrogen yield increased by 46.1% upon adding these NPs at a concentration of 100 mg/L. This improvement was associated with breaking complex chemical compounds into simpler substrates, better enzyme performance, and increased photosynthetic bacterial activity (Zhao and Chen, 2011). Titanium dioxide NPs have demonstrated themselves as adaptable biohydrogen generation enhancers, promoting increases in microbial activity, enzymatic function, and substrate conversion efficiency in various processes.

8.4.7 Iron NPs

The potential of Fe NPs, which are highly reactive, has been widely discussed for their ability to produce biohydrogen from organic waste and carbohydrates. This is due to their exceptional properties. Additionally, the research focused on the role of Fe^0 NPs in microbial dynamics and fermentation in hydrogen production (Yang and Wang, 2018). They found that Fe^0 NPs increased the activity of bacteria and moved the dominant population of microorganisms from *Enterobacter* sp. to *Clostridium* sp. This thereby suggests a more efficient metabolic pathway. The Fe^0 NPs increase Fe^{2+} ions, which is the reason for this. The electron transport between these ions and the ferredoxin/hydrogenase pathway enhances enzyme activity. According to another study, the biohydrogen outputs of anaerobic bacteria were significantly increased by adding lignin-mediated Fe NPs to their conspecifics. Lignin magnetic nanoparticles (LMNPs) and lignin nonmagnetic nanoparticles (LNMNPs) were the two kinds of NPs that were evaluated. When 200 mg/L of these were introduced to the fermentation medium, the yields increased 2.8 and 2.3 times, respectively. Moreover, the hydrogen production increased by 91.0% and 74.3% (do Nascimento Junior et al., 2021). It can be inferred from the studies that Fe NPs have the potential to enhance biohydrogen production when used at appropriate concentrations while highlighting the need to avoid excessive dosages to maintain microbial viability and process efficiency.

Various nanoscale iron oxides have been utilized to enhance biohydrogen production. One example is using hematite or ferric oxide NPs in dark hydrogen fermentation with *Enterobacter aerogenes* (Lin et al., 2016). On a hematite concentration of 200 mg/L, the yield from glucose increased by 17%. In contrast, 400 mg/L suppressed the production, leading to a 5.5% decrease compared to control (Zhang et al., 2018). Another study investigated the effect of hematite concentrations up to 1,600 mg/L using sucrose as the substrate and mixed bacterial culture (Han et al., 2011). Results revealed that at 200 mg/L hematite and pH 8.48, the highest biohydrogen yield was 3.21 mol H_2, a 32.64% improvement over the control test. The potential of magnetite (Fe_3O_4) NPs for biohydrogen production has also been explored (Zhong et al., 2020). Batch experiments showed that a concentration of 50 mg/L Fe_3O_4 NPs with a 40–60 nm particle size yielded 12.97 mL of H_2 per gram of substrate. These studies collectively highlight the critical influence of iron-based NP dosage on biohydrogen production, with concentrations exceeding 400 mg/L generally inhibiting microbial activity and reducing hydrogen yields.

8.4.8 Silica NPs

Silica nanoparticles (SiO_2 NPs) have positively influenced microbial activity and protein function, enhancing biohydrogen production (Patel et al., 2014). In a notable study, porous silica encapsulated NPs, creating silica-NP composites that facilitated glucose fermentation by *C. butyricum* (Beckers et al., 2013). When tested in batch hydrogen production experiments at minimal dosages, cultures containing iron oxide-silica composites produced 58% more hydrogen. Although research on applying SiO_2 NPs in biohydrogen production is limited, existing findings highlight their promising potential. Silica-based NPs benefit biohydrogen generation processes, making them a valuable area for further investigation.

8.4.9 Carbon-Based NPs

Carbon-based NPs, particularly CNTs, have demonstrated significant potential as catalytic supports in fermentative biohydrogen production. Composed of graphene sheets rolled into cylindrical structures, CNTs can exist as single-walled (SWCNTs) or multi-walled (MWCNTs) forms. SWCNTs consist of a single layer of carbon atoms, while MWCNTs comprise multiple graphene layers. Their exceptional ability to facilitate redox reactions and electron transfer dynamics makes them valuable in several applications (Alrebdi et al., 2022; Khilari et al., 2013; Zhong et al., 2019). Moreover, their cost-effectiveness supports their use in anaerobic production systems. The role of CNTs as microbial carriers was also explored. Results demonstrate enhanced microbial activity and biohydrogen production (Liu et al., 2012). In a reactor treated with 100 mg/L CNTs, the hydrogen yield increased to 2.45 mol, with a daily output of 5.55 L. The presence of CNTs accelerated fermentation start-up times and outperformed conventional AC particles, fostering greater coalescence among hydrogen-producing bacteria and significantly boosting hydrogen production efficiency. Similarly, biohydrogen production in a batch process using *E. aerogenes* immobilized on functionalized MWCNTs was also investigated (Boshagh et al., 2019). At a concentration of 1.2 mg/mL, the immobilized *E. aerogenes* achieved higher hydrogen yields, faster production rates, improved consistency, and a reduced lag phase compared to free bacterial cells. These findings underscore the effectiveness of CNTs in enhancing biohydrogen production processes.

8.5 MECHANISMS OF ENHANCED BIOHYDROGEN PRODUCTION

Enzymes are crucial in regulating the overall reaction of biomass conversion to biofuels. Enzymes such as hydrogenase and nitrogenase are crucial in biohydrogen production. These enzymes participate in hydrogen consumption and generation across different biological systems (Koul et al., 2022). Hydrogenase enables the interconversion of hydrogen into electrons and protons. The redox potential of the interacting molecules dictates the direction of the reaction (Akhlaghi and Najafpour-Darzi, 2020).

$$2H^+ + 2e^- \leftrightarrow H_2$$

Nitrogenase generates hydrogen as a by-product during nitrogen fixation without oxygen. This enzyme is commonly found in archaea and bacteria, though its hydrogen production efficiency is lower than that of hydrogenase (Akhlaghi and Najafpour-Darzi, 2020). Efforts to enhance biohydrogen production have focused on improving the performance of these enzymes, which has also led to better electron transfer efficiency in the organisms (Song et al., 2022). Pyruvate ferredoxin oxidoreductase, an essential enzyme that catalyzes the conversion of pyruvate to ferredoxin and acetyl-CoA, is crucial for biohydrogen production by facilitating electron transfer to hydrogenase (Deivayanai et al., 2022). In dark fermentation, biohydrogen production occurs through two main pathways: the formate hydrogen production pathway, where formate hydrogen lyase catalyzes formate decomposition to produce hydrogen yielding 2 mol of hydrogen per mole of glucose theoretically, and the NADH-dependent pathway, which has the potential to produce up to 4 mol of hydrogen for every mole of glucose, facilitated by the activity of hydrogenases (H. Wu et al., 2017) (Figure 8.4a). Recent studies propose that NPs may boost hydrogen production by interacting with specific enzymes. The underlying mechanisms are examined in the discussion that follows.

NPs possess larger surface areas, enabling them to absorb electrons more effectively. The quantum size of these particles influences the electron transfer rate, significantly improving the catalytic efficiency of enzymes responsible for biohydrogen production via particular biological pathways (Shanmugam et al., 2020). When introduced into the fermentation medium, NPs interact with microorganisms, entering cells via the cell membranes. These NPs boost enzyme activity, particularly for enzymes containing metallic clusters at their active sites. They primarily facilitate electron transfer from ferredoxin toward hydrogenase, channeling electrons to hydrogenase to promote hydrogen production (Macrelli et al., 2014; Shanmugam et al., 2020). Metallic ions or NPs, such as Fe^{2+} and Ni^{2+}, are especially effective in enhancing hydrogenase activity (Patel et al., 2018). Yang and Wang (2018) highlighted that Fe^0 NPs facilitate the formation of iron-containing ferredoxin proteins, which are crucial for enhancing electron transfer between ferredoxin and hydrogenase, essential enzymes in the process of biohydrogen production.

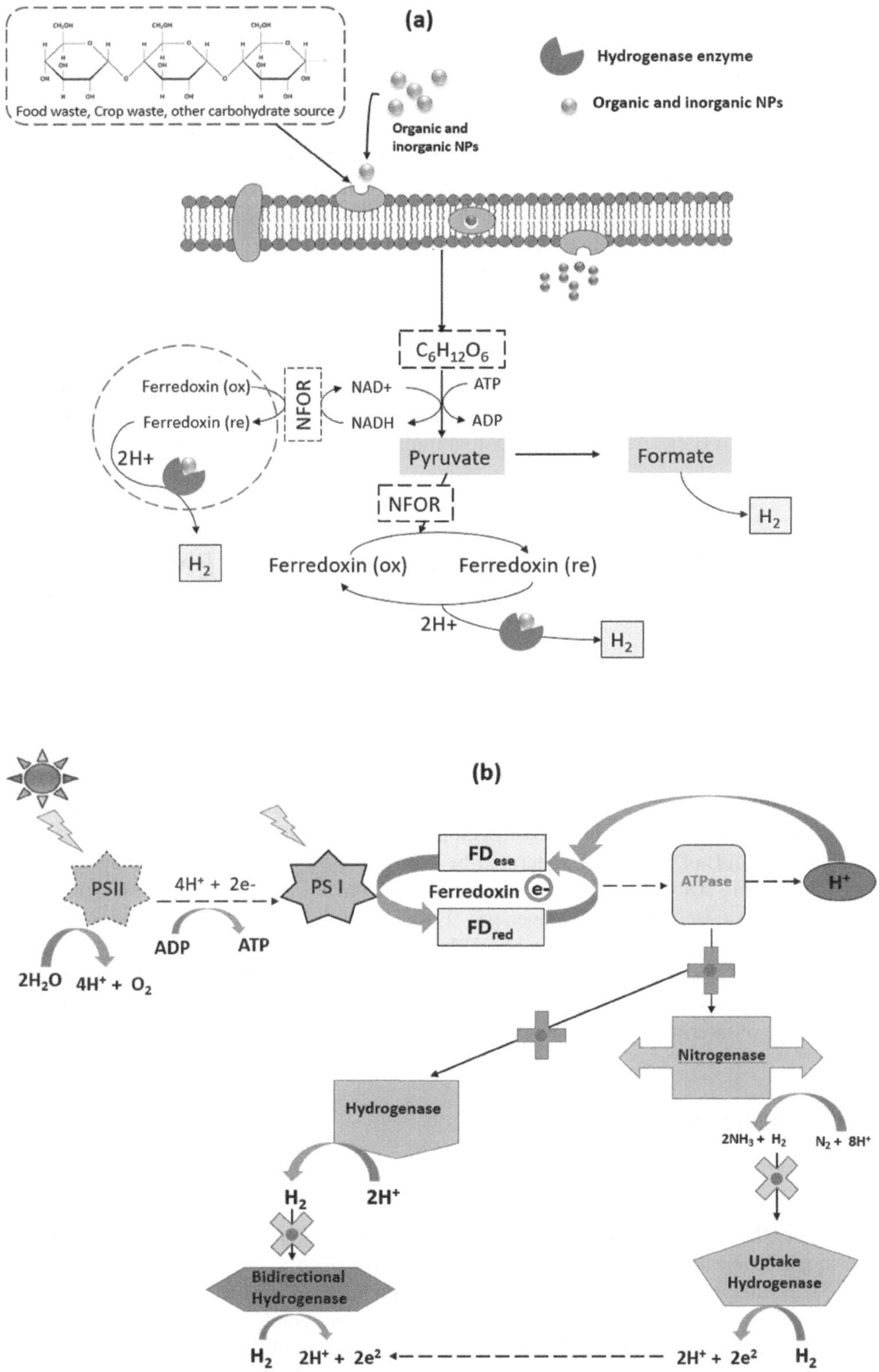

FIGURE 8.4 A proposed mechanism illustrating the role of various nanoparticles in biohydrogen production: (a) dark fermentation and (b) photo-fermentation hydrogen production pathways (Shanmugam et al., 2020).

In photo-fermentation, hydrogen is produced through the action of key enzymes, such as hydrogenases and nitrogenases, which play crucial roles in nitrogen fixation and hydrogen evolution (Sivaramakrishnan et al., 2021) (Figure 8.4b). Zhao and Chen (2011) explored the effect of nanoTiO_2 on hydrogen production by *R. palustris* using dark fermentation liquid derived from waste-activated sludge. Their research revealed that nanoTiO_2 enhanced nitrogenase activity by inhibiting hydrogenase uptake, leading to increased hydrogen production by the photosynthetic bacteria. Additionally, TiO_2 hydrolyzes polysaccharides and supports microbial growth. Additionally, Fe, Ni, and Mo NPs, which are essential cofactors for nitrogenase, are known to influence its activity. Mo plays a key role in regulating nitrogenase genes, thereby boosting hydrogen production in the case of photo-fermentation (Sharma et al., 2020). The relationship between Mo and nitrogenase activity is not yet fully understood and requires additional research. Additionally, NPs enhance biohydrogen production in bio-photolysis by significantly increasing hydrogenase activity. Lin et al. (2016) observed a significant improvement in the activity of *E. aerogenes*, resulting in a 17% increase in biohydrogen production through dark fermentation when supplemented with 200 mg/L of Fe_2O_3 NPs. In addition, immobilizing enzymes onto nanomaterials significantly enhances their thermodynamic stability. For example, Sánchez-Ramírez et al. (2017) demonstrated that cellulases immobilized on chitosan-coated magnetic nanomaterials (Fe_3O_4) exhibited a 1.5-fold improvement in catalytic efficiency.

8.6 FACTORS INFLUENCING THE EFFICIENCY OF NPS IN THE PROCESS OF BIOHYDROGEN PRODUCTION

8.6.1 pH

To optimize the growth of *Clostridium* sp., it is essential to maintain an appropriate pH during the initiation and operation of a bioprocess for hydrogen production using these bacteria (Łukajtis et al., 2018). A pH range of 5–6 is generally ideal for achieving the highest biohydrogen yields. Research indicates that a pH of 5.5 is particularly favorable for biohydrogen production (da Silva Veras et al., 2017). For instance, approximately 2,738 mL of H_2 can be generated by processing 50 g of COD per liter of food waste over 24 hours (Cabrol et al., 2017). The growth of methanogenic bacteria is inhibited by the acid hydrolysis of food waste, which also impedes methanogenesis (Fisgativa et al., 2016). Carbohydrates are most stable between pH 6 and 9, whereas proteins need a pH between 7 and 8 for optimal stability (Walker, 2009). Fermentative bacteria also thrive within a pH range of 4–8.5 (Zhang and Sun, 2014). Furthermore, previous research studies, have indicated that organic acids, like VFAs, can decrease pH levels and reduce the buffering capacity of intermediate products (Kamble et al., 2024; Lee et al., 2014; Mohan et al., 2017).

8.6.2 Temperature

Temperature is a key factor in converting substrates to biohydrogen. Hydrolysis and continuous digestion pre-treatments are classified into three categories based on the temperature: mesophilic, thermophilic, and hyper-thermophilic. Mesophilic bacteria produce hydrogen at temperatures lower than 37°C, while thermophilic bacteria generate lower amounts of hydrogen at elevated temperatures (Alemahdi et al., 2015). Hydrogen production is significantly enhanced by the thermophilic anaerobacterium *Thermoanaerobacterium thermosaccharolyticum*, which produces biohydrogen (Lamolinara et al., 2022). Thermal and biological treatments improve the activity of thermophilic bacteria at 70°C (Lamolinara et al., 2022). In hyper-thermophilic processes, food waste is hydrolyzed between 60°C and 80°C, which is particularly effective for high lipid and protein concentrations (Gadow et al., 2013). Previous research has shown that thermophilic and hyper-thermophilic temperatures can sustain continuous hydrogen production, though thermophilic temperatures generally yield higher hydrogen amounts than hyper-thermophilic ones (S. Sharma et al., 2020). Fang et al. (2006) observed that food waste hydrolysis at temperatures exceeding 80°C leads to the formation of intermediates like polylactide rather than hydrogen gas. Additionally, a decrease in temperature from 65°C to 40°C results in a rapid pH drop from 6.5 to 4.5, a more significant decline in comparison to mesophilic temperatures (35–25°C), which subsequently reduces hydrogen production potential due to the acidic conditions (Abanades et al., 2022).

8.6.3 Metal Ions

Nanoparticles of iron, magnesium, iron, and nickel play a crucial role in biohydrogen production by interacting with hydrogenase enzymes during fermentation. These metal ions are vital for enzyme synthesis, substrate transformation, and the transport processes in microorganisms (Bian et al., 2024). Research has demonstrated that the concentration of metal ions influences the kinetics of microorganisms and enzyme activity (Yogeswari et al., 2016). Iron is crucial for enzyme synthesis in microorganisms. The combination of iron and sulfur supports protein transport and facilitates the metabolic modification of hydrogenase enzymes, resulting in the production of CO_2 and H_2 from the substrate (Wang and Yin, 2021). Excessive concentrations of metal ions can inhibit the rate of biohydrogen production. The addition of extra ions accelerates microbial growth, which in turn leads to greater consumption of feed substrate. As a result, the inhibitory effect is reduced, thereby enhancing microbial production and metabolism. Nickel (Ni) plays a role in electron transfer during biohydrogen production, particularly when hydrogenase enzymes are involved. In fermentation, [Ni-Fe]-hydrogenase utilizes electrons to reduce substrate protons, producing biohydrogen (Hsieh et al., 2016; Lee, 2016).

8.7 APPLICATION OF NANOMATERIALS IN BIOREACTOR DESIGN FOR BIOHYDROGEN PRODUCTION

The biohydrogen production from photosynthetic microorganisms in bioreactors is greatly influenced by several factors. These include the light source, the uniform mixing of gas and nutrients, temperature, and biofilm formation, all of which can impact the overall efficiency of the process (Khoo et al., 2023). While several technologies have been suggested to improve reactor performance, there is a lack of detailed studies focusing on utilizing nanomaterials in bioreactor design for biohydrogen. Among the factors discussed earlier, light illumination plays a crucial role. Light intensity is a key factor influencing microbial growth and metabolism through photosynthesis (Chen et al., 2011). However, the temperature produced during illumination must also be considered, as excessive light intensity can lead to photo-inhibition, negatively impacting the hydrogen production process (Abu-Ghosh et al., 2016). Designing bioreactors for biohydrogen production requires using illuminating materials that are both cost-effective and capable of minimizing heat generation. Nanotechnology is crucial in optimizing nanomaterials to enhance their ability to manipulate light properties effectively. For instance Liao et al. (2014) designed an innovative biofilm photobioreactor (BPBR) that facilitates photosynthetic bacterium *R. palustris* CQK01 HOF adsorption. This system used a GeO_2-SiO_2-chitosan medium to amplify light intensity on the optical fiber surface. Its addition improved the refractive index, which resulted in a substantial 1.5-fold increase in hydrogen production compared to an uncoated optical fiber. Moreover, optical fibers improved light conversion efficiency without excessive heat production (Chan et al., 2023). Sun et al. (2016) devised a strategy to redistribute incident light in a flat plate photobioreactor by integrating planar waveguides embedded with light-scattering organo-silicone NPs measuring 25 cm in width. These NPs enhanced the illuminated surface area by 10.3. As a result, a remarkable 220% increase in biomass production compared to the flat plate without waveguides was observed. Additionally, for the BPBR, SiO_2-chitosan biomaterial was used to modify the surface of its light guiding plate, resulting in a light intensity of 7.9 W/m². This optimization resulted in a 1.12-fold increase in the biohydrogen production rate compared to the BPBR using the traditional polymethyl methacrylate (PMMA) flat plate.

The temperature of the photobioreactor plays a crucial role in influencing light energy, biofilm adhesion characteristics of cells, and metabolism. Traditional methods for temperature control typically rely on indirect heat transfer, which can result in reduced efficiency and increased energy consumption. Photothermal nanomaterials that simultaneously regulate temperature and light illumination have been developed to overcome these limitations. Li et al. (2017) introduced a photothermal nano-biomaterial, GeO_2-SiO_2-Chitosan-Medium-Lanthanumhexaboride (LaB6) [GSCML], capable of absorbing light in the 380–510 nm and 660–780 nm ranges, converting it into heat, and emitting light at approximately 590 nm. This material facilitated biofilm growth at 0.05 mg/cm²/day, with hydrogen production reaching 2.93 mmol/h/m^2. Furthermore, the concentration of LaB6 was adjusted to regulate both temperature and light intensity, highlighting the importance of different NPs in bioreactor design.

8.8 PROSPECTS AND CHALLENGES

Hydrogen is currently one of the most expensive biofuels, with production through photo-fermentation and dark fermentation methods being considered an eco-friendly alternative for sustainable generation. The global demand for biohydrogen is expected to rise substantially, increasing from 71 million metric tons (MMT) in 2017 to 136 MMT by 2040, 287 MMT by 2050, and 519 MMT by 2070, with the transportation sector being a major contributor to its use (Figure 8.5a). Correspondingly, the global market for biohydrogen is predicted to grow from USD 1.20 billion in 2020 to USD 10.05 billion by 2025, USD 37.02 billion by 2028, and USD 89.20 billion by 2030, as shown in Figure 8.5b. However, biologically produced hydrogen faces challenges similar to those of other renewable energy sources, such as high production costs and comparatively lower yields. As a result, different innovative techniques have been explored to enhance biohydrogen production. One such strategy is the use of different kinds of nanomaterials as additives or components in photobioreactor design, aimed at enhancing illumination and temperature efficiency, which has shown improvements in both biohydrogen production rate and yield, as detailed in this chapter. Despite these advancements, some technical challenges remained unresolved in further research.

Numerous studies have emphasized the potential for enhancing biohydrogen production by directly incorporating different NPs into the fermentation broth. However, the ideal dosage of these NPs remains uncertain. Determining the ideal dosage is essential for improving bioavailability while minimizing toxicity to fermentative microbes. The hydrogenases, which are metalloenzymes involved in biohydrogen production, require specific levels of metallic NPs to retain their catalytic activity and avoid inhibition. Moreover, introducing nanomaterials can affect the economic feasibility of the fermentation process, making it a critical area for further research. Investigating innovative, eco-friendly NPs made from low-cost biomaterials with favorable physicochemical characteristics is vital to prevent microbial cell damage and degradation. Additionally, microbial community studies have shown that NPs can boost the abundance of hydrogen-producing microbes during fermentation. As such, it is crucial to assess the effects of NPs on the structure and dynamics of microbial communities to identify microbial populations that enhance hydrogen production.

Research on biohydrogen production from food waste, simple sugars, and other lignocellulosic biomass has been addressed in previous studies. At the same time, the effects of NPs on alternative feedstocks like macroalgae, microalgae, and glycerol have not been extensively explored. Broadening the range of substrates for biohydrogen production should be prioritized to enhance cost-efficiency. So far, NPs have primarily been applied in batch fermentation for hydrogen production, highlighting the need for further investigation into continuous fermentation processes to overcome the technical barriers for large-scale

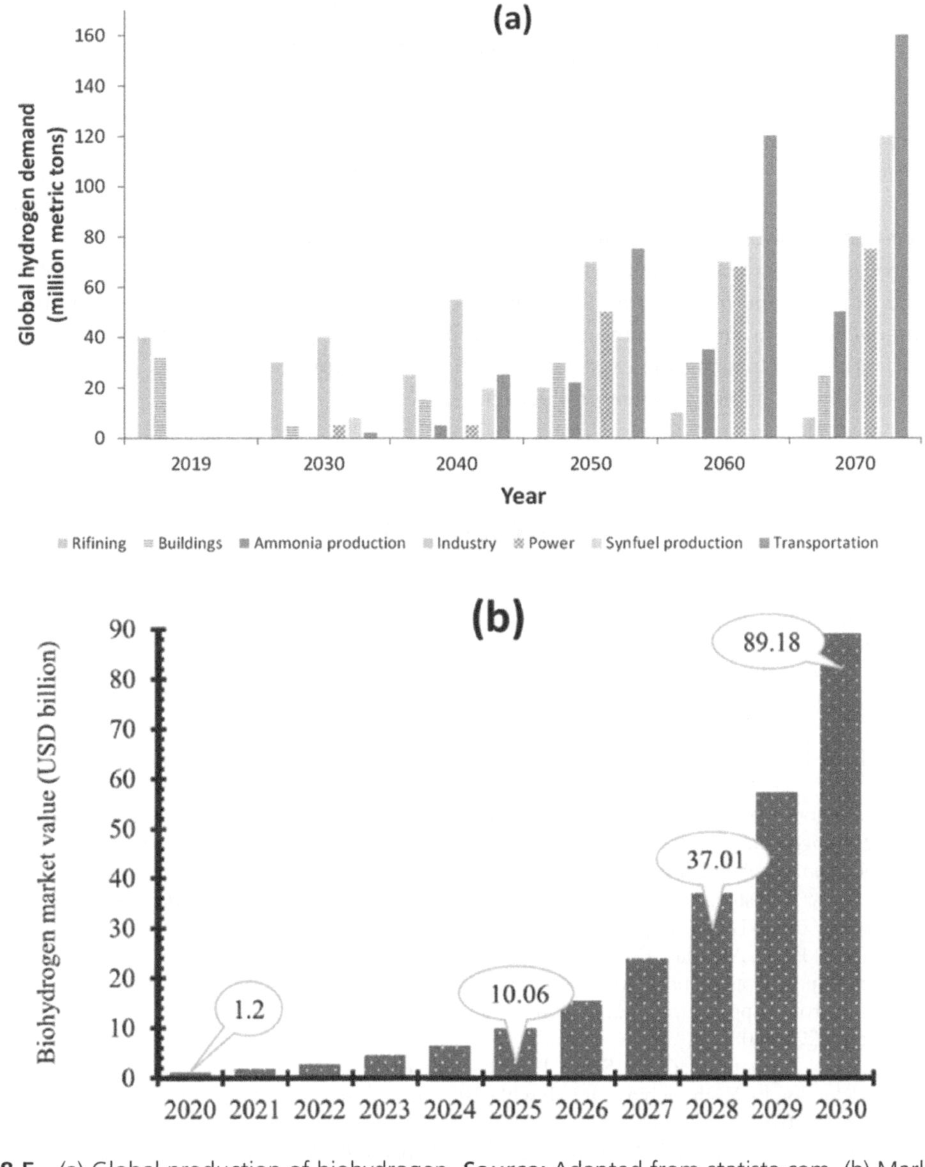

FIGURE 8.5 (a) Global production of biohydrogen. ***Source:*** Adapted from statista.com. (b) Market value of biohydrogen. ***Source:*** Adapted from precedenceresearch.com.

commercial applications. Moreover, replacing powdered NPs with sheets, nanotubes, or hybrid configurations in bioreactors could improve biohydrogen production by supporting microbial or enzyme immobilization, enhancing light absorption, and offering other advantages.

The commercial viability of NPs in biohydrogen production is generally related to integrating multi-stage processes, for instance, dark fermentation combined with biomethane production, as well as utilizing fermentation by-products for the generation of value-added products like polyhydroxyalkanoates (PHA) or polyhydroxy butyrate (PHB). Such multi-step strategies can greatly enhance the economic viability of the system. However, addressing the production of inhibitory by-products during the downstream stages of these processes is crucial. In addition, further research into the toxicity of NPs in fermentation residues is necessary to minimize the associated risks. Future studies should focus on determining the optimal NP concentrations and developing methods for recycling residual NPs to promote their reuse in future operations.

8.9 CONCLUSIONS

Advancements in research and development, coupled with the integration of cutting-edge technologies such as nanotechnology, have established biohydrogen as a promising biofuel for the future. However, several significant challenges remain, including the high costs of feedstock pre-treatment, the sensitivity of oxygen-dependent enzymes, inconsistent light availability for photobiological reactions, and the expense of process intensification methods required for biohydrogen production. This chapter primarily highlights the progress and potential applications of NPs as additives in the fermentation process or as integral components in fermenter design to enhance biohydrogen production through dark or photobiological fermentation. Addressing the existing scientific and technological barriers is crucial to realizing the full potential of NPs in biohydrogen production.

REFERENCES

Abanades, S., Abbaspour, H., Ahmadi, A., Das, B., Ehyaei, M. A., Esmaeilion, F., El Haj Assad, M., Hajilounezhad, T., Jamali, D. H., Hmida, A., Ozgoli, H. A., Safari, S., AlShabi, M., & Bani-Hani, E. H. (2022). A critical review of biogas production and usage with legislations framework across the globe. *International Journal of Environmental Science and Technology*, 19(4), 3377–3400. https://doi.org/10.1007/s13762-021-03301-6

Abu-Ghosh, S., Fixler, D., Dubinsky, Z., & Iluz, D. (2016). Flashing light in microalgae biotechnology. *Bioresource Technology*, 203, 357–363. https://doi.org/10.1016/j.biortech.2015.12.057

Akhlaghi, N., & Najafpour-Darzi, G. (2020). A comprehensive review on biological hydrogen production. *International Journal of Hydrogen Energy*, 45(43), 22492–22512. https://doi.org/10.1016/j.ijhydene.2020.06.182

Alemahdi, N., Che Man, H., Abd Rahman, N., Nasirian, N., & Yang, Y. (2015). Enhanced mesophilic bio-hydrogen production of raw rice straw and activated sewage sludge by co-digestion. *International Journal of Hydrogen Energy*, 40(46), 16033–16044. https://doi.org/10.1016/j.ijhydene.2015.08.106

Alrebdi, T. A., Ahmed, H. A., Alkallas, F. H., Mwafy, E. A., Trabelsi, A. B. G., & Mostafa, A. M. (2022). Structural, linear and nonlinear optical properties of NiO nanoparticles–multi-walled carbon nanotubes nanocomposite for optoelectronic applications. *Radiation Physics and Chemistry*, 195, 110088. https://doi.org/10.1016/j.radphyschem.2022.110088

Alvarado-Cuevas, Z. D., Acevedo, L. G. O., Salas, J. T. O., & De León-Rodríguez, A. (2013). Nitrogen sources impact hydrogen production by Escherichia coli using cheese whey as substrate. *New Biotechnology*, 30(6), 585–590. https://doi.org/10.1016/j.nbt.2013.03.002

Beckers, L., Hiligsmann, S., Lambert, S. D., Heinrichs, B., & Thonart, P. (2013). Improving effect of metal and oxide nanoparticles encapsulated in porous silica on fermentative biohydrogen production by *Clostridium butyricum*. *Bioresource Technology*, 133, 109–117. https://doi.org/10.1016/j.biortech.2012.12.168

Bhatia, S. K., Jagtap, S. S., Bedekar, A. A., Bhatia, R. K., Rajendran, K., Pugazhendhi, A., Rao, C. V., Atabani, A. E., Kumar, G., & Yang, Y.-H. (2021). Renewable biohydrogen production from lignocellulosic biomass using fermentation and integration of systems with other energy generation technologies. *Science of the Total Environment*, 765, 144429. https://doi.org/10.1016/j.scitotenv.2020.144429

Bian, Y., Chen, Y., Zhan, L., Guo, H., Ke, H., Wang, Y., Wang, Q., Gao, Y., & Gao, Y. (2024). Effects of enzyme-induced carbonate precipitation technique on multiple heavy metals immobilization and unconfined compressive strength improvement of contaminated sand. *Science of the Total Environment*, 947, 174409. https://doi.org/10.1016/j.scitotenv.2024.174409

Bolivar, J. M., Woodley, J. M., & Fernandez-Lafuente, R. (2022). Is enzyme immobilization a mature discipline? Some critical considerations to capitalize on the benefits of immobilization. *Chemical Society Reviews*, 51(15), 6251–6290. https://doi.org/10.1039/D2CS00083K

Boshagh, F., Rostami, K., & Moazami, N. (2019). Biohydrogen production by immobilized Enterobacter aerogenes on functionalized multi-walled carbon nanotube. *International Journal of Hydrogen Energy*, 44(28), 14395–14405. https://doi.org/10.1016/j.ijhydene.2018.11.199

Bosu, S., & Rajamohan, N. (2022). Nanotechnology approach for enhancement in biohydrogen production- review on applications of nanocatalyst and life cycle assessment. *Fuel*, 323, 124351. https://doi.org/10.1016/j.fuel.2022.124351

Cabrol, L., Marone, A., Tapia-Venegas, E., Steyer, J.-P., Ruiz-Filippi, G., & Trably, E. (2017). Microbial ecology of fermentative hydrogen producing bioprocesses: Useful insights for driving the ecosystem function. *FEMS Microbiology Reviews*, 41(2), 158–181. https://doi.org/10.1093/femsre/fuw043

Chaganti, S. R., Kim, D.-H., & Lalman, J. A. (2012). Dark fermentative hydrogen production by mixed anaerobic cultures: Effect of inoculum treatment methods on hydrogen yield. *Renewable Energy*, 48, 117–121. https://doi.org/10.1016/j.renene.2012.04.015

Chan, Y. H., Loy, A. C. M., Cheah, K. W., Chai, S. Y. W., Ngu, L. H., How, B. S., Li, C., Lock, S. S. M., Wong, M. K., Yiin, C. L., Chin, B. L. F., Chan, Z. P., & Lam, S. S. (2023). Hydrogen sulfide (H_2S) conversion to hydrogen (H_2) and value-added chemicals: Progress, challenges and outlook. *Chemical Engineering Journal*, 458, 141398. https://doi.org/10.1016/j.cej.2023.141398

Chandra, R., Nikhil, G. N., & Mohan, S. V. (2015). Single-stage operation of hybrid dark-photo fermentation to enhance biohydrogen production through regulation of system redox condition: evaluation with real-field wastewater. *International Journal of Molecular Sciences*, *16*(5), 9540–9556. https://doi.org/10.3390/ijms16059540

Chen, C.-Y., Yeh, K.-L., Aisyah, R., Lee, D.-J., & Chang, J.-S. (2011). Cultivation, photobioreactor design and harvesting of microalgae for biodiesel production: A critical review. *Bioresource Technology*, 102(1), 71–81. https://doi.org/10.1016/j.biortech.2010.06.159

Chen, Z., Chen, L., Khoo, K. S., Gupta, V. K., Sharma, M., Show, P. L., & Yap, P.-S. (2023). Exploitation of lignocellulosic-based biomass biorefinery: A critical review of renewable bioresource, sustainability and economic views. *Biotechnology Advances*, 69, 108265. https://doi.org/10.1016/j.biotechadv.2023.108265

Corneli, E., Adessi, A., Dragoni, F., Ragaglini, G., Bonari, E., & De Philippis, R. (2016). Agroindustrial residues and energy crops for the production of hydrogen and poly-β-hydroxybutyrate via photofermentation. *Bioresource Technology*, 216, 941–947. https://doi.org/10.1016/j.biortech.2016.06.046

da Silva Veras, T., Mozer, T. S., da Costa Rubim Messeder dos Santos, D., & da Silva César, A. (2017). Hydrogen: Trends, production and characterization of the main process worldwide. *International Journal of Hydrogen Energy*, 42(4), 2018–2033. https://doi.org/10.1016/j.ijhydene.2016.08.219

Das, D. (2001). Hydrogen production by biological processes: A survey of literature. *International Journal of Hydrogen Energy*, 26(1), 13–28. https://doi.org/10.1016/S0360-3199(00)00058-6

Das, D., & Veziroglu, T. (2008). Advances in biological hydrogen production processes. *International Journal of Hydrogen Energy*, 33(21), 6046–6057. https://doi.org/10.1016/j.ijhydene.2008.07.098

Deivayanai, V. C., Yaashikaa, P. R., Senthil Kumar, P., & Rangasamy, G. (2022). A comprehensive review on the biological conversion of lignocellulosic biomass into hydrogen: Pretreatment strategy, technology advances and perspectives. *Bioresource Technology*, 365, 128166. https://doi.org/10.1016/j.biortech.2022.128166

Deplanche, K., Caldelari, I., Mikheenko, I. P., Sargent, F., & Macaskie, L. E. (2010). Involvement of hydrogenases in the formation of highly catalytic Pd(0) nanoparticles by bioreduction of Pd(II) using Escherichia coli mutant strains. *Microbiology*, 156(9), 2630–2640. https://doi.org/10.1099/mic.0.036681-0

Dincer, I., & Acar, C. (2015). Review and evaluation of hydrogen production methods for better sustainability. *International Journal of Hydrogen Energy*, 40(34), 11094–11111. https://doi.org/10.1016/j.ijhydene.2014.12.035

do Nascimento Junior, J. R., Zevallos Torres, L. A., Medeiros, A. B. P., Woiciechowski, A. L., Martinez-Burgos, W. J., & Soccol, C. R. (2021). Enhancement of biohydrogen production in industrial wastewaters with vinasse pond consortium using lignin-mediated iron nanoparticles. *International Journal of Hydrogen Energy*, 46(54), 27431–27443. https://doi.org/10.1016/j.ijhydene.2021.06.009

Dudek, M., Dębowski, M., Zieliński, M., Nowicka, A., & Rusanowska, P. (2018). Water from the Vistula Lagoon as a medium in mixotrophic growth and hydrogen production by Platymonas subcordiformis. *International Journal of Hydrogen Energy*, 43(20), 9529–9534. https://doi.org/10.1016/j.ijhydene.2018.04.039

Engliman, N. S., Abdul, P. M., Wu, S.-Y., & Jahim, J. M. (2017). Influence of iron (II) oxide nanoparticle on biohydrogen production in thermophilic mixed fermentation. *International Journal of Hydrogen Energy*, 42(45), 27482–27493. https://doi.org/10.1016/j.ijhydene.2017.05.224

Epelle, E. I., Desongu, K. S., Obande, W., Adeleke, A. A., Ikubanni, P. P., Okolie, J. A., & Gunes, B. (2022). A comprehensive review of hydrogen production and storage: A focus on the role of nanomaterials. *International Journal of Hydrogen Energy*, 47(47), 20398–20431. https://doi.org/10.1016/j.ijhydene.2022.04.227

Fang, H., Li, C., & Zhang, T. (2006). Acidophilic biohydrogen production from rice slurry. *International Journal of Hydrogen Energy*, 31(6), 683–692. https://doi.org/10.1016/j.ijhydene.2005.07.005

Fisgativa, H., Tremier, A., & Dabert, P. (2016). Characterizing the variability of food waste quality: A need for efficient valorisation through anaerobic digestion. *Waste Management*, 50, 264–274. https://doi.org/10.1016/j.wasman.2016.01.041

Gadhe, A., Sonawane, S. S., & Varma, M. N. (2015). Enhancement effect of hematite and nickel nanoparticles on biohydrogen production from dairy wastewater. *International Journal of Hydrogen Energy*, 40(13), 4502–4511. https://doi.org/10.1016/j.ijhydene.2015.02.046

Gadow, S. I., Jiang, H., Hojo, T., & Li, Y.-Y. (2013). Cellulosic hydrogen production and microbial community characterization in hyper-thermophilic continuous bioreactor. *International Journal of Hydrogen Energy*, 38(18), 7259–7267. https://doi.org/10.1016/j.ijhydene.2013.04.041

Giannelli, L., & Torzillo, G. (2012). Hydrogen production with the microalga Chlamydomonas reinhardtii grown in a compact tubular photobioreactor immersed in a scattering light nanoparticle suspension. *International Journal of Hydrogen Energy*, 37(22), 16951–16961. https://doi.org/10.1016/j.ijhydene.2012.08.103

Hallenbeck, P. C. (2009). Fermentative hydrogen production: Principles, progress, and prognosis. *International Journal of Hydrogen Energy*, 34(17), 7379–7389. https://doi.org/10.1016/j.ijhydene.2008.12.080

Han, H., Cui, M., Wei, L., Yang, H., & Shen, J. (2011). Enhancement effect of hematite nanoparticles on fermentative hydrogen production. *Bioresource Technology*, 102(17), 7903–7909. https://doi.org/10.1016/j.biortech.2011.05.089

Hsieh, P.-H., Lai, Y.-C., Chen, K.-Y., & Hung, C.-H. (2016). Explore the possible effect of TiO_2 and magnetic hematite nanoparticle addition on biohydrogen production by Clostridium pasteurianum based on gene expression measurements. *International Journal of Hydrogen Energy*, 41(46), 21685–21691. https://doi.org/10.1016/j.ijhydene.2016.06.197

Jafari, O., & Zilouei, H. (2016). Enhanced biohydrogen and subsequent biomethane production from sugarcane bagasse using nano-titanium dioxide pretreatment. *Bioresource Technology*, 214, 670–678. https://doi.org/10.1016/j.biortech.2016.05.007

Jannat, F. T., Aftab, K., Kalsoom, U., & Baig, M. A. (2024). A bibliometric analysis of the role of nanotechnology in dark fermentative biohydrogen production. *Environmental Science and Pollution Research*, 31(17), 24815–24835. https://doi.org/10.1007/s11356-024-33005-6

Ji, Y., Sultan, M. A., Kim, D. Y., Meeks, N., Hastings, J. T., & Bhattacharyya, D. (2021). Effect of silica-core gold-shell nanoparticles on the kinetics of biohydrogen production and pollutant hydrogenation via organic acid photofermentation over enhanced near-infrared illumination. *International Journal of Hydrogen Energy*, 46(11), 7821–7835. https://doi.org/10.1016/j.ijhydene.2020.11.257

Kamble, S. S., Ram, N. R., & Nikhil, G. N. (2024). Nanoparticle-mediated advancements in volatile fatty acids production through anaerobic fermentation: a comprehensive review. *Chemical Papers*, *78*(10), 5723–5741. https://doi.org/10.1007/s11696-024-03495-2.

Khamtib, S., & Reungsang, A. (2012). Biohydrogen production from xylose by Thermoanaerobacterium thermosaccharolyticum KKU19 isolated from hot spring sediment. *International Journal of Hydrogen Energy*, 37(17), 12219–12228. https://doi.org/10.1016/j.ijhydene.2012.06.038

Khan, I., Anburajan, P., Kumar, G., Yoon, J., Bahuguna, A., de Moura, A. G. L., Pugazhendhi, A., Kim, S., & Kang, S. (2021). Comparative effect of silver nanoparticles (AgNPs) derived from actinomycetes and henna on biohydrogen production by *Clostridium beijerinckii* (KTCC1737). *International Journal of Energy Research*, 45(12), 17269–17278. https://doi.org/10.1002/er.6076

Khan, M. M., Lee, J., & Cho, M. H. (2013). Electrochemically active biofilm mediated bio-hydrogen production catalyzed by positively charged gold nanoparticles. *International Journal of Hydrogen Energy*, 38(13), 5243–5250. https://doi.org/10.1016/j.ijhydene.2013.02.080

Khilari, S., Pandit, S., Ghangrekar, M. M., Das, D., & Pradhan, D. (2013). Graphene supported α-MnO_2 nanotubes as a cathode catalyst for improved power generation and wastewater treatment in single-chambered microbial fuel cells. *RSC Advances*, 3(21), 7902. https://doi.org/10.1039/c3ra22569k

Khoo, K. S., Ahmad, I., Chew, K. W., Iwamoto, K., Bhatnagar, A., & Show, P. L. (2023). Enhanced microalgal lipid production for biofuel using different strategies including genetic modification of microalgae: A review. *Progress in Energy and Combustion Science*, 96, 101071. https://doi.org/10.1016/j.pecs.2023.101071

Kim, D.-H., & Kim, M.-S. (2011). Hydrogenases for biological hydrogen production. *Bioresource Technology*, 102(18), 8423–8431. https://doi.org/10.1016/j.biortech.2011.02.113

Koul, B., Yakoob, M., & Shah, M. P. (2022). Agricultural waste management strategies for environmental sustainability. *Environmental Research*, 206, 112285. https://doi.org/10.1016/j.envres.2021.112285

Kumar, G., Mathimani, T., Rene, E. R., & Pugazhendhi, A. (2019). Application of nanotechnology in dark fermentation for enhanced biohydrogen production using inorganic nanoparticles. *International Journal of Hydrogen Energy*, 44(26), 13106–13113. https://doi.org/10.1016/j.ijhydene.2019.03.131

Kumar, M., Matassa, S., Belloni, C., Pirozzi, F., Esposito, G., & Papirio, S. (2024). Upcycling nitrogen and curbing greenhouse gas emissions from wastewater through H_2-driven assimilatory mixotrophic metabolism. *Chemical Engineering Journal*, 489, 151207. https://doi.org/10.1016/j.cej.2024.151207

Lamolinara, B., Pérez-Martínez, A., Guardado-Yordi, E., Guillén Fiallos, C., Diéguez-Santana, K., & Ruiz-Mercado, G. J. (2022). Anaerobic digestate management, environmental impacts, and techno-economic challenges. *Waste Management*, 140, 14–30. https://doi.org/10.1016/j.wasman.2021.12.035

Lee, C., Lee, S., Han, S.-K., & Hwang, S. (2014). Effect of operational pH on biohydrogen production from food waste using anaerobic batch reactors. *Water Science and Technology*, 69(9), 1886–1893. https://doi.org/10.2166/wst.2014.097

Lee, D.-H. (2016). Cost-benefit analysis, LCOE and evaluation of financial feasibility of full commercialization of biohydrogen. *International Journal of Hydrogen Energy*, 41(7), 4347–4357. https://doi.org/10.1016/j.ijhydene.2015.09.071

Li, H., Riisager, A., Saravanamurugan, S., Pandey, A., Sangwan, R. S., Yang, S., & Luque, R. (2018). Carbon-increasing catalytic strategies for upgrading biomass into energy-intensive fuels and chemicals. *ACS Catalysis*, 8(1), 148–187. https://doi.org/10.1021/acscatal.7b02577

Li, Y., Zhong, N., Liao, Q., Fu, Q., Huang, Y., Zhu, X., & Li, Q. (2017). A biomaterial doped with LaB 6 nanoparticles as photothermal media for enhancing biofilm growth and hydrogen production in photosynthetic bacteria. *International Journal of Hydrogen Energy*, 42(9), 5793–5803. https://doi.org/10.1016/j.ijhydene.2017.01.082

Liao, Q., Zhong, N.-B., Zhu, X., & Chen, R. (2014). High-performance biofilm photobioreactor based on a GeO_2–SiO_2–chitosan-medium-coated hollow optical fiber. *International Journal of Hydrogen Energy*, 39(19), 10016–10027. https://doi.org/10.1016/j.ijhydene.2014.04.130

Lin, R., Cheng, J., Ding, L., Song, W., Liu, M., Zhou, J., & Cen, K. (2016). Enhanced dark hydrogen fermentation by addition of ferric oxide nanoparticles using Enterobacter aerogenes. *Bioresource Technology*, 207, 213–219. https://doi.org/10.1016/j.biortech.2016.02.009

Liu, B., Jin, Y., Wang, Z., Xing, D., Ma, C., Ding, J., & Ren, N. (2017). Enhanced photo-fermentative hydrogen production of *Rhodopseudomonas* sp. nov. strain A7 by the addition of TiO_2, ZnO and SiC nanoparticles. *International Journal of Hydrogen Energy*, 42(29), 18279–18287. https://doi.org/10.1016/j.ijhydene.2017.04.147

Liu, B.-F., Jin, Y.-R., Cui, Q.-F., Xie, G.-J., Wu, Y.-N., & Ren, N.-Q. (2015). Photo-fermentation hydrogen production by Rhodopseudomonas sp. nov. strain A7 isolated from the sludge in a bioreactor. *International Journal of Hydrogen Energy*, 40(28), 8661–8668. https://doi.org/10.1016/j.ijhydene.2015.05.001

Liu, B.-F., Ren, N.-Q., Tang, J., Ding, J., Liu, W.-Z., Xu, J.-F., Cao, G.-L., Guo, W.-Q., & Xie, G.-J. (2010). Bio-hydrogen production by mixed culture of photo- and dark-fermentation bacteria. *International Journal of Hydrogen Energy*, 35(7), 2858–2862. https://doi.org/10.1016/j.ijhydene.2009.05.005

Liu, Z., Lv, F., Zheng, H., Zhang, C., Wei, F., & Xing, X.-H. (2012). Enhanced hydrogen production in a UASB reactor by retaining microbial consortium onto carbon nanotubes (CNTs). *International Journal of Hydrogen Energy*, 37(14), 10619–10626. https://doi.org/10.1016/j.ijhydene.2012.04.057

Łukajtis, R., Hołowacz, I., Kucharska, K., Glinka, M., Rybarczyk, P., Przyjazny, A., & Kamiński, M. (2018). Hydrogen production from biomass using dark fermentation. *Renewable and Sustainable Energy Reviews*, 91, 665–694. https://doi.org/10.1016/j.rser.2018.04.043

Macrelli, S., Galbe, M., & Wallberg, O. (2014). Effects of production and market factors on ethanol profitability for an integrated first and second generation ethanol plant using the whole sugarcane as feedstock. *Biotechnology for Biofuels*, 7(1), 26. https://doi.org/10.1186/1754-6834-7-26

Manish, S., & Banerjee, R. (2008). Comparison of biohydrogen production processes. *International Journal of Hydrogen Energy*, 33(1), 279–286. https://doi.org/10.1016/j.ijhydene.2007.07.026

Megía, P. J., Vizcaíno, A. J., Calles, J. A., & Carrero, A. (2021). Hydrogen production technologies: From fossil fuels toward renewable sources. A mini review. *Energy & Fuels*, 35(20), 16403–16415. https://doi.org/10.1021/acs.energyfuels.1c02501

Mehra, A., Raja Ram, N., Srivastava, N. K., & Nikhil, G. N. (2025). Review of Hydrogen Production Using Organic Waste Materials: Role of Industry 4.0 in Waste Valuation. *Korean Journal of Chemical Engineering*, *42*(3), 455–481. https://doi.org/10.1007/s11814-024-00367-z

Mishra, P., Thakur, S., Mahapatra, D. M., Wahid, Z. A., Liu, H., & Singh, L. (2018). Impacts of nano-metal oxides on hydrogen production in anaerobic digestion of palm oil mill effluent – A novel approach. *International Journal of Hydrogen Energy*, 43(5), 2666–2676. https://doi.org/10.1016/j.ijhydene.2017.12.108

Mohan, S., Mohanakrishna, G., Reddy, S., Raju, B., Rao, K., & Sarma, P. (2008). Self-immobilization of acidogenic mixed consortia on mesoporous material (SBA-15) and activated carbon to enhance fermentative hydrogen production. *International Journal of Hydrogen Energy*, 33(21), 6133–6142. https://doi.org/10.1016/j.ijhydene.2008.07.096

Mohan, S. V., Srikanth, S., & Nikhil, G. N. (2017). Augmentation of bacterial homeostasis by regulating in situ buffer capacity: significance of total dissolved salts over acidogenic metabolism. *Bioresource Technology*, *225*, 34–39. https://doi.org/10.1016/j.biortech.2016.11.027

Mohanraj, S., Anbalagan, K., Kodhaiyolii, S., & Pugalenthi, V. (2014a). Comparative evaluation of fermentative hydrogen production using Enterobacter cloacae and mixed culture: Effect of Pd (II) ion and phytogenic palladium nanoparticles. *Journal of Biotechnology*, 192, 87–95. https://doi.org/10.1016/j.jbiotec.2014.10.012

Mohanraj, S., Anbalagan, K., Rajaguru, P., & Pugalenthi, V. (2016). Effects of phytogenic copper nanoparticles on fermentative hydrogen production by *Enterobacter cloacae* and *Clostridium acetobutylicum*. *International Journal of Hydrogen Energy*, 41(25), 10639–10645. https://doi.org/10.1016/j.ijhydene.2016.04.197

Mohanraj, S., Kodhaiyolii, S., Rengasamy, M., & Pugalenthi, V. (2014b). Phytosynthesized iron oxide nanoparticles and ferrous iron on fermentative hydrogen production using *Enterobacter cloacae*: Evaluation and comparison of the effects. *International Journal of Hydrogen Energy*, 39(23), 11920–11929. https://doi.org/10.1016/j.ijhydene.2014.06.027

Muri, P., Marinšek-Logar, R., Djinović, P., & Pintar, A. (2018). Influence of support materials on continuous hydrogen production in anaerobic packed-bed reactor with immobilized hydrogen producing bacteria at acidic conditions. *Enzyme and Microbial Technology*, 111, 87–96. https://doi.org/10.1016/j.enzmictec.2017.10.008

Nagarajan, D., Lee, D.-J., Kondo, A., & Chang, J.-S. (2017). Recent insights into biohydrogen production by microalgae – From biophotolysis to dark fermentation. *Bioresource Technology*, 227, 373–387. https://doi.org/10.1016/j.biortech.2016.12.104

Nasr, M., Tawfik, A., Ookawara, S., Suzuki, M., Kumari, S., & Bux, F. (2015). Continuous biohydrogen production from starch wastewater via sequential dark-photo fermentation with emphasize on maghemite nanoparticles. *Journal of Industrial and Engineering Chemistry*, 21, 500–506. https://doi.org/10.1016/j.jiec.2014.03.011

Nikolaidis, P., & Poullikkas, A. (2017). A comparative overview of hydrogen production processes. *Renewable and Sustainable Energy Reviews*, 67, 597–611. https://doi.org/10.1016/j.rser.2016.09.044

Nikhil, G. N., Mohan, S. V., & Swamy, Y. V. (2014). Behavior of acidogenesis during biohydrogen production with formate and glucose as carbon source: substrate associated dehydrogenase expression. *International Journal of Hydrogen Energy*, *39*(14), 7486–7495. https://doi.org/10.1016/j.ijhydene.2013.12.169

Nikhil, G. N., Sarkar, O., & Venkata Mohan, S. (2018). Biohydrogen production: an outlook of fermentative processes and integration strategies. In *Optimization and applicability of bioprocesses* (pp. 249–265). Singapore: Springer Singapore. https://doi.org/10.1007/978-981-10-6863-8_12

Niño-Navarro, C., Chairez, I., Christen, P., Canul-Chan, M., & García-Peña, E. I. (2020). Enhanced hydrogen production by a sequential dark and photo fermentation process: Effects of initial feedstock composition, dilution and microbial population. *Renewable Energy*, 147, 924–936. https://doi.org/10.1016/j.renene.2019.09.024

Oey, M., Sawyer, A. L., Ross, I. L., & Hankamer, B. (2016). Challenges and opportunities for hydrogen production from microalgae. *Plant Biotechnology Journal*, 14(7), 1487–1499. https://doi.org/10.1111/pbi.12516

Özgür, E., Mars, A. E., Peksel, B., Louwerse, A., Yücel, M., Gündüz, U., Claassen, P. A. M., & Eroğlu, İ. (2010). Biohydrogen production from beet molasses by sequential dark and photofermentation. *International Journal of Hydrogen Energy*, 35(2), 511–517. https://doi.org/10.1016/j.ijhydene.2009.10.094

Pandey, A., Gupta, K., & Pandey, A. (2015). Effect of nanosized TiO_2 on photofermentation by Rhodobacter sphaeroides NMBL-02. *Biomass and Bioenergy*, 72, 273–279. https://doi.org/10.1016/j.biombioe.2014.10.021

Patel, S. K. S., Kalia, V. C., Choi, J.-H., Haw, J.-R., Kim, I.-W., & Lee, J. K. (2014). Immobilization of laccase on SiO_2 nanocarriers improves its stability and reusability. *Journal of Microbiology and Biotechnology*, 24(5), 639–647. https://doi.org/10.4014/jmb.1401.01025

Patel, S. K. S., Lee, J.-K., & Kalia, V. C. (2018). Nanoparticles in biological hydrogen production: An overview. *Indian Journal of Microbiology*, 58(1), 8–18. https://doi.org/10.1007/s12088-017-0678-9

Priya,K., Naveen, V., Kaur, K., & Sidhu, A. K. (2021). Green synthesis: An eco-friendly route for the synthesis of iron oxide nanoparticles. *Frontiers in Nanotechnology*, 3, 1–16. https://doi.org/10.3389/fnano.2021.655062

Rai, P. K., Singh, S. P., & Asthana, R. K. (2012). Biohydrogen production from cheese whey wastewater in a two-step anaerobic process. *Applied Biochemistry and Biotechnology*, 167(6), 1540–1549. https://doi.org/10.1007/s12010-011-9488-4

Ramprakash, B., & Muthukumar, K. (2018). Influence of sulfuric acid concentration on biohydrogen production from rice mill wastewater using pure and coculture of Enterobacter aerogenes and *Citrobacter freundii*. *International Journal of Hydrogen Energy*, 43(19), 9254–9258. https://doi.org/10.1016/j.ijhydene.2018.03.198

Rashid, N., Lee, K., & Mahmood, Q. (2011). Bio-hydrogen production by *Chlorella vulgaris* under diverse photoperiods. *Bioresource Technology*, 102(2), 2101–2104. https://doi.org/10.1016/j.biortech.2010.08.032

Reddy, K., Nasr, M., Kumari, S., Kumar, S., Gupta, S. K., Enitan, A. M., & Bux, F. (2017). Biohydrogen production from sugarcane bagasse hydrolysate: Effects of pH, S/X, Fe^{2+}, and magnetite nanoparticles. *Environmental Science and Pollution Research*, 24(9), 8790–8804. https://doi.org/10.1007/s11356-017-8560-1

Ren, N., Cao, G., Wang, A., Lee, D., Guo, W., & Zhu, Y. (2008). Dark fermentation of xylose and glucose mix using isolated *Thermoanaerobacterium thermosaccharolyticum* W16. *International Journal of Hydrogen Energy*, 33(21), 6124–6132. https://doi.org/10.1016/j.ijhydene.2008.07.107

Ren, N.-Q., Cao, G.-L., Guo, W.-Q., Wang, A.-J., Zhu, Y.-H., Liu, B., & Xu, J.-F. (2010). Biological hydrogen production from corn stover by moderately thermophile *Thermoanaerobacterium thermosaccharolyticum* W16. *International Journal of Hydrogen Energy*, 35(7), 2708–2712. https://doi.org/10.1016/j.ijhydene.2009.04.044

Sambusiti, C., Bellucci, M., Zabaniotou, A., Beneduce, L., & Monlau, F. (2015). Algae as promising feedstocks for fermentative biohydrogen production according to a biorefinery approach: A comprehensive review. *Renewable and Sustainable Energy Reviews*, 44, 20–36. https://doi.org/10.1016/j.rser.2014.12.013

Sánchez-Ramírez, J., Martínez-Hernández, J. L., Segura-Ceniceros, P., López, G., Saade, H., Medina-Morales, M. A., Ramos-González, R., Aguilar, C. N., & Ilyina, A. (2017). Cellulases immobilization on chitosan-coated magnetic nanoparticles: Application for *Agave Atrovirens* lignocellulosic biomass hydrolysis. *Bioprocess and Biosystems Engineering*, 40(1), 9–22. https://doi.org/10.1007/s00449-016-1670-1

Shanmugam, S., Hari, A., Pandey, A., Mathimani, T., Felix, L., & Pugazhendhi, A. (2020). Comprehensive review on the application of inorganic and organic nanoparticles for enhancing biohydrogen production. *Fuel*, 270, 117453. https://doi.org/10.1016/j.fuel.2020.117453

Sharma, P., Bano, A., Pratap Singh, S., Atkinson, J. D., Shiung Lam, S., Iqbal, H. M. N., & Wah Tong, Y. (2022). Nanomaterials as highly efficient photocatalysts used for bioenergy and biohydrogen production from waste toward a sustainable environment. *Fuel*, 329, 125408. https://doi.org/10.1016/j.fuel.2022.125408

Sharma, S., Basu, S., Shetti, N. P., & Aminabhavi, T. M. (2020). Waste-to-energy nexus for circular economy and environmental protection: Recent trends in hydrogen energy. *Science of the Total Environment*, 713, 136633. https://doi.org/10.1016/j.scitotenv.2020.136633

Shelare, S. D., Belkhode, P. N., Nikam, K. C., Jathar, L. D., Shahapurkar, K., Soudagar, M. E. M., Veza, I., Khan, T. M. Y., Kalam, M. A., Nizami, A.-S., & Rehan, M. (2023). Biofuels for a sustainable future: Examining the role of nano-additives, economics, policy, internet of things, artificial intelligence and machine learning technology in biodiesel production. *Energy*, 282, 128874. https://doi.org/10.1016/j.energy.2023.128874

Shi, X.-Y., & Yu, H.-Q. (2016). Simultaneous metabolism of benzoate and photobiological hydrogen production by Lyngbya sp. *Renewable Energy*, 95, 474–477. https://doi.org/10.1016/j.renene.2016.04.051

Singh, D., Sharma, D., Soni, S. L., Sharma, S., Kumar Sharma, P., & Jhalani, A. (2020). A review on feedstocks, production processes, and yield for different generations of biodiesel. *Fuel*, 262, 116553. https://doi.org/10.1016/j.fuel.2019.116553

Sinharoy, A., Kumar, M., & Pakshirajan, K. (2020). An overview of bioreactor configurations and operational strategies for dark fermentative biohydrogen production. In Bioreactors (pp. 249–288). Elsevier. https://doi.org/10.1016/B978-0-12-821264-6.00014-0

Sivaramakrishnan, R., Shanmugam, S., Sekar, M., Mathimani, T., Incharoensakdi, A., Kim, S.-H., Parthiban, A., Edwin Geo, V., Brindhadevi, K., & Pugazhendhi, A. (2021). Insights on biological hydrogen production routes and potential microorganisms for high hydrogen yield. *Fuel*, 291, 120136. https://doi.org/10.1016/j.fuel.2021.120136

Song, H., Luo, S., Huang, H., Deng, B., & Ye, J. (2022). Solar-driven hydrogen production: Recent advances, challenges, and future perspectives. *ACS Energy Letters*, 7(3), 1043–1065. https://doi.org/10.1021/acsenergylett.1c02591

Stojkovic, D., Torzillo, G., Faraloni, C., & Valant, M. (2015). Hydrogen production by sulfur-deprived TiO_2-encapsulated *Chlamydomonas reinhardtii* cells. *International Journal of Hydrogen Energy*, 40(8), 3201–3206. https://doi.org/10.1016/j.ijhydene.2014.12.115

Strapasson, G. B., Assis, M., Backes, C. W., Corrêa, S. A., Longo, E., & Weibel, D. E. (2021). Microwave assisted synthesis of silver nanoparticles and its application in sustainable photocatalytic hydrogen evolution. *International Journal of Hydrogen Energy*, 46(69), 34264–34275. https://doi.org/10.1016/j.ijhydene.2021.07.237

Sun, Y., Liao, Q., Huang, Y., Xia, A., Fu, Q., Zhu, X., & Zheng, Y. (2016). Integrating planar waveguides doped with light scattering nanoparticles into a flat-plate photobioreactor to improve light distribution and microalgae growth. *Bioresource Technology*, 220, 215–224. https://doi.org/10.1016/j.biortech.2016.08.063

Sun, Y., Wang, Y., Yang, G., & Sun, Z. (2020). Optimization of biohydrogen production using acid pretreated corn stover hydrolysate followed by nickel nanoparticle addition. *International Journal of Energy Research*, 44(3), 1843–1857. https://doi.org/10.1002/er.5030

Taherdanak, M., Zilouei, H., & Karimi, K. (2015). Investigating the effects of iron and nickel nanoparticles on dark hydrogen fermentation from starch using central composite design. *International Journal of Hydrogen Energy*, 40(38), 12956–12963. https://doi.org/10.1016/j.ijhydene.2015.08.004

Tang, D., Tan, G.-L., Li, G.-W., Liang, J.-G., Ahmad, S. M., Bahadur, A., Humayun, M., Ullah, H., Khan, A., & Bououdina, M. (2023). State-of-the-art hydrogen generation techniques and storage methods: A critical review. *Journal of Energy Storage*, 64, 107196. https://doi.org/10.1016/j.est.2023.107196

Tao, Y. (2008). Characteristics of a new photosynthetic bacterial strain for hydrogen production and its application in wastewater treatment. *International Journal of Hydrogen Energy*, 33(3), 963–973. https://doi.org/10.1016/j.ijhydene.2007.11.021

Velvizhi, G., Ranjitha, J., Vijayalakshmi, S., & Nikhil, G. N. (2020). Advancements in nanobiocatalysis for bioenergy and biofuel production. In *Nanobiotechnology for sustainable bioenergy and biofuel production* (pp. 124–149). CRC Press. https://doi.org/10.1201/9780429023194.

Walker, L. (2009). An Investigation into the Bioprocesses of DiCOM®: A Technology Combining Composting and Thermophilic Anaerobic Digestion for the Treatment of Municipal Solid Waste. Murdoch University.

Wang, J., & Yin, Y. (2021). Clostridium species for fermentative hydrogen production: An overview. *International Journal of Hydrogen Energy*, 46(70), 34599–34625. https://doi.org/10.1016/j.ijhydene.2021.08.052

Wen, Z., Fu, Q., Wu, J., & Fan, G. (2020). Ultrafine Pd nanoparticles supported on soft nitriding porous carbon for hydrogen production from hydrolytic dehydrogenation of dimethyl amine-borane. *Nanomaterials*, 10(8), 1612. https://doi.org/10.3390/nano10081612

Wimonsong, P., & Nitisoravut, R. (2014). Biohydrogen enhancement using highly porous activated carbon. *Energy & Fuels*, 28(7), 4554–4559. https://doi.org/10.1021/ef500530v

Wimonsong, P., & Nitisoravut, R. (2015). Comparison of different catalysts for fermentative hydrogen production. *Journal of Clean Energy Technologies*, 3(2), 128–131. https://doi.org/10.7763/JOCET.2015.V3.181

Wu, H., Wang, C., Chen, P., He, A.-Y., Xing, F.-X., Kong, X.-P., & Jiang, M. (2017). Effects of pH and ferrous iron on the coproduction of butanol and hydrogen by Clostridium beijerinckii IB4. *International Journal of Hydrogen Energy*, 42(10), 6547–6555. https://doi.org/10.1016/j.ijhydene.2017.02.094

Wu, X., Li, Q., Dieudonne, M., Cong, Y., Zhou, J., & Long, M. (2010). Enhanced H_2 gas production from bagasse using adhE inactivated Klebsiella oxytoca HP1 by sequential dark-photo fermentations. *Bioresource Technology*, 101(24), 9605–9611. https://doi.org/10.1016/j.biortech.2010.07.095

Yang, G., & Wang, J. (2018). Improving mechanisms of biohydrogen production from grass using zero-valent iron nanoparticles. *Bioresource Technology*, 266, 413–420. https://doi.org/10.1016/j.biortech.2018.07.004

Yao, M., Liang, W., Chen, H., & Zhang, X. (2020). Efficient hydrogen production from formic acid using nitrogen-doped activated carbon supported Pd. *Catalysis Letters*, 150(8), 2377–2384. https://doi.org/10.1007/s10562-020-03141-y

Yildirim, O., Tunay, D., Ozkaya, B., & Demir, A. (2022). Effect of Green synthesized silver oxide nanoparticle on biological hydrogen production. *International Journal of Hydrogen Energy*, 47(45), 19517–19525. https://doi.org/10.1016/j.ijhydene.2021.11.176

Yogeswari, M. K., Dharmalingam, K., Ronald Ross, P., & Mullai, P. (2016). Role of iron concentration on hydrogen production using confectionery wastewater. *Journal of Environmental Engineering*, 142(9), 1–6. https://doi.org/10.1061/(ASCE)EE.1943-7870.0001020

Younas, M., Shafique, S., Hafeez, A., Javed, F., & Rehman, F. (2022). An overview of hydrogen production: Current status, potential, and challenges. *Fuel*, 316, 123317. https://doi.org/10.1016/j.fuel.2022.123317

Yue, M., Lambert, H., Pahon, E., Roche, R., Jemei, S., & Hissel, D. (2021). Hydrogen energy systems: A critical review of technologies, applications, trends and challenges. *Renewable and Sustainable Energy Reviews*, 146, 111180. https://doi.org/10.1016/j.rser.2021.111180

Zagrodnik, R., & Łaniecki, M. (2017). Hydrogen production from starch by co-culture of Clostridium acetobutylicum and Rhodobacter sphaeroides in one step hybrid dark- and photofermentation in repeated fed-batch reactor. *Bioresource Technology*, 224, 298–306. https://doi.org/10.1016/j.biortech.2016.10.060

Zhang, J., Fan, C., Zhang, H., Wang, Z., Zhang, J., & Song, M. (2018). Ferric oxide/carbon nanoparticles enhanced bio-hydrogen production from glucose. *International Journal of Hydrogen Energy*, 43(18), 8729–8738. https://doi.org/10.1016/j.ijhydene.2018.03.143

Zhang, L., & Sun, X. (2014). Changes in physical, chemical, and microbiological properties during the two-stage co-composting of green waste with spent mushroom compost and biochar. *Bioresource Technology*, 171, 274–284. https://doi.org/10.1016/j.biortech.2014.08.079

Zhang, Q., Xu, S., Li, Y., Ding, P., Zhang, Y., & Zhao, P. (2021). Green-synthesized nickel oxide nanoparticles enhances biohydrogen production of *Klebsiella* sp. WL1316 using lignocellulosic hydrolysate and its regulatory mechanism. *Fuel*, 305, 121585. https://doi.org/10.1016/j.fuel.2021.121585

Zhang, Y., & SHEN, J. (2007). Enhancement effect of gold nanoparticles on biohydrogen production from artificial wastewater. *International Journal of Hydrogen Energy*, 32(1), 17–23. https://doi.org/10.1016/j.ijhydene.2006.06.004

Zhao, W., Zhang, Y., Du, B., Wei, D., Wei, Q., & Zhao, Y. (2013). Enhancement effect of silver nanoparticles on fermentative biohydrogen production using mixed bacteria. *Bioresource Technology*, 142, 240–245. https://doi.org/10.1016/j.biortech.2013.05.042

Zhao, W., Zhao, J., Chen, G. D., Feng, R., Yang, J., Zhao, Y. F., Wei, Q., Du, B., & Zhang, Y. F. (2011). Anaerobic biohydrogen production by the mixed culture with mesoporous Fe_3O_4 nanoparticles activation. *Advanced Materials Research*, 306–307, 1528–1531. https://doi.org/10.4028/www.scientific.net/AMR.306-307.1528

Zhao, Y., & Chen, Y. (2011). Nano-TiO_2 enhanced photofermentative hydrogen produced from the dark fermentation liquid of waste activated sludge. *Environmental Science & Technology*, 45(19), 8589–8595. https://doi.org/10.1021/es2016186

Zhong, D., Li, J., Ma, W., & Xin, H. (2020). Magnetite nanoparticles enhanced glucose anaerobic fermentation for bio-hydrogen production using an expanded granular sludge bed (EGSB) reactor. *International Journal of Hydrogen Energy*, 45(18), 10664–10672. https://doi.org/10.1016/j.ijhydene.2020.01.095

Zhong, K., Li, M., Yang, Y., Zhang, H., Zhang, B., Tang, J., Yan, J., Su, M., & Yang, Z. (2019). Nitrogen-doped biochar derived from watermelon rind as oxygen reduction catalyst in air cathode microbial fuel cells. *Applied Energy*, 242, 516–525. https://doi.org/10.1016/j.apenergy.2019.03.050

9 Integrated Biohydrogen Production Strategies Involving Combination of Dark and Photo-Fermentation to Convert Lignocellulosic Biomass to Hydrogen

Nancy Sikodia, Kajal Chauhan, Bindu Battan, and Sulekha Chahal

9.1 INTRODUCTION

Continuously rising energy demand of modern society is the outcome of significant changes in lifestyle. However, due to their limited supply, significant price fluctuations, and detrimental effects on the environment, fossil fuels are unable to entirely satisfy this demand because of their short lifespan (Ibarra-Gonzalez and Rong, 2019). One of the most discussed topics right now is the development and use of renewable energy, which appears to have the potential to displace fossil fuels in the future (Pata and Balcilar, 2024). In this perspective, biofuels are portrayed as an exceptionally potential and environmentally friendly substitute for fossil fuels. They may be able to compete with renewable energy sources and help alleviate the world's energy

DOI: 10.1201/9781003568612-9

crises, which are caused by a variety of fossil fuel restrictions. One of the most useful, renewable, affordable, and carbon-rich resources is biomass, which may be used to generate a variety of biofuels, including biohydrogen, biobutanol, biodiesel, biogas, bioethanol, and biohydrogen (Rodionova et al., 2022). Among them, biohydrogen is a highly esteemed renewable energy source that has garnered significant attention due to its exceptional qualities, including its limitless supply, renewable nature, zero pollution, and affordability (Sikiru et al., 2024). Furthermore, it also has the maximum energy density and produces no carbon dioxide because the only byproduct of the combustion of H_2 is water vapour (Cannone et al., 2021).

Various technologies including biological, chemical, and physical processes are involved in hydrogen production (Raghulchandrana et al., 2020). Due to its many benefits, including its lower cost, ability to employ a variety of organic substrates and microorganisms, and ability to be environmentally friendly and renewable, the biological process of producing hydrogen offers remarkable upper hand over the other methods (Islam et al., 2021).

Due to their considerable availability and high cellulose content, lignocellulosic biomasses can be used as renewable and sustainable substrates for the biological generation of hydrogen (Bhatia et al., 2021). The primary components of lignocellulosic biomass are cellulose, hemicellulose, and lignin; their free form availability is vital for the efficient bioconversion that occurs during dark fermentation (DF) (Liu et al., 2020). The process of producing biohydrogen through DF with sugar hydrolysate has several benefits, including a high rate of hydrogen synthesis and the ability to use a variety of organic substrates, but it also has downside of low yield (Mishra et al., 2019a, 2019b). It is observed that metabolites in the form of fatty acids are produced in large amount because of high H_2 production, consequently the fermentation medium is highly acidic, thus hampering the H_2 production (Silva et al., 2017). These metabolites are the substrates for photo-fermentative (PF) mode of H_2 production but result in low yield. Nevertheless, if both the processes are integrated, the above disadvantage can be conquered (Sekoai et al., 2018).

9.2 PROCESS OVERVIEW: LIGNOCELLULOSIC BIOMASS TO BIOHYDROGEN

For the production of second-generation biofuel, cellulose-containing biomass is considered the best choice due to its upper hand, including availability, non-toxic, eco-friendly, and renewability (Srivastava, et al., 2021). DF is regarded as the most eminent as the process utilizes maximum biomass for the biohydrogen production (Rosales-Calderon and Arantes, 2019). Regarding the DF generation of biohydrogen, agrowaste including wheat straw, rice straw, sugarcane bagasse, corn bran, corn cob, corn stover, corn stalk, sorghum stover, sorghum rusk, sorghum leaves, rice husk, and oat straw is reported (Wei et al., 2016). Several studies on DF and PF employing lignocellulosic biomass (LCB) for the production of biohydrogen are shown in Tables 9.1 and 9.2. The complex structure of lignocellulosic biomass comprised of cellulose, hemicellulose, and lignin disintegrates via pretreatment process (Mankar et al., 2021). Pretreatment of biomass resulted in the removal of hemicellulose and lignin which makes easy passage of cellulose, and disintegration into simple sugar via enzymatic hydrolysis (Galletti and Antonetti, 2012). There are three different ways to pretreat LCB: biological, physical, and chemical (Rezania et al., 2020). To extract the lignin content, the chemical pretreatment method is the most prevalent and effective. However, this procedure is not consistently suitable for industrial usage because of the high cost of the chemicals. For its industrial and commercial application, the cost of the enzymatic hydrolysis of lignocellulosic biomass by cellulase enzyme is of concern in addition to the pretreatment expense. Furthermore, the current method of producing these cellulolytic enzymes involves employing genetically modified strains on synthetic substrate, which raises the total cost conversion process of lignocellulosic biomass to biohydrogen. In this context, the major challenges may be considerably reduced by using cost-effective feasible methods for efficient pretreatment and enzyme manufacturing. The overall production costs associated with biohydrogen using

TABLE 9.1 Research studies on dark fermentative biohydrogen production utilizing distinct types of LCB

FERMENTATION MODE	*MICROORGANISMS*	*SUBSTRATE*	*YIELD*	*REFERENCES*
Dark	***Clostridium termitidis*** ATCC-21846 and ***Clostridium intestinale*** ATCC-BAA 1027	Rice husk	5.59 mL/g dry wt.	Tosuner et al. (2019)
Dark	***Caldicellulosiruptor*** sp. and ***Ralstonia eutropha***	Wheat straw	134.0 mmol H_2/L	Soto et al. (2019)
Dark	***Escherichia coli***	Brewery spent grain	0.75±0.03 mmol H_2/L	Poladyan et al. (2018)
Dark	***Clostridium pasteurianum*** (MTCC116) and ***Bacillus subtilis*** PF_1	Sugarcane bagasse	2,870 mL/L	Srivastava et al. (2018)
Dark	***Clostridium roseum***	Cashew apple bagasse	0.08–1.89 mL H_2/g biomass	Silva et al. (2018)
Dark	***Clostridium guangxiense*** ZGM211T	***Peanut shells***	6.4–39.9 mL H_2/g biomass	Qi et al. (2018)
Dark	***Trichoderma asperellum*** BPLMBT1	Sweet sorghum	402.01 mL	Shanmugam et al. (2020)
Dark	Mixed culture	Sugarcane bagasse	0.874 mol (mol/ glucose)	Reddy et al. (2017)
Dark	Mixed culture	Sugarcane leaf waste	18.6 mL H_2/g	Moodley and Gueguim Kana (2018)
Dark	***Clostridium pasteurianum*** (MTCC116)	Rice straw	2,580 mL/L	Srivastava et al. (2017)
Dark	Mixed culture	Empty fruit bunch of oil palm	275.75 mL H_2/g TS	Ibarra-Gonzalez and Rong (2019)
Dark	***Clostridium butyricum***	Sugarcane bagasse	1.73 mol/mol	Pattra et al. (2016)
Dark	Mixed culture	Rice crop	30.37 mL/VSrem	Sattar et al. (2016)
Dark	***Clostridium butyricum***	Jatropha hull	1.95 mol H_2/ mol-TRS	Jiang et al. (2016)
Dark	***Clostridium thermocellum*** ATCC 27,405	Sugarcane bagasse	109.6 mL/g	Tian et al. (2015)
Dark	Mixed culture	Oat straw	2.39 mol H_2/mol reducing sugars	Arreola-Vargas et al. (2015)
Dark	***Clostridium beijerinckii*** KCTC 1785	Sorghum rusk	1,117 mL/L	Saratale et al. (2015)

LCB may be reduced with appropriate pretreatment procedures along with inexpensive substrate-based efficient cellulase synthesis, either from mesophilic or thermotolerant microorganisms. Numerous studies have looked into the possibility that thermotolerant fungi and chemical pretreatment could help reduce the cost of biomass to biohydrogen production (Nadeem et al., 2020). However, one of the main obstacles to the LCB pretreatment process is the production of inhibitors, which reduce the entire productivity of the enzyme hydrolysis and fermentation process (Srivastava et al., 2020). Biohydrogen production from lignocellulosic biomass also depends on the microorganisms used for fermentation. *Clostridium*, *Bacillus*, and *Enterobacter* are well-known H_2-producing anaerobic bacteria through DF (Banu et al., 2018). It is well studied that separate hydrolysis and fermentation (SHF) is best as it restricts the negative impact of enzyme hydrolysis such as effect of inhibitors, fluctuations in pH and temperature on microorganisms used

TABLE 9.2 Research studies on photo-fermentative biohydrogen production utilizing distinct types of LCB

FERMENTATION MODE	*MICROORGANISMS*	*SUBSTRATE*	*YIELD*	*REFERENCES*
Photo	Mixed culture	Corn stalk	2.6±0.3 mol H_2/mol sugar consumed	Jiang et al. (2016)
Photo	HAU-M1	Corn stalk	119.3 mL/g	Quanguo et al. (2017)
Photo	***Rhodospirillum rubrum*, *Rhodopseudomonas capsulata*, *Rhodopseudomonas palustris*, *Rhodobacter sphaeroides*,** and ***Rhodobacter capsulatus***	Corn stalk	211.9 mL/L medium	Jiang et al. (2020)
Photo	Mixed bacteria strain (HAU-M1)	Corn cob	84.7 mL/g TS	Zhang et al. (2020)
Photo	***Spirulina platensis***	Wet biomass	1.92±0.20 mmol H_2/ mol COD red.	Pandey et al. (2021)

in fermentation (Sudiyani et al., 2019). However, it is typically observed that the rapid rate of H_2 production can quickly cause a DF medium to become extremely acidic because of the significant quantity of acidic metabolites formed such as acetic, malic, butyric, propionic, succinic, and fumaric acid, thus reducing the yield of H_2 generation overall (Uyar et al., 2012). However, given the optimum circumstances, these metabolite effluents, or volatile fatty acids (VFAs), could serve as a substrate for PF bacteria to produce H_2 (Ghosh et al., 2018). Furthermore, by absorbing these VFA effluents, the provocation of low H_2 generation rate regarding PF can be overcome, greatly increasing the overall production of biohydrogen. It is therefore possible to further enhance the biohydrogen production process by utilizing the integrated strategy described in this chapter, which unite DF and PF with LCB as the substrate.

9.3 DARK FERMENTATION

9.3.1 Concept

DF is the biological fermentation mode of H_2 synthesis without light. It is sometimes referred to as the most convenient H_2 production method at ambient settings when using organic substrates (Sarangi and Nanda, 2020). This is a widely utilized biological process that produces H_2 from a variety of cellulosic and organic wastes (Hajizadeh et al., 2021). In addition to the ease of use and adaptability of the substrate, this mode produces H_2 at a far higher rate than other known fermentative methods. It also allows for the possible use of a variety of fermentative bacteria (Jayachandran et al., 2022). Among the well-known possible H_2 producers in the DF process using a variety of substrates are *Enterobacter aerogenes*, *Enterobacter asburiae*, *Clostridium butyricum*, and *Clostridium beijerinckii* (Wang and Yin, 2019). Additionally, similar to other biochemical processes, the DF process of biohydrogen production is significantly impacted by physicochemical variables such as pH, hydraulic retention duration, temperature, and the size and volume of the inoculum (Gopalakrishnan et al., 2019). Furthermore, two primary H_2-producing enzymes, nitrogenase and hydrogenase, are required for the entire synthesis of H_2 (Mishra et al., 2019a, 2019b). Hydrogenases, which are divided into three classes: [FeFe]-hydrogenase, [NiFe]-hydrogenase, and [Fe]-hydrogenase, catalyse the interconversion of hydrogen to protons and electrons (Ogata and Lubitz, 2020). Furthermore, these enzymes are involved in two DF pathways: the butyrate pathway yields 2 M H_2 theoretically from 1 M of glucose, while the acetate pathway yields 4 M H_2 theoretically from 1 M glucose (Sarangi and Nanda, 2020) (Figure 9.1). The process of DF begins with the oxidation of substrate to pyruvate, which produces NADH. Microorganisms that possess ferredoxin oxidoreductase (NFOR)

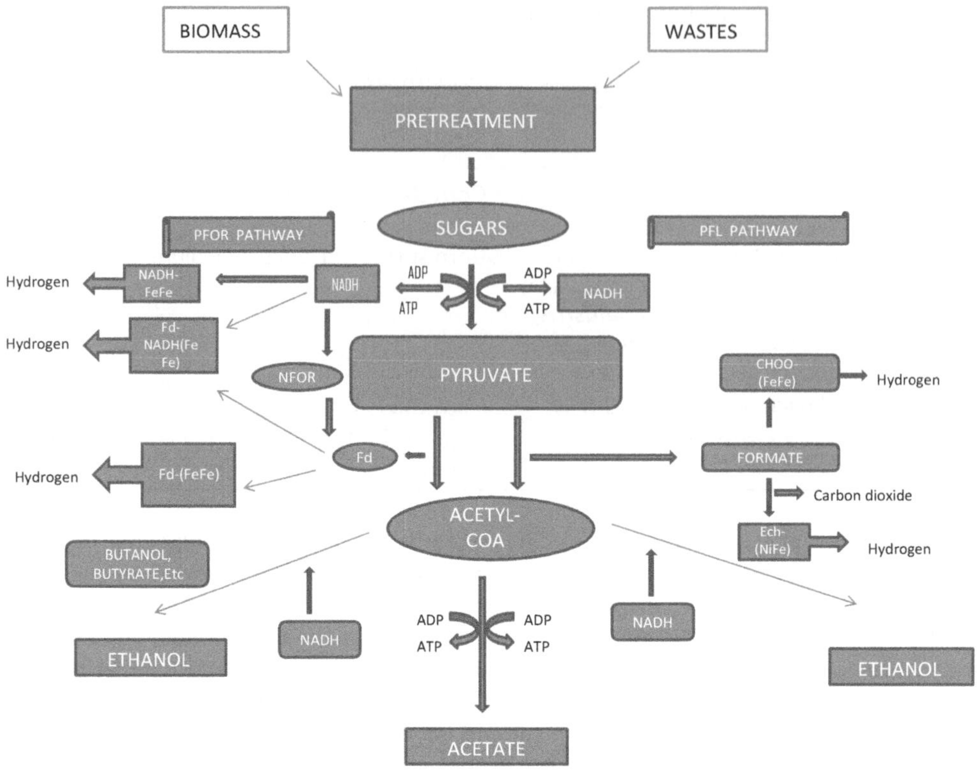

FIGURE 9.1 Biohydrogen production from dark fermentative process. (Adapted from Hallenbeck et al. (2012).)

can use this NADH to create reduced ferredoxin (Pandey et al., 2021). In the second stage, pyruvate can be reduced to acetyl-CoA and formate by ferredoxin via pyruvate-ferredoxin oxidoreductase (PFOR) or acetyl-CoA and formate by pyruvateformate lyase (PFL) to produce H_2 (Sağır and Hallenbeck, 2019). However, a substantial quantity of acid is produced during this process, which reduces the amount of H_2 and is thought to be one of the main issues that needs to be investigated by scientists. Polyhydroxybutyrate (PHB) is the primary end product, although there are also acidic metabolites produced, including ethanol, acetic acid, fumaric, lactic, butyric, and propionic acid. The following equations represent the process of DF reactions (Sarangi and Nanda, 2020):

$$C_6H_{12}O_6 + 2NAD^+ \rightarrow 2CH_3COCOO^- + 4H^+ + 2NADH \tag{9.1}$$

$$CH_3COCOO^- + CoA - H \rightarrow \text{AcetylCoA} + HCOO^- (\text{PFLP}) \tag{9.2}$$

$$HCOO^- + H^+ \rightarrow CO_2 + H_2 \tag{9.3}$$

$$CH_3COCOO^- + CoA + \text{Ferredoxin}(\text{Fd})\text{ox} \rightarrow \text{Acetyl CoA} + \text{Fdred} + CO_2 (\text{FORP}) \tag{9.4}$$

$$\text{Fdred} + 2H^+ \rightarrow \text{Fdox} + 2H_2 \tag{9.5}$$

$$\text{Acetyl CoA} + H_2O \rightarrow CH_3COO^- + H^+ + CoA - H \tag{9.6}$$

$$\text{Acetyl CoA} + 2NADH + 2H^+ \rightarrow CH_3CH_2OH + CoA - H + 2NAD^+ \tag{9.7}$$

$$NADH + H^+ \rightarrow NAD^+ + H_2$$

Reddy et al. (2017) looked at the synthesis of biohydrogen from sugarcane bagasse utilizing DF. Using Enterobacteriaceae and *Clostridium* sp. it was discovered that the addition of magnetite nanoparticles increased biohydrogen generation by around 69.6%. In contrast, the control system (no nanoparticles) had a maximum hydrogen yield (HY) of 0.874 mol (mol/glucose) at the ideal pH of 0.5.0.

9.3.2 Mechanism

Adaptation of the fermentative H_2 generation route is contingent upon the kind of microorganism that carries out the DF operation (Li et al., 2019). Theoretical value of the microorganism PFL pathway was restricted to 2 mol H_2/mol glucose because it was unable to use NADH for H_2 creation (Srivastava et al., 2017). Furthermore, acetyl-CoA eventually transforms into lower VFA salts such as ethanol, butyrate, and acetate during the intermediate stage. Because the acetate path averts microorganisms from consuming more NADH while H_2 is produced, it causes the peak output of 4 mol H_2/mol glucose (4 mol H_2 from 1 mol glucose, 33% of theoretical yield $12H_2$). Osman et al. (2020) conducted research on the synthesis of biohydrogen through DF, utilizing rice straw and *Clostridium pasteurianum* as substrate and fermentative bacterium, respectively. At the initial pH of 7.5 and temperature of 37°C, a maximum hydrogen output of 47.6 mL/g was obtained. Soares et al. (2020) observed biohydrogen generation utilizing rice, maize, and wheat remains, following a similar pattern. The results so found showed that, by mixed culture, biohydrogen generation ranged from 12 to 7,019 mL H_2/L under ideal conditions of 37°C, pH 7.0, and 72 hours of hydraulic retention time. Similarly, Bhatia et al. (2021) found that the acetic acid route produced the maximum specific H_2 yield of 4 mol H_2/mol glucose or 497.8 mL H_2/g glucose.

9.4 PHOTO-FERMENTATION

9.4.1 Concept

After DF, PF is another widely used and popular method of biohydrogen production which is facilitated by the photosynthetic purple non-sulphur (PNS) bacterial group (Rai and Singh, 2016). In the presence of a light source, these bacteria may use both long and short chains of VFAs, which are effectively transformed by the nitrogenase enzyme into H_2 and associated metabolites (Abdelqader, 2021). These bacteria can also effectively create H_2 by utilizing a wide variety of organic substrates and energy in the form of light. In addition *Rhodobacter sphaeroides*, *Rhodopseudomonas palustris*, *Rhodovulum sulfidophilum*, and *Rhodobacter capsulatus* are among the PNS bacteria that are acclaimed for producing H_2 with a high yield through photofermentation on VFAs as a substrate (Kessi and Hörtensteiner, 2018). Potential VFA substrates for the considerable PF H_2 generation have been reported to include organic acids including acetic, butyric, propionic, malic, and lactic (Pandey et al., 2021). However, other studies have also shown that carbohydrates like glucose and sucrose can produce H_2 directly through PF, indicating that this process is potentially used for producing H_2 (Rai et al., 2014). Optimization of physicochemical variables is also crucial for a high H_2 yield in the case of PF, in addition to the substrate and microorganisms (Soto et al., 2019). The bacterial cell growth and hydrogen production of PF occur in the medium temperature range of 30–40°C, whereas pH ranges from 5.0 to 7.0 at normal atmospheric pressure. Different substrates and light sources with wavelengths of 522, 805, and 850 nm are used (Liu et al., 2015). Furthermore, because pH shift is not necessary, the DF effluent can be utilized immediately for PF in the consecutive process (Soto et al., 2019). The type of substrate and conversion efficacy may also have a major impact on how well PF-mediated H_2 generation performs (Zhang and Zhang, 2018). The following equations describe the substrate conversion efficacy of malate, lactate, and acetate in PF reaction (Assawamongkholsiri et al., 2019):

Malate $C_4H_6O_5 + 3H_2O \rightarrow 6H_2 + 4CO_2$

Lactate $C_3H_6O_3 + 3H_2O \rightarrow 6H_2 + 3CO_2$

Acetate $C_2H_4O_2 + 2H_2O \rightarrow 4H_2 + 2CO_2$

9.4.2 Mechanism

Higuchi-Takeuchi and Numata (2019) reported PF H_2 synthesis via PNS through an enzymatic method known as nitrogenase negotiated biochemical process, which may be run employing a number of distinct organic acids as electron donors. When ATP synthase is present, a proton gradient is formed through the membrane by the pumping of protons, which creates ATP, and the electrons produced by the carbon substrate are pumped via a high number of electron carriers. Ferredoxin, an electron acceptor, might receive the extra electrons via the use of the extra ATP produced (Jiang et al., 2020). The electrons from ferredoxin are also utilized by nitrogenase to convert protons into hydrogen when there is no molecular nitrogen present. High theoretical HYs of 8 mol of H_2/mol of glucose have been reported for PF using PNS bacteria (Mishra et al., 2019a, 2019b). However, high energy consumption, poor solar energy conversion efficiency, and low volumetric output rates are the main disadvantages of this technique. The kind of organic acids selected determines the greatest H_2 yield. For instance, acetic acid generates the lowest value of 4 mol H_2/mol substrate, whereas butyric acid yields the theoretical maximum yield (HY) of 10 mol H_2/mol substrate (Zhang et al., 2017) (Figure 9.2).

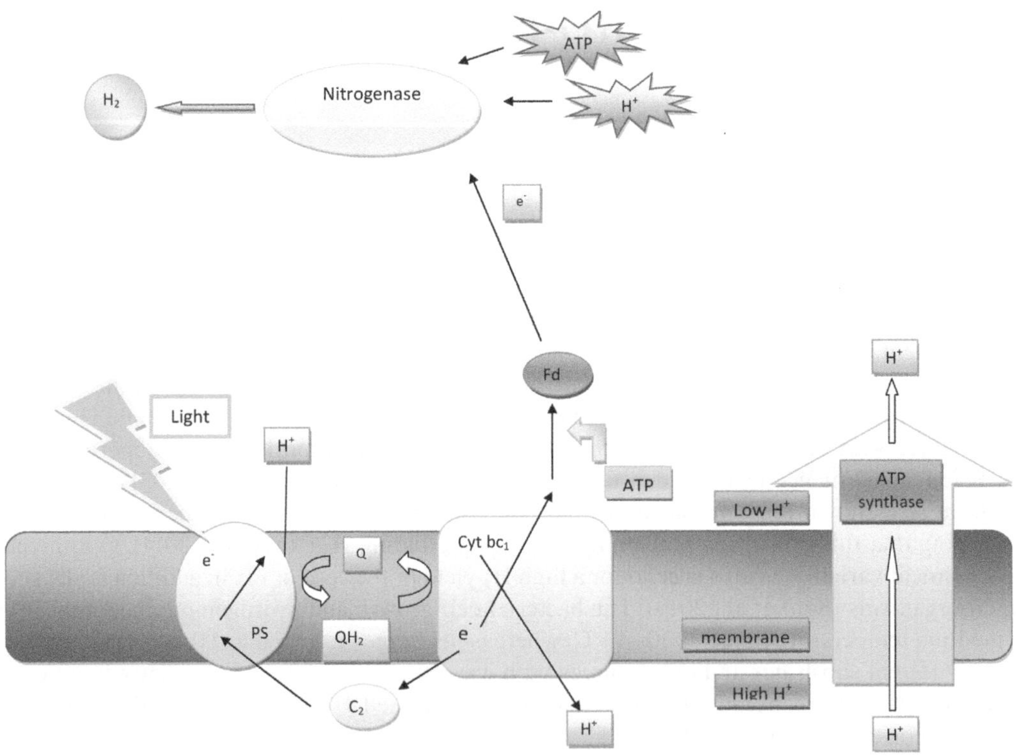

FIGURE 9.2 Photo-fermentation mechanism in purple non-sulphur bacteria. (Adapted from Hädicke et al. (2011) http://creativecommons.org/licenses/by/2.0.)

The following equation explains the chemical reaction for PF hydrogen production from acetate:

$$2CH_3COOH + 2H_2O \rightarrow 4H_2 + 2CO_2, \ \Delta G^\circ = +104 \text{ kJ}$$

Practical HY is always lower than theoretical HY in the PF hydrogen production procedure. This could be because of a variety of environmental factors, including temperature, pH variations, substrate-production inhibition, the viability rate of microorganisms in the fermentative medium, and the influence of intermediate inhibitors (Zhang et al., 2017). Using acetic acid as the substrate, *Rhodopseudomonas* sp. nov. strain A7 bacteria, and an inoculum duration of 14 hours, Liu et al. (2015) reported a HY of 160 mL H_2/60 mL. Using isolated *R. palustris*, Mishra et al. (2016) conducted PF H_2 generation in a batch manner at 30°C and 4 flux light illumination, yielding 2.86 mL H_2/mL palm oil mill effluent (POME).

9.5 INTEGRATED TWO-STAGE SEQUENTIAL TECHNIQUE FOR PRODUCING H_2

Although the most beneficial and sustainable approach, both the fermentative biohydrogen production methods (DF and PF) have major disadvantages that pose significant obstacles to the development of economically viable and sustainable biohydrogen production technology (Ghosh et al., 2018). The primary limitations of the PF process include low H_2 production rate, while the main downsides of the DF process are low yield, high generation costs, and lack of viable substrates. In the present scenario, utilizing LCB as the substrate and integrating the two processes could help to conquer these obstacles (Cheng et al., 2011). Compared to single-stage DF or PF, sequential PF and DF processes are

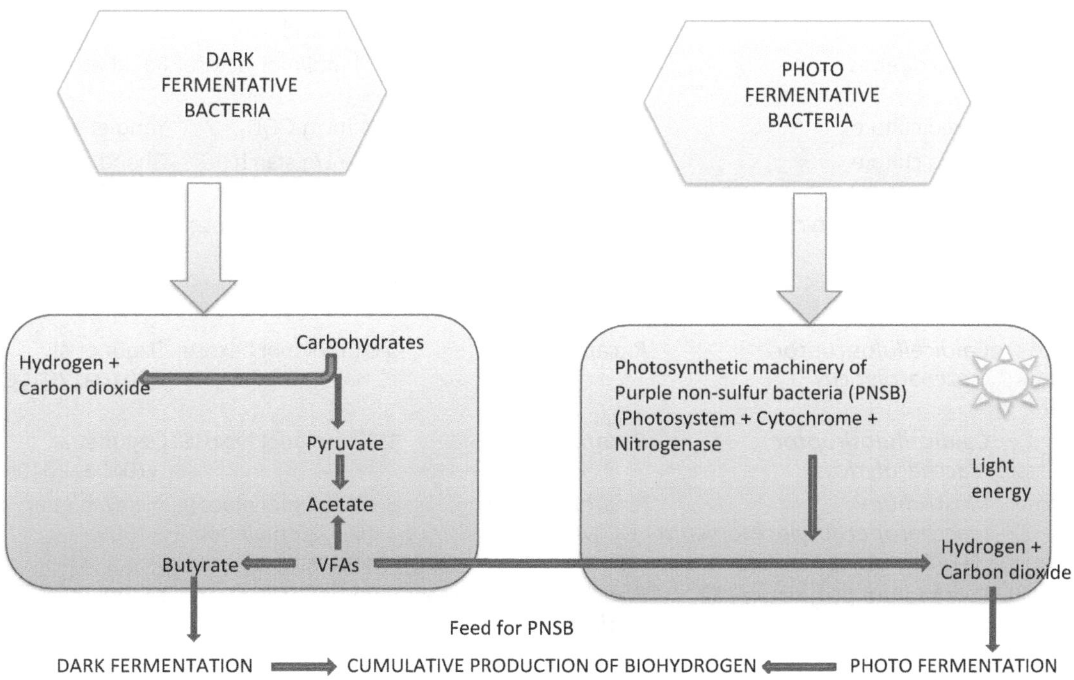

FIGURE 9.3 Integrated biohydrogen production.

thought to be a more competent method for producing hydrogen. In order to operate these processes under their respective ideal conditions, the process of fermentation is sequentially carried out in two different bioreactors in this system (Argun and Kargi, 2011). Because it uses the remaining VFAs in the DF effluents, this procedure produces additional H_2. For example, light-fermentative bacteria use the acetic acid produced during DF as a substrate, resulting in an overall H_2 output of 12 mol/mol glucose. Researchers have previously carried out two-stage sequential dark and light fermentation utilizing glucose as the substrate for the former and the fermented effluent (rich in VFAs) as a substrate for a consortium or PF bacterium (Chen et al., 2008). Numerous industrial and agricultural wastes have been studied thus far for two-stage DF and PF to produce hydrogen (Kumar et al., 2016). Zong et al. (2009) used food waste and cassava as a substrate to show that a two-stage sequential DF and PF method could produce hydrogen. The first stage of fermentation was determined to be 199 mL H_2/g^1 for cassava and 220 mL H_2/g for food waste. However, the subsequent PF of the dark fermented effluent produced 451 and 611 mL H_2/g of hydrogen. Comparing two-stage sequential DF and PF to single-stage DF, the total HY for cassava and food waste was raised by 4.08 and 3.05 times, respectively (Zong et al., 2009). Two-stage sequential DF-PF with pretreatment of corncob has been shown to increase hydrogen output. The greatest HY in the first step (DF) is 120.3 mL H_2/g-corncob; however, following two stages of sequential DF and PF, it rises to 713.6 mL H_2/g-corncob, which accounts for a roughly six-fold increase in hydrogen production (Yang et al., 2010). Using palm oil mill effluent as the substrate, Mishra et al. (2016), in a recent related study, examined the use of two-stage sequential DF and PF. Using *C. butyricum* LS2 as the inoculum, they found a hydrogen output of 0.784 mL H_2/mL POME in the first stage (DF). Under ideal physico-chemical circumstances, DF effluent was

TABLE 9.3 Hydrogen yield from pure carbon sources and agro-industrial effluent utilizing sequential two-stage dark and photo-fermentation

FEED	*DARK FERMENTATIVE INOCULUM*	*PHOTO-FERMENTATIVE INOCULUMS*	*MAXIMUM HYDROGEN YIELD*	*REFERENCES*
Glucose	***E. cloacae*** strain DM11	***R. sphaeroids*** O.U.001	1.86 mol/mol glucose 1.52-1.72 mol/mol acetic acid	Nath et al. (2005)
Cheese whey	***E. aerogenes***	***Rhodopseudomonas*** BHU 01	2.04 mol/mol lactose	Rai et al. (2012)
Corn cob	Mixed culture	***R. sphaeroids***	714 mL/g COD	Yang et al. (2010)
Cassava starch	Mixed culture	***R. palustris***	840 mL/g starch	Cheng et al. (2011)
Starch	***Clostridium butyricum***	***Rhodobacter*** sp. M-19	3.6 mol/mol glucose	Yokoi et al. (1998)
Cheese whey	Mixed culture	***R. palustris***	10 mol/mol lactose	Azbar and Dokgoz (2010)
Potato stem peel	***Caldicellulosiruptor saccharolyticus***	***R. capsulatus***	5.81 mol/mol hexose	Özgür et al. (2010a, 2010b)
Beet molasses	***Caldicellulosiruptor saccharolyticus***	***R. capsulatus***	13.7 mol/mol hexose	Özgür et al. (2010a, 2010b)
Glucose	***Clostridium saccharoperbutylacetonicum*** N14ATCC13564	***R. sphaeroides*** NCIMB8253	3.10 mol/mol glucose 1.10–1.25 mol/mol acetic acid	Alalayah et al. (2009)
Palm oil mill effluent	***Clostridium butyricum LS2***	***Rhodopseudomonas palustris***	3.064 mL H_2/mL POME	Mishra et al. (2016)
Sweet potato starch	Mixed culture (***C. butyricum*** and E.2 ***aerogenes***)	***Rhodobacter sp.***	7 mol/mol hexose	Yokoi et al. (2001)

treated to PF using *R. palustris* as the inoculum, increasing the hydrogen output to 3.064 mL H_2/mL POME (Mishra et al., 2016). Various researchers have adapted the prospect of increased H_2 yield from pure carbon sources and agro-industrial effluent utilizing sequential two-stage DF and PF, as listed in Table 9.3; Figure 3. Though it is difficult to attain in practice, the theoretical H_2 yield utilizing consecutive DF and PF should be 12 mol/mol glucose. One explanation for this invariability is because bacteria use glucose for their metabolic processes (Ren et al., 2011). Potential limitations that prevent the theoretical H_2 yield from being achieved include improper operational procedures (pH, temperature, etc.) and performance metrics (inoculum conversion efficiency).

9.6 SINGLE-STAGE (CO-CULTURE) OR COMBINED FERMENTATION METHOD

Well-known H_2 generators include saccharolytic fermentative bacteria like *Clostridium* sp. However, because they are also very sensitive to oxygen, even a small amount of O_2 in the fermenter inhibits their ability to produce H_2. Researchers have employed L-cysteine as a reducing agent in a medium to address this issue. It does, however, make the process very costly, thus the method that produces H_2 without the reducing agent is preferred (Yokoi et al., 1998). The better choice for a steady and high-yield H_2 production in this search is to use facultative anaerobes as a co-culture, which aid in eating the undesirable O_2 present in the medium (Yokoi et al., 1998). The coupling of PNS with fermentative bacteria is reported to significantly provide high H_2 yield because the PNS photosynthetic bacteria use the organic acids produced as a byproduct of DF (Yokoi et al., 1998). Furthermore, separate reactors are needed for each stage of sequential DF and PF, which raises operational costs. Therefore, co-culturing dark and photosynthetic bacteria in a single reactor using both techniques offers a higher chance to increase H_2 generation in an economical way; compared to two-stage sequential DF and PF, which requires two separate reactors for operation, the single-stage DF and PF (co-culture) mode has the benefit of allowing for simultaneous occurrence in a single bioreactor (Ghosh et al., 2018). Table 9.4 shows the integrated DF and PF employing

TABLE 9.4 Integrated dark and photo-fermentation employing co-cultivation of dark and photo-fermentation using different substrates

FEED	*DARK FERMENTATIVE INOCULUM/ PHOTO-FERMENTATIVE INOCULUM*	*OPERATIONAL PARAMETER PH; TEMPERATURE; LIGHT INTENSITY*	*MAXIMUM HYDROGEN YIELD*	*REFERENCES*
Wheat starch	***Clostridium beijerinckii/ Rhodobacter sphaeroides***	7.0–7.5; 32°C ± 2°C; 10 klux	0.60 mol/ mol hexose	Argun and Kargi (2010)
Wheat starch	***Activated sludge/Rhodobacter sphaeroides***	7.0; 30°C; 5 klux	3.40 mol/ mol hexose	Sagnak et al. (2011)
Sugarcane effluent	***Citrobacter freundii*** and ***E. aerogenes/R. palustris***	7.0; 37°C; 7 klux	2.76 mol/ mol hexose	Vatsala et al. (2008)
Algal biomass (starch)	***Vibrio fluvialis/Rhodobium marinum***	na; 30°C; 330 w/m^2	6.2 mol H_2/ mol hexose	Ike et al. (1999)
Algal biomass (starch)	***Lactobacillus amylovorus/ Rhodobium marinum***	6.5; 30°C; 170 w/m^2	7.2 mol H_2/ mol hexose	Kawaguchi et al. (2001)
Wheat starch	Mixed culture/***Rhodobacter*** sp.	7.3; 30°C; 9.5 klux	176 mL H_2/g starch	Argun et al. (2009)
Starch	***Clostridium butyricum/ Rhodobacter*** sp. M-19	pH6.5;30°C; 5 klux	4.5 mol H_2/ mol glucose	Yokoi et al. (1998)

co-cultivation of DF and PF using different substrates. One of the combination's notable benefits is that the DF inoculum uses organic waste as a substrate and generates hydrogen in addition to VFAs, which are then used by in-situ PF bacteria to produce hydrogen (Pant et al., 2013). The notable benefits of the co-culture system above the two-stage sequential DF and PF production process are also the elimination of the need for external substrate inhibition, pH adjustment, and shorter operating time (because the process is completed in a single stage) (Ozmihci and Kargi, 2010).

9.7 CONCLUSION AND FUTURE PROSPECTS

In developing nations, 38% of primary energy consumption comes from biomass, which is regarded as the world's largest energy source. The growing interest in using different biomasses to produce biological hydrogen is an intriguing area of advanced energy generation technology. This study presents the latest research on fermentative hydrogen production using a variety of techniques, such as DF, PF, and integrated DF-PF systems. As a substrate for DF, PF, and combined DF-PF hydrogen production, organic-rich biomass has been discovered to have a lot of potential. The greatest conversion of substrate to hydrogen (12 mol H_2/mol glucose) is one of the many benefits of the combined DF-PF method, despite the single-stage DF and PF hydrogen. Furthermore, the review indicates that the production of fermentative hydrogen is not limited to the essential operational parameters of pH and temperature. The development of advanced technologies such as immobilization and application nanotechnology has also been perceived to have a positive impact on the production of fermentative hydrogen. When a variety of wastes or wastewaters are exposed to substrates, the immobilized inoculum can effectively improve the stability of the biological hydrogen production process. Furthermore, compared to a non-additive fermentative medium, the addition of nanoparticles of various metal oxides to the fermentative medium resulted in an increase in hydrogen productivity of up to roughly six times. According to the review, a promising method for the single-stage DF and PF hydrogen-generation system is the application of immobilized inoculum and the inclusion of nanoparticles. Additionally, several attempts have been made to increase the rate of hydrogen synthesis as well as the final yield of hydrogen production. The development of genetically modified fermentative microorganisms, advances in bioreactor design, etc., should result in significant improvements and eventually lead to commercially successful sustainable hydrogen production.

REFERENCES

Abdelqader, A. A. (2021). Biohydrogen production from biomass by photofermentation using helium-neon laser irradiation (Doctoral dissertation, Cairo University).

Alalayah, W. M., Kalil, M. S., Kadhum, A. A., Jahim, J. M., Jaapar, S. Z., & Alauj, N. M. (2009). Bio-hydrogen production using a two-stage fermentation process. *Pakistan Journal of Biological Sciences: PJBSciences*: PJBS, *12*(22), 1462–1467.

Argun, H., & Kargi, F. (2010). Bio-hydrogen production from ground wheat starch by continuous combined fermentation using annular-hybrid bioreactor. *International Journal of Hydrogen Energy*, *35*(12), 6170–6178.

Argun, H., & Kargi, F. (2011). Bio-hydrogen production by different operational modes of dark and photo-fermentation: an overview. *International Journal of Hydrogen Energy*, *36*(13), 7443–7459.

Argun, H., Kargi, F., & Kapdan, I. K. (2009). Effects of the substrate and cell concentration on bio-hydrogen production from ground wheat by combined dark and photo-fermentation. *International Journal of Hydrogen Energy*, *34*(15), 6181–6188.

Assawamongkholsiri, T., Reungsang, A., & Sittijunda, S. (2019). Photo-hydrogen and lipid production from lactate, acetate, butyrate, and sugar manufacturing wastewater with an alternative nitrogen source by Rhodobacter sp. KKU–PS1. *PeerJ*, *7*, e6653.

Azbar, N., & Dokgoz, F. T. C. (2010). The effect of dilution and L-malic acid addition on bio-hydrogen production with Rhodopseudomonas palustris from effluent of an acidogenic anaerobic reactor. *International Journal of Hydrogen Energy*, *35*(10), 5028–5033.

Banu, J. R., Kannah, R. Y., Kumar, M. D., Gunasekaran, M., Sivagurunathan, P., Park, J. H., & Kumar, G. (2018). Recent advances on biogranules formation in dark hydrogen fermentation system: mechanism of formation and microbial characteristics. *Bioresource Technology*, *268*, 787–796.

Bhatia, S. K., Jagtap, S. S., Bedekar, A. A., Bhatia, R. K., Rajendran, K., Pugazhendhi, A., ... & Yang, Y. H. (2021). Renewable biohydrogen production from lignocellulosic biomass using fermentation and integration of systems with other energy generation technologies. *Science of the Total Environment*, *765*, 144429.

Cannone, S. F., Lanzini, A., & Santarelli, M. (2021). A review on CO_2 capture technologies with focus on CO_2-enhanced methane recovery from hydrates. *Energies*, *14*(2), 387.

Chen, C. Y., Yang, M. H., Yeh, K. L., Liu, C. H., & Chang, J. S. (2008). Biohydrogen production using sequential two-stage dark and photo fermentation processes. *International Journal of Hydrogen Energy*, *33*(18), 4755–4762.

Cheng, C. L., Lo, Y. C., Lee, K. S., Lee, D. J., Lin, C. Y., & Chang, J. S. (2011). Biohydrogen production from lignocellulosic feedstock. *Bioresource Technology*, *102*(18), 8514–8523.

Cheng, J., Su, H., Zhou, J., Song, W., & Cen, K. (2011). Hydrogen production by mixed bacteria through dark and photo fermentation. *International Journal of Hydrogen Energy*, *36*(1), 450–457.

da Silva Moura, P. M., Ortigueira, J. R., Valdez-Vazquez, I., Saratale, G. D., Saratale, R. G., & da Silva, C. A. M. (2017). Dark fermentative hydrogen production: from concepts to a sustainable production. *Microbial Fuels*, 219–274.

Galletti, A. R., & Antonetti, C. (2012). Biomass pretreatment: separation of cellulose, hemicellulose, and lignin-existing technologies and perspectives. In Biorefinery: From Biomass to Chemicals and Fuels, (pp. 101–121). De Gruyter Brill.

Ghosh, S., Chowdhury, R., & Bhattacharya, P. (2018). A review on single stage integrated dark-photo fermentative biohydrogen production: insight into salient strategies and scopes. *International Journal of Hydrogen Energy*, *43*(4), 2091–2107.

Gopalakrishnan, B., Khanna, N., & Das, D. (2019). Dark-fermentative biohydrogen production. In Biohydrogen (pp. 79–122). Elsevier.

Hädicke, O., Grammel, H., & Klamt, S. (2011). Metabolic network modeling of redox balancing and biohydrogen production in purple nonsulfur bacteria. *BMC Systems Biology*, *5*, 1–18.

Hajizadeh, A., Saady, N. M. C., Zendehboudi, S., Rajagopal, R., & Hung, Y. T. (2021). Biohydrogen production through mixed culture dark anaerobic fermentation of industrial waste. In Integrated Natural Resources Management, (pp. 323–369). Springer.

Ibarra-Gonzalez, P., & Rong, B. G. (2019). A review of the current state of biofuels production from lignocellulosic biomass using thermochemical conversion routes. *Chinese Journal of Chemical Engineering*, *27*(7), 1523–1535.

Ike, A., Murakawa, T., Kawaguchi, H., Hirata, K., & Miyamoto, K. (1999). Photoproduction of hydrogen from raw starch using a halophilic bacterial community. *Journal of Bioscience and Bioengineering*, *88*(1), 72–77.

Islam, A. K., Dunlop, P. S., Hewitt, N. J., Lenihan, R., & Brandoni, C. (2021). Bio-hydrogen production from wastewater: A comparative study of low energy intensive production processes. *Clean Technologies*, *3*(1), 156–182.

Jayachandran, V., Basak, N., De Philippis, R., & Adessi, A. (2022). Novel strategies towards efficient molecular biohydrogen production by dark fermentative mechanism: present progress and future perspective. *Bioprocess and Biosystems Engineering*, *45*(10), 1595–1624.

Jiang, D., Fang, Z., Chin, S. X., Tian, X. F., & Su, T. C. (2016). Biohydrogen production from hydrolysates of selected tropical biomass wastes with Clostridium butyricum. *Scientific reports*, *6*(1), 27205.

Jiang, D., Ge, X., Zhang, T., Chen, Z., Zhang, Z., He, C., ... & Li, Y. (2020). Effect of alkaline pretreatment on photo-fermentative hydrogen production from giant reed: Comparison of NaOH and Ca $(OH)_2$. *Bioresource Technology*, 304, 123001.

Karlilar Pata, S., & Balcilar, M. (2024). Decarbonizing energy: Evaluating fossil fuel displacement by renewables in OECD countries. *Environmental Science and Pollution Research*, *31*(21), 31304–31313.

Kawaguchi, H., Hashimoto, K., Hirata, K., & Miyamoto, K. (2001). H_2 production from algal biomass by a mixed culture of Rhodobium marinum A-501 and Lactobacillus amylovorus. *Journal of Bioscience and Bioengineering*, *91*(3), 277–282.

Kessi, J., & Hörtensteiner, S. (2018). Inhibition of bacteriochlorophyll biosynthesis in the purple phototrophic bacteria Rhodospirillumrubrum and Rhodobacter capsulatus grown in the presence of a toxic concentration of selenite. *BMC Microbiology*, 18, 1–24.

Kumar, G., Mudhoo, A., Sivagurunathan, P., Nagarajan, D., Ghimire, A., Lay, C. H., ... & Chang, J. S. (2016). Recent insights into the cell immobilization technology applied for dark fermentative hydrogen production. *Bioresource Technology*, *219*, 725–737.

Li, Z., Wang, X., Zhang, T., Si, H., Xu, C., Wright, A. D. G., & Li, G. (2019). Heterogeneous development of methanogens and the correlation with bacteria in the rumen and cecum of sika deer (*Cervus nippon*) during early life suggest different ecology relevance. *BMC Microbiology*, *19*, 1–13.

Liu, B. F., Jin, Y. R., Cui, Q. F., Xie, G. J., Wu, Y. N., & Ren, N. Q. (2015). Photo-fermentation hydrogen production by *Rhodopseudomonas* sp. nov. strain A7 isolated from the sludge in a bioreactor. *International Journal of Hydrogen Energy*, *40*(28), 8661–8668.

Liu, Y., Min, J., Feng, X., He, Y., Liu, J., Wang, Y., ... & Sun, Y. (2020). A review of biohydrogen productions from lignocellulosic precursor via dark fermentation: perspective on hydrolysate composition and electron-equivalent balance. *Energies*, *13*(10), 2451.

Mankar, A. R., Pandey, A., Modak, A., & Pant, K. K. (2021). Pretreatment of lignocellulosic biomass: A review on recent advances. *Bioresource Technology*, *334*, 125235.

Mishra, A. K., Kaushik, M. S., & Tiwari, D. N. (2019a). Nitrogenase and hydrogenase: enzymes for nitrogen fixation and hydrogen production in cyanobacteria. In Cyanobacteria (pp. 173–191). Academic Press.

Mishra, P., Krishnan, S., Rana, S., Singh, L., Sakinah, M., & Ab Wahid, Z. (2019b). Outlook of fermentative hydrogen production techniques: An overview of dark, photo and integrated dark-photo fermentative approach to biomass. *Energy Strategy Reviews*, *24*, 27–37.

Mishra, P., Thakur, S., Singh, L., Ab Wahid, Z., & Sakinah, M. (2016). Enhanced hydrogen production from palm oil mill effluent using two stage sequential dark and photo fermentation. *International Journal of Hydrogen Energy*, *41*(41), 18431–18440.

Nadeem, F., Jiang, D., Tahir, N., Alam, M., Zhang, Z., Yi, W., ... & Zhang, Q. (2020). Defect engineering in SnO_2 nanomaterials: pPathway to enhance the biohydrogen production from agricultural residue of corn stover. *Applied Materials Today*, *21*, 100850.

Nath, K., Kumar, A., & Das, D. (2005). Hydrogen production by Rhodobacter sphaeroides strain OU 001 using spent media of Enterobacter cloacae strain DM11. *Applied Mmicrobiology and Bbiotechnology*, *68*, 533–541.

Ogata, H., Lubitz, W., 2020. Hydrogenases structure and function. In Reference Module in Biomedical Sciences: Encyclopedia of Biological ChemistryRef. Module Life Sci. (pp. 562–567). Elsevier.

Osman, A. I., Deka, T. J., Baruah, D. C., & Rooney, D. W. (2020). Critical challenges in biohydrogen production processes from the organic feedstocks. *Biomass Conversion and Biorefinery*, *13*, 8383–84011–19.

Özgür, E., Afsar, N., de Vrije, T., Yücel, M., Gündüz, U., Claassen, P. A., & Eroglu, I. (2010a). Potential use of thermophilic dark fermentation effluents in photofermentative hydrogen production by Rhodobacter capsulatus. *Journal of Cleaner Production*, *18*, S23–S28.

Özgür, E., Mars, A. E., Peksel, B., Louwerse, A., Yücel, M., Gündüz, U., ... & Eroğlu, İ. (2010b). Biohydrogen production from beet molasses by sequential dark and photofermentation. *International Journal of Hydrogen Energy*, *35*(2), 511–517.

Ozmihci, S., & Kargi, F. (2010). Bio-hydrogen production by photo-fermentation of dark fermentation effluent with intermittent feeding and effluent removal. *International Journal of Hydrogen Energy*, *35*(13), 6674–6680.

Pandey, A., Sinha, P., & Pandey, A. (2021). Hydrogen production by sequential dark and photofermentation using wet biomass hydrolysate of Spirulina platensis: rResponse surface methodological approach. *International Journal of Hydrogen Energy*, *46*(10), 7137–7146.

Pant, D., Arslan, D., Van Bogaert, G., Gallego, Y. A., De Wever, H., Diels, L., & Vanbroekhoven, K. (2013). Integrated conversion of food waste diluted with sewage into volatile fatty acids through fermentation and electricity through a fuel cell. *Environmental Technology*, 34(13–14), 1935–1945.

Poladyan, A., Trchounian, K., Vassilian, A., & Trchounian, A. (2018). Hydrogen production by Escherichia coli using brewery waste: optimal pretreatment of waste and role of different hydrogenases. *Renewable Energy*, *115*, 931–936.

Raghulchandrana, R., Tamizhini, A., Snehya, A. V., & Pandey, R. P. (2020). Pretreatment studies of biohydrogen production from agro-industrial waste.

Rai, P. K., & Singh, S. P. (2016). Integrated dark-and photo-fermentation: Recent advances and provisions for improvement. *International Jjournal of Hydrogen Eenergy*, *41*(44), 19957–19971.

Rai, P. K., Singh, S. P., & Asthana, R. K. (2012). Biohydrogen production from cheese whey wastewater in a two-step anaerobic process. *Applied Biochemistry and Biotechnology*, *167*, 1540–1549.

Rai, P. K., Singh, S. P., Asthana, R. K., & Singh, S. (2014). Biohydrogen production from sugarcane bagasse by integrating dark-and photo-fermentation. *Bioresource Technology*, *152*, 140–146.

Reddy, K., Nasr, M., Kumari, S., Kumar, S., Gupta, S. K., Enitan, A. M., & Bux, F. (2017). Biohydrogen production from sugarcane bagasse hydrolysate: effects of pH, S/X, Fe^{2+}, and magnetite nanoparticles. *Environmental Science and Pollution Research*, *24*, 8790–8804.

Ren, N., Guo, W., Liu, B., Cao, G., & Ding, J. (2011). Biological hydrogen production by dark fermentation: challenges and prospects towards scaled-up production. *Current Opinion in Biotechnology*, *22*(3), 365–370.

Rezania, S., Oryani, B., Cho, J., Talaiekhozani, A., Sabbagh, F., Hashemi, B., …... & Mohammadi, A. A. (2020). Different pretreatment technologies of lignocellulosic biomass for bioethanol production: An overview. *Energy*, *199*, 117457.

Rodionova, M. V., Bozieva, A. M., Zharmukhamedov, S. K., Leong, Y. K., Lan, J. C. W., Veziroglu, A., …... & Allakhverdiev, S. I. (2022). A comprehensive review on lignocellulosic biomass biorefinery for sustainable biofuel production. *International Journal of Hydrogen Energy*, *47*(3), 1481–1498.

Rosales-Calderon, O., & Arantes, V. (2019). A review on commercial-scale high-value products that can be produced alongside cellulosic ethanol. *Biotechnology for Biofuels*, *12*(1), 240.

Sağır, E., & Hallenbeck, P. C. (2019). Photofermentative hydrogen production. In Biohydrogen (pp. 141–157). Elsevier.

Sagnak, R., & Kargi, F. (2011). Photo-fermentative hydrogen gas production from dark fermentation effluent of acid hydrolyzed wheat starch with periodic feeding. *International Journal of Hydrogen Energy*, *36*(7), 4348–4353.

Sarangi, P. K., & Nanda, S. (2020). Biohydrogen production through dark fermentation. *Chemical Engineering & Technology*, *43*(4), 601–612.

Sekoai, P. T., Yoro, K. O., Bodunrin, M. O., Ayeni, A. O., & Daramola, M. O. (2018). Integrated system approach to dark fermentative biohydrogen production for enhanced yield, energy efficiency and substrate recovery. *Reviews in Environmental Science and Bio/Technology*, *17*, 501–529.

Shanmugam, S., Hari, A., Pandey, A., Mathimani, T., Felix, L., & Pugazhendhi, A. (2020). Comprehensive review on the application of inorganic and organic nanoparticles for enhancing biohydrogen production. *Fuel*, *270*, 117453.

Sikiru, S., Oladosu, T. L., Amosa, T. I., Olutoki, J. O., Ansari, M. N. M., Abioye, K. J., …... & Soleimani, H. (2024). Hydrogen-powered horizons: Transformative technologies in clean energy generation, distribution, and storage for sustainable innovation. *International Journal of Hydrogen Energy*, *56*, 1152–1182.

Soares, J. F., Confortin, T. C., Todero, I., Mayer, F. D., & Mazutti, M. A. (2020). Dark fermentative biohydrogen production from lignocellulosic biomass: technological challenges and future prospects. *Renewable and Sustainable Energy Reviews*, *117*, 109484.

Soto, L. R., Byrne, E., van Niel, E. W., Sayed, M., Villanueva, C. C., & Hatti-Kaul, R. (2019). Hydrogen and polyhydroxybutyrate production from wheat straw hydrolysate using *Caldicellulosiruptor* species and *Ralstonia eutropha* in a coupled process. *Bioresource Technology*, *272*, 259–266.

Srivastava, N., Srivastava, M., Kushwaha, D., Gupta, V. K., Manikanta, A., Ramteke, P. W., & Mishra, P. K. (2017). Efficient dark fermentative hydrogen production from enzyme hydrolyzed rice straw by *Clostridium pasteurianum* (MTCC116). *Bioresource Technology*, *238*, 552–558.

Srivastava, N., Srivastava, M., Mishra, P. K., Kausar, M. A., Saeed, M., Gupta, V. K., …... & Ramteke, P. W. (2020). Advances in nanomaterials induced biohydrogen production using waste biomass. *Bioresource Technology*, *307*, 123094.

Srivastava, R. K., Shetti, N. P., Reddy, K. R., Kwon, E. E., Nadagouda, M. N., & Aminabhavi, T. M. (2021). Biomass utilization and production of biofuels from carbon neutral materials. *Environmental Pollution*, *276*, 116731.

Sudiyani, Y., Dahnum, D., Burhani, D., & Putri, A. M. H. (2019). Evaluation and comparison between simultaneous saccharification and fermentation and separated hydrolysis and fermentation process. In Second and Third Generation of Feedstocks (pp. 273–290). Elsevier.

Uyar, B., Kars, G., Eroğlu, İ., Yücel, M., & Gündüz, U. (2012). Hydrogen production via photofermentation. In State of the Art and Progress in Production of Biohydrogen (pp. 54–77)State of the art and progress in production of biohydrogen. Bentham Science Publishers, United Arab Emirates, 54–77.

Vatsala, T. M., Raj, S. M., & Manimaran, A. (2008). A pilot-scale study of biohydrogen production from distillery effluent using defined bacterial co-culture. *International Journal of Hydrogen Eenergy*, *33*(20), 5404–5415.

Wang, J., & Yin, Y. (2019). Progress in microbiology for fermentative hydrogen production from organic wastes. *Critical Reviews in Environmental Science and Technology*, *49*(10), 825–865.

Wei, H., Junhong, T., & Yongfeng, L. (2016). Utilization of food waste for fermentative hydrogen production. *Physical Sciences Reviews*, *1*(10), 20160050.

Yang, H., Guo, L., & Liu, F. (2010). Enhanced bio-hydrogen production from corncob by a two-step process: dark-and photo-fermentation. *Bioresource Technology*, *101*(6), 2049–2052.

Yokoi, H., Mori, S., Hirose, J., Hayashi, S., & Takasaki, Y. (1998). H_2 production from starch by a mixed culture of Clostridium butyricum and Rhodobacter sp. M [h] 19. *Biotechnology Letters*, *20*, 895–899.

Yokoi, H., Saitsu, A., Uchida, H., Hirose, J. U. N., Hayashi, S., & Takasaki, Y. (2001). Microbial hydrogen production from sweet potato starch residue. *Journal of Bioscience and Bioengineering*, *91*(1), 58–63.

Zhang, Q., & Zhang, Z. (2018). Biological hydrogen production from renewable resources by photofermentation. In Advances in Bioenergy (Vol. 3, pp. 137–160). Elsevier.

Zhang, Q., Zhang, T., Zhang, Z., Zhou, X., Wang, Y., & He, C. (2017). Experiment of integrated fermentation hydrogen production by photosynthetic bacteria cooperating with Enterobacter aerogenes. *Transactions of the Chinese Society of Agricultural Engineering*, *33*(9), 243–249.

Zhang, Z., Zhang, H., Li, Y., Lu, C., Zhu, S., He, C., ... & Zhang, Q. (2020). Investigation of the interaction between lighting and mixing applied during the photo-fermentation biohydrogen production process from agricultural waste. *Bioresource Technology*, *312*, 123570.

Zong, W., Yu, R., Zhang, P., Fan, M., & Zhou, Z. (2009). Efficient hydrogen gas production from cassava and food waste by a two-step process of dark fermentation and photo-fermentation. *Biomass and Bioenergy*, *33*(10), 1458–1463.

Microbial Biohydrogen Production Using Organic Wastes

Current Advances and Future Opportunities

10

Prasann Kumar

10.1 INTRODUCTION

Biohydrogen has marked itself as a clean and sustainable energy option, as it offers vast potential to supply the world's energy demands without compromising the environmental landscape. Burning hydrogen to produce energy is highly efficient and releases only water as a by-product, so hydrogen is often regarded as a suitable replacement for fossil fuels (Musharavati et al., 2024; Salazar-Batres & Moreno-Andrade, 2022; Salem et al., 2018). Biohydrogen is unique among hydrogen production methodologies in that it is an environmentally friendly process that harnesses biological resources to shoot organic substrates to hydrogen. Biohydrogen outperforms other hydrogen production technologies, such as steam methane reforming and electrolysis, in many aspects and outperforms them. Biohydrogen technology uses renewable resources such as organic waste as supplements, which is one crucial feature. As a result, biohydrogen thus accomplishes a dual goal: it generates energy and transforms waste into highly advantageous materials. As nations and industries worldwide move closer to significant commitments to reduce greenhouse gas (GHG) emissions and achieve Paris Accord objectives, biohydrogen production technology is crucial in enabling trading conditions to become more sustainable (Jiménez-Ocampo et al., 2021; Saravanan et al., 2021; Weide et al., 2019).

This makes it an important area of research; using organic wastes in biohydrogen production only increases its importance. Organic waste, including agricultural residues, food waste, municipal solid waste, and industrial by-products, poses a significant environmental problem due to the high volume and improper disposal methods. When not properly managed, millions of tons of organic

DOI: 10.1201/9781003568612-10

waste generated yearly through food and animal manure contributes to soil degradation, methane emissions, and water pollution. Utilising these wastes to produce biohydrogen mitigates their environmental ramifications and taps into their hidden energy potential. Each of these substrates can be considered a viable candidate for microbial fermentation processes, and examples such as wheat straw, rice husk, and sugarcane bagasse are all rich in carbohydrates. Likewise, food and kitchen waste, usually thrown in landfills, can be effectively transformed into hydrogen via dark fermentation (Ghimire et al., 2018; Pomdaeng et al., 2024; Zhazhkov et al., 2024). In recent years, biohydrogen derived from organic waste have emerged as promising solutions for sustainable energy generation derived from diverse organic waste materials. Moreover, the distributed characteristics of organic waste availability are suited to decentralised hydrogen production facilities, minimising transportation expenses and promoting local energy independence. That, in turn, lays the foundational context for the rest of this book because biohydrogen's role in a rapidly changing energy landscape is what this section is all about. Then, the chapter addresses the need to use organic wastes, considering their availability and environmental impacts, and as a renewable substrate for hydrogen generation. The review highlights applicable waste types for biohydrogen production and discusses the potential for waste-to-energy systems and their usefulness for waste valorisation. Microorganisms are the driving force of this process, with different species of bacteria and algae utilising metabolic pathways, including dark fermentation, photofermentation and biophotolysis, to generate hydrogen. Extracting insights about the diversity and versatility of the biological systems making up these communities can be achieved by analysing which type of microorganisms are represented by the datasets and their metabolic roles. The focus is on engineering microbial strains and consortia to improve hydrogen yield and substrate utilisation efficiency. One of the most studied processes, dark fermentation refers to microbes breaking down organic material anaerobically, where they can then produce hydrogen and organic acids and alcohols (Aboudi et al., 2021; Budhraja, 2024; Pasupuleti et al., 2014). This pathway is especially appropriate for lignocellulosic biomass and food waste. Conversely, photofermentation utilises photosynthetic bacteria to produce hydrogen with light energy and has complementary benefits when combined with dark fermentation. Potential hybrid approaches integrating multiple pathways are also discussed to optimise substrate conversion and hydrogen yield. The chapter, therefore, discussed substrate optimisation, bioreactor design innovations, and pretreatment techniques that enhance the bioavailability of complex organic materials. It also reviews the role of genetic engineering in producing high-performance microbial strains, highlighting achievements that can help overcome existing shortcomings in biohydrogen production systems. These studies will come in case studies and examples of recent research related to these strategies (Anand et al., 2024; Chandra et al., 2024; Roslan et al., 2024).

While acknowledging the merits of biohydrogen production, this chapter does not focus on challenges, including low yield, process inefficiency, and scalability. Although these challenges may seem daunting, they also present valuable innovation and interdisciplinary collaboration opportunities. By highlighting the barriers, the chapter seeks to offer a fair perspective, encouraging more research and policy development to close the gaps that still exist.

The chapter ends with a discussion of the future of biohydrogen production from organic wastes. We highlight emerging trends, including nanotechnology advancements, next-generation bioreactor platforms, and the coupling of biohydrogen to sustainable energy ecosystems. It also covers the role that policy frameworks and economic incentives can play in scaling up biohydrogen production, highlighting the importance of collaborative efforts between governments, industries, and researchers.

This chapter provides a comprehensive outline of biohydrogen production through organic wastes, covering its scientific, technological, and environmental aspects. The connection between biohydrogen's clean energy potential and the sustainable handling of organic wastes showcases the transformative nature of this technology towards a sustainable circular energy future (Table 10.1 and Figure 10.1).

TABLE 10.1 Biohydrogen production from organic wastes

MICROORGANISM TYPE	*MICROBIAL SPECIES/ STRAINS*	*SUBSTRATES USED*	*HYDROGEN PRODUCTION PATHWAY*	*KEY ENZYMES INVOLVED*	*PRODUCTION YIELD*	*PROCESS ENHANCEMENTS*	*CHALLENGES*	*FUTURE PROSPECTS*
Fermentative microbes	***Clostridium butyricum***, ***Clostridium acetobutylicum***, ***Enterobacter aerogenes***	Agricultural residues (e.g., corn stover, wheat straw), food waste (e.g., glucose, sucrose), industrial by-products	Dark fermentation	Hydrogenase (NiFe hydrogenase, FeFe hydrogenase), pyruvate formate lyase	1–2 moles H_2 per mole of glucose	Genetic engineering of strains, metabolic pathway optimisation, co-culture systems	Low yield, slow hydrogen production rate, accumulation of by-products (e.g., organic acids)	Improvement in yield through engineered strains, enhanced fermentation efficiency, integration with other pathways
Photosynthetic microbes	***Rhodobacter sphaeroides***, ***Chlamydomonas reinhardtii***, ***Anabaena cylindrica***	Organic acids, wastewater, agricultural residues	Photofermentation	Hydrogenase, photosystem II, ferredoxin	1–3 moles H_2 per mole of substrate	Light intensity optimisation, metabolic engineering of photosynthetic pathways, use of immobilised cells	Dependence on light conditions, low efficiency under natural light, oxygen inhibition	Enhancement in light utilisation, photosystem optimisation, use of photobioreactors for improved hydrogen production
Syntrophic microbes	***Methanobacterium formicicum***, ***Syntrophomonas wolfei***, ***Desulfovibrio vulgaris***	Organic wastes (e.g., food waste, brewery waste, cellulose)	Syntrophic fermentation coupled with hydrogen production	Hydrogenase, formate dehydrogenase, acetate kinase	0.5–1 mole H_2 per mole of substrate	Development of consortia systems, co-cultivation with fermentative or methanogenic microbes	Complex metabolic interactions, slow rate of hydrogen production, high dependency on partner microorganisms	Use of engineered consortia, optimisation of syntrophic interactions, enhancement of electron transfer mechanisms
Archaeal microbes	***Methanococcus maripaludis***, ***Pyrococcus furiosus***, ***Thermococcus litoralis***	Lignocellulosic biomass, waste gases (e.g., CO, CO_2)	Thermophilic fermentation, hydrogen production via CO_2 reduction	[NiFe] hydrogenase, formate dehydrogenase, CO dehydrogenase	2–4 moles H_2 per mole of substrate	Genetic modification of hydrogenases, co-cultures for enhanced synergy, use of thermophilic reactors	High-temperature sensitivity, the requirement for high-temperature reactors, substrate specificity	Thermophilic bioreactor optimisation, development of heat-resistant strains, expansion to industrial waste streams

(Continued)

TABLE 10.1 (*Continued*) Biohydrogen production from organic wastes

MICROORGANISM TYPE	*MICROBIAL SPECIES/ STRAINS*	*SUBSTRATES USED*	*HYDROGEN PRODUCTION PATHWAY*	*KEY ENZYMES INVOLVED*	*PRODUCTION YIELD*	*PROCESS ENHANCEMENTS*	*CHALLENGES*	*FUTURE PROSPECTS*
Bacteria for biophotolysis	***Synechocystis*** sp., ***Anabaena*** PCC 7120, ***Chlamydomonas reinhardtii***	Water (H_2O), organic waste containing nitrogen and phosphorus	Biophotolysis (light-driven water splitting)	Photosystem II, hydrogenase, photoreceptor proteins	0.5–2 moles H_2 per mole of water	Genetic engineering to enhance photosynthetic hydrogenase activity, optimisation of light harvesting systems	Low efficiency in natural conditions, oxygen inhibition, limited natural sunlight availability	Integration into photobioreactors, engineering for enhanced photolytic activity, scaling up systems for commercial applications
Algae for hydrogen production	***Chlamydomonas reinhardtii***, ***Scenedesmus obliquus***, green algae	Organic waste, water, CO_2	Photoautotrophic production and biophotolysis	Hydrogenase, pyruvate dehydrogenase, ferredoxin	1–4 moles H_2 per mole of substrate	Genetic engineering, optimised light and nutrient conditions, closed-loop systems	Oxygen inhibition, growth dependence on light and CO_2, low productivity under controlled conditions	Algae strain improvement, use of dual-purpose algae strains for biofuels and hydrogen, optimisation of culture systems
Hybrid pathways	***Rhodobacter sphaeroides*** (combined with ***Clostridium acetobutylicum***), ***Chlamydomonas reinhardtii*** with syntrophic bacteria	Wastewater, organic acids, glucose, food waste	Hybrid (dark fermentation + photofermentation + biophotolysis)	Hydrogenase, pyruvate dehydrogenase, nitrogenase	Up to 5 moles H_2 per mole of substrate	Co-culturing of different microorganisms, hybrid system design, synergy of multiple pathways	System complexity, high operational costs, difficulty in maintaining optimal conditions for multiple pathways	Integration of multi-stage bioreactors, use of engineered hybrid systems, scalability improvements

Source: Aboudi et al. (2021), Alavi-Borazjani et al. (2021), Kovalev et al. (2020), Pomdaeng et al. (2024), Roslan et al. (2024), Zhazhkov et al. (2024)

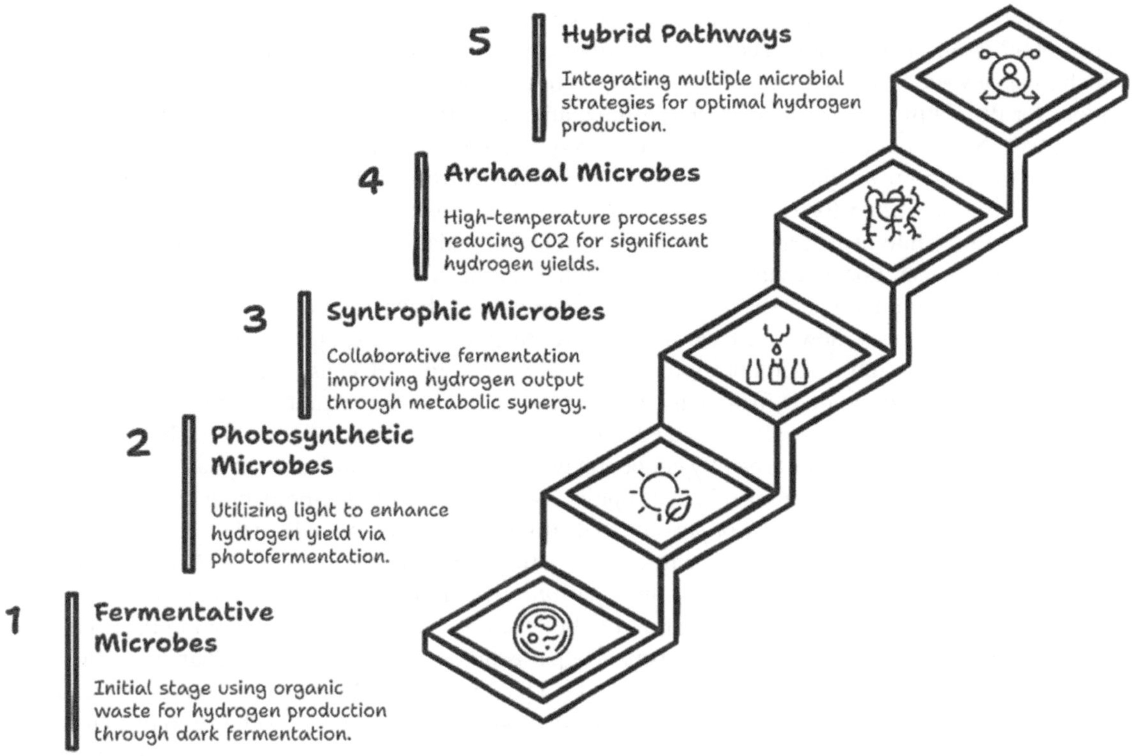

FIGURE 10.1 Advancing biohydrogen production.

10.2 TYPES OF MICROORGANISMS INVOLVED IN BIOHYDROGEN PRODUCTION

Biohydrogen production using microbes is an emerging scientific practice that plays a key role in renewable energy generation. This process utilises organic waste materials to produce hydrogen gas, offering a clean and sustainable alternative fuel source. The following vignette explores its significance and potential. Different fermentative, photosynthetic, and syntrophic microorganisms mediate biohydrogen production (Mogili et al., 2022; Plakar et al., 2025; Reyna-Gómez et al., 2021). Depending on the organic substrates used, their physiological properties, and environmental circumstances, microorganisms involved in distinct metabolic pathways can provide heterogeneous benefits and disadvantages regarding hydrogen production efficiencies. In anaerobic fermentation, the hydrogen is produced by *Clostridium* species and various other fermentative microorganisms (e.g., *Escherichia coli*) by degradation of organic substrates (e.g., glucose, fatty acids, and different types of sugar). Hydrogen production in anaerobic environments relies on this group of bacteria to use hydrogenases to mediate the H_2 production reaction. However, during fermentation, it is also possible to produce other by-products, such as acids or coproduced alcohol and gaseous products, in addition to hydrogen, thus reducing net hydrogen production. There are such challenges; however, fermentative microorganisms remain the foci of biohydrogen production as they can utilise various organic waste materials, enabling biohydrogen production from food, agricultural, and industrial effluents (El Bari et al., 2022; Shomope et al., 2025; Srivastava et al., 2024).

Microbial hydrogen production is performed in the photosynthesising bacteria, including cyanobacteria and purple non-sulphur bacteria, which induce hydrogen production using light. Two classes of photosynthetic organisms can carry out this process: oxygenic and anoxygenic photosynthetic bacteria. When photosynthesising, cyanobacteria, for example, harness light energy to apply to water molecules, which frees oxygen. Still, the by-product oxygen can inhibit hydrogen production, so light energy has limitations for those kinds of organisms (Alavi-Borazjani et al., 2021; Kovalev et al., 2020; Martínez-Fraile et al., 2024; Nabgan et al., 2021). In contrast, anoxygenic photosynthetic bacteria (e.g., members of the genus *Rhodobacter*) use light energy to reduce protons to hydrogen gas without O_2 generation, making them an intriguing group for biohydrogen production. Hydrogenase-contributing bacteria are required to convert protons into hydrogen gas. However, hurdles still exist in improving conditions for photosynthetic hydrogen generation, such as increasing light harvesting efficiency, activating hydrogen production through the upstream hydrogen gas branches, and overcoming downstream oxygen inhibition problems. Photosynthetic microbes, however, have their environmental demands, such as sufficient sunlight and water, which are the aim of outdoor systems or photobioreactors. Because these microorganisms can harness solar energy to produce hydrogen, they present a unique opportunity for sustainable hydrogen production. However, further studies are needed to improve their effectiveness and scalability.

These syntrophic microbes typically cooperate with other microorganisms and play a significant role in hydrogen production. For example, in syntrophic associations, hydrogen-producing bacteria cooperate with methanogenic archaea or other organisms to degrade organic compounds such as fatty acids, alcohols, and volatile acids. Hydrogen scavengers co-existing with methanogens (e.g., *Methanococcus* and *Methanosarcina*) include *Syntrophomonas* and *Syntrophus* sp. In this way, the methanogens eat away the hydrogen by-product of the bacteria, avoiding the build-up of hydrogen gas that would otherwise make the activity difficult. Some of these organisms are parasitic on other organisms (Chandra et al., 2024; Jiménez-Ocampo et al., 2021; Pasupuleti et al., 2014; Zhazhkov et al., 2024). A metabolic partnership of this sort is known as syntrophy, where a new organism is added to a metabolic pathway so that an organic compound (theoretically, an inorganic one could be digested, too), which could not be oxidised previously under anaerobic conditions, can be successfully digested; syntrophy, in this case, is going on in anaerobic systems (e.g., landfills, anaerobic digesters, wastewater treatment systems) in which an extremely high load of organic waste can be metabolically transformed to biohydrogen. So, syntrophic hydrogen production can benefit large-scale waste-to-energy technologies because it can mediate the degradation of complex organic molecules while producing hydrogen. However, obstacles that include establishing stable microbial communities, optimising hydrogen transport to microorganisms, and regulating hydrogen production efficiency must be overcome to scale up the syntrophic biohydrogen system (Budhraja, 2024; Roslan et al., 2024; Weide et al., 2019).

Hydrogen production by fermentative bacteria and archaea microbial hydrogen production in the biological production of hydrogen is critical because both groups of hydrogen producers exploit unusual biochemical pathways to ferment various organic substrates, thus producing hydrogen. Some bacteria, such as *Clostridium acetobutylicum* and *Enterobacter cloacae*, are classified as fermentative hydrogen producers. Archaea such as *Methanococcus* and *Thermococcus* are also essential for hydrogen consumption and generation in extreme environments. Hydrogen tends for methanogenic archaea to consume and generate methane, and certain types of archaea, called thermophilic archaea, can produce hydrogen from protons (Budhraja, 2024; Roslan et al., 2024; Salem et al., 2018; Weide et al., 2019). Hydrogenases were significant hydrogen-generating enzymes from the kingdom of bacteria and archaea. Hydrogenases belong to two general classes: [FeFe]-hydrogenases and [NiFe]-hydrogenases, which catalyse the reduction of protons to produce molecular hydrogen. According to the type of organic substrates at their disposal and the described parameters, these microbial processes exhibit different efficiencies, pH ranges, temperature, and co-factor requirements for their hydrogenase-mediated activity. Bacteria and archaea have a distinction in their metabolic pathways and act as excellent pathways for organic waste to biohydrogen production, thus making them a good option for energy production and industrial ecology (Budhraja, 2024; Chandra et al., 2024; Srivastava et al., 2024; Zhazhkov et al., 2024).

Significant progress has been made in developing microbial strains for hydrogen production in recent years. This will involve methods of genetic engineering, adaptive evolution, and metabolic engineering. This can involve genetic modifications such as introducing hydrogenase-producing or increased favourable fermentation pathway-upregulating genes to enhance the microbes' hydrogen generation capabilities. For instance, genetically modified *E. coli* has been designed to produce higher hydrogen yields by overexpressing hydrogenases or repressing competing metabolic pathways. Microorganism strains adapted for hydrogen production are more likely obtained through continuous cultivations in an anaerobic environment or with a selected substratum (Alavi-Borazjani et al., 2021; El Bari et al., 2022; Kovalev et al., 2020; Reyna-Gómez et al., 2021). This has effectively developed *Clostridium* species to create more hydrogen in various ecological conditions. Further, metabolic engineering can rewire the microbial metabolic network to divert carbon and energy flow to hydrogen production. One application area of metabolic engineering is to increase the levels of biohydrogen production in microorganisms by maximising hydrogen-producing pathways while minimising competing by-product production and upregulating critical enzymes for those pathways (Table 10.2 and Figure 10.2).

Promising strategies to increase the hydrogen yields and enhance the efficiency of the biohydrogen systems have been the significant advances made in engineering microbial strains. Continuous research to fine-tune these microbial systems utilising genomics, adaptive evolution, and metabolic engineering will make biohydrogen generation a more practical and resource-efficient renewable energy technology. Knowing these metabolic routes in both groups will open new doors for hydrogen generation from organic waste materials that would assist a clean, environmentally friendly energy future. Thus, strategies targeting the generation of high-performance microbial strains will be a key topic to overcome the actual boundaries of biohydrogen production, which is expected to enable the utilisation of this renewable energy carrier for large-scale applications (Anand et al., 2024; Budhraja, 2024; Mogili et al., 2022; Shomope et al., 2025).

10.3 DIVERSE PATHWAYS FOR BIOHYDROGEN PRODUCTION

Biohydrogen can be produced via dark fermentation, photofermentation, and biophotolysis processes, using diverse metabolic pathways that convert organic matter or light energy into hydrogen gas. In particular, dark fermentation is an anaerobic pathway mediated by various microorganisms, including *Clostridium*, *Enterobacter*, and *E. coli*, that convert complex organic substrates, such as carbohydrates, fatty acids, and proteins, into simpler molecules. These microorganisms deploy enzymes, including hydrogenases and pyruvate decarboxylases, to catalyse organic material into hydrogen gas, organic acids, alcohols, and other metabolites during this process (El Bari et al., 2022; Mogili et al., 2022; Pasupuleti et al., 2014; Zhazhkov et al., 2024). Hydrogen production occurs at the hydrogenase site, where the enzymes reduce protons to molecular hydrogen. Hydrogenases have a pivotal role in dark fermentation, and two predominant types of the hydrogenase class of enzymes drive proton reduction: [FeFe] and [NiFe] hydrogenases. The activity of these enzymes can be affected by substrate concentration, pH, temperature, and co-factors like iron or nickel and thus requires fine regulation. Glycolysis, acetogenesis, and hydrogen production are integral aspects of the metabolic process of dark fermentation. Fermentative bacteria use an anaerobic respiration pathway in anaerobic conditions to produce ATP and other cellular functions, with the last step in the electron transport chain being hydrogen production. Dark fermentation occurs optimally in controlled anaerobic environments, meaning hydrogen production can be maximised by avoiding oxidative environments that inhibit hydrogenase activity. Dark fermentation, in particular, is a favourable pathway for biohydrogen generation; however, there are still hurdles to overcome regarding improving hydrogen production, reducing by-products that impact performance, and optimising metabolic pathways (Chandra et al., 2024; El Bari et al., 2022; Nabgan et al., 2021; Reyna-Gómez et al., 2021).

Unlike dark fermentation, photofermentation uses light energy to promote hydrogen production. Photosynthetic microorganisms mainly use purple non-sulphur bacteria (*Rhodobacter sphaeroides*, *Rhodopseudomonas palustris*, and *Chromatium* sp.) that use light to produce hydrogen gas from

TABLE 10.2 Different microorganisms and their types for hydrogen production

MICROORGANISM TYPE	*SPECIES/STRAINS*	*HYDROGEN PRODUCTION PATHWAY*	*SUBSTRATES USED*	*METABOLIC PROCESS*	*KEY ENZYMES INVOLVED*	*HYDROGEN YIELD*	*ADVANTAGES*	*CHALLENGES*	*FUTURE PROSPECTS*
Bacteria	***Clostridium acetobutylicum***	Dark fermentation	Organic waste (e.g., glucose, cellulose, starch)	Anaerobic fermentation	Hydrogenase, NADH dehydrogenase	Up to 2 moles of H_2 per mole of glucose	High yield, substrate versatility, rapid growth	Low yield from complex substrates, low scalability	Genetic engineering to enhance yield, co-culture systems
	Clostridium butyricum	Dark fermentation	Lignocellulosic biomass, food waste	Anaerobic fermentation	Hydrogenase, formate dehydrogenase	1.8 moles of H_2 per mole of glucose	High substrate flexibility, efficient degradation of organic wastes	Sensitive to environmental conditions, low yield from complex substrates	Optimisation of fermentation conditions, metabolic engineering
	Rhodobacter sphaeroides	Photofermentation	Organic acids, sugars	Photosynthesis under anaerobic conditions	Hydrogenase, nitrogenase	1–3 moles of H_2 per mole of acetate	Utilises light energy, environmentally sustainable	Requires light, low yield under anaerobic conditions	Improvement in light utilisation, genetic modifications for higher efficiency
	Enterobacter cloacae	Dark fermentation	Wastewater, food waste	Anaerobic fermentation	Hydrogenase, NADH dehydrogenase	1.5–2 moles of H_2 per mole of glucose	Can utilise wastewater as a substrate, high growth rate	Low yield requires optimised environmental conditions	Bioreactor optimisation, strain improvement
Archaea	***Methanobacterium formicicum***	Hydrogenotrophic methanogenesis	Formate, carbon dioxide	Hydrogenotrophic methanogenesis	Hydrogenase, methyl coenzyme M reductase	1 mole of H_2 per mole of formate	Can utilise low-energy substrates, produces methane as a co-product	Low hydrogen yield, sensitive to oxygen	Co-culture with other microorganisms to enhance hydrogen production
	Thermococcus barophilus	Hydrogen production from water	Inorganic compounds (e.g., CO_2, H_2O)	Chemosynthesis	Hydrogenase, carbon monoxide dehydrogenase	2 moles of H_2 per mole of CO_2	High-temperature tolerance, high efficiency	High energy requirements, low scalability	Genetic engineering to improve efficiency at lower temperatures

(Continued)

TABLE 10.2 (*Continued*) Different microorganisms and their types for hydrogen production

MICROORGANISM TYPE	*SPECIES/STRAINS*	*HYDROGEN PRODUCTION PATHWAY*	*SUBSTRATES USEDÇ*	*METABOLIC PROCESS*	*KEY ENZYMES INVOLVED*	*HYDROGEN YIELD*	*ADVANTAGES*	*CHALLENGES*	*FUTURE PROSPECTS*
Algae	***Chlamydomonas reinhardtii***	Biophotolysis	Water, light	Photosynthesis (biophotolysis)	Hydrogenase, ATP synthase	2 moles of H_2 per mole of water	Efficient use of sunlight, renewable and clean	Low yield requires specific environmental conditions	Genetic modification to improve hydrogenase activity, bioreactor optimisation
	Chlorella vulgaris	Biophotolysis	Water, light	Photosynthesis (biophotolysis)	Hydrogenase	Up to 1.5 moles of H_2 per mole of water	Sustainable, utilise sunlight	Limited hydrogen production under normal conditions	Genetic enhancement, bioreactor design for optimised light exposure
Fungi	***Neurospora crassa***	Dark fermentation	Agricultural residues, food waste	Anaerobic fermentation	Hydrogenase	1.2 moles of H_2 per mole of glucose	Can degrade complex organic substrates	Limited hydrogen production, slow growth rate	Genetic engineering for increased hydrogenase activity
Co-cultures	***Clostridium*** and ***Rhodobacter*** spp.	Hybrid pathway (dark fermentation + photofermentation)	Organic wastes, food waste, wastewater	Fermentation + photosynthesis	Hydrogenase, nitrogenase	3–4 moles of H_2 per mole of substrate	Synergistic effects, enhanced yield from organic waste	Complex interactions, difficult to scale	Co-culture optimisation, bioreactor design for effective synergy
Cyanobacteria	***Synechocystis*** sp. PCC 6803	Biophotolysis	Water, light	Photosynthesis	Hydrogenase	1 mole of H_2 per mole of water	It uses sunlight; there is no need for external organic substrates	Low yield, limited hydrogenase activity	Genetic engineering, bioreactor optimisation for light capture

Source: Aboudi et al. (2021), Anand et al. (2024), Chandra et al. (2024), Kovalev et al. (2020), Nabgan et al. (2021), Plakar et al. (2025), Saravanan et al. (2021)

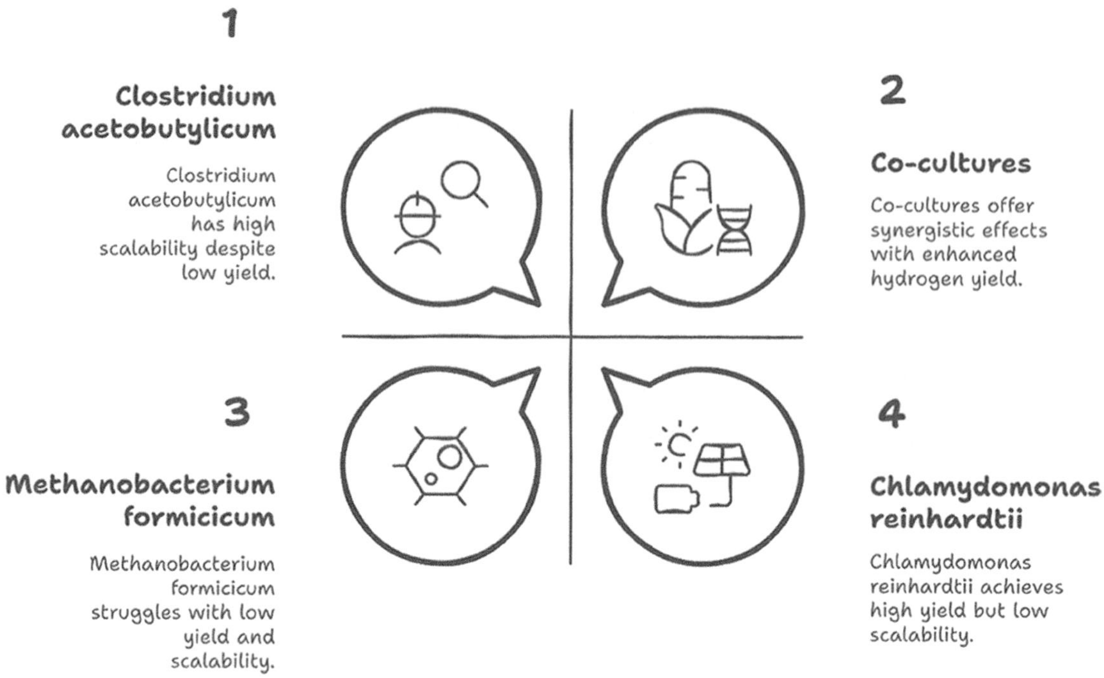

FIGURE 10.2 Microorganisms for hydrogen production.

organic compounds. In photofermentation, photosynthetic pigments like bacteriochlorophyll capture the energy from light, which is used to drive the reduction of protons to molecular hydrogen through hydrogenase enzymes. Contrasting dark fermentation, photofermentation does not depend exclusively on complex organic substrates to achieve hydrogen production; this process employs light energy as a main driving force to transform organic acids in hydrogen production, including acetate, butyrate, and lactate (Pasupuleti et al., 2014; Roslan et al., 2024; Srivastava et al., 2024; Zhazhkov et al., 2024). These microorganisms conduct anoxygenic photosynthesis, but thus, they do not form oxygen as waste material. Instead, they lower protons with energy from light, spewing hydrogen gas as a waste product of their metabolism. Photofermentation is beneficial because it can source solar energy directly, making it a more sustainability-friendly approach for large-scale hydrogen production. Nonetheless, photofermentation is limited by light penetration, the requirement for a specialised environment to prevent oxygen inhibition, and the challenge of optimising light harvesting for photosynthetic bacteria. Moreover, photofermentation processes contribute to a relatively low productive hydrogen output because the bacteria compete for light energy on many metabolic pathways. Nonetheless, photofermentation is a promising photo-biohydrogen production method, especially when using advanced systems like photobioreactors that can optimise light exposure and maximise microbial growth. Further exploration of these pathways, combined with the investigation of novel amendments to optimise photosynthetic hydrogen production and stability of photofermentative bacteria, should significantly enhance the efficiency of light energy conversion by these organisms and hence unlock more significant potential from this process as a renewable energy source (Aboudi et al., 2021; Alavi-Borazjani et al., 2021; Mogili et al., 2022; Shomope et al., 2025).

Another method to obtain hydrogen gas is called biophotolysis, where the photonic energy directly splits water molecules. It is an advanced mechanism of light energy capture and water photolysis found in photosynthetic microorganisms, e.g., cyanobacteria and algae. The cyanobacteria (e.g., *Synechocystis* sp., *Anabaena*, and *Nostoc*) and algae (e.g., *Chlamydomonas reinhardtii*) use light-driven reactions to oxidise water molecules and generate oxygen, electrons, and protons. Electrons are then passed to hydrogenase enzymes that reduce the protons to hydrogen gas. Biophotolysis is key to utilising water as an abundant and renewable feedstock to obtain a large-scale hydrogen fuel. This uranium acid-base tetraborate polymer, which could exist in anionic (uranyl dodecahydrate) and cationic forms (uranyl dimaleate), is synthesised through two significant processes: direct water dissociation (photolysis) performed by cyanobacteria and algae and, as the result of indirect water disintegration, in the reduction of protons to hydrogen gas by photosynthetic organisms that produce intermediate electron carriers (Chandra et al., 2024; Mogili et al., 2022; Shomope et al., 2025). Biophotolysis is advantageous over other methods because water (rather than organic substrates) is the only feedstock, making this an environmentally sustainable process. Biophotolysis, however, is also challenging, including low hydrogen production rates, the sensitivity of hydrogenase enzymes to oxygen, and the need to optimise the rates of light absorption and energy conversion efficiency. By altering their metabolic pathways or introducing more efficient hydrogenase enzymes, the hydrogen production capacity of cyanobacteria and algae can be up-scaled. In addition, genetic engineering tools, including clustered regularly interspaced short palindromic repeats (CRISPR)-Cas9 and synthetic biology, can potentially enhance the biophotolysis process and provide solutions to its current challenges (Aboudi et al., 2021; Kovalev et al., 2020; Plakar et al., 2025).

Combining dark fermentation, photofermentation, and biophotolysis (hybrid pathways) is an excellent approach to increasing biohydrogen production. To date, hydrogen-producing processes have been combined into hybrid systems, where the limitations of individual processes can be overcome, leading to higher overall hydrogen yields and efficiency. One such approach could be a hybrid dark fermentation-photofermentation system that uses organic waste as the substrate for dark fermentation and solar energy for photofermentation to harness the benefits of both processes at two levels. Dark fermentation produces various organic acids that can be used as substrates for photofermentative bacteria, establishing a synergistic relationship that increases hydrogen production. Likewise, combining biophotolysis with dark or photofermentation could enhance a more sustainable strategy for hydrogen production with a co-catalytic approach, whereby high hydrogen production rates from photosynthetic systems would complement the organic substrate consumption of fermentation-based systems (Table 10.3).

This combines pathways that allow the use of organic waste and solar power, resulting in a more stable and scalable biohydrogen production process. Therefore, hybrid systems proved advantageous since they can utilise the particular benefits of each process, for instance, the high hydrogen yield in dark fermentation and when light is used as a renewable energy source in photofermentation and biophotolysis. In addition, hybrid-type approaches can overcome some limitations of single pathways, such as oxygen inhibition of photofermentation or limitations by the light penetration of biophotolysis. Thus, with optimal process conditions and efficient energy and electron flow, these biohybrid systems may represent the key to realising large, sustainable biohydrogen production systems. Research on hybrid system development drives much work to the nascent stage. Nevertheless, this technology makes great strides towards making biohydrogen a commercially available clean and renewable energy source (Anand et al., 2024; Kovalev et al., 2020; Reyna-Gómez et al., 2021).

10.4 STRATEGIES FOR ENHANCED BIOHYDROGEN PRODUCTION

It is vital to optimise substrates and pretreatment technologies to maximise the efficiency of biohydrogen production processes. For H_2 production, the substrate type has a significant impact, leading to final yield and cost benefit. Agricultural residues, food waste, and industrial by-products are abundant and of low cost, which makes them more popular as feedstock. Nonetheless, those substrates containing pretreatment must be imparted, making them against those characteristics like acidity and other traits associated with their biodegradability, which would be somewhat yin and yang regarding microbial conversion. Physical, chemical, and biological methods are commonly used as a treatment to change complex organic structures like cellulose, hemicellulose, and lignin into more digestible forms. Examples of physical pretreatments

TABLE 10.3 Diverse pathways for biohydrogen production

PATHWAY TYPE	MICROORGANISMS INVOLVED	SUBSTRATES USED	METABOLIC PROCESS	KEY ENZYMES	HYDROGEN PRODUCTION YIELD	ADVANTAGES	CHALLENGES	PROCESS IMPROVEMENTS	FUTURE PROSPECTS
Dark fermentation	***Clostridium*** sp., ***Enterobacter*** sp., ***Bacteroides*** sp.	Organic wastes (e.g., glucose, starch, food waste)	Anaerobic fermentation: fermentation of sugars to produce hydrogen	Hydrogenases, pyruvate ferredoxin oxidoreductase	~1–2 moles of H_2 per mole of glucose	It utilises organic wastes; no light is required, and it is fast	Low yield, high substrate specificity, limited scalability	Metabolic engineering to improve hydrogenase activity and fermentation efficiency	Potential for scaling with advancements in metabolic engineering and bioreactors
Photofermentation	***Rhodobacter*** sp., ***Rhodopseudomonas*** sp., ***Bradyrhizobium*** sp.	Organic acids (e.g., butyric acid, acetic acid)	Light-driven hydrogen production through fermentation	Hydrogenases, nitrogenase	~2–3 moles of H_2 per mole of organic acid	The light-driven process utilises organic acids with a higher yield than dark fermentation	It requires light, has low yields, and costs maintaining photobioreactors	Genetic optimisation of hydrogenase enzymes improved light utilisation	Promising with optimised light harvesting systems and photobioreactor designs
Biophotolysis	***Chlamydomonas reinhardtii***, ***Anabaena*** sp., ***Synechocystis*** sp.	Water	Photolysis of water molecules into hydrogen and oxygen using light	Hydrogenase, photosystem II	~1–3 moles of H_2 per mole of water	Direct hydrogen production from water means there is no need for organic substrates	Low hydrogen production efficiency and sensitivity to environmental conditions	Enhancements in algal strain development and genetic engineering of photosystems	Advances in synthetic biology and optimised algal strains for better efficiency
Hybrid pathways	***Clostridium acetobutylicum***, ***Rhodobacter*** sp., ***Synechocystis*** sp.	Organic waste, water	Combination of dark fermentation and photofermentation	Hydrogenases, nitrogenases, hydrogenase-like enzymes	Varies depending on the combination used	It combines the strengths of both dark fermentation and photofermentation	Complexity in system integration and optimisation of microbial interactions	Integrated bioreactor systems and improved microbial consortia for better synergy	High potential with technological integration for optimised performance
Electrohydrogenesis	***Geobacter*** sp., ***Shewanella*** sp., ***Desulfovibrio*** sp.	Organic waste, CO_2, wastewater	Electrochemical conversion of organic waste or CO_2 into hydrogen	Hydrogenases, formate dehydrogenase	Varies, up to 2.5 mol H_2 per mol of organic substrate	It can use waste or CO_2 and combines electrochemical methods with microbial systems	High energy demand, expensive electrochemical setup, scalability issues	Development of cost-effective electrochemical cells and integration with renewable energy	Emergence of microbial-electrochemical systems to improve cost and efficiency
Nitrogen fixation	***Rhizobium*** sp., ***Azotobacter*** sp., ***Clostridium*** sp.	Nitrogenous compounds (e.g., ammonia)	Hydrogen production via nitrogenase enzymes in nitrogen-fixing microorganisms	Nitrogenase, hydrogenase	~0.3–0.5 moles of H_2 per mole of nitrogen	It can be integrated with nitrogen-fixing crops, a low-cost process	A low yield of hydrogen requires nitrogen fixation	Engineering of nitrogenase and hydrogenase activities, co-cultivation strategies	Future integration with agricultural practices for sustainable hydrogen production

(Continued)

TABLE 10.3 (*Continued*) Diverse pathways for biohydrogen production

PATHWAY TYPE	*MICROORGANISMS INVOLVED*	*SUBSTRATES USED*	*METABOLIC PROCESS*	*KEY ENZYMES*	*HYDROGEN PRODUCTION YIELD*	*ADVANTAGES*	*CHALLENGES*	*PROCESS IMPROVEMENTS*	*FUTURE PROSPECTS*
Dark and photofermentation hybrid	***Clostridium*** sp., ***Rhodobacter*** sp., ***Synechocystis*** sp.	Glucose, organic acids, wastewater	A hybrid system combining dark fermentation for substrate breakdown and photofermentation for hydrogen production	Hydrogenases, pyruvate ferredoxin oxidoreductase, nitrogenases	Varies, enhanced yields compared to individual pathways	The potential for higher yields uses multiple substrates	In high complexity, control of microbial systems is challenging	Optimisation of co-cultures, enhanced nutrient and light utilisation	Development of more robust hybrid bioreactors for improved scalability
Biological water-splitting	***Chlamydomonas reinhardtii***, ***Synechocystis*** sp.	Water, CO_2	Splitting of water molecules via photosynthetic pathways to produce hydrogen	Hydrogenase, photosystem II	~1.5–2 moles of H_2 per mole of water	Direct conversion of water to hydrogen is environmentally friendly	Low hydrogen yield, unstable systems	Genetic modification of cyanobacteria, optimised light conditions	Promising advances in synthetic biology and optimisation of photosystem genes.

Source: Budhraja (2024), Chandra et al. (2024), Kovalev et al. (2020), Mogili et al. (2022), Pasupuleti et al. (2014), Salem et al. (2018), Saravanan et al. (2021), Srivastava et al. (2024), Weide et al. (2019)

are mechanical grinding, milling, and heating, which make more surface area of the substrate available to microbial enzymes. Pretreatment (chemical), for instance, acid and alkali treatment, decomposes the lignocellulosic matrix, releasing fermentable sugars, thus allowing them to be accessible to microbial degradation (El Bari et al., 2022; Martínez-Fraile et al., 2024; Plakar et al., 2025; Srivastava et al., 2024). Alternatively, enzymatic pretreatment deploys specific enzymes to hydrolyse cellulose and hemicellulose to simple sugars, offering a more directed pathway in decreasing substrate availability. Pretreatment methods are broad and consist of mechanical/biological/thermal/chemical processes. The type of process that is applied relates to both substrate type and the necessary hydrogen yield and economic feasibility. Appropriate pretreatment will guarantee optimum hydrogen yield, as during fermentation, the feedstock must be hydrolysed efficiently into fermentable sugars digestible by microorganisms. Thus, substrate optimisation considers the organic waste matter most pertinent to their composition, fermentable sugars, and tolerance to inhibitory products. When used with appropriate substrate selection, these pretreatment approaches can be highly effective in increasing the biohydrogen levels and providing cheaper processes.

Recently, promising strategies for enhancing biohydrogen production emerged from genetic engineering and synthetic biology approaches. Ultimately, these processes promote the development of strains with enhanced hydrogen production rates, enabling systems to select for greater system efficiency and stability via microbial genome duplication. Application of genetic engineering approaches (gene knock-outs, over-expression, alterations of metabolic pathways) enables the design of microbial strains with heightened hydrogenase activity, broadened substrate utilisation potential, and resistance to various stressors. Hydrogenase enzymes (which catalyse hydrogen production) are key to biohydrogen production, and genetic modifications that yield higher levels of hydrogenase or increased stability, to name a few changes, can vastly increase hydrogen yields (Anand et al., 2024; Budhraja, 2024; Plakar et al., 2025). The metabolic pathways associated with biohydrogen production, glycolysis and the fermentation pathways, can similarly be entirely engineered to reduce by-product generation and improve hydrogen production yield. Molecular biological approaches including designing and constructing novel biological parts, devices, and systems can yield novel, biohydrogen-optimised microbial strains. To enable the design of microorganisms with novel functionalities, such as increased hydrogen production rates, enhanced substrate degradation, and operation across broader environmental conditions, advanced synthetic biology techniques (including CRISPR-Cas9) are used for genetic engineering applications. To be pretentious, it allows one to design microbial consortia with complementary metabolic functions, paving the way for biohydrogen production. These engineered microbial communities improve substrate utilisation and hydrogen production by coupling metabolic pathways. Genetically engineered and synthetic biology microorganisms have the potential to probe the potential of biohydrogen production because these engineered microorganisms can perform superiorly and develop scalable production systems compared to natural organisms (El Bari et al., 2022; Pasupuleti et al., 2014; Plakar et al., 2025).

Innovations in bioreactor designs and operational parameters are crucial for optimising biohydrogen production. The bioreactor that provides the best conditions for microbial growth and hydrogen production is one of the most critical components in large-scale biohydrogen production systems. Different bioreactors could be employed in the biohydrogen generation according to biohydrogen production requirements (e.g., batch, continuous, and fed-batch reactors). Furthermore, the parameters of the bioreactor, including scale-up, mixing, temperature, and pH, can directly affect the hydrogen production efficiency. For example, continuous bioreactors have been proposed for large-scale biohydrogen production (e.g., continuous addition of substrate and withdrawal of by-products for constant hydrogen production) such that operational parameters, such as temperature, pH, and nutrient concentration, should be further optimised to maintain optimal conditions for microbial proliferation and hydrogenase activity. Temperature is a critical factor since extreme temperatures have been reported to inhibit microbial activity, while lower temperatures may also restrict hydrogen production due to inhibited metabolic activity (Anand et al., 2024; Ghimire et al., 2018; Plakar et al., 2025). Providing the optimal pH for the various enzymes, as low and high pH states, may reduce enzyme activity or microbial performance. In addition to optimisation of these traditional parameters, novel improvements in bioreactor construction have also occurred recently: for example, using more advanced techniques such as membrane bioreactors to recover hydrogen from the fermentation broth or gas-tight bioreactors that minimise hydrogen losses and enhance yield overall. In addition, the latest bioreactors feature sophisticated sensors and automatic control systems that facilitate monitoring and optimising operational conditions in real time, leading to increased efficiency and stability during hydrogen generation. The future success of biohydrogen production systems will rely not only on the engineering of the metabolic pathways that generate hydrogen but also on the development of innovative bioreactor designs and process innovations that optimise the economics and scalability of the process (Aboudi et al., 2021; Chandra et al., 2024; El Bari et al., 2022; Reyna-Gómez et al., 2021).

A circular bioeconomy approach for biohydrogen production, utilising co-products, appears promising for increasing sustainability and economic viability. Such a circular bioeconomy allows waste products from biohydrogen production (organic acids, biomass, and residual fermentation liquids) to be employed as valuable co-products, reducing waste and maximising the sustainable impact of the biohydrogen production process. The produced organic acids can be further utilised as feedstock for subsequent processes, such as for the generation of biopolymers or as precursors for the biosynthesis of biofuels. Similarly, the residual biomass left after hydrogen generation can be processed into other value-added products such as biochar, animal feed, or fertilisers, which further mitigate the ecological impact of the process while contributing to the circular economy (Aboudi et al., 2021; Roslan et al., 2024; Srivastava et al., 2024; Zhazhkov et al., 2024). Co-product enrichment integration is driven by the desire to enrich biohydrogen-producing microorganisms with nutrients such as nitrogen or phosphorus through the utilisation of nutrient-rich wastes as feedstock, which in turn promotes biohydrogen production while minimising the acquisition of nutrient supplements from distant supply chains, thus increasing the overall economic feasibility of the process. On the other hand, through co-product integration, we mean assessing potential alternative and innovative purposes for co-products formed in the biohydrogen production process. This can help develop further income streams from biohydrogen production, closing the resource utilisation loop and making the process sustainable. Additionally, using co-products helps improve the efficiency of the entire biohydrogen production process, as these by-products can be utilised to lower hydrogen production costs. These examples show good potential for process integration and new designs of biohybrids using H_2 production as either a bioprocess integrated with another one or as a hybrid bioprocess itself through the evolution of

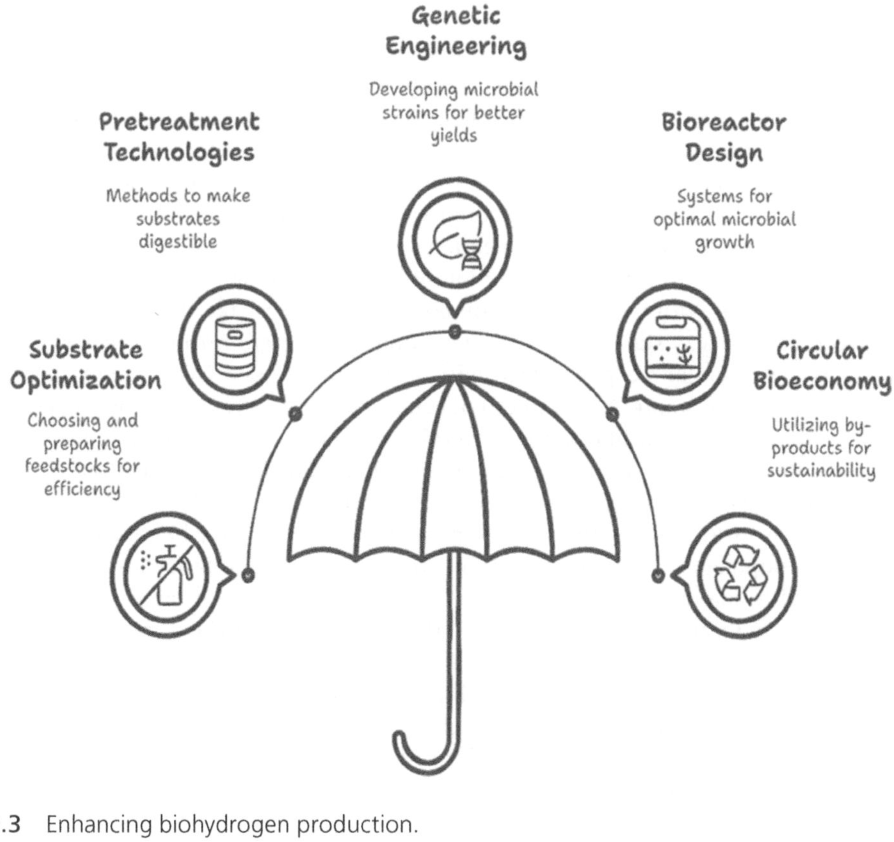

FIGURE 10.3 Enhancing biohydrogen production.

H_2 production through metabolic engineering-integrated bioprocesses with improved yield and sustainability (Anand et al., 2024; Mogili et al., 2022; Srivastava et al., 2024). It will also be essential to pair biohydrogen production in biorefineries with the production of other high-added-value commodities and materials to fully take advantage of its potential use in the context of a circular bioeconomy. Biohydrogen production can thus significantly contribute to the future circular bioeconomy and transition to greener energy by valorising co-products and minimising waste (Figure 10.3).

10.5 CHALLENGES IN BIOHYDROGEN PRODUCTION

Biohydrogen production is considered a clean and renewable home appliance, but certain limitations often hinder its large-scale application. The relatively low yield in existing biohydrogen production systems is one of the main limitations that prevent the process from being economically viable at an industrial level. Biohydrogen yield is primarily limited by several properties, such as the low efficiency of microbial hydrogen production pathways, substrate specificity, and the inefficiency in utilising several organic materials (Alavi-Borazjani et al., 2021; Martínez-Fraile et al., 2024; Nabgan et al., 2021). For instance, although dark fermentation, photofermentation, and biophotolysis offer prospects for hydrogen production, all three are constrained in the rate of hydrogen produced per unit of substrate. Moreover, abundant and cheap raw materials, mainly complicated organic waste, often need pretreatment to be suitable for microbial transformation. Pretreatments can make it even more complex and expensive. In addition, the substrate specificity of some microorganisms may also restrict the feedstock candidates for biohydrogen production. Despite the adaptation of different microbial strains to degrade specific substrates, these groups show limited efficiency in what they convert to hydrogen. The diversity of the organic waste material also brings variability into the process and thus makes it challenging to achieve reproducible yields (El Bari et al., 2022; Kovalev et al., 2020; Shomope et al., 2025). Therefore, improving different microbial strains by broadening the substrate specificity or enhancing the conversion efficiency of fermentation pathways is continuously a focus of research. Thus, although biohydrogen production of organic waste is promising, the low yield and the substrate specificity remain significant barriers to broad applications.

Commercial production of biohydrogen also faces economic and technological constraints that this process must overcome. From a challenging standpoint, the current bioreactor configurations do not function at high levels of efficiency concerning gas-liquid separation, which is a critical element in isolating and purifying hydrogen gas. Conventional bioreactors may not be designed to maximise hydrogen production rates or to limit the formation of by-products, and novel reactor designs can aid in these areas. In addition, with modern biotechnological processes like dark fermentation or photosynthetic bacteria, the process stability is often low, and the operation times are long, which makes these technologies inefficient and expensive (Mogili et al., 2022; Reyna-Gómez et al., 2021; Srivastava et al., 2024). Additionally, to improve overall efficiencies, the continued process optimisation of aseptic hydrogen production is crucial in fine-tuning several parameters, such as temperature, pH, substrate concentration, and microbial strain selection. Economically, biohydrogen production is still expensive compared to standard hydrogen production methods (namely, steam methane reforming or electrolysis). Economic challenges are proven through the need for high-efficiency bioreactors, the high cost of microbial strain development, and the pretreatment costs of feedstocks. Since large-scale biohydrogen production is needed to make it economically feasible, the initial cost of large-scale plants is one of the major bottlenecks in its application. In addition, there are limited infrastructures for the large-scale deployment of biohydrogen production systems, further intensifying the economic challenges. Hence, enhancing process efficiencies, lowering feedstock pretreatment costs, and providing inexpensive bioreactor designs are essential to surpassing these technological and financial challenges. If biohydrogen is to be a commercial competitor to fossil fuel-derived hydrogen, these barriers must be surmounted (Chandra et al., 2024; Plakar et al., 2025; Roslan et al., 2024).

Biohydrogen production faces other significant challenges, including environmental impacts and scalability-related problems. Although biohydrogen is a clean energy by nature, environmental impacts typical for the production process itself challenge its overall sustainability. Biohydrogen was produced via pretreatment of lignocellulosic biomass, which typically involves the chemical of acids, alkalis, or enzymes, and all of these techniques can have environmental impacts (Aboudi et al., 2021; Anand et al., 2024; Pasupuleti et al., 2014). Moreover, large-scale biohydrogen technology may be characterised by its high water and energy consumption, especially when (i) energy-extensive pre-treatment methods are employed or (ii) optimal growth conditions (e.g., temperature, pH, and osmotic pressure) need to be sustained for complex microbial strains. Such resource inputs can negate biohydrogen production's environmental benefits, mainly when they depend on non-renewable energy sources or increase water consumption. Proper waste management is crucial in industrial-scale processes, as leftover biomass, by-products, and effluents can cause environmental issues if not handled correctly. Similarly, waste generated during biohydrogen production must be treated effectively to prevent pollution of water, land, and air. Biohydrogen production from organic waste materials is still at an early production phase concerning their scalability; thus, several logistical, technical, and financial barriers stand as hurdles to scaling up biohydrogen production from laboratory or pilot-scale systems to industrial-scale operations. The heterogeneity of feedstocks is one of the main concerns related to scalability. Microbial fermentation of organic waste materials varies depending on the initial feedstock, which causes inconsistencies in the fermentation process and can reduce hydrogen yields. This requires strain robustness and plasticity to utilise a variety of feedstock types while maintaining high hydrogen production rates. The scalability of biohydrogen production requires advanced gas recovery and purification systems that operate with high efficiency and throughput (Budhraja, 2024; Ghimire et al., 2018; Zhazhkov et al., 2024). This approach enables higher biohydrogen production due

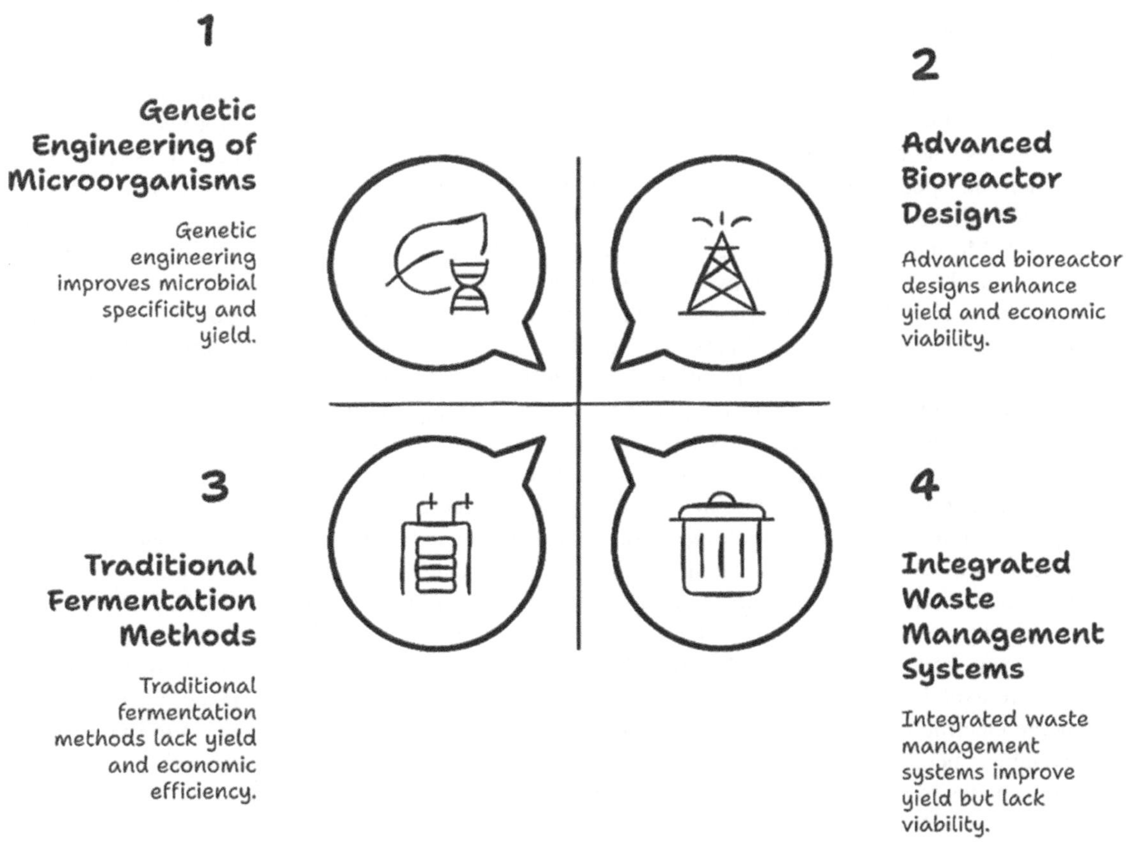

FIGURE 10.4 Challenges and solutions in biohydrogen production.

to its higher value and versatility than other biological end-products. Also, it promotes a more sustainable and economically feasible progression compared to scenarios where biohydrogen is produced in separate plants since existing infrastructure and systems for organic waste collection, pretreatment, and waste-to-energy conversion could be utilised.

Though the potential of biohydrogen generation from organic waste as renewable and clean energy is significant, serious challenges persist in low yield, substrate specificity, techno-economics, and scalability. Every challenge can be overcome with breakthroughs in microbial strain, bioreactor and pretreatment technologies, such as improving substrate utilisation and hydrogen production rates while enhancing process efficiency via more optimised bioreactor processes or pretreatment methods to make the feedstocks more accessible for microbial conversion (Figure 10.4). However, addressing the environmental costs of resource inputs and waste disposal is essential for achieving sustainable biohydrogen production in the long run. Finally, the key to realising biohydrogen as a new mainstream energy source relies on developing large-scale production infrastructure by integrating biohydrogen systems into the whole energy circular economy (Pomdaeng et al., 2024; Saravanan et al., 2021; Weide et al., 2019).

10.6 FUTURE PROSPECTS AND EMERGING TRENDS

Microbial consortia and engineered microbes are essential for addressing some key limitations of biohydrogen production, including microbial efficiency, substrate specificity, and hydrogen yield. Microbial consortia, specific examples being the few different types of microorganisms working together, have emerged as an appealing approach to increase biohydrogen production. Such consortia often act synergistically, with different microbes exerting complementary functions. For example, fermentative bacteria break down complexes into simple organic matter, and hydrogen-producing bacteria use them as substrates to convert to hydrogen. Consortia metabolise a broader range of feedstocks (for instance, complex organic wastes) and result in better hydrogen yields, taking advantage of the metabolic versatility of different microbial species. Moreover, synthetic biology and genetic engineering strategies can finally result in advanced microorganism strains with enhanced biohydrogen-producing capabilities (Jiménez-Ocampo et al., 2021; Musharavati et al., 2024; Salazar-Batres & Moreno-Andrade, 2022; Salem et al., 2018). By characterising the enzyme and pathways for hydrogen production and substrate utilisation ability, you can create strains with better enzymatic behaviours and pathways for H_2 production. Microbes can be genetically engineered to address suppressing challenges, such as building and creating better resistive strains, thus reducing substrate competition with industrial production processes typically happening in environments that vary across pH or temperature. Synthetic biology strategies, which allow for designing completely new metabolic routes or inserting foreign genes into microbial genomes, can also significantly increase hydrogen production in microbial systems. This study establishes a distinct model providing evidence that integrating natural microbial consortia with engineered microbes could yield novel proofs of concept to overcome specific technological bottlenecks associated with efficient biohydrogen production and pave the way to foster biohydrogen as a sustainable and scalable energy carrier.

In addition, some of the challenges in biohydrogen production have been highlighted and are being addressed with the help of nanotechnology. Nanomaterials, including nanoparticles and nanostructured materials, have been extensively studied to enhance biohydrogen production. Exploiting the use of nanomaterials to exploit gas-liquid interfaces can enable improved bioreactor performance, for instance, by enhancing electron transfer processes and increasing the overall surface area available to the microbes. For example, metallic nanoparticles (platinum or palladium) may act as bio-catalytic enhancers of enzymes relative to H_2 production or electron transfer between organisms (pathogens) in microbiological electrolysis cells (MEC). They can also be employed to develop more sensitive sensors and monitoring systems that can offer real-time insights into microbial activity, substrate consumption, and hydrogen production rates. These advancements provide nanotechnology for higher process control of biohydrogen production, more yield, and operational efficiency. Moreover, nanotechnology does not only represent a group of techniques practical for biohydrogen production in microbial processes. Nanomaterials facilitate the downstream process of hydrogen gas recovery and purification, which is essential for the commercial viability of biohydrogen production. For example, nanoporous materials are used for gas separation and storage. This also applies to nanofiltration technology, which improves the purity of the final hydrogen product for use in fuel cells and other energy applications. So, incorporating nanotechnology into biohydrogen production systems is a promising method for reducing their production costs and environmental impacts while enhancing their efficiency and scaling up biohydrogen systems.

The effect of policy support and economic stimulations in expanding biohydrogen production should also be well-noted. Governments and regulatory bodies drive the expansion of the biohydrogen sector by gearing financial incentives and supportive policies towards an atmosphere where research and development can be supported. Many countries have recognised the potential of biohydrogen as a clean and sustainable energy carrier, including through research, development, and commercialisation programmes. Incentives for renewable energy technology research (biohydrogen renewable energy included) can give power to universities and industries to work on the technical issues related to biohydrogen production. Financial instruments such as tax credits, grants, or low-interest loans can also help lower the capital cost of biohydrogen production plants and pilot projects. Such incentives are critical for designing the early-stage biohydrogen industry, where economics still do not favour the new technology versus other forms of hydrogen production, like steam methane reforming or electrolysis. Governments can also implement carbon pricing or cap-and-trade systems that penalise burning fossil fuels and make biohydrogen comparatively cheaper. Policies facilitating biohydrogen integration into existing infrastructure, like injection as opposed to natural gas pipelines or transportation infrastructure, can alleviate part of the logistical–economic challenges to large-scale deployment. The full potential of biohydrogen can be realised only through collaborative efforts between policymakers and industry stakeholders to formulate long-term scalability strategies for production, infrastructure, and system sustainability.

The future success of biohydrogen will depend significantly on the extent to which it disconnects systems of renewable energy grids and the broader bioeconomy. Biomass and biogas can be suspended on the way to their joint conversion into biohydrogen, which may be a final fermentation product. It is considered one of the promising bioenergy sources in the interaction between biohydrogen in the family of renewable energy sources, including solar and wind. It is applied to sufficient organic waste materials, usually considered waste products of several industries for biohydrogen production. It allows the conversion of these residuals into biohydrogen and the generation of a circular bioeconomy where residual products are processed into valuable products such as clean energy. Biohydrogen has the potential to be incorporated into renewable energy grids as an energy carrier and energy storage. Specific renewable sources such as wind or solar are intermittent and can produce excess energy, and biohydrogen (or hydrogen from renewable sources) can be used to harness that excess energy. This hydrogen stored can be used in a situation when renewable energy is lacking, which is not only stable but also sustainable. Diversification of biohydrogen is potentially integrated into energy grids that aim to reduce fossil fuels and improve energy security. Thus, biohydrogen can be a crucial player in the journey to future energy without sacrificing the continuous growth of global energy demand and reducing total GHG emissions. Integrating policies with existing and emerging improvements in microbial and nanotechnology-enabled biohydrogen technologies could see, on a relatively large and adjunct scale of our implementation, the upcoming biohydrogen energisation of energy production supports renewable energy and biofuel need within the global energy basket while also accounting for the integration with a circular bioeconomy.

Biohydrogen production represents a sustainable alternative to conventional hydrogen production methods, and large-scale implementation of biohydrogen technologies is contingent upon successfully solving numerous technical, economic, and policy challenges. Using microbial consortia or engineered microbes as a potential microbial platform for biohydrogen is an auspicious approach to enhance the biohydrogen yield and substrate utilisation combined with nanotechnology to increase the reactor's performance and hydrogen purification. To realise its full potential, the authors say this pathway requires appropriate policy frameworks and economic incentives to address biohydrogen's current cost competitiveness gap and, more generally, enable its large-scale production. Finally, integrating biohydrogen production into renewable energy grids and the larger bioeconomy can facilitate the transition towards a more sustainable energy future and contribute to closing the loop of the circular economy. If, with the technological advancements brought on by addressing these complex problems, biohydrogen will become an essential player in the global transition to clean and renewable energy, this will be a world-changing result.

10.7 CONCLUSION

Biohydrogen production from organic wastes is being focused as an alternative because of the increasing demand for clean and sustainable energy. The consequences of this domain illuminate its auspicious outlook as a durable alternative to the current fossil fuel energy framework. Biohydrogen is produced via metabolically active microorganisms (fermentative, photosynthetic, and syntrophic) that can use organic waste substrates as feedstock. Microbial pathways offer a low-carbon approach for hydrogen production through biohydrogen production. In this respect, biohydrogen is a potential green energy carrier to mitigate climate change and GHG emissions. Biohydrogen from organic wastes is a viable solution for a circular bioeconomy since they are abundant, and their generation is often an industrial waste stream. Therefore, one of the present-day methods of H_2 receiving describes processes of anaerobic treatment of organic wastes resulting in the formation of molecule biohydrogen (H_2); it is an ecologically safe way of obtaining this molecule compared with classical methods and against a background of waste management resources (or matter) recovery as a method of its effective exploitation of existing biomass. The fermentation process is still one of the most optimal avenues for biohydrogen production, where particular microbes in fermentative bacteria, photosynthetic bacteria, and the syntrophic consortium have been exploited to treat organic substrates for hydrogen production. By optimising microbial activity and metabolic pathways accompanying hydrogen production, researchers have borrowed excess hydrogen and increased the overall production of hydrogen for use in biohydrogen production processes. These optimisations have only been augmented with recent advances in genetic engineering and synthetic biology by developing strains of microorganisms that produce higher quantities of hydrogen and can utilise a broader range of organic substrates.

Despite the bioproduction of hydrogen initiatives, several barriers still exist for large-scale applications. One of the significant challenges is the low yield of hydrogen generated per mole of organic waste due to bioenergetics inefficiencies and substrate specificity. Additional research is ongoing, and several promising approaches, such as metabolic engineering, co-culture systems, and novel reactor designs, are being used to optimise microbial activity while increasing hydrogen yields. Genetic modifications of microorganisms and microbial consortia have been shown to improve substrate utilisation capacity and increase hydrogen production rates while minimising environmental stressors likely limiting microbial activity. Nanotechnology has also been developed to enhance the performance of biohydrogen production systems in terms of bioreactors, electron transfer, and, therefore, the overall yield of hydrogen. For biohydrogen to be a product on the market, new bioreactor designs and operating parameters need to be developed to maximise the production process. Finally, biohydrogen generation is integral to the bio-succinic acid and biofuel process. It will create additional value by becoming part of the circular bioeconomy with cost repayment to the overall process and improved economic efficiency of biohydrogen production due to products of higher value than the feedstock. Such techniques encounter barriers regarding technology and finances, particularly the high capital cost of building the infrastructure and, subsequently, the logistical problems of industrialising biohydrogen production. In summary, according to these constraints, they conclude that more studies and inverters are essential to overcome the remaining obstacles to biohydrogen becoming a more cost-effective commercial energy source.

Sustainable biohydrogen production from organic wastes: The idea is to create an ecological, manageable scale, and economically viable model that can influence global energy culture. Making this vision a reality will involve overcoming key barriers in several critical areas, such as enhanced microbial performance, improved reactor design, and decreased production costs. This underscores the potential benefits of implementing genetic engineering, synthetic biology, and microbial consortia to reach higher yields and better process efficiency. Moreover, introducing nanotechnology into biohydrogen systems could enhance the system's performance, facilitate electron transfer, and optimise hydrogen recovery and purification. Regular optimisation of metabolic pathways and developing unconventional bioprocess techniques shall be pivotal to advancing large-scale biohydrogen generation. At the same time, supportive policies, favourable economic measures, and regulatory structures will also be necessary to generate research and investments in biohydrogen technologies. To this end, governments can participate in the transfer dynamic, creating favourable conditions for preparing biohydrogen production infrastructure, contributing to the transition from fossil fuels to a new renewable energy source, and decreasing GHG emissions in a sustainable energy future. Another critical challenge in biohydrogen production will be to couple biohydrogen production with renewable energy grids and other energy storage systems (including battery and cell technologies) to maintain biohydrogen economically competitive as a green energy carrier in the long term.

Biohydrogen production from organic wastes aligns with a circular bioeconomy, in which the waste substances are maximised as resources. Due to the utilisation of organic wastes that would otherwise go to waste or be thrown in landfills, biohydrogen production acts as a clean energy source, alleviates environmental pollution, and performs waste management actions simultaneously. This technology is beneficial, particularly in areas that yield an enormous amount of biodegradable waste from the agricultural, industrial, and municipal sectors. The latter can facilitate the potential sustainable production of biohydrogen at a level of local energy systems positioned near the waste generation sites to reduce fuel consumption and carbon value for transportation and regional sustainability-driven benefits. In addition, biohydrogen can complement other renewable sources like solar energy and wind energy by serving as a stable energy carrier and storage medium, which is vital in the wake of increased demand for clean energy. Developments in this field and financial commitment to infrastructure and education can effectively make biohydrogen production one of the building blocks that push us towards a renewable, low-emission energy economy, thanks to the combined efforts of researchers, lawmakers, and industrial leaders.

In conclusion, producing biohydrogen based on organic wastes is promising for clean, renewable, and sustainable energy generation. We may have removed all further roadblocks to achieving sustainable production of excess elemental hydrogen at the process scale based on a metagenomic understanding of high-efficiency microbial consortium communities coupled with metabolic engineer genetics and advances in nanotechnology and process innovation. Given the progressive advances in biohydrogen and the resolution of challenges, biohydrogen can play a significant role in the global energy transition and circular bioeconomy by combining waste management, reducing GHG emissions, and generating sustainable power. So this idea of cell-free sustainable biohydrogen produced from biowastes outperforms an environmentally friendly reliable energy source using the asset in the card yet linking the advantages of behavioural renewable energy systems with the best practice of microbial processes. Biohydrogen can become the prosthetic organs of the future clean energy economy by bridging them into today's energy grids and systems and mediating the transition to a circular bioeconomy. Driven by research, innovation, and collaboration, biohydrogen can play a fundamental role in unlocking a sustainable and resilient energy future for the world.

REFERENCES

Aboudi, K., Fernández-Güelfo, L. A., Álvarez-Gallego, C. J., & Romero-García, L. I. (2021). 11 – Biogas, biohydrogen, and polyhydroxyalkanoates production from organic waste in the circular economy context. In R. C. Ray (Ed.), *Applied Biotechnology Reviews* (pp. 305–343). Academic Press. https://doi.org/10.1016/B978-0-12-820297-5.00002-5

Alavi-Borazjani, S. A., Tarelho, L. A. da C., & Capela, M. I. (2021). Parametric optimization of the dark fermentation process for enhanced biohydrogen production from the organic fraction of municipal solid waste using Taguchi method. *International Journal of Hydrogen Energy*, *46*(41), 21372–21382. https://doi.org/10.1016/j.ijhydene.2021.04.017

Anand, A., Umesh, & Moholkar, V. S. (2024). Chapter 3 – Biohydrogen production from microbial fermentation of organic wastes. In S. Nanda, A. K. Dalai, & V. V. B. T.- E. B. Goud (Eds.) *Emerging Biofuels* (pp. 27–52). Elsevier. https://doi.org/10.1016/B978-0-323-99547-4.00011-3

Budhraja, N. (2024). Simulation and optimization for biohydrogen production potential of various organic waste via anaerobic digestion. *Fuel, 360*, 130563. https://doi.org/10.1016/j.fuel.2023.130563

Chandra, S., Pandit, S., Deb, S., Mohan, C., Rajeev, M., Ranjan, N., Kumar, A., & Dikshit, P. K. (2024). Utilization of water chestnut waste for biohydrogen production and enhanced power generation by stacked microbial fuel cell. *Biocatalysis and Agricultural Biotechnology, 62*, 103425. https://doi.org/10.1016/j.bcab.2024.103425

El Bari, H., Lahboubi, N., Habchi, S., Rachidi, S., Bayssi, O., Nabil, N., Mortezaei, Y., & Villa, R. (2022). Biohydrogen production from fermentation of organic waste, storage and applications. *Cleaner Waste Systems, 3*, 100043. https://doi.org/10.1016/j.clwas.2022.100043

Ghimire, A., Trably, E., Frunzo, L., Pirozzi, F., Lens, P. N. L., Esposito, G., Cazier, E. A., & Escudié, R. (2018). Effect of total solids content on biohydrogen production and lactic acid accumulation during dark fermentation of organic waste biomass. *Bioresource Technology, 248*, 180–186. https://doi.org/10.1016/j.biortech.2017.07.062

Jiménez-Ocampo, U. E., Santiago, S. G., Vargas, A., & Moreno-Andrade, I. (2021). Feedback control strategy for optimizing biohydrogen production from organic solid waste in a discontinuous process. *International Journal of Hydrogen Energy, 46*(72), 35831–35839. https://doi.org/10.1016/j.ijhydene.2021.01.062

Kovalev, A. A., Kovalev, D. A., Litti, Y. V., & Katraeva, I. V. (2020). Biohydrogen production in the two-stage process of anaerobic bioconversion of organic matter of liquid organic waste with recirculation of digister effluent. *International Journal of Hydrogen Energy, 45*(51), 26831–26839. https://doi.org/10.1016/j.ijhydene.2020.07.124

Martínez-Fraile, C., Muñoz, R., Teresa Simorte, M., Sanz, I., & García-Depraect, O. (2024). Biohydrogen production by lactate-driven dark fermentation of real organic wastes derived from solid waste treatment plants. *Bioresource Technology, 403*, 130846. https://doi.org/10.1016/j.biortech.2024.130846

Mogili, N. V., Murugesan, N., Ayothiraman, S., Gautam, R., Deshavath, N. N., & Reddy Erva, R. (2022). Chapter 6 – Biohydrogen production from wastewater and organic solid wastes. In C. M. Hussain, S. Singh, & L. Goswami (Eds.), *Waste-to-Energy Approaches towards Zero Waste* (pp. 165–195). Elsevier. https://doi.org/10.1016/B978-0-323-85387-3.00009-4

Musharavati, F., Ahmad, A., Javed, M. H., Sajid, K., & Nizami, A.-S. (2024). Advancing biohydrogen production from organic fraction of municipal solid waste through thermal liquefaction. *International Journal of Hydrogen Energy*. https://doi.org/10.1016/j.ijhydene.2024.05.144

Nabgan, W., Tuan Abdullah, T. A., Nabgan, B., Jalil, A. A., Nordin, A. H., Ul-Hamid, A., Hassan, N. S., Hussain, I., Coelho, A., Amin, A., & Ikram, M. (2021). Catalytic biohydrogen production from organic waste materials: a literature review and bibliometric analysis. *International Journal of Hydrogen Energy, 46*(60), 30903–30925. https://doi.org/10.1016/j.ijhydene.2021.04.100

Pasupuleti, S. B., Sarkar, O., & Venkata Mohan, S. (2014). Upscaling of biohydrogen production process in semi-pilot scale biofilm reactor: evaluation with food waste at variable organic loads. *International Journal of Hydrogen Energy, 39*(14), 7587–7596. https://doi.org/10.1016/j.ijhydene.2014.02.034

Plakar, G. P. Z., Kovo, A. S., Oguzie, K. L., & Oguzie, E. E. (2025). Valorization of mixed blackwater/agricultural wastes for bioelectricity and biohydrogen production: a microbial treatment pathway. *Heliyon, 11*(1), e41126. https://doi.org/10.1016/j.heliyon.2024.e41126

Pomdaeng, P., Chu, C.-Y., Masa-ad, A., Ngamnurak, P., Setthapun, W., & Sintuya, H. (2024). An innovative acid-resistant elephant dung biochar powder for fermentative biohydrogen production improvement from food waste slurry. *International Journal of Hydrogen Energy*. https://doi.org/10.1016/j.ijhydene.2024.10.123

Reyna-Gómez, L. M., Cruz-López, A., Alfaro, J. M., & Suárez-Vázquez, S. I. (2021). Evaluation of the production of biohydrogen during the co-digestion of organic wastes in an upflow hybrid anaerobic reactor. *Chemical Engineering Journal, 425*, 129235. https://doi.org/10.1016/j.cej.2021.129235

Roslan, E., Mohamed, H., Abu Hassan, S. H., Carrere, H., & Trably, E. (2024). Coupling lactic acid fermentation of food waste at various concentrations as storage strategy with dark fermentation for biohydrogen production. *International Journal of Hydrogen Energy, 88*, 358–368. https://doi.org/10.1016/j.ijhydene.2024.09.134

Salazar-Batres, K. J., & Moreno-Andrade, I. (2022). Effect of nickel concentration on biohydrogen production: organic solid waste vs. glucose. *International Journal of Hydrogen Energy, 47*(70), 30097–30106. https://doi.org/10.1016/j.ijhydene.2022.05.281

Salem, A. H., Brunstermann, R., Mietzel, T., & Widmann, R. (2018). Effect of pre-treatment and hydraulic retention time on biohydrogen production from organic wastes. *International Journal of Hydrogen Energy, 43*(10), 4856–4865. https://doi.org/10.1016/j.ijhydene.2018.01.114

Saravanan, A., Senthil Kumar, P., Khoo, K. S., Show, P.-L., Femina Carolin, C., Fetcia Jackulin, C., Jeevanantham, S., Karishma, S., Show, K.-Y., Lee, D.-J., & Chang, J.-S. (2021). Biohydrogen from organic wastes as a clean and environment-friendly energy source: production pathways, feedstock types, and future prospects. *Bioresource Technology*, *342*, 126021. https://doi.org/10.1016/j.biortech.2021.126021

Shomope, I., Tawalbeh, M., Al-Othman, A., & Almomani, F. (2025). Predicting biohydrogen production from dark fermentation of organic waste biomass using multilayer perceptron artificial neural network (MLP–ANN). *Computers & Chemical Engineering*, *192*, 108900. https://doi.org/10.1016/j.compchemeng.2024.108900

Srivastava, N., Singh, R., Lal, B., & Haque, S. (2024). Evaluation of bioprocess parameters for pilot scale fermentative biohydrogen production using organic waste: environmental remediation, techno-economic challenges & future solutions. *International Journal of Hydrogen Energy*. https://doi.org/10.1016/j.ijhydene.2024.05.211

Weide, T., Brügging, E., Wetter, C., Ierardi, A., & Wichern, M. (2019). Use of organic waste for biohydrogen production and volatile fatty acids via dark fermentation and further processing to methane. *International Journal of Hydrogen Energy*, *44*(44), 24110–24125. https://doi.org/10.1016/j.ijhydene.2019.07.140

Zhazhkov, V. V, Politaeva, N. A., Velmozhina, K. A., Shinkevich, P. S., & Norov, B. K. (2024). Production of biogas from organic waste at landfills by anaerobic digestion and its further conversion into biohydrogen. *International Journal of Hydrogen Energy*, *70*, 779–785. https://doi.org/10.1016/j.ijhydene.2024.03.062

11 Enhanced Biohydrogen Production from Microalgae

Strategies and Advances

Deepti Verma, Taranjit Kaur Gargeria, Janhavee Asthana, Chaitanya Gatkal, and Ambuj Kumar Kushwaha

11.1 INTRODUCTION

To address the increasing global demand for clean, renewable energy sources, it is essential to explore and develop alternative energy solutions that are both viable and environmentally friendly. Among the most promising options is the production of hydrogen from microalgae, a process that holds significant potential in mitigating the energy crisis. Microalgae, due to their ability to undergo photosynthesis, can efficiently convert sunlight and carbon dioxide into oxygen and organic compounds, making them ideal candidates for biohydrogen production (Masojídek et al. 2021). This process of biological hydrogen production in microalgae is both energy-efficient and environmentally beneficial. Specific microalgal species can generate hydrogen under anaerobic conditions when subjected to stress factors, such as sulfur deprivation or low light (Yu et al. 2024). The ability of these microorganisms to produce hydrogen in such conditions offers a renewable energy source that can significantly contribute to reducing greenhouse gas emissions and combating global warming. Additionally, this process can be integrated into existing systems for carbon sequestration, further promoting its role in environmental sustainability.

Microalgae are not only capable of serving as a natural carbon sink, capturing CO_2 from the atmosphere and contributing to the reduction of global warming, but they also create a wide range of commercially significant and valuable products. These include biofuels, pharmaceuticals, and high-value nutraceuticals, which are gaining increasing interest in various industries. The diverse applications of microalgae make them highly versatile, adding another layer of appeal to their potential as a hydrogen

DOI: 10.1201/9781003568612-11

source. Beyond their energy production capabilities, microalgae can also play a pivotal role in environmental remediation. Since they can flourish in nutrient-rich wastewater, they present an opportunity for eco-efficient wastewater treatment (Ahmad and Ashraf 2024). This capability enables microalgae to treat effluent in an environmentally responsible manner while simultaneously serving as a valuable source of hydrogen. Additionally, microalgae do not compete with food crops for arable land, allowing them to be cultivated in diverse environments, including seawater and non-arable land. During their growth process, microalgae remove harmful substances such as CO_2 and phosphorous, reducing pollutants while generating energy (Sundaram et al. 2023). This ability to grow in wastewater treatment systems without negative environmental impacts further enhances the potential of microalgae for large-scale, low-cost, sustainable hydrogen production. The production of hydrogen by microalgae is driven by several natural biological pathways. These pathways, including the hydrogenase, nitrogenase, and photosynthetic hydrogen production pathways, enable microalgae to produce hydrogen under defined and optimal conditions (Kamshybayeva et al. 2022). Additionally, the photosynthetic hydrogen production pathway utilizes light energy to drive the production of hydrogen in the presence of light and certain environmental stresses, such as nutrient deprivation or altered metabolic conditions. These natural pathways set microalgae apart as an efficient and promising platform for large-scale hydrogen production, which is further strengthened by the fact that it offers a more cost-effective alternative to conventional hydrogen production methods, such as steam methane reforming (SMR) and electrolysis. These traditional methods often require significant energy inputs and resources, making them less efficient and environmentally harmful (Dębowski et al. 2021). On the other hand, microalgae cultivation requires significantly fewer land and water resources, and their ability to grow on non-arable land further reduces competition with food crops (Ammar et al. 2022). Additionally, microalgae-based hydrogen production can be integrated into other environmental initiatives, such as carbon capture and wastewater treatment, allowing for synergistic benefits that support broader sustainability goals (Hamid Nour et al. 2024). The metabolic pathways of microalgae, particularly the hydrogenase and nitrogenase pathways, enable them to produce hydrogen under specific conditions; however, while these unique metabolic processes make microalgae highly valuable for biohydrogen production, they also present challenges in optimizing yield (Hippler and Khosravitabar 2024). By manipulating the algal genome and using synthetic biology, researchers can improve hydrogenase enzyme expression and optimize environmental conditions for higher hydrogen output, making biohydrogen more commercially viable. This approach offers environmental, economic, and technological benefits, moving us closer to sustainable, hydrogen-based energy production (Scranton et al. 2016).

11.1.1 Key Pathways for Biohydrogen Production in Microalgae

Microalgae have several natural pathways for hydrogen production, each of which depends on key enzymes or proteins that play vital roles. The hydrogenase, nitrogenase, and photosynthetic hydrogen production pathways offer promising opportunities for biohydrogen production, each accompanied by its own set of unique challenges (Figure 11.1).

11.1.1.1 Hydrogenase Pathway

The hydrogenase enzyme plays a crucial role in the hydrogenase pathway by facilitating the conversion of protons and electrons to hydrogen gas (H_2). For instance, when exposed to sulfur starvation, *Chlamydomonas reinhardtii* activates hydrogenase, enabling anaerobic hydrogen production (Ahmad and Ashraf 2024). Nonetheless, hydrogenase is sensitive to oxygen levels, which hinders its activity and makes the process unfeasible for large-scale production. Additionally, hydrogenase tends to degrade over time, reducing its efficiency and stability for extended periods (Bingham, Smith, and Swartz 2012).

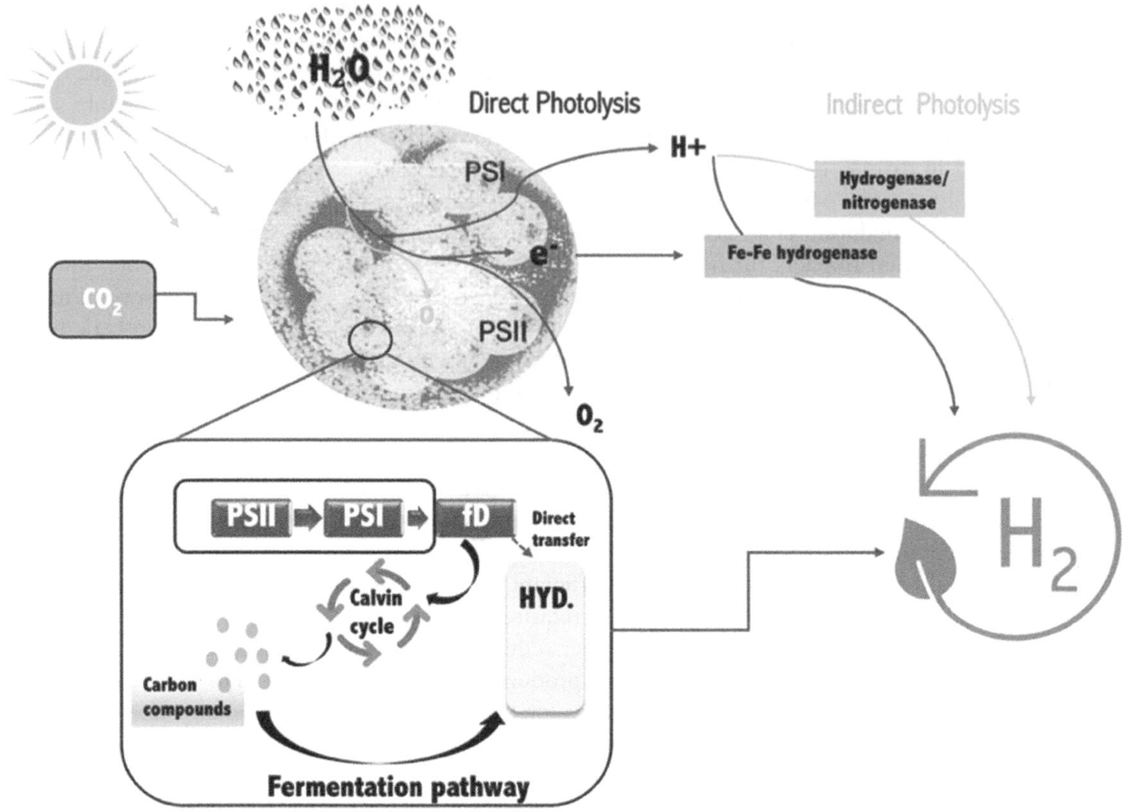

FIGURE 11.1 Metabolic pathways for natural hydrogen production in microalgae. Overview of metabolic processes (e.g., photosynthesis, fermentative pathways) involved in natural hydrogen production.

11.1.1.2 Nitrogenase Pathway

The nitrogenase enzyme is central to the nitrogenase pathway, where it fixes nitrogen and produces hydrogen gas (H_2) as a by-product. Nitrogenase is present in nitrogen-fixing microalgae, such as *Anabaena* sp. (Xuan et al. 2023). Like hydrogenase, nitrogenase is highly sensitive to oxygen, which diminishes its efficiency. This issue is mitigated by heterocyst, specialized cells in nitrogen-fixing microalgae that provide an anaerobic environment for nitrogenase. However, nitrogenase is an energy-intensive enzyme, and hydrogen production competes with nitrogen fixation, leading to lower hydrogen yields (Redding et al. 2022).

11.1.1.3 Photosynthetic Hydrogen Production Pathway

In the photosynthetic hydrogen production pathway, light-driven water splitting is facilitated by the photosystem II (PSII) protein complex, along with associated hydrogenase enzymes. In species like *C. reinhardtii*, light energy can be converted into hydrogen in the presence of nutrient starvation (Allakhverdiev et al. 2010). However, the concurrent production of oxygen through photosynthesis inhibits hydrogenase activity, which diminishes the overall hydrogen yield. The primary challenge lies in optimizing the conditions that control the balance between hydrogen production and oxygen evolution (Zidan, Javed, and Hassan 2024).

11.1.2 Challenges Across Pathways

Across all pathways, the key challenges include the oxygen sensitivity of critical enzymes like hydrogenase and nitrogenase, low hydrogen yields, and the need for optimized cultivation conditions (Javed et al. 2022). The efficiency and scalability of biohydrogen production can be enhanced by engineering more robust enzymes, such as oxygen-tolerant variants of hydrogenase, and by developing improved cultivation systems (Chandrasekhar, Lee, and Lee 2015). These pathways provide a renewable and sustainable source of hydrogen through microalgae, but further research is needed to address challenges related to enzyme oxygen sensitivity, low yields, and process optimization. Advances in genetic engineering and optimized cultivation techniques will be crucial for scaling up biohydrogen production to meet industrial demands.

11.2 MICROALGAL STRAINS

The selection of an algal strain plays a crucial role in determining the efficiency, sustainability, and economic feasibility of biohydrogen production processes. Choosing an optimal strain can drive significant progress in renewable energy generation while promoting environmental sustainability. Some of the ideal properties desired in algal strains for biohydrogen production include a high hydrogen production rate, rapid growth rate, robustness to environmental conditions, high biomass yield, metabolic flexibility, tolerance to stress, effective carbon fixation, genetic manipulability, low nutrient requirements, compatibility with bioreactor systems, and optimized photosynthetic efficiency (Shetty et al. 2019; Xuan et al. 2023). Algal strains with ideal characteristics can be isolated from various natural sources (Table 11.1).

11.2.1 Freshwater Species

The microalgae extracted from freshwater sources are highly promising feedstock for biofuel production due to their rapid growth rates, elevated biomass yields (for example, *Chlorella vulgaris* recognized for its substantial lipid production and notable capacity for hydrogen generation), and substantial carbon dioxide sequestration capabilities (*Scenedesmus obliquus* exhibits strong potential for hydrogen production and excels in wastewater treatment, contributing to both energy generation and environmental sustainability) (da Silva et al. 2023). One of the most notable freshwater algal species is *C. reinhardtii* (with its rich research background and well-established genetic tools, it serves as a valuable model for studying algal metabolism, presenting a promising avenue for biohydrogen production) (Vargas et al. 2018).

11.2.2 Marine Species

In contrast to freshwater microalgae, the variety of saltwater species is relatively diminished, as the majority of algae present in marine environments tend to be macroalgae (for example, *Padina tetrastromatica*, a seaweed that can generate substantial biomass in a short time) and numerous saltwater microalgae exhibit toxic or dangerous characteristics (Radha and Murugesan 2017). Examples of nontoxic microalgae include *Geitlerinema* sp. RMK-SH10, which is known for its ability to thrive in marine environments and contribute to biohydrogen production through photosynthetic processes. Its filamentous structure may enhance biomass accumulation and facilitate the harvesting process (Tinpranee, Incharoensakdi, and Phunpruch 2018). *Chlorella* sp. LSD-W2 is a well-studied microalga recognized for its high growth rate and lipid content, which is utilized in biohydrogen production, especially under specific conditions that promote hydrogenase enzyme activity (Tinpranee, Incharoensakdi, and Phunpruch 2016). The utilization of marine algae presents several advantages for applications in renewable energy. Marine algae facilitate

TABLE 11.1 Sources of algal strain along with cultivation condition and hydrogen gas yield

CATEGORY	*SPECIES*	*CONDITIONS*	*YIELD/ PRODUCTION*	*REFERENCE*
Freshwater species	***Chlamydomonas reinhardtii***	Grown in TAP medium (pH 7.2) aerobically to mid-exponential phase, then transferred to sulfur-deprived anaerobic photobioreactors.	Yield: 5.95 μmol/mg	Vargas et al. (2018)
	Chlorella vulgaris	Sulfur-deprived immobilized cells, nitrogen-purged, exposed to light for 24 hours, then kept in darkness for 48 hours.	Total H_2 production: 530±5 mL/L	Rashid, Lee, and Mahmood (2011)
	Scenedesmus obliquus	Grown in urban artificial wastewater under anaerobic, sulfur-deprived conditions (pH 7.5, 30°C) with 140 μE/m²/s purple light.	Yield: 128 mL H_2/L; Productivity rate: 204.8 mL H_2/L/day	Ruiz-Marin, Canedo-Lopez, and Chavez-Fuentes (2020)
	Chlorella pyrenoidosa	Sulfur-deprived TAP medium (pH 7), continuously illuminated (100 μE/m²/s), 25°C, with constant shaking.	Total H_2 production: 43.2 mL/L	Wang et al. (2011)
Marine water species	***Padina tetrastromatica*** (macroalgae)	Pretreated with dilute sulfuric acid; media containing 2 g $(NH_4)_2SO_4$, 0.5 g $NaCl_2$, 0.05 g $MgSO_4$, 0.1 g K_2HPO_4; pH 6±0.5.	Yield: 78±2.9 mL/0.05 g	Radha and Murugesan (2017)
	Geitlerinema sp. RMK-SH10 (microalgae)	Cultivated under nitrogen deprivation in ASN III medium; agitated at 120 rpm; illuminated with white light at 30 μmol photons/m²/s for 14 days.	Maximum H_2 yield: 0.271±0.013 μmol H_2/mg^{-1} dry weight h^{-1}	Tinpranee, Incharoensakdi, and Phunpruch (2018)
	Chlorella sp. LSD-W2 (microalgae)	NH_4Cl-deprived TAP medium; light intensity of 30 μmol photons/m²/s; temperature 30°C.	Photohydrogen production rate: 4.25 mL H_2/L/h	Tinpranee, Incharoensakdi, and Phunpruch (2016)
	Codium fragile	Sludge heat-pretreated at 90°C for 20 min, then incubated with marine algae and tap water.	Yield: 3.70±0.17 mL H_2/g COD added	Jung, Kim, and Shin (2011)
	Laminaria japonica	A 1 mL inoculum, 1 mL nutrient stock, and 0.07 M phosphate buffer (pH 7.5) were mixed, purged with nitrogen, and incubated at 35°C with 150 rpm agitation.	Total amount of H_2 produced 444 mL (from 10 g/L of dry algae in 62 h)	Lee et al. (2010)
Wastewater	***Chlorella pyrenoidosa***	Cultivated in domestic wastewater with 10% seed culture from BG11 medium; controlled temperature 25±2°C; light intensity 120 μmol/m²/s.	Maximum H_2 production: 45.50±0.32 mL H_2/g VS	Satheesh et al. (2023)
	Scenedesmus obliquus	Cultivated in domestic wastewater with 10% seed culture from BG11 medium; controlled temperature 25±2°C; light intensity 120 μmol/m²/s.	Maximum H_2 production: 38.43±0.42 mL H_2/g VS	Satheesh et al. (2023)
	Chlorella sorokiniana	Cultivated in domestic wastewater with 10% seed culture from BG11 medium; controlled temperature 25±2°C; light intensity 120 μmol/m²/s.	Maximum H_2 production: 34.83±1.82 mL H_2/g VS	Satheesh et al. (2023)
	Tetraselmis subcordiformis	Sulfur-free, chloride-substituted medium in deionized water (pH 8, 25°C); cultivated under a 14:10 h light-dark cycle with 5 klux intensity.	Maximum H_2 production: 116±7 mL to 162±7 mL	Dudek, Dębowski, et al. (2022)

the swift synthesis of biomass through the assimilation of natural resources, the production yields of algal biomass per unit area significantly exceed those associated with terrestrial biomass, marine algae exhibit a more elevated rate of carbon dioxide fixation relative to terrestrial biomass, suggesting a potentially greater capacity for the remediation of carbon dioxide (Gao and Mckinley 1994). Additionally, marine algae are devoid of hemicellulose and lignin, hence they can be depolymerized with relative ease compared to lignocellulosic biomass. Ultimately, marine algae can be cultivated in diverse marine environments such as freshwater, saltwater, or municipal wastewater (Wei, Quarterman, and Jin 2013).

11.2.3 Wastewater Species

Besides biohydrogen production, algae have also become a significant organism for the biological purification of wastewater since they can accumulate plant nutrients, heavy metals, pesticides, organic/inorganic, and toxic substances in their cells showing excellent bioaccumulation abilities. Microalgae are utilized to reduce biochemical oxygen demand (BOD) and remove nitrogen and phosphorus, making them effective for water purification, remediation, and the reduction of global greenhouse gases. The selected species must be capable of degrading color-causing compounds and tolerating high nutrient concentrations (Randrianarison and Ashraf 2017). Synthetic wastewater treatment using algal species has been optimized under varying conditions. For instance, *Micractinium reisseri* achieved an 89% reduction in total nitrogen and a 95% reduction in total phosphorus in tertiary wastewater effluents after 8 days of cultivation (Abou-Shanab et al. 2014).

Wastewater is rich in nutrients, phosphorus, and nitrates, which support microalgae growth for biohydrogen production while enabling simultaneous wastewater treatment. This approach reduces the economic constraints of biomass-derived biofuel production. Additionally, industrial wastewater, such as dairy manure used as a nutrient supplement, can further enhance algae-based biofuel yields (Dudek, Debowski, et al. 2022).

11.3 HARVESTING

Algal harvesting involves concentrating a diluted algal suspension into a thick algal paste, resulting in a slurry with at least 2%–7% algal suspension based on dry matter. The process presents a major challenge in algal biofuel production due to the small size of algal cells and their extremely low concentration in photobioreactors. As a result, a large volume of suspension must be processed (Gudin and Thépenier 1986). Estimates indicate that biomass recovery from the culture suspension accounts for approximately 20%–30% of the total biomass production costs (Chini Zittelli et al. 2006). Typically, a two-step concentration method is used, where the suspension is first thickened to form a slurry with around 2%–7% total suspended solids (TSS), achieving a concentration factor of 100–200. This is followed by further dewatering to produce a cake with 15%–25% TSS, reaching a concentration factor of 2–10 (Vandamme, Foubert, and Muylaert 2013). The efficacy of a harvesting technique is contingent upon various factors, including the specific microalgal strain, the dimensions of the cells, and the targeted products (Table 11.2). Common harvesting techniques include the following steps:

a. **Screening**

    Screening is a physical separation technique where microalgae are collected by passing the culture through mesh screens or sieves. The process relies on the difference in size between algal cells and the screen pores, allowing water and smaller particles to pass through while retaining the algae. Vibrating screens employ mechanical vibrations to separate algae from water, enhancing the efficiency of biomass recovery. In contrast, microstrainers utilize finely woven stainless steel fabrics to filter out algae from water, they are attached on top of a partially submerged drum which rotates slowly in a trough of suspended algal particles. Water spray is used to dislodge the algal cake attached to the filters, resulting in high removal rates (Sethunge, Manage, and Jayasinghe 2012; Gupta and Wakudkar 2023).

TABLE 11.2 Different algal harvesting techniques along with their advantages, disadvantages, and recovery efficiency

HARVESTING TECHNIQUE	*MICROALGAE SPECIES*	*RECOVERY*	*ADVANTAGES*	*DISADVANTAGES*	*REFERENCE*
Cross flow filtration	***Chlorella*** sp.	100%	High energy efficiency, no need for chemical usage	Susceptible to membrane fouling and damage to fragile materials	Hwang et al. (2013)
Ultrafiltration	***Chlorella*** sp.	94%	Quick and straightforward processing; reliable for small-scale algal cultures; low water content in the product; minimal cell damage, energy, and chemical use	Requires significant capital investment	Hwang et al. (2015)
Electro-flocculation	***Scenedesmus almeriensis***	99%	Cost-effective, suitable for diverse types of microalgal biomass, chemical-free	High energy usage; reliant on ionic bond strength and the culture medium's pH	Vasistha et al. (2021)
Plant bio-flocculation	***Chlorella vulgaris***	90.80%	Affordable, biodegradable, and low toxicity	Risk of product contamination	Maia et al. (2023)
Microbial bio-flocculation	***Scenedesmus*** sp.	99.40%	Economical and eco-friendly; allows reuse of culture medium and non-toxic to algal biomass	Alterations in cell composition and potential contamination of algal products	Indahsari et al. (2022)
Chemical flocculation	***Chlorella vulgaris***	85%	Cost-efficient, easy to scale up, no energy requirements	Use of harmful chemicals; limited recycling potential of culture medium	Jiang et al. (2021)
Buoy-bead flotation	***Chlorella vulgaris***	89.90%	Chemical-free with high reusability	Expensive	Wen et al. (2019)
Magnesium coagulation-dissolved air flotation	***Chlorella zofingiensis***	95.20%	Does not require external coagulants; coagulants and biomass are highly recyclable	Involves the use of toxic chemicals	Zhang et al. (2012)
Electrolytic flotation	***Chlorella vulgaris***	99%	Free from chemicals and supports continuous systems	High operational and initial costs	Koley et al. (2017)
Foam flotation	***Chlorella vulgaris***	96%	Highly scalable, requires minimal space and short processing times	High operational and initial costs	Al-Humairi et al. (2023)
Ozone flotation	***Chlorella vulgaris***	68.62%	Improved recovery of bioactive compounds from microalgae	Needs on-site specialized equipment for ozone generation	Kadir et al. (2021)

(*Continued*)

TABLE 11.2 (*Continued*) Different algal harvesting techniques along with their advantages, disadvantages, and recovery efficiency

HARVESTING TECHNIQUE	*MICROALGAE SPECIES*	*RECOVERY*	*ADVANTAGES*	*DISADVANTAGES*	*REFERENCE*
Gravity sedimentation	***Scenedesmus dimorphus***	80%	Simple, cost-effective method; avoids chemical use and requires little energy	Time-intensive, ineffective for small-sized algae, and low efficiency, potential degradation of biomass, low concentration of recovered algal slurry and cake	Wang et al. (2014)
Centrifugation	***Chlorella*** sp.	90%	Rapid method, high recovery rates, and compatible with nearly all algal species	High cost, energy-intensive, primarily suited for recovering high-value products, and possible cell damage caused by high shear forces	Ma et al. (2022)

b. Filtration

The filtration process utilizes a semipermeable membrane that can retain microalgae on the membrane while allowing the liquid media to pass through, leaving the algae biomass behind to be collected. Filters are chosen based on the size of the algal cells. This method is efficient for harvesting larger algal species and high-density cultures. In surface filtration, solids are deposited on the filter medium in the form of a paste or cake, whereas in deep bed filtration, solids are deposited within the filter bed matrix (Velten et al. 2024).

c. Centrifugation

Centrifugation separates microalgae cells from the culture medium by utilizing centrifugal force, relying on differences in density and particle size. This method is significantly more efficient than natural gravity sedimentation. Various centrifugal systems are employed in the industry, including perforated basket centrifuges, which are disc bowl centrifuges designed for suspensions with low solid content (ranging from 0.01% to 20% algal dry weight). In contrast, decanter centrifuges are used for suspensions with a higher solid content, typically between 10% and 50% algal dry weight (Najjar and Abu-Shamleh 2020). Among various centrifugation techniques, hydrocyclone technology may represent the most promising avenue for large-scale economic application because of its simplicity and minimal moving components. In a hydrocyclone system, the walls of the chamber remain fixed, while algal cells within the chamber traverse in a spiral motion, generating centripetal forces that facilitate the removal of denser particles from the liquid medium (Shakeel Syed et al. 2017).

d. Flocculation

Flocculation aggregates free-floating microalgae cells into larger particles (flocs) by adding a flocculating agent to neutralize surface charges. Polymer flocculants promote flocculation by forming bridges between particles, with positively charged functional groups aiding interaction with negatively charged algal cells. Increasing pH in algal cultures enhances auto-flocculation, while co-culturing with yeast or bacteria also improves floc formation. Common flocculants include inorganic (iron chloride, aluminum sulfate) and organic (chitosan, acrylic acid) compounds. This process concentrates the suspension by 20–100 times, with Fe_3O_4-embedded carbon microparticles achieving 99% *Chlorella* recovery within a minute (Seo et al. 2015).

e. Flotation

Flotation, an "inverted" sedimentation process, uses air bubbles to lift microalgae to the surface, forming a scum that is skimmed off. In dissolved air flotation, air-saturated water is mixed into the culture under high pressure, producing fine bubbles (~40 μm) for effective separation, though with higher energy demand. Dispersed air flotation generates bubbles by releasing pressurized air through a nozzle or porous media. Dispersed ozone flotation replaces air with ozone, which modifies the algal cell wall and reduces surface charge, enhancing cell adhesion to bubbles (Zhang et al. 2014; Valeriano Gonzalez et al. 2021).

11.4 CULTIVATION OF MICROALGAE FOR BIOHYDROGEN PRODUCTION

Biohydrogen is considered the cleanest fuel and can be produced through various methods. While thermochemical processes exist, biological methods are preferred for their eco-friendliness and cost-effectiveness. Among them, microalgae-based biohydrogen production is commercially viable for large-scale use. Microalgae generate biohydrogen *via* biophotolysis, where sunlight splits water (H_2O) into hydrogen (H_2). The released H_2 is collected from the bioreactor for energy production. Efficient microalgae cultivation methods are key to optimizing biohydrogen yield.

11.4.1 Methods for Cultivating Microalgae for Biohydrogen Production

Microalgae grow rapidly and require no fertile land, excessive freshwater, herbicides, or pesticides, avoiding competition for these resources. They can be cultivated using wastewater from domestic sewage and palm oil mills, facilitating bioremediation while simultaneously producing biomass. Most microalgae species thrive in nutrient-rich wastewater, efficiently absorbing nutrients and metals, making them a cost-effective and sustainable wastewater treatment option. Additionally, microalgae contribute to reducing atmospheric CO_2 through photosynthesis, helping mitigate global warming and the greenhouse effect (Tan et al. 2020; Alam et al. 2022). Cultivation methods primarily include open pond systems and photobioreactors (Figure 11.2).

11.4.1.1 Open Pond System

The simplest culture system for the mass production of microalgae is an open pond, in which the microalgae are cultivated under conditions that are comparable to those found in the outside world (Yu, Lee, and Sim 2022). Open raceway ponds, typically lined with polyvinyl chloride (PVC) sheets, operate at a culture depth of 20–30 cm, with paddle wheels or pumps facilitating mixing and gas exchange. Their performance is influenced by climatic factors such as air temperature and solar radiation, though light availability, determined by design, is the most critical factor (Venkata Subhash et al. 2022; Barboza-Rodriguez et al. 2024). Key advantages of open raceway ponds include low initial investment costs, a low-shear environment, and effective nutrient removal from municipal wastewater. However, they also have limitations, such as lower biomass productivity, susceptibility to bacterial contamination, sensitivity to rainwater affecting culture media, and water loss due to evaporation (Barboza-Rodriguez et al. 2024; Ali et al. 2022; Lu, Xiao, and Chen 2021).

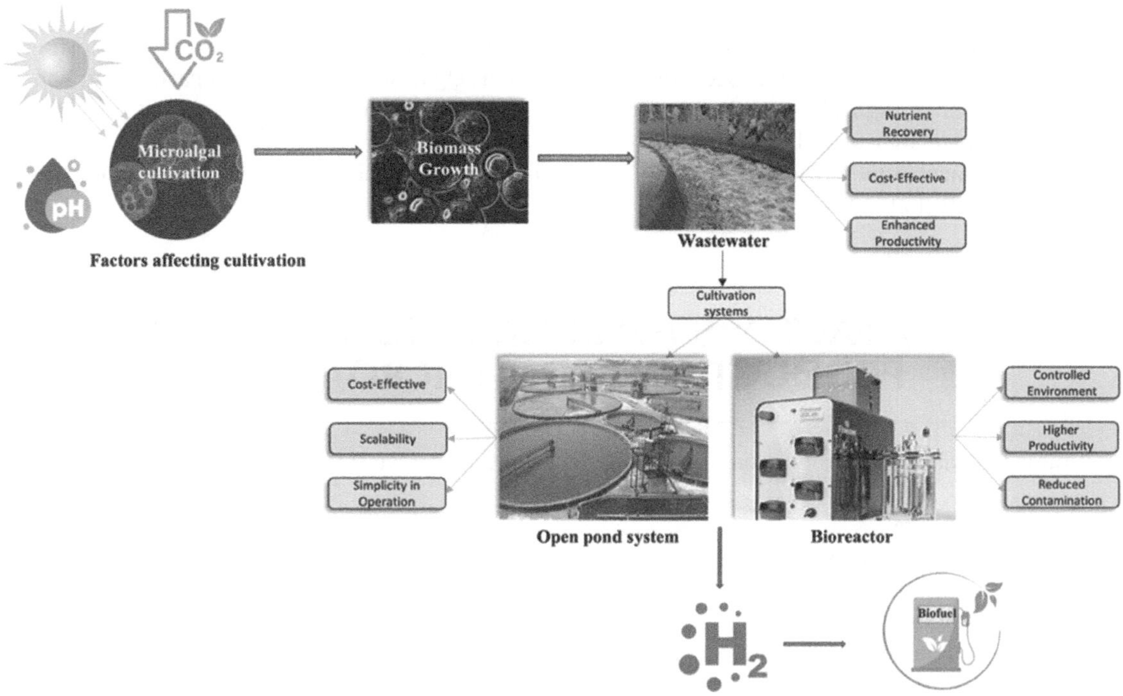

FIGURE 11.2 Factors and cultivation systems for optimized H_2 production in microalgae. Diagram showing key growth factors (light, temperature, nutrients, pH) and cultivation systems (photobioreactors, open ponds) that enhance biohydrogen production.

11.4.1.2 Photobioreactors

A photobioreactor is a closed system that prevents direct material exchange between the culture and the environment, making it ideal for cultivating phototrophic organisms like microalgae. The photosynthetic rate of microalgae in these systems is closely linked to environmental conditions, directly influencing nutrient and metabolite production (Tan et al. 2020; Manu et al. 2024).

Closed-system photobioreactors offer several advantages, including higher product yields, reduced water loss, and improved photosynthetic efficiency due to controlled conditions, though at a higher operational cost. These systems also support optimal microalgal growth, contribute to CO_2 sequestration—a key factor in mitigating global warming—and facilitate the production of valuable compounds. Notably, photobioreactors can achieve CO_2 removal efficiencies ranging from 16% to 58% (Morgado et al. 2024; Benner et al. 2022). Various types of bioreactors exist, including:

a. **Stirred Tank Bioreactor**

Stirred tank photobioreactors' primary benefits include the ability to guarantee effective heat and mass transmission as well as uniform nutrient dispersion when the stirring system is operating at its best (Aslanbay Guler et al. 2020).

b. **Horizontal Tubular**

Horizontal tube photobioreactors are made of transparent materials and installed outdoors in areas that receive sunlight. The enormous surface area per unit volume of these cultivation vessels maximizes the amount of sunlight that microalgae receive (Abdur Razzak et al. 2024). The advantages of this photobioreactor arrangement include a large surface-to-volume ratio, prevented shading, improved air resistance time, and enhanced dissolved oxygen (Deruyck et al. 2019). However, it also has certain limitations such as the requirement of a cooling system when doing scale-up to prevent overheating.

c. Flat Panel

Flat panel photobioreactor is a suitable option for research with the primary goal of testing how different types of light affect the growth of microalgae and its constituent parts. Their benefits include high photosynthetic efficiency, high gas-liquid mass transfer, ease of construction and relative affordability, ease of cleanup and temperature control, proper mixing through mechanical rotation or air bubbling, appropriate light path, and high biomass productivity (Tayebati et al. 2024).

11.5 FACTORS AFFECTING BIOHYDROGEN PRODUCTION FROM MICROALGAE

Several factors affect the production of biohydrogen from microalgae (Figure 11.2).

a. Temperature

Temperature is crucial for microalgae growth, directly influencing biochemical processes like photosynthesis. Non-optimal temperatures, especially in outdoor systems, lead to high biomass losses. In large-scale production, particularly open pond cultures, temperature control is essential. Elevated temperatures hinder photosynthesis by deactivating essential proteins and disrupting energy balance, while lower temperatures impair carbon assimilation. The key photosynthetic enzyme, Rubisco, is highly temperature-sensitive and functions as either an oxygenase or carboxylase depending on the O_2 and CO_2 levels in chloroplasts (Batyrova et al. 2015).

b. pH

pH is a key factor affecting microalgae growth, metabolism, and biohydrogen production. Optimal pH varies by species, with marine and freshwater microalgae having different requirements. Biohydrogen generation is most effective between pH 4.5 and 8.0, while ideal microalgae growth typically occurs between 5 and 9 (Sathinathan et al. 2023; Min Woon et al. 2023; Arimbrathodi et al. 2023; Tang et al. 2022). Higher pH can reduce lag time, boosting hydrogen production (Chowdury, Nahar, and Deb 2020). pH fluctuations arise from nitrate absorption and carbon fixation, while volatile fatty acid (VFA) accumulation can lower pH below 4.5, shifting metabolism toward solventogenesis and reducing biohydrogen yield.

c. Light

Light intensity is a key factor limiting microalgae growth, directly influencing photosynthesis, biochemical composition, and biomass yield. In bioreactors or open pond systems, light levels vary within the culture and decrease with depth, requiring careful consideration. Different algae species have specific light requirements for optimal growth and biomass accumulation (Srivastava et al. 2020).

Microalgae growth is highly dependent on light intensity, with both excessively low and high levels hindering development. Beyond a certain threshold, photosynthesis stabilizes as photorespiration and photoinhibition counterbalance its rate. Algal photosynthesis requires specific light/dark cycles, as ATP and NADPH production, critical for carbon skeleton formation in dark reactions, relies on light (González-Camejo et al. 2019; Cheirsilp and Torpee 2012). Growth correlates with light intensity and duration up to saturation, with most studies indicating that a 16-hour light and 8-hour dark cycle is optimal. While fluorescent tubes are effective, light-emitting diode (LED) lights offer a superior alternative (Carvalho et al. 2011). For most microalgae species, ideal light intensities range between 200 and 400 μmol photons/m²/s, though some species thrive at 100 μmol/m²/s, as demonstrated experimentally (Wu 2016).

d. Nutrients

Essential nutrients must be supplied to enhance cell division and hydrogen production in microalgae. The nutrient requirements vary depending on the microalgae's growth phase—mixotrophy, heterotrophy, or autotrophy. For instance, *Chlorella* sp. utilizes both internal and external substrates for hydrogen production in its heterotrophic metabolism (Khetkorn et al. 2017; Morales-Sánchez, Martinez-Rodriguez, and Martinez 2017). External carbon sources, such as acetate, glucose, sucrose, and fructose, are crucial for growth and can significantly boost hydrogen yields (Min Woon et al. 2023). Starch, an important internal carbon source, is produced during early photosynthesis and later serves as an electron donor for hydrogen generation. The availability of carbon directly influences metabolic pathways, biohydrogen yield, and fermentation by-products. An optimal glucose concentration of 30 mM has been found to reduce hydrogen production time by approximately 10 hours. Microalgal growth and hydrogen production can be further enhanced by supplying trace elements, nitrogen, and phosphorus in appropriate amounts (Goswami et al. 2021). Nitrogen regulates protein synthesis and growth metabolites, while phosphorus is vital for cellular functions and metabolic activities (Su 2021). Additionally, Fe^{3+} is essential due to its role in hydrogenase enzyme activity. Sulfur compounds must be excluded from the culture medium, as they inhibit protein synthesis and partially deactivate the PSII. This suppression reduces photosynthetic oxygen production, preventing oxygen-induced inhibition of hydrogenase and supporting sustained hydrogen generation (Veeravalli et al. 2019).

11.6 PRETREATMENT

11.6.1 Inoculum Treatment

Inoculum pretreatment procedures are employed to enhance hydrogen (H_2) production yield. The pretreatment of sludge mitigates methane generation and suppresses the activity of hydrogen-consuming bacteria, thereby increasing H_2 production efficiency. Studies have shown that pretreatment of the inoculum or substrate can lead to a 43%–69% increase in hydrogen content (Xie et al. 2008). The primary objective of pretreatment is the liquefaction and saccharification of microalgal cells, which not only reduces processing costs but also minimizes the formation of fermentation inhibitors (Nagarajan, Chang, and Lee 2020).

Pretreatment of anaerobic inoculum facilitates hydrolysis, thereby overcoming the rate-limiting step and enhancing process stability. This optimization of anaerobic digestion significantly improves H_2 generation efficiency. Various inoculum pretreatment techniques are employed to enhance hydrogen production, including heat pretreatment, acid/alkali treatment, ultrasonic waves, chloroform, and iodopropane. Among these, heat pretreatment, acid/alkali treatment, and ultrasonic treatment are the most effective for H_2 production (Singh et al. 2022).

11.6.2 Biomass Pretreatment

Pretreatment is a crucial step in the processing and utilization of microalgal biomass for applications such as biofuel production, pharmaceuticals, nutraceuticals, and wastewater treatment. It aids in breaking down the robust cell walls of microalgae, which are composed of cellulose, pectin, or sporopollenin that acts as a barrier to the recovery of intracellular compounds like lipids, proteins, and carbohydrates. By disintegrating cell wall components, pretreatment increases pore size, surface area, and substrate digestibility, enhancing the accessibility to valuable compounds. This, in turn, improves downstream processing efficiency and optimizes resource recovery.

11.6.3 Physical Pretreatment Technologies

Several biomass and inoculum treatment methods are available, each contributing to enhanced H_2 production (Figure 11.3).

11.6.3.1 Physical Treatment

Physical pretreatment techniques provide better cell wall disintegration because of shear and impact stresses; however, these techniques are very energy-intensive, expensive, and take a long time to extract oil from microalgae. The physical treatment methods include thermal treatment, freezing and thawing, microwave radiation, ultrasonication (US), radiation, and milling.

11.6.3.2 Thermal Treatment

Thermal pretreatment facilitates the proliferation of hydrogen-producing bacteria (HPB) while inhibiting methanogenic activity. The presence of spores serves as an indicator of HPB, even after the removal of non-spore-forming microorganisms, such as methanogens and non-sporulating HPB, during the pretreatment process. Heat-shock assays are employed to assess the hydrogen production potential of mixed and anaerobic microbial populations (Singh et al. 2022).

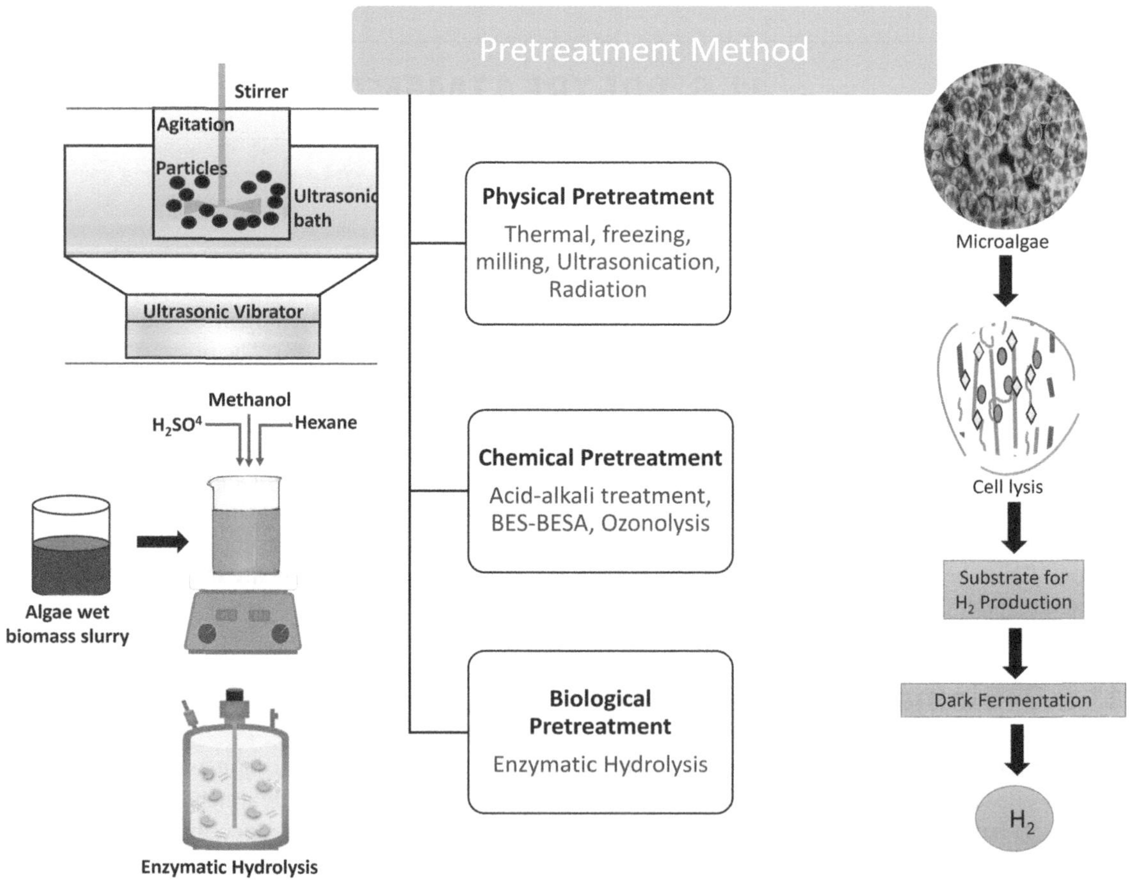

FIGURE 11.3 Pretreatment methods for enhanced biohydrogen production. Overview of physical, chemical, and biological strategies to optimize microalgal hydrogen production.

According to Tukey's test, the hydrogen yield follows the trend: acid treatment>heat shock>alkali treatment>control in lactic acid (LA)-based cultures. It was observed that the heat-shocked and acid-treated cultures exhibited no methanogenic activity, unlike the control cultures, which received no treatment. However, alkali-treated cultures displayed some methanogenic activity. Additionally, both alkali-treated cells (600 mg/L) and control cultures (720 mg/L) exhibited LA production (Chaganti, Kim, and Lalman 2012).

A study by Luo et al. (2010) demonstrated that anaerobic granular sludge, subjected to heat shock and load treatment, produced hydrogen yields of 64.4 mL H_2/g VS when used as inoculum with cassava stillage as the substrate (Luo et al. 2010). Similarly, Xie et al. (2008) utilized heat-shock-treated activated sludge to obtain 271.2 mL H_2/g VS from enzymatically treated potatoes (Xie et al. 2008). These variations in hydrogen yields are likely attributed to differences in inoculum sources and operational conditions, as bacterial metabolism plays a crucial role in biohydrogen synthesis (Ghimire et al. 2015).

11.6.3.3 Freezing Treatment

The recurrent development of massive intracellular ice crystals through freezing treatment or freeze-thaw cycles promotes the growth of pores in cell walls and membranes, which break cells and release intracellular chemicals (Dixon and Wilken 2018).

11.6.3.4 Microwave Radiation

Microwave pretreatment significantly enhances lipid extraction from microalgae. For example, *Botryococcus braunii* yielded 56.42% lipids under microwave exposure, while *C. vulgaris* achieved 41.31% lipids, surpassing ultrasonic methods. Additionally, in *Chlorella pyrenoidosa*, microwave treatment stimulated microalgal growth and oxygen production. Notably, chlorophyll-a and oxygen levels increased by 12.4% and 10.7%, respectively, at 400 W for 25 seconds, improving the efficiency of algal-bacterial wastewater treatment systems (He et al. 2022).

11.6.3.5 Ultrasonication

The process of US uses ultrasonic wave energy at frequencies between 10 kHz and 20 MHz to disturb cells (Nagarajan, Chang, and Lee 2020). Sonication pretreatment of algal biomass produced more hydrogen (0.42 m^3/m^3) than autoclaved biomass (0.100 m^3/m^3). This might be because sonication produces smaller starch molecules, which are easier and more suited to convert to H_2 than pure starch (Roy et al. 2014).

US pretreatment enhances biohydrogen production by improving the anaerobic digestion of complex food waste through increased solubilization of organic materials. The highest hydrogen yield from sonicated food waste is significantly boosted by raising the total solids (TS) concentration from 2% to 8% and extending ultrasonic treatment from 5 to 15 minutes (Gadhe, Sonawane, and Varma 2014). Hydrogen production rates can increase 1.30-fold when ultrasound is applied directly to digested sludge and 1.48-fold when used on a solution containing all hydrogen-producing components, with optimal ultrasonic duration of 10 seconds and intensity of 130 W/L (Guo et al. 2010). While sonication, combined with thorough solvent mixing, is a simple and effective method, a resting period in a sealed vessel under refrigeration is crucial for maximizing extraction yield.

11.6.3.6 Milling

One of the first methods for disrupting microalgal cells is bead milling, which involves rapidly stirring the biomass of microalgae using steel, ceramic, glass, or quartz beads to create shear forces (Nagarajan, Chang, and Lee 2020).

11.6.3.7 Pulse Electric Field

Pulsed electric field (PEF) is a technique that disrupts cells using electric forces. In this method, a cell suspension is exposed to an electric field, and a high-voltage current is applied to alter the transmembrane electric potential, leading to pore formation (Nagarajan, Chang, and Lee 2020). PEF is a relatively mild cell disruption technique, as it electroporates algal cells without generating cell debris. This allows for the selective enhancement of small ionic compounds and carbohydrate extraction, achieving yields comparable to high-pressure homogenization. However, protein extraction efficiency remains low (Carullo et al. 2018).

11.6.4 Chemical Pretreatment Technologies

11.6.4.1 Acid and Alkali Pretreatment

Heat and acid treatments facilitate the breakdown of complex carbohydrates into simpler sugars. For instance, when algal starch undergoes HCl heat pretreatment, it hydrolyzes into various soluble sugars, predominantly glucose, cellobiose, and dextrin, depending on the HCl concentration used (Roy et al. 2014). Among the commonly employed chemicals, sulfuric acid (H_2SO_4) is the preferred acid, while sodium hydroxide (NaOH) is the most widely used base (Nagarajan, Chang, and Lee 2020).

11.6.4.2 BES and BESA

BES (2-bromoethanesulfonate) is a potent and selective inhibitor of methanogens. At a concentration of 286 μM, it suppresses methane production by 64%, and at 572 μM, it nearly eliminates methane generation, thereby increasing total biohydrogen yield (Chae et al. 2010). Additionally, BES is introduced into the methanogenic column influent at concentrations ranging between 1×10^{-4} and 6×10^{-6} M to inhibit methanogenic activity (Bouwer and McCarty 1983).

11.6.4.3 Ozonolysis

Ozone pretreatment oxidizes, dissolves, and degrades the cell wall, improving hydrolysis efficiency. This method selectively degrades lignin while minimizing effects on sugar polymers, avoiding the formation of inhibitory compounds and producing weak carboxylic acids (Sulfahri, Langford, and Tassakka 2020). Ozonolysis enhances carbohydrate accessibility for microbial fermentation, which correlates with higher biohydrogen production (Pal et al. 2025).

11.6.4.4 Chloroform

Chloroform ($CHCl_3$) has been utilized in dark fermentation to inhibit methanogenic activity and enhance biohydrogen yield (Singh et al. 2022). It works by inhibiting methyl-coenzyme M reductase and blocking corrinoid enzyme activity, thereby effectively reducing methanogenesis and increasing hydrogen production (Hu and Chen 2007).

11.6.4.5 Iodopropane

Iodopropane acts as a solvent, disrupting the rigid cell walls of microalgae, thereby facilitating the release of intracellular contents for improved fermentation efficiency (Giang et al. 2019).

11.6.5 Biological Pretreatment Technologies

Biological treatment primarily involves enzymatic pretreatment, which ensures high product purity, free from contaminating chemicals, while maintaining simple reaction conditions that do not require complex instrumentation (Nagarajan, Chang, and Lee 2020). In microalgal pretreatment, hydrolytic microorganisms facilitate hydrolysis by secreting specific enzymes that degrade intracellular biopolymer cell walls (Zabed et al. 2019). Studies have shown that enzymatic pretreatment under thermophilic conditions achieved a maximum hydrogen yield of 135±3.11 mL H_2 g VS^{-1} at a substrate concentration of 10 g dry algae. Furthermore, enzymatically pretreated *C. vulgaris* exhibited a 7-fold increase in hydrogen production compared to untreated samples (Srivastava et al. 2020).

11.7 GENETIC ENGINEERING-BASED STRATEGIES FOR ENHANCED DIHYDROGEN PRODUCTION

Genetic modifications through metabolic pathway optimization and synthetic biology application present a great potential to harness the dihydrogen (H_2)-producing capability of various microorganisms (Table 11.3 and Figure 11.4).

11.7.1 Engineering of Metabolic Pathways

11.7.1.1 Optimizing Hydrogenase Activity

Green algae, such as *C. reinhardtii*, harness sunlight to split water into protons, electrons, and oxygen. Under specific conditions, they can also generate hydrogen (H_2) using hydrogenase enzymes (HYDA1 and HYDA2) within their chloroplasts. However, in their natural state, hydrogen production is limited because a significant portion of electrons, which could be directed to hydrogenase for H_2 generation, are instead

TABLE 11.3 Various strategies for optimizing hydrogen production in different microalgal strains

MICROALGAL SPECIES	*TARGETED MECHANISM/ APPLICATION*	*KEY FINDINGS*	*REFERENCES*
Chlamydomonas reinhardtii	Improving electron flow to hydrogenase (HYDA1, HYDA2).	Increased H_2 production by 3–4 times by optimizing electron flow to hydrogenase.	Weiner et al. (2018)
Chlamydomonas reinhardtii	Enhancing hydrogenase activity with HydA-SOD fusion protein.	Sustained H_2 production for 14 days, slower growth.	Ben-Zvi et al. (2019)
Parachlorella kessleri	Using protonophores to boost H_2 production.	Protonophores increased H_2 yield by 3-fold by creating anaerobic conditions.	Manoyan et al. (2019)
Chlamydomonas reinhardtii	Removing the FDP pathway to boost H_2 production.	Removal of FDP pathway increased H_2 production by redirecting electrons to hydrogenase.	Jokel et al. (2019)
Chlamydomonas reinhardtii	Building biomimetic electron transfer chain with thermophilic enzymes.	Synthetic pathways boosted H_2 output six times faster than traditional methods.	Kim et al. (2016)
Chlorella sp.	Phosphorus deprivation to optimize H_2 production.	Phosphorus deprivation redirected resources to H_2 production.	Batyrova and Hallenbeck (2017)
Chlamydomonas reinhardtii	Removing the FDP pathway to increase electron flow to hydrogenase.	Increased H_2 production by elimination of the FDP pathway.	Jokel et al. (2019)

(Continued)

TABLE 11.3 (*Continued*) Various strategies for optimizing hydrogen production in different microalgal strains

MICROALGAL SPECIES	*TARGETED MECHANISM/ APPLICATION*	*KEY FINDINGS*	*REFERENCES*
Chlamydomonas reinhardtii	Enhancing transgene expression and preventing silencing.	Strong promoters reduced silencing and increased hydrogenase expression.	McQuillan et al. (2022)
Chlamydomonas reinhardtii	Optogenetic regulation to enhance H_2 production.	Blue light-inducible system improved H_2 production without affecting growth.	Rumpel et al. (2014)
Chlamydomonas reinhardtii	Precise genome editing for hydrogen production.	CRISPR edits improved photosynthesis, boosting H_2 output.	Baek et al. (2016)
Chlamydomonas reinhardtii	CRISPR methods for gene insertion and editing.	CRISPR/Cas9 methods enhanced biofuel production.	Patel et al. (2019)
Chlamydomonas reinhardtii	RNAi to silence genes and boost H_2 production.	Silencing light-harvesting proteins increased H_2 production by 180%.	Oey et al. (2013)
Chlamydomonas reinhardtii	Targeting miRNAs to enhance hydrogen production.	Upregulation of miR1166.1 significantly improved H_2 yields.	Kao and Ng (2017)

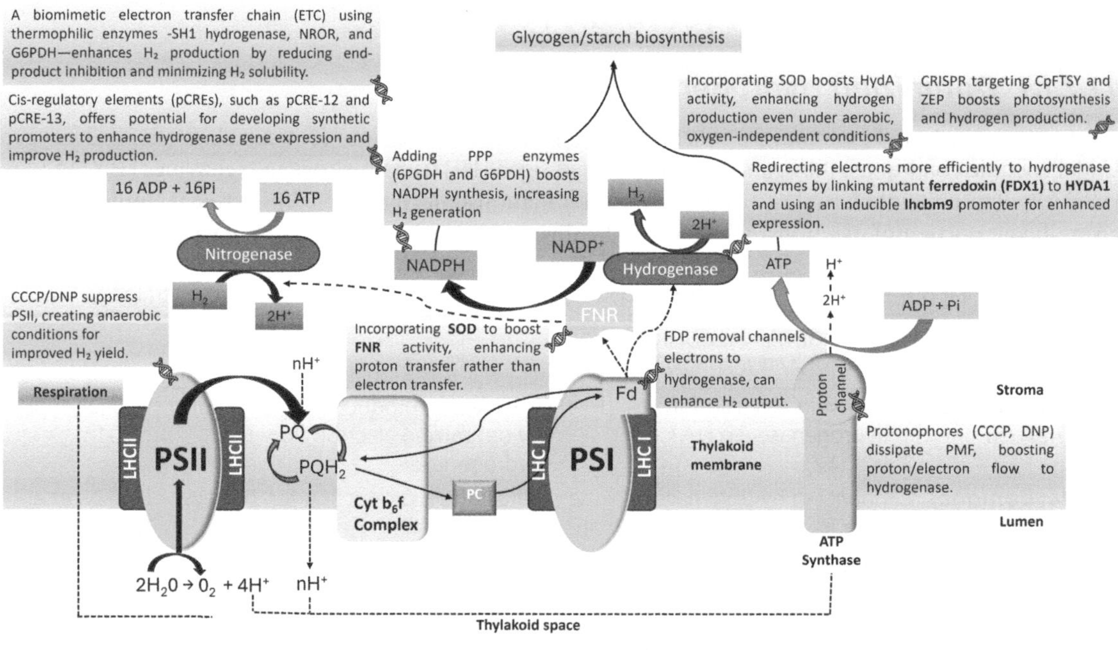

FIGURE 11.4 Genetic engineering for dihydrogen production in microalgae. Illustration of genetic modifications and metabolic strategies to enhance hydrogenase expression and optimize conditions for improved hydrogen production in microalgae.

utilized in other cellular processes, such as ferredoxin-NADP+ reductase (FNR), an enzyme involved in electron transfer (Siitonen et al. 2023).

To enhance hydrogen production, researchers have employed genetic modifications to improve enzyme activity and expression. One study optimized electron flow to hydrogenase enzymes (HYDA1 and HYDA2), leading to increased H_2 production (Xiong et al. 2021). In wild-type algae, most electrons divert to FNR, limiting hydrogen generation. To address this, scientists engineered a fusion protein by linking ferredoxin (FDX1) with a mutant form (D19A, D58A) that has a reduced affinity for FNR and fused it to HYDA1. To enhance expression, they employed a highly inducible *lhcbm9* promoter and incorporated introns to boost nuclear expression. The fusion constructs were introduced into a HYDA1/

HYDA2 knockout strain and tested under sulfur-deprived, illuminated conditions. The modified algae exhibited a 3- to 4-fold increase in H_2 production compared to the wild type, demonstrating more efficient electron transfer to hydrogenase (Xiong et al. 2021).

Further studies revealed that high-expressing clones (OP68 and R2D2) had lower levels of active hydrogenase but still produced more hydrogen, likely due to improved enzyme performance. These clones also contained substantial amounts of inactive hydrogenase, suggesting that enhancing hydrogenase maturation could further increase the active enzyme pool and hydrogen output (Weiner et al. 2018).

Additionally, the research highlighted the role of superoxide dismutase (SOD) in protecting and enhancing [FeFe]-hydrogenase (HydA) for hydrogen production, even under oxygen-rich conditions. SOD was found to boost HydA activity and increase FNR function, facilitating improved proton transfer and, consequently, better hydrogen generation. By fusing HydA with SOD, hydrogen production increased by 300%, achieving a record rate of 700 mmol H_2/mg chlorophyll/h (Ben-Zvi et al. 2019).

11.7.1.2 Pathway Modification

As an alternative to directly optimizing hydrogenase activity, modifying metabolic pathways can also enhance hydrogen (H_2) production. In green algae, H_2 is generated through a process known as "anaerobic oxygenic photosynthesis," where the oxygen produced is consumed by the algae, creating anaerobic conditions that activate the [FeFe]-hydrogenase enzyme. Despite its sensitivity to oxygen, this enzyme still facilitates hydrogen production under low O_2 conditions (Manoyan et al. 2019).

In *Parachlorella kessleri*, the use of protonophores such as CCCP and DNP was tested to improve H_2 yield, leading to a 3-fold increase. These compounds dissipate the proton motive force across the thylakoid membrane, increasing the availability of protons and electrons for hydrogenase while simultaneously inhibiting PSII. This reduction in O_2 production fosters anaerobic conditions, further enhancing H_2 generation (Manoyan et al. 2019).

In *C. reinhardtii*, researchers have investigated the role of alternative electron transport pathways, particularly ferredoxin-dependent proteins (FDPs), which act as electron sinks to create microoxic conditions that activate [FeFe]-hydrogenase. The deletion of FDPs, as seen in the *flv 208* mutant, redirected electrons to hydrogenase, significantly increasing H_2 production. Additionally, the study found that adjusting light pulse duration influenced H_2 photoproduction—shorter pulses favored H_2 generation by limiting CO_2 fixation, while longer pulses promoted CO_2 fixation at the expense of hydrogen output (Jokel et al. 2019). These findings highlight that optimizing light conditions and oxygen management can enhance both hydrogen production and biomass accumulation, offering valuable strategies for scalable biohydrogen production.

11.7.2 Engineering Approaches in Synthetic Biology

11.7.2.1 Constructing Synthetic Pathways

Researchers have constructed a biomimetic electron transfer chain (ETC) incorporating thermophilic enzymes like SH1 hydrogenase, NROR, and G6PDH to enhance hydrogen (H_2) production. This system achieved a volumetric H_2 productivity of 93 mmol H_2/L/h at 80°C by reducing end-product inhibition and minimizing H_2 solubility. The use of hyper-thermostable enzymes optimized reaction rates, and the addition of PPP enzymes (6PGDH and G6PDH) further boosted NADPH synthesis, increasing H_2 generation to 164 mmol H_2/L/h. Incorporating 6PGL and electron mediators like NROR and BV raised the H_2 output to 310 mmol H_2/L/h, setting a new benchmark for biological hydrogen production—over six times faster than traditional methods (González-Camejo et al. 2019).

However, improving hydrogen production through synthetic biological systems remains challenging. Despite advancements in NAD(P)H-dependent pathways in *Escherichia coli*, hydrogen yields remain constrained due to limited NADP(H) turnover, thermodynamic limitations, cofactor imbalances (NADPH/NADP+ and NADH/NAD+), and end-product inhibition. Additionally, the accumulation of hydrogen within culture containers further impedes industrial-scale production, highlighting the need for strategies to overcome these bottlenecks (Agapakis et al. 2010).

11.7.2.2 Metabolic Burden Reduction

Reducing metabolic load by limiting non-essential by-products enhances the efficiency of ATP and reducing equivalents, which are crucial for hydrogen production. In *Chlorella* sp., phosphorus deprivation promotes starch accumulation, redirecting metabolic resources away from wasteful by-products toward hydrogen production. This strategy optimizes hydrogenase activity under anaerobic conditions, while further improvements in growth conditions and CO_2 supply reduce by-product formation, ensuring the availability of more resources for hydrogen production (Ye et al. 2012).

Similarly, in *C. reinhardtii*, hydrogen production can be enhanced by decreasing metabolic burden by removing the FDP pathway. The FDP pathway reroutes excess electrons from photosynthesis to oxygen reduction, limiting the electron availability for hydrogen production. In *flv 208* mutants, which lack this pathway, electrons are redirected toward hydrogenase activity, significantly increasing hydrogen yields in both anaerobic and microoxic conditions. By eliminating the FDP pathway, by-product formation is reduced, optimizing ATP and reducing equivalents, and improving the overall efficiency of hydrogen production (Jokel et al. 2019).

11.7.3 Manipulation of Regulatory Element

In *Chlamydomonas*, hydrogen production is primarily driven by hydrogenase enzymes, which catalyze anaerobic hydrogen gas production, particularly under sulfur-depleted conditions. However, the efficiency of genetic modifications aimed at enhancing hydrogenase activity is often hindered by transgene silencing, which affects the stable expression of hydrogenase genes. Although upregulation of these genes can boost hydrogen production, transgene silencing remains a challenge.

Promoter engineering offers a promising solution by improving genetic transformation efficiency in *C. reinhardtii*, particularly by enhancing transgene expression and minimizing silencing (Schroda, Beck, and Vallon 2002).

Researchers have successfully increased gene expression and prevented silencing by utilizing strong promoters such as HSP70A and modifying the RBCS2 promoter. These strategies are crucial for achieving stable and high-level transgene expression, which can be directly applied to optimizing hydrogenase gene expression for increased hydrogen production.

Additionally, the identification of cis-regulatory elements (pCREs), such as pCRE-12 and pCRE-13, has opened new avenues for developing synthetic promoters to further enhance hydrogenase gene expression. By integrating these elements with strong native promoters like *HSP70A* or *RBCS2*, more efficient transcriptional activation can be achieved, ultimately boosting biohydrogen production. The study also emphasizes the advantage of using smaller promoter constructs incorporating pCREs, which simplifies genetic engineering. Future research could focus on developing modular promoters that optimize hydrogenase expression, thereby maximizing hydrogen production in *C. reinhardtii* (McQuillan et al. 2022).

11.7.4 Transcription Factor Modification

Significant potential exists for optimizing hydrogen production through transcription factor manipulation in algae, particularly under conditions of regulated growth. Researchers can improve hydrogen generation by fine-tuning metabolic pathways through sophisticated genetic engineering techniques such as targeted gene expression and optogenetic systems, which allow for exact control over the circumstances that encourage hydrogen creation.

11.7.4.1 Optogenetics

C. reinhardtii can produce more hydrogen by implementing a blue light-inducible method that promotes targeted gene expression without impairing algal growth (Cheirsilp and Torpee 2012). This approach offers a more effective way to regulate genes, especially in large-scale cultures, by using heterodimerizing proteins that trigger transcription in response to light.

11.7.4.2 Inducible Expression Systems

By sustaining hydrogenase activity for extended periods, strains such as *C. reinhardtii* cy6Nac2.49 show that regulated expression of PSII can improve hydrogen generation (Batyrova and Hallenbeck 2017). Even though these developments are promising, scaling these methods for real-world uses is still difficult, and further study is needed to increase the sustainability and effectiveness of producing biohydrogen from microalgae.

11.7.5 CRISPR and Genome Editing

CRISPR/Cas9 technology has revolutionized precise genome editing, particularly in microorganisms like bacteria and microalgae. This tool allows researchers to disable genes that inhibit hydrogen production or introduce beneficial genes to enhance the process. In *C. reinhardtii*, CRISPR/Cas9 has been used to improve photosynthetic efficiency by targeting *CpFTSY* and *ZEP* genes. Knocking out *CpFTSY* reduces chlorophyll antenna size, enhancing light penetration in dense cultures, while *ZEP* knockout increases zeaxanthin accumulation, boosting photosynthetic efficiency. These modifications enhance biomass production and create optimal conditions for hydrogenase activity, improving hydrogen output (Baek et al. 2016). CRISPR technology has progressed in recent years, and three methods have been developed globally to date for the insertion of target genes in algae cells:

i. **Integrative Plasmid Based**

   This approach involves the use of plasmids that integrate into the algal genome. The plasmid carries the CRISPR/Cas9 components along with the desired gene, which is inserted into the target site through homologous recombination. This method allows for stable integration of the gene, enabling long-term expression and functional studies (Moosburner et al. 2020).

ii. **Replicative Plasmid Based**

   In this method, plasmids replicate independently within the algal cells. The CRISPR/Cas9 system is delivered *via* these plasmids, which do not integrate into the genome but can express the Cas9 protein and guide RNA transiently (Poliner et al. 2018; Patel et al. 2019). This approach is useful for transient gene expression studies and can be advantageous for applications where stable integration is not required.

iii. **Ribonucleoprotein (RNP) Based**

   The RNP-based approach involves the direct delivery of the Cas9 protein and guide RNA into algal cells. This method is advantageous because it bypasses the need for plasmid DNA, reducing the risk of off-target effects and potential regulatory issues associated with genetically modified organisms (GMOs). The RNP complex can efficiently induce double-strand breaks at the target site, leading to gene editing through the cell's natural repair mechanisms (Patel et al. 2023).

   These approaches enhance CRISPR's versatility in microalgae, driving advancements in metabolic engineering, biofuel production, and biotechnology.

11.7.6 RNA Interference for Enhanced H_2 Production

Gene silencing *via* RNA interference (RNAi) has emerged as a powerful tool in algal biotechnology, particularly for enhancing hydrogen (H_2) production. Research has demonstrated that the application of RNAi in microalgae, such as *C. reinhardtii*, can significantly improve H_2 output by targeting specific genes involved in photosynthesis and metabolic pathways. For instance, researchers reported that knocking down light-harvesting complex proteins (LHCMB1, 2, and 3) resulted in a remarkable 180% increase in H_2 production efficiency (Oey et al. 2013). This approach leverages the natural RNAi machinery in various algal species, including key components like Dicer and Argonaute proteins that process double-stranded RNA into small interfering RNAs (siRNAs). Furthermore, studies have shown that transient expression of artificial microRNAs (amiRNAs) can enhance gene silencing efficiency, allowing for more precise control over metabolic pathways that influence H_2 production (Li et al. 2015). The integration of RNAi technology with metabolic engineering strategies holds great promise for optimizing algal strains for biohydrogen production, potentially leading to sustainable energy solutions.

11.7.7 MicroRNA Regulation in Hydrogen Optimization

Fine-tuning gene expression through miRNAs has emerged as a promising strategy to enhance hydrogen production in *C. reinhardtii*. A study identified miR1166.1 as a key regulator of hydrogen biosynthesis, with its upregulation significantly boosting hydrogen yields. Using a heat-inducible system to control miR1166.1 expression enabled sustained hydrogen production without medium replacement, demonstrating the potential of miRNA-based regulation in improving biohydrogen efficiency (Wang et al. 2018).

Additionally, CRISPRi-mediated repression of the *CrPEPC1* gene achieved up to 94% suppression, leading to a 74.4% increase in lipid content and productivity compared to the wild type (Kao and Ng 2017). This, along with miRNA regulation, underscores the importance of optimizing multiple genetic targets to enhance hydrogen biosynthesis. Furthermore, RNAi-based downregulation of ferredoxin-NADP$^+$ reductase (FNR) redirected metabolism away from CO_2 fixation, resulting in a 2.5-fold increase in hydrogen production under sulfur-deprived conditions (Sun et al. 2013). These findings highlight the potential of precise gene regulation in advancing microalgal biofuel production and renewable energy technologies.

11.8 INTEGRATED APPROACHES FOR PROCESS OPTIMIZATION

A holistic approach integrating multiple key strategies is essential for enhancing the efficiency of biohydrogen production from microalgal biomass. One of the most critical steps is the selection of an optimal microalgal strain, as different species exhibit varying hydrogen production capabilities. Among the most promising candidates are *C. reinhardtii* and *S. obliquus*, known for their high growth rates and significant hydrogen production potential (Figueroa-Torres et al. 2020; Chader, Hacene, and Agathos 2009; Ban et al. 2019).

Following strain selection, environmental modifications play a crucial role in maximizing both biomass yield and hydrogen production. Key factors such as light intensity, temperature, and nutrient availability—particularly nitrogen—must be carefully optimized. Studies have shown that nitrogen supplementation in the growth medium can significantly enhance hydrogen production (Chanquia, Vernet, and Kara 2022; Show et al. 2019). Additionally, efficient harvesting techniques, including flocculation and centrifugation, help reduce energy costs associated with biomass recovery, making the process more economically viable (Lafarga 2019; Hoh, Watson, and Kan 2016).

Another important aspect is biomass pretreatment, which improves the accessibility of carbohydrates for fermentation. Methods such as enzymatic hydrolysis and chemical treatments break down complex biomasses into simpler sugars, thereby increasing hydrogen production efficiency (Rajendran et al. 2021). Furthermore, advanced genetic engineering tools, including CRISPR/Cas9, offer exciting possibilities for optimizing metabolic pathways to achieve even higher hydrogen yields. By integrating these strategies—strain selection, environmental optimization, efficient harvesting, biomass pretreatment, and genetic modifications—biohydrogen production can be significantly enhanced, paving the way for a more efficient and sustainable system (Shin et al. 2016).

11.9 FUTURE PERSPECTIVES AND CHALLENGES

Despite its potential, large-scale commercialization of biohydrogen production from microalgal biomass faces several challenges. Profitability remains a major hurdle, as the high costs associated with cultivation, harvesting, and downstream processing limit the scalability of this technology (Wang and Yin 2018; Jiao et al. 2024).

Beyond economic constraints, technological limitations also pose significant obstacles. Photobioreactor efficiency is often suboptimal, leading to lower-than-expected yields. Optimizing reactor design to enhance light penetration, gas exchange, and overall performance is crucial for improving efficiency (Peter et al. 2022).

Microalgal growth and hydrogen production are further restricted by metabolic constraints, including the oxygen sensitivity of hydrogenase enzymes, low carbon-to-hydrogen conversion efficiency, and metabolic shifts favoring growth over hydrogen production (Yu, Lee, and Sim 2022). Other reducing factors include by-products, energy transfer limitations, and deprivation effects due to sulfur, all of which tend to lower the overall yield. Additional factors such as energy transfer limitations, by-product accumulation, and nutrient deprivation, particularly sulfur depletion, further reduce overall yields (Navarro et al. 2009). Moreover, inefficient photosynthesis and the absence of effective hydrogen storage mechanisms hinder productivity, making large-scale, reproducible hydrogen output challenging (Wei and Ma 2024).

Despite these barriers, advancements in genetic engineering, metabolic optimization, and bioreactor design continue to drive progress in this field. Research efforts are focused on developing algal strains with enhanced hydrogen yields, improved hydrogenase activity, and greater photosynthetic efficiency (Omer et al. 2025). Integrating complementary technologies such as wastewater treatment and carbon capture could further enhance sustainability, paving the way for a viable future in microalgal biohydrogen production.

11.10 CONCLUSION

Microalgal biomass holds great potential as a renewable source for biohydrogen production. Significant improvements in hydrogen yields can be achieved through a combination of strategies, including optimal strain selection, cultivation optimization, advanced harvesting techniques, effective pretreatments, and genetic engineering. However, economic and technological challenges must be addressed to enable large-scale implementation. Continued innovation and research will be essential to unlocking the full potential of microalgal biomass as a sustainable energy source.

ACKNOWLEDGMENTS

AKK is grateful to the University Grants Commission, New Delhi, for the Start-Up Research Grant (Letter No. F.30–544/2021), the Institute of Eminence (IoE) Seed Grant, Banaras Hindu University (Dev. Scheme No. 6031), and Science and Engineering Research Board Start-Up Research Grant (SRG/2021/001732) for providing financial assistance. DV is thankful to the Banaras Hindu University for UGC Research Fellowship. TK, JA, and CG are thankful to the Department of Biotechnology (DBT) for the student fellowship. Financial support from the DBT to the School of Biotechnology, BHU, is also acknowledged.

ABBREVIATIONS

amiRNA	Artificial MicroRNA
ASN III	Artificial Seawater Nutrient Medium III
BES	2-bromoethanesulfonate
BESA	Bioenergy and Environmental Sustainability Assessment
BG II	Blue Green II
BHP	Biohydrogen Production Potential
BOD	Biochemical Oxygen Demand
BV	Biogas Volume
CCCP	Carbonyl Cyanide M-Chlorophenyl Hydrazone
CRISPR	Clustered Regularly Interspaced Short Palindromic Repeats
DNP	Dinitrophenol
ETC	Electron Transport Chain
FDP	Flavodiiron protein
FNR	Ferredoxin-NADP+ Reductase
FP-PBR	Flat Panel Photobioreactor
GMO	Genetically Modified Organism
HPB	Hydrogen-Producing Bacteria
LHCMB	Light-Harvesting Chlorophyll A/B-Binding
PBR	Photobioreactor
PEF	Pulsed Electric Field
PPP	Pentose Phosphate Pathway
PVC	Polyvinyl Chloride
RNAi	RNA Interference
RNP	Ribonucleoprotein
siRNA	Small Interfering RNA
SMR	Steam Methane Reforming
SOD	Superoxide Dismutase
TAP	Total Algal Productivity
TS	Total Solids
TSS	Total Suspended Solids
UASB	Upflow Anaerobic Sludge Blanket
VFA	Volatile Fatty Acids
ZEP	Zeaxanthin Epoxidase

REFERENCES

Abdur Razzak, S., K. Bahar, K. M. Oajedul Islam, A. K. Haniffa, M. O. Faruque, S. M. Zakir Hossain, and M. M. Hossain. 2024. Microalgae cultivation in photobioreactors: Sustainable solutions for a greener future, Green Chemical Engineering, 5: 418–39.

Abou-Shanab, R. A. I., M. M. El-Dalatony, M. M. EL-Sheekh, M. K. Ji, E. S. Salama, A. N. Kabra, and B. H. Jeon. 2014. Cultivation of a new microalga, *Micractinium reisseri*, in municipal wastewater for nutrient removal, biomass, lipid, and fatty acid production, Biotechnology and Bioprocess Engineering, 19: 510–18.

Agapakis, C. M., D. C. Ducat, P. M. Boyle, E. H. Wintermute, J. C. Way, and P. A. Silver. 2010. Insulation of a synthetic hydrogen metabolism circuit in bacteria, Journal of Biological Engineering, 4: 3.

Ahmad, A., and S. S. Ashraf. 2024. Harnessing microalgae: Innovations for achieving UN Sustainable Development Goals and climate resilience, Journal of Water Process Engineering, 68: 106506-06.

Al-Humairi, S. T., J. G. M. Lee, A. P. Harvey, A. D. Salman, T. Juzsakova, B. Van, P. C. Le, D. D. La, A. K. Mungray, P. L. Show, and D. D. Nguyen. 2023. A foam column system harvesting freshwater algae for biodiesel production: An experiment and process model evaluations, Science of the Total Environment, 862: 160702.

Alam, S. N., Z. Khalid, B. Singh, and A. Guldhe. 2022. Biodiesel and an overview of waste utilization at the various production stages. In Waste and Biodiesel: Feedstocks and Precursors for Catalysts (pp. 1–16), Elsevier.

Ali, S. S., S. G. Mastropetros, M. Schagerl, M. Sakarika, T. Elsamahy, M. El-Sheekh, J. Z. Sun, and M. Kornaros. 2022. Recent advances in wastewater microalgae-based biofuels production: A state-of-the-art review, Energy Reports, 8: 13253–80.

Allakhverdiev, S. I., V. Thavasi, V. D. Kreslavski, S. K. Zharmukhamedov, V. V. Klimov, S. Ramakrishna, D. A. Los, M. Mimuro, H. Nishihara, and R. Carpentier. 2010. Photosynthetic hydrogen production, Journal of Photochemistry and Photobiology C-Photochemistry Reviews, 11: 101–13.

Ammar, E. E., A. A. A. Aioub, A. E. Elesawy, A. M. Karkour, M. S. Mouhamed, A. A. Amer, and N. A. El-Shershaby. 2022. Algae as bio-fertilizers: Between current situation and future prospective, Saudi Journal of Biological Sciences, 29: 3083–96.

Arimbrathodi, S. P., M. A. Javed, M. A. Hamouda, A. A. Hassan, and M. E. Ahmed. 2023. BioH production using microalgae: Highlights on recent advancements from a bibliometric analysis, Water, 15: 185–85.

Aslanbay Guler, B., I. Deniz, Z. Demirel, S. S. Oncel, and E. Imamoglu. 2020. Computational fluid dynamics modelling of stirred tank photobioreactor for *Haematococcus pluvialis* production: Hydrodynamics and mixing conditions, Algal Research, 47: 101854–54.

Baek, K., D. H. Kim, J. Jeong, S. J. Sim, A. Melis, J. S. Kim, E. Jin, and S. Bae. 2016. DNA-free two-gene knockout in *Chlamydomonas reinhardtii* via CRISPR-Cas9 ribonucleoproteins, Scientific Reports, 6: 30620.

Ban, S., W. Lin, Z. Luo, and J. Luo. 2019. Improving hydrogen production of *Chlamydomonas reinhardtii* by reducing chlorophyll content via atmospheric and room temperature plasma, Bioresource Technology, 275: 425–29.

Barboza-Rodriguez, R., R. M. Rodriguez-Jasso, G. Rosero-Chasoy, M. L. Rosales Aguado, and H. A. Ruiz. 2024. Photobioreactor configurations in cultivating microalgae biomass for biorefinery, Bioresource Technology, 394: 130208.

Batyrova, K., A. Gavrisheva, E. Ivanova, J. Liu, and A. Tsygankov. 2015. Sustainable hydrogen photoproduction by phosphorus-deprived marine green microalgae Chlorella sp, International Journal of Molecular Sciences, 16: 2705–16.

Batyrova, K., and P. C. Hallenbeck. 2017. Hydrogen production by a *Chlamydomonas reinhardtii* strain with inducible expression of photosystem II, International Journal of Molecular Sciences, 18: 647.

Ben-Zvi, O., E. Dafni, Y. Feldman, and I. Yacoby. 2019. Re-routing photosynthetic energy for continuous hydrogen production in vivo, Biotechnology for Biofuels and Bioproducts, 12: 266.

Benner, P., L. Meier, A. Pfeffer, K. Kruger, J. E. Oropeza Vargas, and D. Weuster-Botz. 2022. Lab-scale photobioreactor systems: Principles, applications, and scalability, Bioprocess and Biosystems Engineering, 45: 791–813.

Bingham, A. S., P. R. Smith, and J. R. Swartz. 2012. Evolution of an [FeFe] hydrogenase with decreased oxygen sensitivity, International Journal of Hydrogen Energy, 37: 2965–76.

Bouwer, E. J., and P. L. McCarty. 1983. Effects of 2-bromoethanesulfonic acid and 2- chloroethanesulfonic acid on acetate utilization in a continuous-flow methanogenic fixed-film column, Applied and Environmental Microbiology, 45: 1408–10.

Carullo, D., B. D. Abera, A. A. Casazza, F. Donsi, P. Perego, G. Ferrari, and G. Pataro. 2018. Effect of pulsed electric fields and high pressure homogenization on the aqueous extraction of intracellular compounds from the microalgae, Algal Research-Biomass Biofuels and Bioproducts, 31: 60–69.

Carvalho, A. P., S. O. Silva, J. M. Baptista, and F. X. Malcata. 2011. Light requirements in microalgal photobioreactors: An overview of biophotonic aspects, Applied Microbiology and Biotechnology, 89: 1275–88.

Chader, S., H. Hacene, and S. N. Agathos. 2009. Study of hydrogen production by three strains of isolated from the soil in the Algerian Sahara, International Journal of Hydrogen Energy, 34: 4941–46.

Chae, K. J., M. J. Choi, K. Y. Kim, F. F. Ajayi, I. S. Chang, and I. S. Kim. 2010. Selective inhibition of methanogens for the improvement of biohydrogen production in microbial electrolysis cells, International Journal of Hydrogen Energy, 35: 13379–86.

Chaganti, S. R., D. H. Kim, and J. A. Lalman. 2012. Dark fermentative hydrogen production by mixed anaerobic cultures: Effect of inoculum treatment methods on hydrogen yield, Renewable Energy, 48: 117–21.

Chandrasekhar, K., Y. J. Lee, and D. W. Lee. 2015. Biohydrogen production: Strategies to improve process efficiency through microbial routes, International Journal of Molecular Sciences, 16: 8266–93.

Chanquia, S. N., G. Vernet, and S. Kara. 2022. Photobioreactors for cultivation and synthesis: Specifications, challenges, and perspectives, Engineering in Life Sciences, 22: 712–24.

Cheirsilp, B., and S. Torpee. 2012. Enhanced growth and lipid production of microalgae under mixotrophic culture condition: Effect of light intensity, glucose concentration and fed-batch cultivation, Bioresource Technology, 110: 510–6.

Chini Zittelli, G., L. Rodolfi, N. Biondi, and M. R. Tredici. 2006. Productivity and photosynthetic efficiency of outdoor cultures of Tetraselmis suecica in annular columns, Aquaculture, 261: 932–43.

Chowdury, K. H., N. Nahar, and U. K. Deb. 2020. The growth factors involved in microalgae cultivation for biofuel production: A review, Computational Water, Energy, and Environmental Engineering, 09: 185–215.

da Silva, L., I. A. Donato, C. A. C. Goncalves, J. R. Scherf, H. S. Dos Santos, E. Mori, H. D. M. Coutinho, and F. A. B. da Cunha. 2023. Antibacterial potential of chalcones and its derivatives against Staphylococcus aureus, 3 Biotech, 13: 1.

Dębowski, M., M. Dudek, M. Zieliński, A. Nowicka, and J. Kazimierowicz. 2021. Microalgal hydrogen production in relation to other biomass-based technologies—A review, Energies, 14: 6025.

Deruyck, B., K. H. T. Nguyen, E. Decaestecker, and K. Muylaert. 2019. Modeling the impact of rotifer contamination on microalgal production in open pond, photobioreactor and thin layer cultivation systems, Algal Research-Biomass Biofuels and Bioproducts, 38: 101398–98.

Dixon, C., and L. R. Wilken. 2018. Green microalgae biomolecule separations and recovery, Bioresources and Bioprocessing, 5: 1–24.

Dudek, M., M. Dębowski, J. Kazimierowicz, M. Zieliński, P. Quattrocelli, and A. Nowicka. 2022. The cultivation of biohydrogen-producing tetraselmis subcordiformis microalgae as the third stage of dairy wastewater aerobic treatment system, Sustainability, 14: 12085.

Figueroa-Torres, G., E. B. Padilla, J. K. Pittman, and C. Theodoropoulos. 2020. Microalgae strain catalogue. In, 1–58. Zenodo.

Gadhe, A., S. S. Sonawane, and M. N. Varma. 2014. Ultrasonic pretreatment for an enhancement of biohydrogen production from complex food waste, International Journal of Hydrogen Energy, 39: 7721–29.

Gao, K., and K. R. Mckinley. 1994. Use of macroalgae for marine biomass production and CO2 remediation - a review, Journal of Applied Phycology, 6: 45–60.

Ghimire, A., L. Frunzo, L. Pontoni, G. d'Antonio, P. N. Lens, G. Esposito, and F. Pirozzi. 2015. Dark fermentation of complex waste biomass for biohydrogen production by pretreated thermophilic anaerobic digestate, J Environ Manage, 152: 43–8.

Giang, T. T., S. Lunprom, Q. Liao, A. Reungsang, and A. Salakkam. 2019. Improvement of hydrogen production from Chlorella sp. biomass by acid-thermal pretreatment, PeerJ, 7: e6637.

González-Camejo, J., A. Viruela, M. V. Ruano, R. Barat, A. Seco, and J. Ferrer. 2019. Effect of light intensity, light duration and photoperiods in the performance of an outdoor photobioreactor for urban wastewater treatment, Algal Research, 40: 101511–11.

Goswami, R. K., S. Mehariya, P. K. Obulisamy, and P. Verma. 2021. Advanced microalgae-based renewable biohydrogen production systems: A review, Bioresource Technology, 320: 124301.

Gudin, C., and C. Thępenier. 1986. Bioconversion of solar energy into organic chemicals by microalgae.

Guo, Y. P., S. H. Kim, S. H. Sung, and P. H. Lee. 2010. Effect of ultrasonic treatment of digestion sludge on bio-hydrogen production from sucrose by anaerobic fermentation, International Journal of Hydrogen Energy, 35: 3450–55.

Gupta, M., and H. Wakudkar. 2023. Microalgae harvesting strategies for biofuel production, EAI/Springer Innovations in Communication and Computing, Part F1482: 153–68.

Hamid Nour, A., A. A. Bin Mokaizh, M. Y. D. Alazaiza, M. J. K. Bashir, S. E. Mustafa, and A. O. Baarimah. 2024. Innovative strategies for microalgae-based bioproduct extraction in biorefineries: current trends and future solutions integrating wastewater treatment, Sustainability, 16: 10565–65.

He, Z. Q., Y. T. Qu, W. B. Jin, X. Zhou, W. Han, K. Song, S. H. Gao, Y. D. Chen, S. Y. Yin, and G. M. Jiang. 2022. Effect of microwave on biomass growth and oxygen production of microalgae cultured in real wastewater, Process Safety and Environmental Protection, 161: 22–33.

Hippler, M., and F. Khosravitabar. 2024. Light-driven H_2 production in *Chlamydomonas reinhardtii*: Lessons from engineering of photosynthesis, Plants (Basel), 13: 2114–14.

Hoh, D., S. Watson, and E. Kan. 2016. Algal biofilm reactors for integrated wastewater treatment and biofuel production: A review, Chemical Engineering Journal, 287: 466–73.

Hu, B., and S. L. Chen. 2007. Pretreatment of methanogenic granules for immobilized hydrogen fermentation, International Journal of Hydrogen Energy, 32: 3266–73.

Hwang, T., M. R. Kotte, J. I. Han, Y. K. Oh, and M. S. Diallo. 2015. Microalgae recovery by ultrafiltration using novel fouling-resistant PVDF membranes with in situ PEGylated polyethyleneimine particles, Water Research, 73: 181–92.

Hwang, T., S. J. Park, Y. K. Oh, N. Rashid, and J. I. Han. 2013. Harvesting of Chlorella sp. KR-1 using a cross-flow membrane filtration system equipped with an anti-fouling membrane, Bioresource Technology, 139: 379–82.

Indahsari, H. S., A. C. M. A. R. Tassakka, E. N. Dewi, M. Yuwono, and E. A. Suyono. 2022. Effects of salinity and bioflocculation during *Euglena* sp. harvest on the production of lipid, chlorophyll, and carotenoid with *Skeletonema* sp. as a bioflocculant, Journal of Pure and Applied Microbiology, 16: 2901–11.

Javed, M. A., A. M. Zafar, A. A. Hassan, A. A. Zaidi, M. Farooq, A. El Badawy, T. Lundquist, M. M. A. Mohamed, and S. Al-Zuhair. 2022. The role of oxygen regulation and algal growth parameters in hydrogen production via biophotolysis, Journal of Environmental Chemical Engineering, 10: 107003–03.

Jiang, J., W. Jin, R. Tu, S. Han, Y. Ji, and X. Zhou. 2021. Harvesting of microalgae chlorella pyrenoidosa by bio-flocculation with bacteria and filamentous fungi, Waste and Biomass Valorization, 12: 145–54.

Jiao, H., K. Tsigkou, T. Elsamahy, K. Pispas, J. Sun, G. Manthos, M. Schagerl, E. Sventzouri, R. Al-Tohamy, M. Kornaros, and S. S. Ali. 2024. Recent advances in sustainable hydrogen production from microalgae: Mechanisms, challenges, and future perspectives, Ecotoxicology and Environmental Safety, 270: 115908.

Jokel, M., V. Nagy, S. Z. Toth, S. Kosourov, and Y. Allahverdiyeva. 2019. Elimination of the flavodiiron electron sink facilitates long-term H_2 photoproduction in green algae, Biotechnology for Biofuels and Bioproducts, 12: 280.

Jung, K. W., D. H. Kim, and H. S. Shin. 2011. Fermentative hydrogen production from Laminaria japonica and optimization of thermal pretreatment conditions, Bioresource Technology, 102: 2745–50.

Kadir, W. N. A., M. K. Lam, Y. Uemura, J. W. Lim, P. L. Kiew, S. Lim, S. S. Rosli, C. Y. Wong, P. L. Show, and K. T. Lee. 2021. Simultaneous harvesting and cell disruption of microalgae using ozone bubbles: Optimization and characterization study for biodiesel production, Frontiers of Chemical Science and Engineering, 15: 1257–68.

Kamshybayeva, G. K., B. D. Kossalbayev, A. K. Sadvakasova, B. K. Zayadan, A. M. Bozieva, D. Dunikov, S. Alwasel, and S. I. Allakhverdiev. 2022. Strategies and economic feasibilities in cyanobacterial hydrogen production, International Journal of Hydrogen Energy, 47: 29661–84.

Kao, P. H., and I. S. Ng. 2017. CRISPRi mediated phosphoenolpyruvate carboxylase regulation to enhance the production of lipid in *Chlamydomonas reinhardtii*, Bioresource Technology, 245: 1527–37.

Khetkorn, W., R. P. Rastogi, A. Incharoensakdi, P. Lindblad, D. Madamwar, A. Pandey, and C. Larroche. 2017. Microalgal hydrogen production – A review, Bioresource Technology, 243: 1194–206.

Kim, E. J., C. H. Wu, M. W. Adams, and Y. P. Zhang. 2016. Exceptionally high rates of biological hydrogen production by biomimetic in vitro synthetic enzymatic pathways, Chemistry, 22: 16047–51.

Koley, S., S. Prasad, S. K. Bagchi, and N. Mallick. 2017. Development of a harvesting technique for large-scale microalgal harvesting for biodiesel production, RSC Advances, 7: 7227–37.

Lafarga, Tomás. 2019. Cultured microalgae and compounds derived thereof for food applications: Strain selection and cultivation, drying, and processing strategies, Food Reviews International, 36: 559–83.

Lee, J-H, D-G Lee, J-I Park, and J-Y Kim. 2010. Bio-hydrogen production from a marine brown algae and its bacterial diversity, Korean Journal of Chemical Engineering, 27: 187–92.

Li, H., L. Zhang, L. F. Shu, X. S. Zhuang, Y. M. Liu, J. Chen, and Z. L. Hu. 2015. Sustainable photosynthetic H-production mediated by artificial miRNA silencing of OEE2 gene in green alga, International Journal of Hydrogen Energy, 40: 5609–16.

Lu, Q., Y. Xiao, and F. Chen. 2021. Application of microalgae biotechnology for the sustainable development of aquaculture, Advances in Bioenergy, 6: 117–63.

Luo, G., L. Xie, Z. H. Zou, W. Wang, Q. Zhou, and H. J. Shim. 2010. Anaerobic treatment of cassava stillage for hydrogen and methane production in continuously stirred tank reactor (CSTR) under high organic loading rate (OLR), International Journal of Hydrogen Energy, 35: 11733–37.

Ma, W., C. Feng, F. Guan, D. Ma, and J. Cai. 2022. Effective *Chlorella vulgaris* biomass harvesting through sulfate and chloride flocculants, Journal of Marine Science and Engineering, 11: 47–47.

Maia, C., C. A. Sousa, H. Sousa, F. Vale, and M. Simões. 2023. Parabens removal from wastewaters by microalgae – Ecotoxicity, metabolism and pathways, Chemical Engineering Journal, 453: 139631–31.

Manoyan, J., L. Gabrielyan, N. Kozel, and A. Trchounian. 2019. Regulation of biohydrogen production by protonophores in novel green microalgae *Parachlorella kessleri*, J Photochem Photobiol B, 199: 111597.

Manu, Lusia, J. F. Mokolensang, W. B. Gunawan, A. Setyawardani, N. Salindeho, R. A. Syahputra, M. Iqhrammullah, and F. Nurkolis. 2024. Photobioreactors are beneficial for mass cultivation of microalgae in terms of areal efficiency, climate implications, and metabolites content, Journal of Agriculture and Food Research, 18: 101282–82.

Masojídek, J., K. Ranglová, G. E. Lakatos, A. M. S. Benavides, and G. Torzillo. 2021. Variables governing photosynthesis and growth in microalgae mass cultures, Processes, 9: 820–20.

McQuillan, J. L., A. J. Berndt, A. E. Sproles, S. P. Mayfield, and J. Pandhal. 2022. Novel cis-regulatory elements as synthetic promoters to drive recombinant protein expression from the *Chlamydomonas reinhardtii* nuclear genome, New Biotechnology, 68: 9–18.

Min Woon, J., K. S. Khoo, M. Akermi, M. M. Alanazi, J. W. Lim, Y. J. Chan, P. S. Goh, B. S. Chidi, M. K. Lam, J. Zaini, M. R. Bilad, Y. Zhou, N. T. Sahrin, and F. M. Ardo. 2023. Reviewing biohydrogen production from microalgal cells through fundamental mechanisms, enzymes and factors that engendering new challenges and prospects, Fuel, 346: 128312–12.

Moosburner, M. A., P. Gholami, J. K. McCarthy, M. Tan, V. A. Bielinski, and A. E. Allen. 2020. Multiplexed knockouts in the model diatom phaeodactylum by episomal delivery of a selectable Cas9, Frontiers in Microbiology, 11: 5.

Morales-Sánchez, D., O. A. Martinez-Rodriguez, and A. Martinez. 2017. Heterotrophic cultivation of microalgae: Production of metabolites of commercial interest, Journal of Chemical Technology & Biotechnology, 92: 925–36.

Morgado, D., A. Fanesi, T. Martin, S. Tebbani, O. Bernard, and F. Lopes. 2024. Exploring the dynamics of astaxanthin production in Haematococcus pluvialis biofilms using a rotating biofilm-based system, Biotechnology and Bioengineering, 121: 991–1004.

Nagarajan, D., J. S. Chang, and D. J. Lee. 2020. Pretreatment of microalgal biomass for efficient biohydrogen production - Recent insights and future perspectives, Bioresource Technology, 302: 122871.

Najjar, Y. S. H., and A. Abu-Shamleh. 2020. Harvesting of microalgae by centrifugation for biodiesel production: A review, Algal Research-Biomass Biofuels and Bioproducts, 51: 102046–46.

Navarro, E., A. Montagud, P. F. de Córdoba, and J. F. Urchueguía. 2009. Metabolic flux analysis of the hydrogen production potential in sp PCC6803, International Journal of Hydrogen Energy, 34: 8828–38.

Oey, M., I. L. Ross, E. Stephens, J. Steinbeck, J. Wolf, K. A. Radzun, J. Kugler, A. K. Ringsmuth, O. Kruse, and B. Hankamer. 2013. RNAi knock-down of LHCBM1, 2 and 3 increases photosynthetic H_2 production efficiency of the green alga Chlamydomonas reinhardtii, PLoS ONE, 8: e61375.

Omer, S. N., P. Saravanan, P. Kumar, R. R. Kannan, M. Rajasimman, and V. Shanmugam. 2025. Insights into renewable biohydrogen production from algal biomass: Technical hurdles and economic analysis, Biomass Conversion and Biorefinery, 15: 3281–300.

Pal, D., S. Banerjee, S. Chandra, D. Das, S. Pandit, A. Roy, M. Hasan, M. U. Khandaker, and M. A. Alreshidi. 2025. The generation of biohydrogen from pretreated algal biomass in batch fermentation mode, International Journal of Hydrogen Energy, 107: 321–31.

Patel, V. K., A. Das, R. Kumari, and S. Kajla. 2023. Recent progress and challenges in CRISPR-Cas9 engineered algae and cyanobacteria, Algal Research-Biomass Biofuels and Bioproducts, 71: 103068–68.

Patel, V. K., N. Soni, V. Prasad, A. Sapre, S. Dasgupta, and B. Bhadra. 2019. CRISPR-Cas9 system for genome engineering of photosynthetic microalgae, Molecular Biotechnology, 61: 541–61.

Peter, A. P., A. K. Koyande, K. W. Chew, S.-H. Ho, W.-H. Chen, J.-S. Chang, R. Krishnamoorthy, F. Banat, and P. L. Show. 2022. Continuous cultivation of microalgae in photobioreactors as a source of renewable energy: Current status and future challenges, Renewable and Sustainable Energy Reviews, 154: 111852–52.

Poliner, E., J. A. Pulman, K. Zienkiewicz, K. Childs, C. Benning, and E. M. Farre. 2018. A toolkit for Nannochloropsis oceanica CCMP1779 enables gene stacking and genetic engineering of the eicosapentaenoic acid pathway for enhanced long-chain polyunsaturated fatty acid production, Plant Biotechnology Journal, 16: 298–309.

Radha, M., and A. G. Murugesan. 2017. Enhanced dark fermentative biohydrogen production from marine macroalgae Padina tetrastromatica by different pretreatment processes, Biofuel Research Journal, 4: 551–58.

Rajendran, N., B. Gurunathan, J. Han, S. Krishna, A. Ananth, K. Venugopal, and R. B. Sherly Priyanka. 2021. Recent advances in valorization of organic municipal waste into energy using biorefinery approach, environment and economic analysis, Bioresource Technology, 337: 125498.

Randrianarison, G., and M. A. Ashraf. 2017. Microalgae: A potential plant for energy production, Geology, Ecology, and Landscapes, 1: 104–20.

Rashid, N., K. Lee, and Q. Mahmood. 2011. Bio-hydrogen production by *Chlorella vulgaris* under diverse photoperiods, Bioresource Technology, 102: 2101–4.

Redding, K. E., J. Appel, M. Boehm, W. Schuhmann, M. M. Nowaczyk, I. Yacoby, and K. Gutekunst. 2022. Advances and challenges in photosynthetic hydrogen production, Trends in Biotechnology, 40: 1313–25.

Roy, S., K. Kumar, S. Ghosh, and D. Das. 2014. Thermophilic biohydrogen production using pre-treated algal biomass as substrate, Biomass & Bioenergy, 61: 157–66.

Ruiz-Marin, A., Y. Canedo-Lopez, and P. Chavez-Fuentes. 2020. Biohydrogen production by *Chlorella vulgaris* and *Scenedesmus obliquus* immobilized cultivated in artificial wastewater under different light quality, AMB Express, 10: 191.

Rumpel, S., J. F. Siebel, C. Fares, J. F. Duan, E. Reijerse, T. Happe, W. Lubitz, and M. Winkler. 2014. Enhancing hydrogen production of microalgae by redirecting electrons from photosystem I to hydrogenase, Energy & Environmental Science, 7: 3296–301.

Satheesh, S., A. Pugazhendi, B. A. Al-Mur, and R. Balasubramani. 2023. Biohydrogen production coupled with wastewater treatment using selected microalgae, Chemosphere, 334: 138932.

Sathinathan, P., H. M. Parab, R. Yusoff, S. Ibrahim, V. Vello, and G. C. Ngoh. 2023. Photobioreactor design and parameters essential for algal cultivation using industrial wastewater: A review, Renewable and Sustainable Energy Reviews, 173: 113096–96.

Schroda, M., C. F. Beck, and O. Vallon. 2002. Sequence elements within an HSP70 promoter counteract transcriptional transgene silencing in Chlamydomonas, The Plant Journal, 31: 445–55.

Scranton, M. A., J. T. Ostrand, D. R. Georgianna, S. M. Lofgren, D. Li, R. C. Ellis, D. N. Carruthers, A. Dräger, D. L. Masica, and S. P. Mayfield. 2016. Synthetic promoters capable of driving robust nuclear gene expression in the green alga, Algal Research-Biomass Biofuels and Bioproducts, 15: 135–42.

Seo, J. Y., K. Lee, R. Praveenkumar, B. Kim, S. Y. Lee, Y.-K. Oh, and S. B. Park. 2015. Tri-functionality of Fe_3O_4-embedded carbon microparticles in microalgae harvesting, Chemical Engineering Journal, 280: 206–14.

Sethunge, S. S., P. M. Manage, and S. Jayasinghe. 2012. Microstrainer for removal of algae in drinking water treatment, in *Proceedings of International Forestry and Environment Symposium*, 14. University of Sri Jayewardenepura, Sri Lanka.

Shakeel Syed, M., M. Rafeie, R. Henderson, D. Vandamme, M. Asadnia, and M. Ebrahimi Warkiani. 2017. A 3D-printed mini-hydrocyclone for high throughput particle separation: Application to primary harvesting of microalgae, Lab Chip, 17: 2459–69.

Shetty, P., I. Z. Boboescu, B. Pap, R. Wirth, K. L. Kovács, T. Bíró, Z. Futó, R. A. White, and G. Maróti. 2019. Exploitation of algal-bacterial consortia in combined biohydrogen generation and wastewater treatment, Frontiers in Energy Research, 7.

Shin, S. E., J. M. Lim, H. G. Koh, E. K. Kim, N. K. Kang, S. Jeon, S. Kwon, W. S. Shin, B. Lee, K. Hwangbo, J. Kim, S. H. Ye, J. Y. Yun, H. Seo, H. M. Oh, K. J. Kim, J. S. Kim, W. J. Jeong, Y. K. Chang, and B. R. Jeong. 2016. CRISPR/Cas9-induced knockout and knock-in mutations in Chlamydomonas reinhardtii, Scientific Reports, 6: 27810.

Show, K. Y., Y. Yan, C. Zong, N. Guo, J. S. Chang, and D. J. Lee. 2019. State of the art and challenges of biohydrogen from microalgae, Bioresource Technology, 289: 121747.

Siitonen, V., A. Probst, G. Tóth, R. Kourist, M. Schroda, S. Kosourov, and Y. Allahverdiyeva. 2023. Engineered green alga as a whole-cell photosynthetic biocatalyst for stepwise photoproduction of H and ε-caprolactone, Green Chemistry, 25: 5945–55.

Singh, H., S. Tomar, K. A. Qureshi, M. Jaremko, and P. K. Rai. 2022. Recent advances in biomass pretreatment technologies for biohydrogen production, Energies, 15: 999–99.

Srivastava, N., M. Srivastava, P. K. Mishra, M. A. Kausar, M. Saeed, V. K. Gupta, R. Singh, and P. W. Ramteke. 2020. Advances in nanomaterials induced biohydrogen production using waste biomass, Bioresource Technology, 307: 123094.

Su, Y. 2021. Revisiting carbon, nitrogen, and phosphorus metabolisms in microalgae for wastewater treatment, Science of the Total Environment, 762: 144590.

Sulfahri, S. M., A. Langford, and A. C. M. A. R. Tassakka. 2020. Ozonolysis as an effective pretreatment strategy for bioethanol production from marine algae, Bioenergy Research, 13: 1269–79.

Sun, Y. L., M. Chen, H. M. Yang, J. Zhang, T. Y. Kuang, and F. Huang. 2013. Enhanced H photoproduction by down-regulation of ferredoxin-NADP reductase (FNR) in the green alga, International Journal of Hydrogen Energy, 38: 16029–37.

Sundaram, T., S. Rajendran, L. Gnanasekaran, N. Rachmadona, J. J. Jiang, K. S. Khoo, and P. L. Show. 2023. Bioengineering strategies of microalgae biomass for biofuel production: Recent advancement and insight, Bioengineered, 14: 2252228.

Tan, J. S., S. Y. Lee, K. W. Chew, M. K. Lam, J. W. Lim, S. H. Ho, and P. L. Show. 2020. A review on microalgae cultivation and harvesting, and their biomass extraction processing using ionic liquids, Bioengineered, 11: 116–29.

Tang, T., Y. Chen, M. Liu, Y. Du, and Y. Tan. 2022. Effect of pH on the performance of hydrogen production by dark fermentation coupled denitrification, Environmental Research, 208: 112663.

Tayebati, H., F. P. Shariati, N. Soltani, and H. Sepasi Tehrani. 2024. Effect of various light spectra on amino acids and pigment production of Arthrospira platensis using flat-plate photobioreactor, Preparative Biochemistry & Biotechnology, 54: 1028–39.

Tinpranee, N., A. Incharoensakdi, and S. Phunpruch. 2016. Hydrogen production by unicellular green alga chlorella sp. LSD-W2 isolated from seawater in Thailand, Asia-Pacific Journal of Science and Technology, 21: 256–66.

Tinpranee, N., A. Incharoensakdi, and S. Phunpruch. 2018. Screening cyanobacteria from marine coastal waters of Thailand for biohydrogen production, Journal of Applied Phycology, 30: 3471–81.

Valeriano Gonzalez, M. T., M. T. Orta Ledesma, S. B. Velasquez-Orta, and I. Monje Ramirez. 2021. Harvesting microalgae using ozone-air flotation for recovery of biomass, lipids, carbohydrates, and proteins, Environ Technol, 42: 3267–77.

Vandamme, D., I. Foubert, and K. Muylaert. 2013. Flocculation as a low-cost method for harvesting microalgae for bulk biomass production, Trends in Biotechnology, 31: 233–9.

Vargas, S. R., P. V. D. Santos, L. A. Giraldi, M. Zaiat, and M. D. C. Calijuri. 2018. Anaerobic phototrophic processes of hydrogen production by different strains of microalgae Chlamydomonas sp, FEMS Microbiology Letters, 365.

Vasistha, S., A. Khanra, M. Clifford, and M. P. Rai. 2021. Current advances in microalgae harvesting and lipid extraction processes for improved biodiesel production: A review, Renewable & Sustainable Energy Reviews, 137: 110498–98.

Veeravalli, S. S., S. R. Shanmugam, S. Ray, J. A. Lalman, and N. Biswas. 2019. Biohydrogen production from renewable resources, in Advanced Bioprocessing for Alternative Fuels, Biobased Chemicals, and Bioproducts: Technologies and Approaches for Scale-Up and Commercialization (pp. 289–312). Woodhead Publishing.

Velten, H., D. Krahe, N. Hasport, T. Fundneider, U. Grabbe, L. Knorr, and U. Theilen. 2024. Pile cloth media filtration for harvesting microalgae used for wastewater treatment, Fermentation-Basel, 10: 325–25.

Venkata Subhash, G., M. Rajvanshi, G. R. K. Kumar, U. S. Sagaram, V. Prasad, S. Govindachary, and S. Dasgupta. 2022. Challenges in microalgal biofuel production: A perspective on techno economic feasibility under biorefinery stratagem, Bioresource Technology, 343: 126155.

Wang, H., X. Fan, Y. Zhang, D. Yang, and R. Guo. 2011. Sustained photo-hydrogen production by *Chlorella pyrenoidosa* without sulfur depletion, Biotechnology Letters, 33: 1345–50.

Wang, J., and Y. Yin. 2018. Fermentative hydrogen production using pretreated microalgal biomass as feedstock, Microbial Cell Factories, 17: 22.

Wang, Y., X. Zhuang, M. Chen, Z. Zeng, X. Cai, H. Li, and Z. Hu. 2018. An endogenous microRNA (miRNA1166.1) can regulate photobio-H_2 production in eukaryotic green alga Chlamydomonas reinhardtii, Biotechnology for Biofuels and Bioproducts, 11: 126.

Wang, Z. W., J. Hou, D. Bowden, and J. M. Belovich. 2014. Evaluation of an inclined gravity settler for microalgae harvesting, Journal of Chemical Technology and Biotechnology, 89: 714–20.

Wei, L. Z., and W. M. Ma. 2024. Photosynthetic H_2 production: Lessons from the regulation of electron transfer in microalgae, Global Change Biology Bioenergy, 16: e13118.

Wei, N., J. Quarterman, and Y. S. Jin. 2013. Marine macroalgae: An untapped resource for producing fuels and chemicals, Trends in Biotechnology, 31: 70–7.

Weiner, I., N. Shahar, Y. Feldman, S. Landman, Y. Milrad, O. Ben-Zvi, M. Avitan, E. Dafni, S. Schweitzer, H. Eilenberg, S. Atar, A. Diament, T. Tuller, and I. Yacoby. 2018. Overcoming the expression barrier of the ferredoxin-hydrogenase chimera in supports a linear increment in photosynthetic hydrogen output, Algal Research-Biomass Biofuels and Bioproducts, 33: 310–15.

Wen, H., X. Zou, K. Xu, Z. Shen, X. Ren, and Y. Li. 2019. Buoy-bead flotation application for the harvesting of microalgae and mechanistic analysis of significant factors, Bioprocess and Biosystems Engineering, 42: 391–400.

Wu, H. 2016. Effect of different light qualities on growth, pigment content, chlorophyll fluorescence, and antioxidant enzyme activity in the red alga Pyropia haitanensis (Bangiales, Rhodophyta), BioMed Research International, 2016: 7383918.

Xie, B., J. Cheng, J. Zhou, W. Song, J. Liu, and K. Cen. 2008. Production of hydrogen and methane from potatoes by two-phase anaerobic fermentation, Bioresource Technology, 99: 5942–6.

Xiong, D. K., T. Happe, B. Hankamer, and I. L. Ross. 2021. Inducible high-level expression of a variant δD19A, D58A-ferredoxin-hydrogenase fusion increases photohydrogen production efficiency in the green alga Chlamydomonas reinhardtii, Algal Research-Biomass Biofuels and Bioproducts, 55: 102275–75.

Xuan, J., L. He, W. Wen, and Y. Feng. 2023. Hydrogenase and nitrogenase: Key catalysts in biohydrogen production, Molecules, 28: 1392–92.

Ye, C. P., M. C. Zhang, Y. F. Yang, and G. Thirumaran. 2012. Photosynthetic performance in aquatic and terrestrial colonies of (Cyanophyceae) under aquatic and aerial conditions, Journal of Arid Environments, 85: 56–61.

Yu, B. S., S. Y. Lee, and S. J. Sim. 2022. Effective contamination control strategies facilitating axenic cultivation of Haematococcus pluvialis: Risks and challenges, Bioresource Technology, 344: 126289.

Yu, B. S., S. Pyo, J. Lee, and K. Han. 2024. Microalgae: A multifaceted catalyst for sustainable solutions in renewable energy, food security, and environmental management, Microbial Cell Factories, 23: 308.

Zabed, H. M., S. Akter, J. H. Yun, G. Y. Zhang, F. N. Awad, X. H. Qi, and J. N. Sahu. 2019. Recent advances in biological pretreatment of microalgae and lignocellulosic biomass for biofuel production, Renewable & Sustainable Energy Reviews, 105: 105–28.

Zhang, X., P. Amendola, J. C. Hewson, M. Sommerfeld, and Q. Hu. 2012. Influence of growth phase on harvesting of Chlorella zofingiensis by dissolved air flotation, Bioresource Technology, 116: 477–84.

Zhang, X., J. C. Hewson, P. Amendola, M. Reynoso, M. Sommerfeld, Y. Chen, and Q. Hu. 2014. Critical evaluation and modeling of algal harvesting using dissolved air flotation, Biotechnology and Bioengineering, 111: 2477–85.

Zidan, A., M. A. Javed, and A. A. Hassan. 2024. Optimization of hydrogen production using a coculture of Chlamydomonas reinhardtii and activated sludge bacteria, Chemosphere, 369: 143789.

12 Unlocking the Potential of Microalgae for Sustainable Biohydrogen Production

Nikita Gupta, Deepshikha Gupta,
Pradeep Kumar Singh, and Rajveer Singh Chauhan

12.1 INTRODUCTION

12.1.1 Background on Bio-hydrogen Production

The process of producing hydrogen gas (H_2) through biological processes is known as "biohydrogen production," which provides a sustainable and eco-friendly substitute for traditional hydrogen production techniques (Ntaikou et al., 2010). In contrast to conventional methods like steam methane reforming or electrolysis, which are dependent on non-renewable resources and need a lot of energy, biohydrogen is produced from organic materials by biological processes like photosynthesis or fermentation. These methods are appealing in the global transition to greener energy solutions since they not only use renewable resources but also help to reduce greenhouse gas emissions (Ramprakash et al., 2022).

The flexibility of biohydrogen generation to use a variety of organic feedstocks that are otherwise regarded as trash is one of the main reasons it has gained momentum. These feedstocks include organic waste from industries, wastewater, and agricultural wastes (Singh & Dhar 2020). These organic molecules are converted into hydrogen gas by microorganisms, such as bacteria, archaea, and microalgae, through processes like fermentation or photosynthesis. This type of waste-to-energy recovery helps manage trash disposal, which promotes environmental sustainability, in addition to meeting the requirement

DOI: 10.1201/9781003568612-12

for renewable energy. By converting waste materials into useful energy resources, the method closes the loop in a circular bioeconomy, minimizes the environmental impact, and lowers greenhouse gas emissions (Wang et al., 2021).

There are various categories of biological hydrogen production processes, such as photofermentation, biophotolysis, and dark fermentation. In dark fermentation, anaerobic microbes use organic substrates like carbohydrates or organic acids to create hydrogen in the absence of light. The method's adaptability to different environments and very straightforward process requirements have attracted interest (Bhagwat et al., 2024). The inadequate conversion of substrates, which frequently results in the creation of byproducts such as organic acids and alcohols, is one of the drawbacks of dark fermentation. Because of this inefficiency, which lowers the total output of hydrogen, researchers are looking into hybrid systems that combine dark fermentation with other processes to increase yield and efficiency (Kumar & Singh, 2020).

Another viable method for producing biohydrogen is photofermentation, which uses light energy to transform organic compounds into hydrogen using photosynthetic bacteria. In contrast to dark fermentation, photofermentation takes place in the presence of light and can use a variety of organic substrates because the bacteria involved, such as purple non-sulfur bacteria, are able to do so (Chandrasekhar & Anuprasanna, 2023). Compared to dark fermentation, photofermentation has the benefit of creating hydrogen more effectively and with fewer byproducts. Nevertheless, this approach is very reliant on light availability, which may restrict its scalability and usefulness, particularly in areas with little sunlight. Furthermore, it can be expensive to maintain ideal lighting conditions for large-scale photofermentation, which calls for greater investigation into more economical lighting options or different photo-bioreactor designs (Lee et al., 2019).

Biophotolysis, which uses light energy to divide water molecules into hydrogen and oxygen in microalgae or cyanobacteria, is the third important biological mechanism for producing hydrogen. With just three inputs—light, water, and CO_2—this process can produce hydrogen in an extremely clean and sustainable manner, much like natural photosynthesis. Because of their high photosynthetic efficiency, adaptability to a variety of conditions, and capacity to sequester carbon dioxide during the hydrogen production process, microalgae in particular have garnered attention as a dual-benefit system for energy generation and carbon sequestration (Chandrasekhar et al., 2015). Biophotolysis is still in its infancy, though, and there are still obstacles to overcome, such as low rates of hydrogen production, the sensitivity of the enzymes involved, and the requirement for microalgae strains that have been tuned for optimal performance in practical settings (Zhou & Fang, 2020).

The ability of biohydrogen production to be integrated with waste management systems is one of its main benefits. For instance, organic waste can be converted into hydrogen and the water can be treated at the same time by including microorganisms that produce hydrogen into wastewater treatment plants (Velmozhina et al., 2023). This integration increases the sustainability and economic viability of the wastewater treatment process by producing a valuable energy resource while also offsetting the energy expenses normally associated with the operation. These integrated systems, which turn trash into clean energy and useful byproducts like treated water, are very appealing to those who want to create circular bioeconomies, which decrease waste generation and maximize resource efficiency (Wang et al., 2021).

Biohydrogen production has a lot of promise, but before it can be extensively used, a few obstacles need to be removed. The low efficiency of hydrogen synthesis in many biological processes is one of the main obstacles. For example, photofermentation systems are limited by their reliance on light, while the hydrogen yields from dark fermentation are sometimes restricted by the partial breakdown of substrates (Vaishnav et al., 2023). Furthermore, the sensitivity of many of the microorganisms involved in the synthesis of biohydrogen to environmental factors including pH, temperature, and substrate concentration can further restrict the scalability and efficiency of these processes. It will take developments in metabolic engineering, bioprocess optimization, and the creation of resilient microbial strains with increased hydrogen yields under industrial settings to overcome these obstacles (Ramprakash et al., 2022).

The high cost of infrastructure and operational requirements is another significant barrier to the generation of biohydrogen (Yadav et al., 2021). The development of specialized bioreactors, microorganism-optimized growth medium, and mechanisms for maintaining optimal environmental conditions are necessary for the large-scale production of biohydrogen (Boboescu et al., 2014). These expenses make the generation of biohydrogen less economically viable when compared to conventional techniques, especially when considering the relatively low hydrogen yields currently feasible with current technology. However, the prices of producing biohydrogen are anticipated to drop as research and technology progress, making it a more attractive option going forward (Kumar & Singh, 2020).

Apart from the scientific obstacles, the commercial feasibility of producing biohydrogen is also heavily influenced by regulatory and legislative frameworks. Subsidies, tax breaks, and research money are examples of government policies that promote the development of renewable energy and can give biohydrogen technologies the boost they need to go from lab trials to commercial implementations. Making biohydrogen a major force in the world energy scene will also require the development of hydrogen markets and infrastructure for distribution and storage. Biohydrogen technology development is already at the forefront in nations like Germany and Japan, which have robust renewable energy policies and investments in hydrogen infrastructure (Wang et al., 2021).

The generation of biohydrogen has a bright future ahead of it, but overcoming current obstacles will need a coordinated effort from academics, decision-makers in government, and business executives (Oey et al., 2016). The efficiency of microorganisms that produce hydrogen may be increased by biotechnology advancements such as genetic engineering and synthetic biology, while production costs may be reduced by advances in bioreactor design and process integration. Realizing the full potential of this technology will also depend on the creation of circular bioeconomy models that combine the production of biohydrogen with other sustainable practices like waste management. Biohydrogen is anticipated to play a bigger part in the move toward greener, more sustainable energy systems as the world's energy transition progresses. It provides a way to cut carbon emissions while opening up business prospects in the green energy industry (Zhou & Fang, 2020).

12.2 ROLE OF MICROALGAE IN BIOHYDROGEN PRODUCTION

Microalgae play a crucial role in biohydrogen production due to their high photosynthetic efficiency, ability to thrive in diverse conditions, and capacity for carbon sequestration (Subhash & Mohan 2019). These microorganisms utilize sunlight to convert water into hydrogen through biological processes such as direct photolysis and photofermentation (Kumar & Singh, 2020). A key advantage of microalgae-based hydrogen production is its dual benefit: renewable energy generation and CO_2 absorption, making it an environmentally sustainable option (Lee et al., 2019).

In addition to producing biohydrogen, microalgae-based systems can integrate with waste management processes, utilizing organic waste streams like industrial byproducts, wastewater, and agricultural residues. This not only enhances sustainability but also lowers production costs by repurposing waste materials (Wang et al., 2021 & Touloupakis et al., 2021).

Biohydrogen production through microalgae involves three primary biological pathways:

1. **Biophotolysis**: Uses light energy to split water molecules into hydrogen and oxygen, mimicking photosynthesis. While highly sustainable, it faces challenges related to low hydrogen yields and enzyme sensitivity (Zhou & Fang, 2020).
2. **Photofermentation**: Uses photosynthetic bacteria to convert organic substrates into hydrogen. It offers higher yields than dark fermentation but is light-dependent, affecting scalability (Lee et al., 2019).

3. **Dark Fermentation**: Anaerobic microbes break down organic substrates to produce hydrogen without light. Though simple and flexible, incomplete substrate breakdown leads to lower yields (Kumar & Singh, 2020).

12.2.1 Economic and Technological Considerations

The feasibility of large-scale biohydrogen production depends on several factors, including costs, scalability, and efficiency (Table 12.1). While co-producing biofuels and bioplastics improves profitability, high initial investments remain a challenge (Gendy & El-Temtamy 2013). Revenue generation through hydrogen sales, carbon credits, and byproduct utilization can enhance long-term viability (Gholkar et al., 2021).

12.2.2 Challenges in Microalgae-Based Biohydrogen Production

Despite its potential, microalgae-based hydrogen production faces significant challenges that hinder scalability and cost-effectiveness (Table 12.2). Key limitations include (Nagarajan et al., 2021 & Sharma et al., 2021)[]:

TABLE 12.1 Role of microalgae in biohydrogen production

FACTOR	*DESCRIPTION*	*IMPACT ON BIOHYDROGEN PRODUCTION*
Photosynthetic efficiency	Converts sunlight into energy for hydrogen production	Enhances hydrogen yield
Carbon sequestration	Absorbs CO_2, reducing greenhouse gases	Supports sustainable production
Multiple product approaches	Co-generation of biofuels, bioplastics, and hydrogen	Boosts economic feasibility
Waste utilization	Uses organic waste and wastewater for growth	Reduces production costs
Scalability challenges	Requires optimization for large-scale operations	Impacts commercial viability

TABLE 12.2 Challenges in microalgae-based biohydrogen production

BOTTLENECK	*DESCRIPTION*	*IMPACT ON BIOHYDROGEN PRODUCTION*
Hydrogenase enzyme sensitivity	Oxygen sensitivity limits hydrogen synthesis efficiency	Reduces overall hydrogen yield
High energy demand	Energy-intensive processes, especially for cultivation	Increases operational costs
Bioreactor design limitations	Inefficient designs lead to low productivity and scalability	Restricts large-scale adoption
Substrate availability	Limited nutrient sources affect microalgae growth	Impacts continuous hydrogen production
Process optimization challenges	Requires precise conditions for maximum hydrogen yield	Limits efficiency and feasibility

- **Hydrogenase Enzyme Sensitivity**: Oxygen exposure reduces hydrogen yield
- **High Energy Demand**: Cultivation and processing are energy-intensive.
- **Bioreactor Inefficiencies**: Poor designs limit productivity and scalability.
- **Substrate Availability**: Nutrient limitations can impact continuous production.
- **Process Optimization**: Precise environmental conditions are needed for efficiency.

12.3 STATE-OF-THE-ART IN MICROALGAE-BASED HYDROGEN PRODUCTION

The production of hydrogen from microalgae has gained momentum as a sustainable energy source since it can produce carbon-neutral biohydrogen. The main goals of current developments are to integrate hybrid systems, optimize growing conditions, and increase efficiency through genetic engineering. Important advancements consist of:

1. **Bioengineering of Microalgae Strains**: Significant promise has been shown by genetic alterations to boost the activity of the hydrogenase enzyme, control photosynthetic pathways, and improve the efficiency of light-to-hydrogen conversion. Methods like CRISPR-Cas9 are employed to block oxygen-sensitive pathways that hinder the synthesis of hydrogen (Velmozhina et al., 2023).
2. **Optimization of Cultivation Systems**: Improvements are being made to photobioreactors (PBRs) to improve their ability to distribute light, absorb CO_2, and deliver nutrients. Under controlled conditions, PBRs that are flat-plate or tubular are commonly used to improve microalgae growth and hydrogen output (Srirangan & Young, 2019 & Ahmed et al., 2022).
3. **Hybrid Systems for Enhanced Yield**: In order to more effectively capture light energy, microalgae are being combined with other renewable technologies, such as solar photovoltaic (PV) systems. The goal of these systems is to stabilize the environment for continuous hydrogen generation while addressing the sporadic nature of solar energy.
4. **Co-cultivation and Symbiotic Relationships**: It has been demonstrated that co-cultivating microalgae with bacteria or other microbes can boost the total hydrogen output. The symbiotic relationship aids in lowering oxygen buildup, which is frequently an inhibitor of the process that produces hydrogen (Ghimire et al., 2017).
5. **Metabolic Pathway Engineering**: Thanks to developments in the understanding of microalgae metabolism, energy fluxes can now be redirected toward hydrogen synthesis. The goals of pathway engineering are to increase ATP synthesis for hydrogenase activity and decrease energy losses in the photosynthetic chain (Ahmad et al., 2024).
6. **Bioreactor Scaling and Commercialization Challenges**: While successful experiments with small-scale laboratory setups have been shown, the industrialization of these technologies is still a hurdle. For commercial viability, problems including bioreactor cost, pollution, and nutrient supply efficiency must be resolved (Goswami et al., 2021 & Li et al., 2022)[].

12.4 FUTURE PROSPECTS AND R&D CHALLENGES

12.4.1 Future Prospects

Integrating future microalgae-based hydrogen production systems with other renewable energy sources, such as solar and wind power, is anticipated to yield considerable benefits. The main benefit of this integration is that it can help overcome solar energy's intermittent nature, perhaps allowing for continuous hydrogen production. Greater dependability and efficiency will be provided by hybrid systems that integrate solar, wind, and microalgae technologies. This will result in a more resilient framework for energy generation that can function in all-weather scenarios (Yaashikaa et al., 2022).

The creation of multi-product systems is a possible avenue to achieve the economic viability of microalgae-based hydrogen production. Microalgae in these systems would yield high-value byproducts like biofuels, medicines, and bioplastics in addition to hydrogen. By minimizing waste and obtaining numerous products from a single source, the circular economy method has the potential to lower production costs and provide new revenue streams (Chisti, 2018). This would make the commercialization of microalgae-based systems more feasible and attractive to investors.

Automated bioreactors will probably become more common in the future of large-scale microalgae hydrogen generation. With real-time sensors and artificial intelligence (AI)-powered platforms, these cutting-edge systems will be able to continuously optimize growing conditions by modifying elements like light, temperature, and fertilizer supply for optimal effectiveness (Okeke et al., 2022). Automation will increase overall yields while decreasing the risk of contamination and operational faults by reducing the need for human intervention. This development in technology has the potential to greatly increase the scalability and affordability of hydrogen production using microalgae (Banu et al., 2020).

The development of synthetic biology and bioengineering will be crucial to increasing the effectiveness of microalgae-based hydrogen production. Developments in these areas may result in the production of microalgae strains with enhanced photosynthetic efficiency and the capacity to generate hydrogen in conditions rich in oxygen through the development of hydrogenase enzymes that are tolerant of oxygen. These genetically modified strains might assist in overcoming the present constraints in the generation of biohydrogen and greatly increase hydrogen yields (Faraloni & Torzillo 2023).

The production of hydrogen from microalgae needs to overcome a number of obstacles before it can be widely commercialized. It will be essential to make significant infrastructure investments, such as creating sophisticated bioreactors and supply lines. Furthermore, government policies that are favorable as well as technological advancements will be necessary to guarantee that microalgae-based systems can rival more traditional hydrogen generation techniques. Microalgae-based hydrogen generation has the potential to play a significant role in the sustainable energy landscape as attention turns to lowering carbon emissions and implementing green energy solutions on a global scale, but it will need ongoing innovation and support (Ahmed et al., 2022).

12.4.2 R&D Challenges

1. **Low Hydrogen Yield**: The primary obstacle in the generation of hydrogen from microalgae is its comparatively low yield in contrast to other renewable energy sources. The oxygen sensitivity of the hydrogenase enzyme, which produces hydrogen, is quite high, and photosynthetic efficiency is not suited for large-scale production. Research and development is concentrated on enhancing microalgae's photosynthetic efficiency, creating genetically resistant strains of the organism, and maximizing hydrogenase activity to boost production (Al-Dailami et al., 2022).
2. **Bioreactor Design and Efficiency**: Although PBR technology is a promising field, there are several obstacles to overcome, including inadequate light dispersion, subpar CO_2 absorption, and difficulty with scaling up. Innovative PBR designs, such as flat-plate, tubular, and hybrid reactors, are the focus of R&D efforts because they improve light penetration and nutrient delivery while lowering the danger of contamination (Woon et al., 2023).
3. **Resource and Energy Demands**: Growing microalgae need a lot of water, nutrients, and energy. For large-scale production, managing these resources responsibly is a major task. Researchers are working on low-energy harvesting methods and recycling strategies to minimize the amount of water and nutrients used.

4. **Oxygen Inhibition**: One of the biggest obstacles still standing in the way of hydrogen generation is its sensitivity to oxygen. Oxygen buildup during photosynthesis inhibits the hydrogenase enzyme, hence halting the generation of hydrogen. Research is still being done on creating novel strains of microalgae and altering their metabolic pathways to reduce or tolerate oxygen interference.

12.5 COMMERCIALIZATION ASPECTS

1. **High Production Costs:** Commercialization is challenging due to the high cost of the infrastructure needed for hydrogen production and microalgae growing, particularly PBRs. While R&D has made significant progress in streamlining production processes, lower operating and capital costs (CapEx and OpEx) are still necessary for economic viability. For large-scale commercialization, improvements are required in the areas of byproduct valorization, biomass processing, and cost-effective harvesting (Singh & Gu, 2010).
2. **Lack of Industrial-Scale Demonstrations:** There aren't many large-scale demonstrations of microalgae-based hydrogen generation systems; most are still operated at the pilot or lab scale. Maintaining consistent growth conditions and preventing contamination are only two of the many technical and financial obstacles that come with scaling up from controlled laboratory conditions to industrial levels.
3. **Policy and Market Support:** Microalgae-based hydrogen production is not yet commercially viable without significant policy incentives or subsidies. Research funding, carbon credits, and green hydrogen-promoting policies from the government can be quite helpful in bridging the gap between R&D and commercial acceptance. The market for green hydrogen is growing, while more conventional processes like electrolysis and steam methane reforming continue to face fierce competition (Su et al., 2017).
4. **Economic Bottlenecks:** Financial barriers including expensive inputs (nutrients, light sources, PBR materials) and intricate downstream processing techniques make the commercialization of microalgae-based hydrogen production difficult, even with advancements in technology. These expenses can be mitigated by co-using microalgae for biofuels, nutraceuticals, or bioplastics; however, finding the best possible balance is currently being investigated (Khan et al., 2023).

Table 12.3 lists the critical advancements that are required to advance the production of hydrogen from microalgae and their implications for future commercialization. Investing heavily in infrastructure, such as cutting-edge bioreactors and effective supply chains, is essential to becoming market-ready and facilitating more productive manufacturing methods. Furthermore, robust policy support—such as government grants and carbon credit programs—can improve market competitiveness and increase

TABLE 12.3 Aspects of hydrogen derived from microalgae for commercialization

ASPECT	*REQUIRED DEVELOPMENT*	*COMMERCIALIZATION POTENTIAL*
Investments in infrastructure	Sophisticated bioreactors and effective supply networks	Essential for being prepared for the market
Support for policy	Government grants and programs for carbon credits	Increases the competitiveness of the market
Innovations in technology	Improved strains of microalgae and automation	Improves efficiency and scalability

the technology's attractiveness to stakeholders and investors (Dębowski et al., 2022). Finally, greater scalability and efficiency will be made possible by technological advancements in the development of improved microalgae strains and automation, which will eventually enable a wider acceptance of microalgae-based hydrogen production in the renewable energy market. Taken as a whole, these features highlight how development and commercialization are linked to maximizing the potential of this novel energy solution (Anwar et al., 2019).

12.6 SUCCESS AND FAILURE STORIES

12.6.1 Success Stories

1. **Flat-Plate PBR Systems:** It has been possible for certain researchers to create flat-plate PBR systems with higher light utilization and hydrogen output. These systems' minimal space needs, simplicity, and ease of maintenance present a promising avenue for future growth (Anwar et al., 2019).
2. **Genetically Engineered Strains:** The creation of microalgae strains with increased hydrogenase activity and resistance to oxygen inhibition has proven to be a noteworthy accomplishment. The efficiency of hydrogen production in lab conditions has been greatly increased by this genetic engineering method, indicating the possibility for additional advancements.

12.6.2 Failure Stories

1. **Scaling Challenges:** Although the production of hydrogen using microalgae has proven successful in controlled settings, scaling up has resulted in difficulties sustaining steady productivity. Large reactor growth conditions, insufficient light penetration, and contamination have all contributed to the failure of numerous attempts at industrial-scale operations (Rashid et al., 2013).
2. **High Operational Costs:** Numerous attempts to increase the generation of hydrogen from microalgae have been hindered by high operating expenses, such as the energy needed to maintain ideal growth conditions, which rendered the projects unfeasible from an economic standpoint.

12.7 INTEGRATED MICROALGAE-BASED PROCESSES: AN OVERVIEW

Several biological and technical techniques are combined in integrated microalgae-based systems to increase the efficiency of hydrogen production (Figure 12.1). These procedures frequently include using waste products as fuel for hydrogen production or combining microalgae cultivation with fermentation (Wang et al., 2021). Combining several approaches maximizes the use of available resources, lowers production costs, and increases the total amount of hydrogen produced. These strategies also include the utilization of modified microalgae strains for increased efficiency, improvements in bioreactor design, and nutrient recycling (Zhou & Fang, 2020).

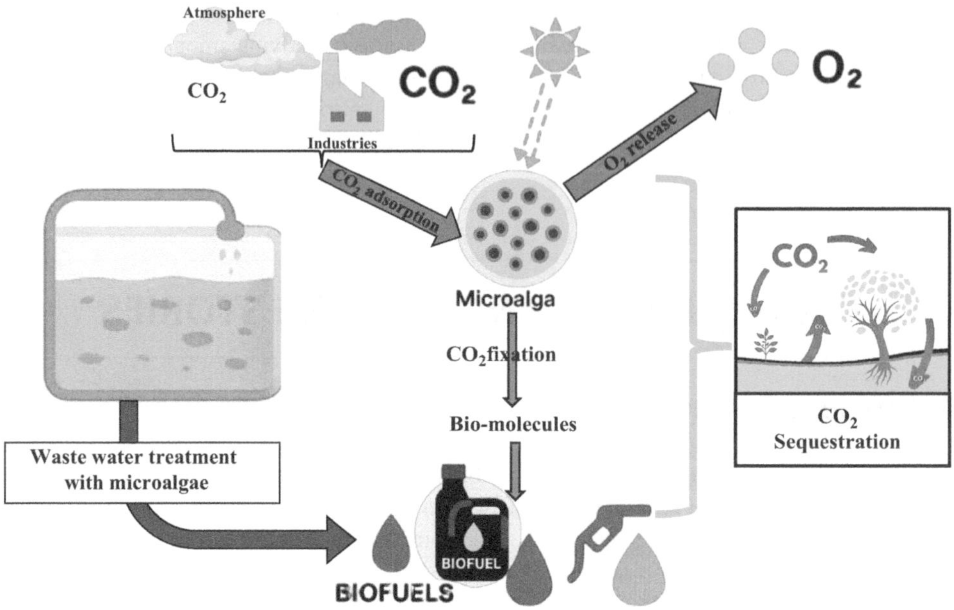

FIGURE 12.1 Wastewater treatment, CO_2 sequestration, and biomass valorization in an integrated microalgae biorefinery.

Combining fermentation and microalgae culture is one of the main tactics in integrated microalgae-based processes. Anaerobic conditions can be achieved through the biological process of fermentation, which breaks down organic materials and produces hydrogen. Researchers can maximize the use of available resources and minimize waste by combining microalgae culture with fermentation. This allows researchers to use the biomass produced by microalgae as feedstock for fermentation. This method produces more hydrogen overall and offers a more sustainable alternative by turning organic matter and light into useful energy. For example, anaerobic bacteria that break down organic materials in dark fermentation can be coupled with photosynthetic bacteria to further transform fermentation products into hydrogen when exposed to light. Comparing this two-stage approach to traditional one-stage techniques can greatly increase the production of hydrogen.

Moreover, waste materials can be used as feedstock for integrated microalgae-based processes that produce hydrogen. Wastewater, for instance, is an economical and ecologically beneficial resource since it contains organic materials that may be utilized to feed fermentation and microalgae activities. Wastewater can be an effective growing medium for microalgae, which can remove nitrogen and phosphorus and produce biomass that can be fermented to produce hydrogen. This takes care of the problem of getting rid of waste and also lessens the demand for expensive inputs like fertilizers. Furthermore, incorporating waste usage into the hydrogen production process reduces pollution in the environment and promotes a more sustainable and circular bioeconomy. In large-scale applications, where resource efficiency and waste minimization are essential for economic sustainability, using waste products as feedstock is very beneficial.

An additional essential element of integrated microalgae-based processes is the development of bioreactor designs. Systems called bioreactors are used to grow microalgae and enable the controlled synthesis of hydrogen. There are a number of drawbacks to using traditional open pond systems for microalgae growing, including their vulnerability to contamination, ineffective use of light, and restricted ability to regulate environmental parameters. Closed bioreactor solutions, including PBRs, have been created to overcome these difficulties. More control over elements like temperature, light intensity, and food availability—all critical for maximizing microalgae growth and hydrogen production—is provided by these systems. Better biomass production and thus better hydrogen yields can be achieved by optimizing light exposure and ensuring effective mixing of the microalgal culture within PBRs. Additionally, these systems can be combined with gas exchange mechanisms to absorb and use the oxygen generated during photosynthesis, which, if not controlled, can prevent the generation of hydrogen. Scaling up and commercializing microalgae-based hydrogen production require the integration of such cutting-edge bioreactor systems.

An additional crucial component of integrated microalgae-based systems is nutrient recycling. For optimum growth, microalgae need a variety of nutrients, such as nitrogen, phosphorus, and trace elements. The constant delivery of these nutrients in large-scale agricultural systems can become expensive and unsustainable for the ecosystem. Integrated processes that concentrate on recycling nutrients within the system have been developed to address this problem. For instance, the waste materials produced during the fermentation process can be cleaned up and utilized again as a supply of nutrients for the growth of microalgae (Javed et al., 2019). As a result, there is less waste and less demand for outside inputs, increasing the process's overall sustainability. In order to sustain the microalgal culture's equilibrium and guarantee continuous development and hydrogen production over time, nutrient recycling is also helpful.

12.8 SIGNIFICANCE OF HYDROGEN AS A CLEAN ENERGY SOURCE

Since burning hydrogen releases just water as a byproduct and emits no greenhouse gases, hydrogen is regarded as one of the cleanest energy sources (IEA, 2022). Because hydrogen has the ability to decarbonize industries like power production, transportation, and industry, it is becoming more and more important as a component of the renewable energy mix. By employing renewable biological resources and adhering to international climate targets to cut carbon emissions and combat climate change, the use of biohydrogen further maximizes this potential (Wang et al., 2021).

The main reasons hydrogen is good for the environment are that it only produces heat and water vapor when it burns or combines with oxygen in a fuel cell. This is in stark contrast to the conventional burning of fossil fuels, which produces significant amounts of carbon dioxide emissions (CO_2) and other dangerous contaminants into the air (Ayub et al., 2024). Because of this, hydrogen is being pushed as a clean energy alternative, especially in industries where it is challenging to decarbonize using alternative renewable energy sources. For instance, hydrogen-powered fuel cells can be employed in long-distance transportation, where the energy density of batteries is insufficient for long-haul vehicles, ships, or airplanes, and in heavy industries like steel manufacture, where high temperatures are necessary (IEA, 2022 & Agyekum et al., 2022).

The potential of hydrogen to dramatically cut carbon emissions and its wide range of applications have led to an increased attention on hydrogen as a component of the renewable energy mix on a worldwide scale. The transportation industry is one of the most potential applications for hydrogen (Ardo et al., 2022). Even though electric vehicles (EVs) have received a lot of attention as a way to cut emissions, they aren't necessarily the best option for all types of transportation. Compared to battery EVs (BEVs), hydrogen-powered vehicles, especially fuel cell electric vehicles (FCEVs), have benefits including greater driving ranges and quicker refueling periods. Because of this, hydrogen is a desirable alternative for industries including freight transportation, buses, and trains, where long-range capacity and low downtime are crucial (Wang et al., 2021).

12.9 CONCLUSION

In conclusion, biohydrogen production presents a promising avenue for sustainable energy generation by utilizing organic materials through biological processes such as fermentation and photosynthesis. This method not only reduces greenhouse gas emissions but also integrates with waste management, converting organic waste into valuable hydrogen and treated water. While various processes, including photofermentation and dark fermentation, exhibit distinct advantages and challenges, the scalability of these technologies remains hindered by low efficiencies, high production costs, and infrastructural demands. Advancements in genetic engineering, bioreactor design, and hybrid systems could enhance the productivity and economic feasibility of biohydrogen production. As regulatory support and market infrastructure evolve, biohydrogen is set to play a crucial role in the transition to greener energy systems, aiding in the reduction of carbon emissions and aligning with global sustainability targets. Ultimately, ongoing innovations and collaborations among researchers, policymakers, and industry leaders will be essential to realize the full potential of biohydrogen as a clean energy source.

REFERENCES

Agyekum, E. B., Nutakor, C., Agwa, A. M., & Kamel, S. (2022). A critical review of renewable hydrogen production methods: Factors affecting their scale-up and its role in future energy generation. *Membranes*, 12(2), 173.

Ahmad, A., Yadav, A. K., & Hasan, S. (2024). Enhanced production of methane enriched biogas through intensified co-digestion process and its effective utilization in a biodiesel/biohydrogen fueled engine with duel injection strategies: ML-RSM based an efficient optimization approach. *International Journal of Hydrogen Energy*, 65, 671–686.

Ahmed, S. F., Mofijur, M., Nahrin, M., Chowdhury, S. N., Nuzhat, S., Alherek, M., ... & Mahlia, T. M. I. (2022). Biohydrogen production from wastewater-based microalgae: Progresses and challenges. *International Journal of Hydrogen Energy*, 47(88), 37321–37342.

Al-Dailami, A., Ahmad, I., Abdullah, N., Koji, I., & Yuzir, A. (2022, April). Feasibility and viability of procuring biohydrogen from microalgae: An emerging and sustainable energy resource technology. *Journal of Physics: Conference Series*, 2259(1), 012014.

Anwar, M., Lou, S., Chen, L., Li, H., & Hu, Z. (2019). Recent advancement and strategy on bio-hydrogen production from photosynthetic microalgae. *Bioresource Technology*, 292, 121972.

Ardo, F. M., Lim, J. W., Ramli, A., Lam, M. K., Kiatkittipong, W., Abdelfattah, E. A., ... & Sahrin, N. T. (2022). A review in redressing challenges to produce sustainable hydrogen from microalgae for aviation industry. *Fuel*, 330, 125646.

Ayub, H. M. U., Nizami, M., Qyyum, M. A., Iqbal, N., Al-Muhtaseb, A. A. H., & Hasan, M. (2024). Sustainable hydrogen production via microalgae: Technological advancements, economic indicators, environmental aspects, challenges, and policy implications. *Environmental Research*, 244, 117815.

Banu, J. R., Kavitha, S., Gunasekaran, M., & Kumar, G. (2020). Microalgae based biorefinery promoting circular bioeconomy-techno economic and life-cycle analysis. *Bioresource Technology*, 302, 122822.

Bhagwat, P., Permaul, K., Singh, S., & Pillai, S. (2024). Feedstocks and cases III: Biohydrogen from algal biomass (macro and microalgae). In *Biohydrogen-Advances and Processes* (pp. 337–364). Springer Nature Switzerland.

Boboescu, I. Z., Ilie, M., Gherman, V. D., Mirel, I., Pap, B., Negrea, A., ... & Maróti, G. (2014). Revealing the factors influencing a fermentative biohydrogen production process using industrial wastewater as fermentation substrate. *Biotechnology for Biofuels*, 7(1), 139.

Chandrasekhar, K., Lee, Y. J., & Lee, D. W. (2015). Biohydrogen production: Strategies to improve process efficiency through microbial routes. *International Journal of Molecular Sciences*, 16(4), 8266–8293.

Chandrasekhar, T., & Anuprasanna, V. (2023). Shedding light on the production of biohydrogen from algae. In *Solar Fuels* (pp. 161–184). Wiley Online Library.

Chisti, Y. (2018). Biodiesel from microalgae: Processes and challenges. *Biotechnology Advances*, 29(3), 564–575. https://doi.org/10.1016/j.biotechadv.2018.04.006

Dębowski, M., Świca, I., Kazimierowicz, J., & Zieliński, M. (2022). Large scale microalgae biofuel technology—Development perspectives in light of the barriers and limitations. *Energies*, 16(1), 81.

Faraloni, C., & Torzillo, G. (2023). Interplay between photobiological hydrogen production by microalgae and bioeconomy. In *A Sustainable Green Future: Perspectives on Energy, Economy, Industry, Cities and Environment* (pp. 541–559). Springer International Publishing.

Gendy, T. S., & El-Temtamy, S. A. (2013). Commercialization potential aspects of microalgae for biofuel production: An overview. *Egyptian Journal of Petroleum*, 22(1), 43–51.

Ghimire, A., Kumar, G., Sivagurunathan, P., Shobana, S., Saratale, G. D., Kim, H. W., … & Munoz, R. (2017). Bio-hythane production from microalgae biomass: Key challenges and potential opportunities for algal bio-refineries. *Bioresource Technology*, 241, 525–536.

Gholkar, P., Shastri, Y., & Tanksale, A. (2021). Renewable hydrogen and methane production from microalgae: A techno-economic and life cycle assessment study. *Journal of Cleaner Production*, 279, 123726.

Goswami, R. K., Mehariya, S., Obulisamy, P. K., & Verma, P. (2021). Advanced microalgae-based renewable biohydrogen production systems: A review. *Bioresource Technology*, 320, 124301.

IEA (2022). *Global Hydrogen Energy Report: Transitioning to a Hydrogen Economy*. International Energy Agency.

Javed, F., Aslam, M., Rashid, N., Shamair, Z., Khan, A. L., Yasin, M., … & Bazmi, A. A. (2019). Microalgae-based biofuels, resource recovery and wastewater treatment: A pathway towards sustainable biorefinery. *Fuel*, 255, 115826.

Khan, S., Das, P., Abdul Quadir, M., Thaher, M. I., Mahata, C., Sayadi, S., & Al-Jabri, H. (2023). Microalgal feedstock for biofuel production: Recent advances, challenges, and future perspective. *Fermentation*, 9(3), 281.

Kumar, P., & Singh, A. (2020). Advances in bio hydrogen production technologies: Microalgae and beyond. *Biotechnology Advances*, 38(2), 25–35. https://doi.org/10.1016/j.biotechadv.2019.101524

Lee, J., Park, H., & Kim, S. (2019). The potential of microalgae in sustainable energy production: A review. *Energy Reports*, 5(1), 100–110. https://doi.org/10.1016/j.egyr.2019.01.009

Li, S., Li, F., Zhu, X., Liao, Q., Chang, J. S., & Ho, S. H. (2022). Biohydrogen production from microalgae for environmental sustainability. *Chemosphere*, 291, 132717.

Nagarajan, D., Dong, C. D., Chen, C. Y., Lee, D. J., & Chang, J. S. (2021). Biohydrogen production from microalgae—Major bottlenecks and future research perspectives. *Biotechnology Journal*, 16(5), 2000124.

Ntaikou, I., Antonopoulou, G., & Lyberatos, G. (2010). Biohydrogen production from biomass and wastes via dark fermentation: A review. *Waste and Biomass Valorization*, 1(1), 21–39.

Oey, M., Sawyer, A. L., Ross, I. L., & Hankamer, B. (2016). Challenges and opportunities for hydrogen production from microalgae. *Plant Biotechnology Journal*, 14(7), 1487–1499.

Okeke, E. S., Ejeromedoghene, O., Okoye, C. O., Ezeorba, T. P. C., Nyaruaba, R., Ikechukwu, C. K., … & Orege, J. I. (2022). Microalgae biorefinery: An integrated route for the sustainable production of high-value-added products. *Energy Conversion and Management:* X, 16, 100323.

Ramprakash, B., Lindblad, P., Eaton-Rye, J. J., & Incharoensakdi, A. (2022). Current strategies and future perspectives in biological hydrogen production: A review. *Renewable and Sustainable Energy Reviews*, 168, 112773.

Rashid, N., Rehman, M. S. U., Memon, S., Rahman, Z. U., Lee, K., & Han, J. I. (2013). Current status, barriers and developments in biohydrogen production by microalgae. *Renewable and Sustainable Energy Reviews*, 22, 571–579.

Sharma, A. K., Ghodke, P. K., Manna, S., & Chen, W. H. (2021). Emerging technologies for sustainable production of biohydrogen production from microalgae: A state-of-the-art review of upstream and downstream processes. *Bioresource Technology*, 342, 126057.

Singh, A., & Dhar, D. (2020). Microalgal hydrogen production: Recent progress, challenges, and future perspectives. *Biofuels Journal*, 13(4), 670–680. https://doi.org/10.1080/17597269.2020.1723062

Singh, J., & Gu, S. (2010). Commercialization potential of microalgae for biofuels production. *Renewable and Sustainable Energy Reviews*, 14(9), 2596–2610.

Srirangan, K., & Young, M. (2019). Optimization of bio hydrogen production through metabolic engineering of microalgae. *Trends in Biotechnology*, 37(7), 555–568. https://doi.org/10.1016/j.tibtech.2019.02.008

Su, Y., Song, K., Zhang, P., Su, Y., Cheng, J., & Chen, X. (2017). Progress of microalgae biofuel's commercialization. *Renewable and Sustainable Energy Reviews*, 74, 402–411.

Subhash, G., & Mohan, R. (2019). Hydrogen production through photobiological processes: Microalgae integration and advancements. *Bioresource Technology*, 244(2), 120–128. https://doi.org/10.1016/j.biortech.2018.11.052

Touloupakis, E., Faraloni, C., Silva Benavides, A. M., & Torzillo, G. (2021). Recent achievements in microalgal photobiological hydrogen production. *Energies*, 14(21), 7170.

Vaishnav, S., Saini, T., Chauhan, A., Gaur, G. K., Tiwari, R., Dutt, T., & Tarafdar, A. (2023). Livestock and poultry farm wastewater treatment and its valorization for generating value-added products: Recent updates and way forward. *Bioresource Technology*, 382, 129170.

Velmozhina, K., Shinkevich, P., Zhazhkov, V., Politaeva, N., Korablev, V., Vladimirov, I., & Morales, T. C. (2023). Production of biohydrogen from microalgae biomass after wastewater treatment and air purification from CO_2. *Processes*, 11(10), 2978.

Wang, Y., Zhang, X., & Li, L. (2021). Economic analysis of bio hydrogen production through microalgae integration. *Renewable and Sustainable Energy Reviews*, 64(1), 59–71. https://doi.org/10.1016/j.rser.2020.109859

Woon, J. M., Khoo, K. S., Akermi, M., Alanazi, M. M., Lim, J. W., Chan, Y. J., … & Ardo, F. M. (2023). Reviewing biohydrogen production from microalgal cells through fundamental mechanisms, enzymes and factors that engendering new challenges and prospects. *Fuel*, 346, 128312.

Yaashikaa, P. R., Devi, M. K., & Kumar, P. S. (2022). Algal biofuels: Technological perspective on cultivation, fuel extraction and engineering genetic pathway for enhancing productivity. *Fuel*, 320, 123814.

Yadav, G., Shanmugam, S., Sivaramakrishnan, R., Kumar, D., Mathimani, T., Brindhadevi, K., … & Rajendran, K. (2021). Mechanism and challenges behind algae as a wastewater treatment choice for bioenergy production and beyond. *Fuel*, 285, 119093.

Zhou, M., & Fang, Y. (2020). Technological advancements in bio hydrogen production: A focus on microalgae-based processes. *Bioenergy Research*, 13(3), 432–446. https://doi.org/10.1007/s12155-020-10156-2

Integrated Dark and Photo-Fermentation of Lignocellulosic Biomass for Hydrogen Production

13

Prerana Limbani, Kandasamy Yazhini, Ananya Sood, Aadharshini Pradeep, Janani Balaji, and Pasupathi Rathinasabapathi

13.1 INTRODUCTION

Over the past two centuries, the world's primary energy sources have shifted from wood to coal, then to oil, followed by natural gas, and, more recently, to renewable energy sources. Human civilization has been shaped by the availability of these natural resources (T. Singh et al., 2021). Energy sources are generally classified into two categories: renewable and non-renewable. The non-renewable resources are coal, oil, nuclear energy, and natural gas; these resources take millions of years to be replenished once the stocks are depleted. On the other hand, renewable resources like solar energy, the energy from the wind, and hydropower are resources that are self-generated and, therefore, replenish themselves. The level of fossil fuels, coal, oil, and natural gas, remains high and takes approximately 80% of energy consumption (Hitam & Jalil, 2023). Even though these fuels have contributed to industrialization and economic development, they are associated with high costs to the environment and geopolitics. Their combustion releases carbon dioxide gases, thereby releasing greenhouse gases that cause climate change (Bhatia et al., 2021). This has a central role in explaining why there is a need to search for environmentally friendly energy sources and those that are renewable. Bioenergy can be in the form of biodiesel, bioethanol, and even biohydrogen if it is derived from biomass (Abraham et al., 2020). Biohydrogen stands out as a promising renewable energy source due to its high energy density and environmentally friendly nature (Roque et al., 2025). When burned, biohydrogen produces only water, making it a carbon-neutral fuel (Hitam & Jalil, 2023). Various biological methods offer efficient and cost-effective ways to produce biohydrogen, making

DOI: 10.1201/9781003568612-13

it a sustainable alternative to traditional energy sources. Biomass includes not only plant and animal matter but also biopolymers extracted from these sources and microorganisms. This makes biomass a crucial component of renewable energy. Common biodegradable biopolymers include cellulose, starch, silk, and collagen, which find applications in various industries. Mostly made up of cellulose, hemicellulose, and lignin, lignocellulosic biomass (LCB) is an important source of polymers (Vertés et al., 2020). Forestry and agricultural waste can be converted into energy and materials, promoting sustainability and renewability (Kumar et al., 2024). Shifting toward biomass and other clean energy sources demonstrates a growing awareness of the need to minimize the environmental impact of energy production, ensuring a more sustainable future.

13.2 LCB: A VERSATILE FEEDSTOCK

LCB is an incoming raw material that is available in plenty, and its use is an attempt to minimize the use of petroleum products in energy production. Bio-refineries, mainly those that process LCB, use thermochemical processing and mechanical and biochemical conversion of biomass for energy end-products. Some major technologies for converting LCB through heat include burning and reaction with carbon dioxide, cracking, hydrothermal synthesis, thermal decarboxylation, pyrolysis, as well as gasification. Among them, gasification is most efficient for the conversion of LCB to syngas, which is a mixture of gases suitable for use as a fuel or a starting material for chemical production. Biomass gasification involves heating biomass in the presence of a gasifying agent at high temperatures (Richardson et al., 2015). This process produces a gas mixture with reduced amounts of ash, tar, and char (Zhao et al., 2022).

Energy system development plays an essential role in supporting most of the societal objectives, ranging from sustainable economic growth attainment of the millennium development goals to the prevention of disastrous climate change. However, great challenges are associated with the current energy system that centralizes energy around non-renewable, oil-based fuels, which severely impacts climate, air, and water quality. For instance, polyethylene (PE), polypropylene (PP), polystyrene (PS), polyvinyl alcohol (PVA), and polyvinyl chloride (PVC) form the manufacturing base of the plastic industry and are generated from petroleum products. These plastics result in long-standing pollution effects since they are not biodegradable, forming much permanent waste. To address these challenges, natural polysaccharides like cellulose, starch, carrageenan, alginate, chitin, and chitosan have gained attention. These biopolymers are biodegradable, non-toxic, and versatile, finding applications in various industries, including chemical processing, textiles, food, agriculture, and medicine (Ren et al., 2016).

Recent developments in biotechnology have greatly improved the engineering of the metabolic pathway of microbial cells based on LCB, thus improving the efficiency of their conversion into biopolymers or monomers. All these have enhanced the bioconversion processes, making them more plausible and effective (Hitam & Jalil, 2023), opening opportunities for environment-friendly industrial uses of resourceful bioresources. This chapter focuses on the composition, formation, and utilization in biohydrogen generation as a resource for renewable energy of LCB. It also analyzes the use of photo-fermentation (PF) in LCB as well as the factors and prospects for biomass catalyst production. Also, the chapter discloses the socio-economic and environmental implications for LCB gasification and contingency for sustainable energy sources.

Biohydrogen, with a high energy density of approximately 120 MJ/kg, offers nearly three times the energy of conventional fossil fuels like gasoline, making it an efficient energy carrier. Lignocellulosic feedstocks offer significant advantages over other biomass sources because they come from the non-edible parts of plants and do not compete with food supplies (T. Singh et al., 2021). Large quantities of lignocellulosic waste from forestry, agriculture, and agro-industries accumulate yearly and can be repurposed to produce valuable products, contributing to a circular economy. Economically, LCB is faster and cheaper to produce compared to biofuel feedstocks like corn starch, soybeans, and sugar cane, and it is also much less expensive than crude oil (Soares et al., 2020). Although it has numerous advantages, LCB can be challenging to process due to its complex structure and high crystallinity (T. Singh et al., 2021). This could be a significant problem if it is not handled properly, such as through burning or improper disposal. Thermochemical methods like pyrolysis and hydrothermal carbonization, as well as activation processes, convert carbon-based materials from biomass. However, lignin hinders the biochemical conversion process, making biofuel production less commercially viable. To address this, biomass undergoes pretreatment using physical, chemical, or biochemical methods to enhance biodegradability. Enzymatic hydrolysis breaks down cellulose and hemicellulose into oligosaccharides and monosaccharides, which are then converted into bioproducts via microbial fermentation (Soares et al., 2020). While enzymatic hydrolysis is crucial for breaking down biomass, it is expensive, often requiring large amounts of enzymes to achieve high yields. Integrated bioprocessing—combining pretreatment, enzyme production, and fermentation in one reactor—helps reduce costs and energy consumption. LCB offers significant potential for biohydrogen production, which can help address global energy challenges. Although pretreatment and processing hurdles remain, advancements in biotechnology, enzyme development, and process integration can pave the way for a sustainable biohydrogen economy.

13.3 PRETREATMENTS OF LCB

LCB is rich in cellulose, hemicellulose, and lignin. The components of LCB are tightly bound, which makes it resistant to microbial attack and accessible to enzymes. Without pretreatment, these structural components limit the release of fermentable sugars, hindering efficient fermentation. Biomass pretreatment is the first and most important step in biohydrogen production (T. Singh et al., 2021). This can be achieved through various physical, chemical, and biological methods. It ensures the removal of the lignin component and degradation of the complex structure of lignocellulose into fermentable units, which exposes it to enzymatic hydrolysis. Integrating dark fermentation (DF) and PF processes with effective pretreatment ensures both the hydrogen production rate and yield (T. Singh et al., 2021).

13.3.1 Physical Pretreatment

Physical pretreatment includes mechanical disruption, sonication, exposure to radiation, high temperature, and pressure (Soares et al., 2020). The main aim of mechanical disruption is particle size reduction, which increases the surface area-to-volume ratio. This will increase the chemical and enzymatic hydrolysis, resulting in more yield (Cheng et al., 2011). After undergoing physical pretreatment, the structure that is tightly packed will be relaxed, which will increase its accessibility. Chemical agents and enzymes can permeate them more effectively and further break them down as a result. Hemicellulose and cellulose are transformed into fermentable sugars such as xylose, arabinose, glucose, and other sugars through the process of hydrolysis, which can be either chemical or enzymatic (Soares et al., 2020). Studies have demonstrated that ultrafine grinding can effectively disrupt the rigid cellulose structure in Paulownia stems and corn stover, significantly increasing biohydrogen production. Combining mechanical disruption with other pretreatment methods can further enhance efficiency (Q. Zhang et al., 2021). A pilot-scale study on the steam explosion of softwood found that smaller particle sizes (2–5 mm and 5–8 mm) resulted in greater hemicellulose hydrolysis compared to larger particles (8–12 mm) at 190°C. Increasing the temperature to 210°C allowed larger particles to achieve maximum hydrolysis (Ballesteros et al., 2000). Additionally, raising the pressure from 8 to 12 MPa improved reducing sugar yield in wheat straw (Alinia et al., 2010). Ultrasonication is another method that uses high-frequency sound waves to generate cavitation, which disrupts the cell walls of LCB, increasing the availability of fermentable sugars for microbial fermentation (Ong & Wu, 2020). A study showed a 60% increase in hydrogen production in maize stalks compared to untreated biomass using ultrasonication (Carrère et al., 2010). Radiation, like UV rays, gamma rays, and microwaves, is another way to break the concrete structure of the LCB. Microwaves have the property to break hydrogen bonds and cause evaporation of water. Gamma radiation and electron beams can disrupt 1,4-glycosidic bonds in LCB (Kucharska et al., 2018). Studies on rice straw exposed to varying irradiation doses (7.6–90 kGy) and durations (4–36 minutes) found that glucose yield and production rate increased with higher electron beam current. A dose of 80 kGy and 1 MeV yielded the highest glucose yield for all timeframes (Bak et al., 2009). To increase hydrogen yield, it is vital to optimize both the light intensity and the wavelength. According to research, a fourfold increase in light intensity might result in a threefold rise in the creation of hydrogen gas, while at the same time resulting in a threefold decrease in the efficiency of light conversion. At a relatively modest light intensity of 3.75 W/m^2 (2,500 lux), the best light conversion efficiency (0.51%) was obtained. This exemplifies the trade-off that exists between the production of hydrogen and efficiency (I. Eroglu et al., 2010).

DF relies solely on anaerobic microorganisms. Physical pretreatments in DF are used to break down complicated structures in LCB. This process will enhance microbial hydrolysis by mechanical grinding or milling. Steam explosion involves rapidly reducing the pressure after it has been heated to high temperatures and pressures. On the other hand, PF usually works with more aqueous or liquid substrates, such as sugars and organic acids, which don't need to be thoroughly mechanically processed. PF processes are simpler and do not require significant energy input, making steam explosion and mechanical pretreatment often unnecessary. It is a process where photosynthetic bacteria utilize sunlight to convert organic substrates into hydrogen and carbon dioxide. Since PF bacteria rely on light to produce hydrogen, exposure to light or illumination becomes a crucial physical pretreatment step in PF (Hitam & Jalil, 2023).

13.3.2 Chemical Pretreatment

Chemical pretreatment includes acid, alkali hydrolysis, treatment with oxidizers, and carbon dioxide. It is usually combined with physical methods like temperature. Oxidizers like peroxides and alcohol are used to remove the lignin content, exposing the cellulose (Sołowski et al., 2020). A study comparing CaO_2 and CaO pretreatments for biohydrogen production from wheat straw found that increasing the oxidizer concentration from 0% to 6% led to higher hydrogen yields. However, further increases caused lignin aggregation and precipitation, reducing yield (J. Zhang et al., 2020). Another study on wheat straw demonstrated that a supercritical CO_2 explosion resulted in higher reduced sugar yields compared to a steam explosion (Alinia et al., 2010). The most commonly used chemical pretreatment is hydrolysis by acid and alkali. Hydrolysis depends on many factors like temperature, pH, the concentration of both biomass and acid or alkali, time of exposure, production of inhibitory compound formation, etc. Alkali treatment removes lignin, increases porosity and internal surface area, and exposes cellulose and hemicellulose for enzymatic hydrolysis. A study on giant reed biomass found that NaOH and $Ca(OH)_2$ pretreatments improved hydrogen yield, with 20% NaOH achieving the highest yield. Similar results were observed with rice straw, corn stalk, sugarcane bagasse, and oat straw (Jiang et al., 2020). Acid pretreatment can decompose cellulose and hemicellulose into smaller, fermentable sugar molecules, leading to higher reducing sugar yields compared to alkaline pretreatment. A study on *Platanus orientalis* leaves showed that 4% H_2SO_4 pretreatment produced 38.74% more biohydrogen than alkali (Q. Zhang et al., 2021). However, high acid concentrations can produce inhibitory compounds like hydroxymethylfurfural (HMF) and furfural, which can damage DNA and prevent RNA and protein production (El-Sayed Nasr, 2017). A study on sugarcane bagasse found that furfural concentration increased with higher H_2SO_4 concentrations (Pattra et al., 2008). pH is another factor that affects LCB pretreatment. As the pH of the sugarcane bagasse was raised from 3.0 to 5.0, the amount of hydrogen that was produced increased, but it decreased when the pH reached 9.0. It has been established that the pH range of 7.0–9.0 is the most favorable for the production of hydrogen (Reddy et al., 2017). The creation of hydrogen by photosynthetic bacteria like *Rhodobacter sphaeroides* is accomplished by the utilization of enzymes such as nitrogenase. To function at their highest potential, these enzymes require particular metal cofactors such as Mo and Fe. It is possible to considerably improve the efficiency of the production of hydrogen during PF by adjusting the quantities of metal ions in the medium (Bundhoo et al., 2015).

Glycosidic linkages in cellulose and hemicellulose are broken by acid hydrolysis, and lignin is eliminated by alkaline pretreatment, increasing the accessibility of polysaccharides to DF bacteria. Thermochemical techniques combine chemicals and heat to improve the decomposition of biomass even further. Therefore, acid hydrolysis, alkaline pretreatment, and thermochemical pretreatment are employed in DF to convert complicated lignocellulosic material into sugars that may be fermented. On the other hand, PF uses chemical treatments such as metal ion supplements like molybdenum and iron to increase the activity of important enzymes like hydrogenase and nitrogenase, buffer solutions for stringent pH control, and chemical reduction agents like sodium sulfide to maintain anaerobic conditions. These are peculiar to the light-dependent processes of PF.

13.3.3 Enzymatic Pretreatment

The breakdown of LCB is accomplished through the utilization of enzymes, which are biological catalysts, in the process of enzymatic pretreatment (Tornyie et al., 2024). These enzymes are naturally produced by cellulolytic organisms such as bacteria from the genera *Bacillus*, *Streptomyces*, *Candida*, and *Aspergillus*, as well as fungi like *Phanerochaete chrysosporium*, *Gloeophyllum trabeum*, and *Trichoderma reesei*. Commercially available enzymes are also widely used for this purpose. One of the most frequent enzymes, cellulases, is responsible for targeting the β-1,4-glycosidic links that are present in cellulose. Endoglucanases, exoglucanases, and β-glucosidases are the three forms of cellulases that are considered to be the most prominent. The production of long-chain oligomers is accomplished by endoglucanases by the cleavage of internal glycosidic linkages in the amorphous sections of cellulose (Ummalyma et al., 2019). These are further broken down into short-chain oligomers by exoglucanases, and finally, β-glucosidases hydrolyze the oligomers into fermentable sugars (Soares et al., 2020). Within a heterogeneous set of linear and branching hemicelluloses that are connected by hydrogen bonds, hemicellulases are responsible for their action (Ummalyma et al., 2019). Lignin-degrading enzymes like laccases and peroxidases break down the complex aromatic lignin to make cellulose and hemicellulose more accessible for enzymatic hydrolysis (Kumar et al., 2024). Pectinases like polygalacturonases and pectin lyases break down pectin, a component of the plant cell wall. Accessory enzymes such as acetyl xylan esterases and ferulic acid esterases play a vital role in removing side chains from hemicellulose (Ummalyma et al., 2019). It has been preferred in recent years since it is less time-consuming, can be reused, and inhibitory compounds are not produced (Ferdeş et al., 2020). Using a combination of Agropect Pomace and Celustar XL enzymes, scientific research was conducted on sugar beet pulp and wasted hops to evaluate the process of enzymatic hydrolysis (Ziemiński et al., 2012). Higher enzyme doses increased the reducing sugar concentration in the hydrolysates, which eventually reached levels that were much higher than before (Ziemiński et al., 2012). Enzymatic hydrolysis is often the final step in LCB pretreatment, following physical and chemical treatments to make the substrate more accessible to enzymes. Numerous studies have demonstrated the effectiveness of enzymatic hydrolysis in achieving high saccharification yields. Enzymatic hydrolysis enhanced the release of sugars and significantly increased hydrogen production by 30% due to improved substrate availability for microbial fermentation (Wang & Wan, 2008). Similarly, hemicellulases can break down hemicellulose into pentose sugars like xylose and arabinose. The hydrogen yield from hemicellulose-treated bagasse was 40% higher than untreated biomass (Hawkes et al., 2002). Some enzymes like nitrogenase, typically used for nitrogen fixation, which can also catalyze hydrogen production as a side reaction, superoxide dismutase, and catalase, are sometimes used to detoxify reactive oxygen species (ROS) that could inhibit photosynthetic hydrogen production and lipases. These compounds are poorly utilized in DF but are efficiently converted to hydrogen in PF under light conditions (Z. Zhu et al., 2010).

At the beginning of the DF process, the major objective of the pretreatment is to break down the complicated structure of the LCB (Soares et al., 2020). Enzymatic pretreatments involving cellulases and hemicellulases hydrolyze LCB into fermentable sugars, while pectinases target plant cell wall components. By breaking down lignin, enzymes that degrade lignin, such as laccases and peroxidases, further increase the levels of accessibility to cellulose and hemicellulose (Kumar et al., 2024). In contrast, these enzymes are generally not required for PF, which utilize simpler substrates like organic acids and monosaccharides. PF hydrogen production is light-dependent and driven by photosynthetic enzymes, primarily nitrogenase and hydrogenase, whose activity is enhanced by metal cofactors such as iron and molybdenum. Following pretreatment, LCB undergoes fermentation to produce biohydrogen through two main pathways: DF and PF. In the absence of light, anaerobic bacteria are utilized in the process of DF, which results in the production of hydrogen gas from organic materials. On the contrary, PF is a process that involves the utilization of sunlight by photosynthetic bacteria to produce hydrogen (Soares et al., 2020).

13.4 BIOHYDROGEN PRODUCTION THROUGH DF

The process of DF involves oxidation and reduction, and it is the process that is responsible for the transformation of organic molecules into both inorganic and organic compounds (Bastidas-Oyanedel et al., 2015). DF begins with bacterial hydrolysis, where complex organic materials are broken down into soluble compounds, making them accessible to microorganisms (Figure 13.1). Afterward, acidogenic bacteria are responsible for the transformation of these hydrolysis products into alcohols, carbon dioxide, ammonia,

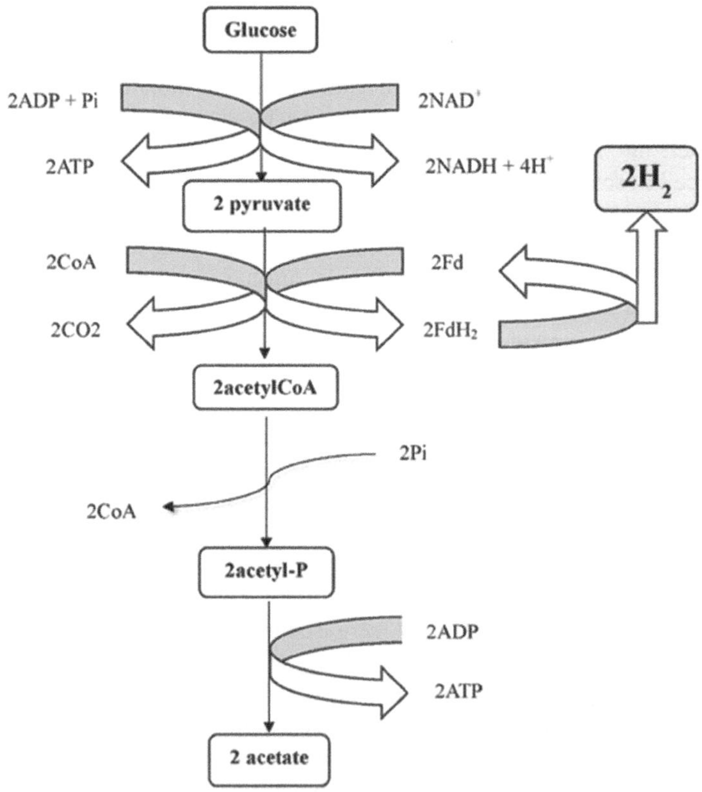

FIGURE 13.1 Biohydrogen production through dark fermentation: The image illustrates the process of hydrogen production occurring alongside acetate formation through glycolysis. This metabolic route results in the production of NADH as a by-product, and it is during the process of glycolysis that glucose is broken down. It is possible to manufacture more molecular hydrogen by utilizing the NADH that is produced during glycolysis, which is produced when acetate is produced (Wang & Yin, 2017a). This simultaneous production of hydrogen and acetate highlights a key aspect of biohydrogen generation, where NADH plays a crucial role in facilitating the release of extra hydrogen molecules in the presence of acetate formation (Wang & Yin, 2017a).

hydrogen, and a variety of organic acids (Kongjan, 2017). There are several benefits that DF provides in comparison to other biological processes that produce hydrogen (Abreu, 2011). Unlike PF, it does not require light, enabling the use of a wide range of raw materials as carbon sources. The microorganisms involved typically exhibit rapid growth and high hydrogen production rates, making the process efficient and scalable (Ramprakash & Karuppan, 2019). All of the microorganisms that are involved in the process generally demonstrate rapid growth and high hydrogen generation rates, which makes the process both efficient and scalable. Both stringent and facultative anaerobic microorganisms can be utilized in the process of microbial development. These bacteria can be used either as pure cultures, such as *Escherichia coli*, *Clostridium*, and *Enterobacter*, or as mixed cultures, which are produced from sources such as organic compost, anaerobic sludge, and bovine dung. Both types of cultures are available for use (Soares et al., 2020). Pure cultures require a sterile environment to prevent contamination, which is difficult and expensive to create on an industrial scale. Mixed cultures, on the other hand, are often preferred for applications that are carried out on a large scale (Bakonyi et al., 2014). In DF, both pure sugars—such as glucose, xylose, lactose, galactose, and sucrose—and various waste-derived materials—like food waste, industrial wastewater, glycerol, municipal solid waste, and LCB—are commonly used as substrates (Hay et al., 2013). Many elements can affect the production and yield of biohydrogen. These factors include the inoculum source, the kind of substrate, the pH, the temperature, the retention time of the hydraulic system, and the gas partial pressure (Soares et al., 2020).

13.4.1 Microbes Involved in DF

Numerous microbes can be used to produce hydrogen through fermentation, and each has unique preferences for substrates, pH, and temperature. These parameters not only affect microbial growth but also critically influence the metabolic pathways used, hence considerably affecting the overall hydrogen production of the process. Producing hydrogen can be accomplished through the utilization of both mixed acidogenic microbial cultures and pure cultures of particular bacteria that produce hydrogen (Koukios & Panagiotopoulos, 2018). It is common practice to obtain mixed cultures from natural sources such as soil, sludge from wastewater treatment plants, and compost. Bacteria that produce hydrogen can be classified as mesophilic (25°C–40°C), thermophilic (40°C–65°C), severe thermophilic (65°C–80°C), or hyperthermophilic (above 80°C) according to the temperatures at which they thrive (Ntaikou et al., 2010). The particular types of microorganisms that should be used, as well as the type of culture that should be used (mixed, pure, or co-culture), should be chosen by the requirements of each process (Ntaikou et al., 2010).

Pure Culture: It has proven possible to produce hydrogen by the use of pure cultures of specific species that can produce hydrogen (Abreu, 2011). Pure cultures provide several advantages, including substrate selectivity, enhanced metabolic manipulation through altered growing conditions, increased hydrogen yields due to less by-products, and consistent process results. Pure cultures, on the other hand, are extremely susceptible to contamination and often require aseptic conditions, which might raise costs. Diverse microorganisms capable of hydrogen production by DF, including strict anaerobes (e.g., *Clostridia*, methylotrophs, rumen bacteria, methanogens, archaea), facultative anaerobes (such as *E. coli*, *Enterobacter*, and *Citrobacter*), and aerobes (including *Alcaligenes* and *Bacillus*) (Monlau et al., 2013). *Clostridium* sp., such as *C. paraputrificum*, *C. thermocellum*, *C. tyrobutyricum*, *C. thermolacticum*, *C. beijerinckii*, *C. acetobutyricum*, and *C. butyricum*, have been the subject of the most comprehensive research because of their ability to produce hydrogen during their exponential growth phase (Kumar Sharma et al., 2022). Recent studies have focused on hydrogen production at high temperatures using thermophilic and hyperthermophilic bacteria, which can enhance reaction kinetics (Abreu, 2011). Thermophiles, including *Caldicellulosiruptor saccharolyticus*, *Thermoanaerobacterium* sp. such as *T. thermosaccharolyticum*, and *Thermotoga* sp. including *T. maritima* and *T. elfii* have been studied in this context. Hyperthermophilic bacteria are attracting interest for their ability to flourish at extremely high temperatures (up to 110°C), which reduces contamination risk and enhances resilience to high hydrogen partial pressures (Ntaikou et al., 2010).

Mixed Cultures: Microbiomes that are phylogenetically diverse play specific roles in the process of breaking down complex organic substrates and facilitating the production of biohydrogen from organic wastes in mixed culture systems (Cabrol et al., 2017). Mixed cultures are widely used for hydrogen production via DF due to their greater stability, efficiency, and cost-effectiveness compared to pure cultures (Kumar et al., 2024). Their diverse microbial communities are capable of rapidly degrading a broad range of substrates. Mixed cultures are widely used for hydrogen production via DF due to their greater stability, efficiency, and cost-effectiveness compared to pure cultures (Karimi, 2015). Their diverse microbial communities are capable of rapidly degrading a broad range of substrates (Pachapur et al., 2019). For full-scale applications, mixed cultures are generally preferred from an engineering perspective. They simplify process control and operation by eliminating the need for medium sterilization, which reduces overall costs and allows for a wider selection of feedstocks. Mixed consortia can be sourced from various natural environments. The utilization of mixed cultures concerns the danger of non-hydrogen-producing species, including lactic acid bacteria, homoacetogens, and methanogens, potentially dominating and affecting the viability of the process (Monlau et al., 2013). To minimize this risk, strategies include pretreating the seed to eliminate undesired microorganisms and maintaining conditions that favor hydrogen-producing species. Common pretreatment methods involve heat treatment, where high temperatures (75°C, 100°C, or even 121°C) for 15 minutes–2 hours ensure that only spore-forming acidogenic microorganisms survive. On the other hand, acid/base treatments can be used to maintain seed microorganisms in extreme pH conditions, effectively removing methanogens that cannot tolerate such environments (Ntaikou et al., 2010).

Genetically Modified Organisms: The production of hydrogen through fermentation, whether it be through pure or mixed cultures, is subject to a great deal of limitations (Ntaikou et al., 2010). The poor conversion of substrates, the formation of by-products, and a low tolerance to oxygen levels and high hydrogen partial pressures are all contributing factors (Abreu, 2011), and a theoretical upper limit of hydrogen yield. To address these challenges, researchers have proposed using genetically modified microbes with improved properties created by mutagenesis or genetic engineering (Abreu, 2011). Key areas of modification include introducing or modifying genes to overexpress cellulolytic enzymes (cellulases, hemicellulases, and lignases) or enzymes for sugar uptake to improve substrate availability and conversion, eliminating hydrogen-consuming hydrogenases, and overexpressing hydrogen-producing hydrogenases that are also hydrogen-tolerant. *E. coli*, a frequently examined facultative anaerobe for hydrogen production, has been extensively utilized for genetic alterations (Ntaikou et al., 2010).

13.4.2 Mechanism and Pathway

The synthesis of biohydrogen by DF is primarily carried out by bacteria that are either anaerobic or facultatively anaerobic (Ntaikou et al., 2010). Although a wide variety of organic compounds, including carbohydrates, sugars, proteins, and lipids, have the potential to serve as substrates, the bioconversion of glucose to acetate is frequently regarded as the standard reference for determining theoretical yields of fermentative hydrogen (Yagmur Goren et al., 2023). It is possible to get a maximum yield of 4 mol H_2/mol of glucose through the process of glucose metabolism, as shown by the reaction (13.1) (Ghimire et al., 2015):

$$C_6H_{12}O_6 + 4H_2O \rightarrow 2CH_3COO^- + 2HCO_3^- + 4H^+ + 4H_2 \quad (13.1)$$

Alternatively, the maximum theoretical yield of 4 mol H_2/mol of glucose can be achieved through a two-step process: the fermentation of glucose to acetate and formate, as represented by reactions 13.2a and 13.2b (Ntaikou et al., 2010):

$$C_6H_{12}C_6 + 2H_2O \rightarrow 2CH_3COO^- + 2HCOO^- + 4H^+ + 2H_2 \tag{13.2a}$$

$$2HCOOH \rightarrow 2CO_2 + 2H_2 \tag{13.2b}$$

In the event where butyrate as opposed to acetate is the sole by-product, the maximum potential hydrogen yield will reduce to 2 mol H_2/mol of glucose, as shown in reaction (13.3) (Ntaikou et al., 2010):

$$C_6H_{12}C_6 + 2H_2O \rightarrow CH_3CH_2CH_2COO^- + 2HCO_3^- + 3H^+ + 2H_2 \tag{13.3}$$

The conversion of glucose to pyruvate is the major step in these processes (13.4). This conversion, which potentially results in the creation of 2 mol of hydrogen through NADH regeneration, is the primary step. Most bacteria utilize the Embden–Meyerhof–Parnas (EMP) pathway (Fang et al., 2015) for this metabolic conversion:

$$C_6H_{12}O_6 + 2NAD^+ \rightarrow 2CH3COCOO^- + 4H^+ + 2NADH \tag{13.4}$$

Acetyl-CoA, which is produced from pyruvate, is an essential component in the production of hydrogen (Fang et al., 2015). Its subsequent metabolic fate determines not only whether the theoretical maximum yield is 4 or 2 mol H_2/mol of glucose, but also whether the actual yield approaches this maximum. Acetyl-CoA can be formed via reactions (13.5) or (13.6) (Ntaikou et al., 2010):

$$\text{pyruvate} + \text{CoA} + \text{Fdox} \leftrightarrow \text{acetyl CoA} + CO_2 + \text{Fdred} \tag{13.5}$$

$$\text{pyruvate} + \text{CoA} \leftrightarrow \text{acetyl CoA} + \text{formate} \tag{13.6}$$

It is pyruvate-ferredoxin oxidoreductase that is responsible for catalyzing reaction (13.5), which results in the production of acetyl-CoA, with ferredoxin serving as an electron receiver. This enzyme can be found in a wide variety of bacteria, including cyanobacteria, facultatively anaerobic bacteria, and bacteria that are totally anaerobic. A further metabolization of acetyl-CoA can result in the production of either acetate or butyrate. The hydrogenase enzyme is responsible for creating 1 mol of hydrogen for every mole of reduced ferredoxin (Ntaikou et al., 2010). In contrast, the formation of butyrate results in the consumption of NADH during the conversion of acetoacetyl-CoA to butyrate. This limitation places the amount of hydrogen produced at a maximum of 2 mol H_2/mol of glucose (Karimi, 2015). In contrast, when butyrate is produced, NADH is consumed in the reduction of acetoacetyl-CoA to butyrate, limiting hydrogen production to 2 mol H_2/mol of glucose (Karimi, 2015). When specific strains of *Clostridium*, such as *C. pasteurianum* and *C. butyricum*, are utilized, the hydrogen yields often fall within the range of 2–4 mol H_2/mol of glucose. This is the case when the growth conditions are combined (Ghimire et al., 2015).

In the second pathway for the creation of acetyl-CoA, which is referred to as reaction (13.6), the enzyme pyruvate formate lyase is involved. This enzyme is responsible for the formation of formate. This particular pathway is frequently reported in enterobacteria, including *Enterobacter aerogenes* and *E. coli*, among others (Ntaikou et al., 2010), and is regulated by environmental oxygen levels. Some *Clostridium* species also possess pyruvate formate lyase (Ghimire et al., 2015). In addition to hydrogen, the fermentation of glucose can result in a wide range of other metabolic products. These products include lactate, propionate, succinate, and 2,3-butanediol, all of which are derived directly from pyruvate. Additionally, ethanol, butanol, and isopropanol are produced as a result of the further metabolism of acetyl-CoA (Ntaikou et al., 2010). These compounds do not contribute to hydrogen production and instead reduce the overall yield, and are therefore considered undesired by-products. It occurs more frequently with germs that have been grown in mixed cultures or under different culture conditions. *E. coli*, for instance, has a propensity to create lactate at pH values that are lower than 7, which can result in decreased hydrogen yields. When dealing with situations like these, it is of the utmost importance to carefully select the optimal growth conditions in order to direct the metabolism toward the creation of hydrogen and to favor microorganisms that produce hydrogen, particularly when employing mixed cultures (Ghimire et al., 2015).

13.4.3 Factors Influencing DF

The production of biohydrogen through DF is a complicated process that is influenced by a number of different factors, including the inoculum, temperature, pH, hydraulic retention time (HRT), gas partial pressure, and fermentation end-products (Soares et al., 2020).

1. **Inoculum**: The production of biohydrogen has been accomplished through the utilization of a variety of bacterial species, including strict anaerobes (*Clostridium*), facultative anaerobes (*Citrobacter*, *Rhodopseudomonas*, *E. coli*, and *Enterobacter*), and specific aerobic bacteria (*Bacillus*). *Clostridium* species are the most extensively researched due to their significant hydrogen production capabilities. Facultative anaerobes produce up to 2 mol H_2/mol of glucose, while strict anaerobes can reach a theoretical maximum of 4 mol H_2/mol of glucose. However, strict anaerobes are particularly sensitive to oxygen; even a brief encounter can entirely stop hydrogen generation. In contrast, facultative anaerobes consume oxygen quickly, helping to restore anaerobic conditions, making them more practical and cost-effective for large-scale biohydrogen production. Mixed microbial cultures have additional benefits, such as the ability to digest fermentation inhibitors such as furfural and partially break down 5-HMF during DF. However, effective pretreatment is required to exclude hydrogen-consuming microbes while selectively enriching hydrogen producers such as *Clostridium* sp. (Soares et al., 2020). Pretreatment methods include thermal, chemical, aeration, and microwave, with heat treatment being the most common. The typical condition is heating to 100°C for 30 minutes, which effectively enriches H_2-producing microorganisms while inhibiting hydrogen consumers.
2. **Temperature**: Temperature plays a crucial role in anaerobic fermentation, directly affecting the activity of hydrogen-producing bacteria. Thermophilic bacteria typically produce more biohydrogen, whereas mesophilic bacteria exhibit higher volumetric production rates due to slower growth and lower cell densities (Łukajtis et al., 2018). The ideal temperature range for this process is 37°C–55°C. The possibility of producing hydrogen rises with a temperature between 20°C and 40°C but falls between 40°C and 55°C. At 40°C, the highest potential for producing hydrogen was found to be 269.9 mL (Wang & Yin, 2017b).
3. **pH**: A key element influencing hydrogen production is pH. It has an impact on metabolic pathways and hydrogenase activity. Higher pH levels can reduce hydrogen generation, while rising pH can enhance it within a certain range. The ideal starting pH for fermentative hydrogen production cannot be fixed due to variations in the inoculum, substrate, and pH range; 4.2 was the ideal pH for fermentative hydrogen generation; however, 7.0 was also mentioned in some studies. When the initial pH was raised from 4.0 to 7.0, the hydrogen production rate increased; however, when the initial pH was raised from 7.0 to 10.0, the production rate decreased (Wang & Yin, 2017b).

4. **HRT**: HRT is a critical operational parameter in anaerobic fermentation, affecting microbial metabolism and biohydrogen production. Different HRTs yield varying results. For instance, hydrogen production from acid-hydrolyzed wheat waste was highest (1.75 L H_2/L.d) at an HRT of 2 hours (Kirli & Karapinar, 2018). Shorter HRTs generally enhance biohydrogen production, but the optimal HRT depends on the specific reactor configuration and feedstock.
5. **H_2 Partial Pressure**: The partial pressure of hydrogen is an essential component in the production of hydrogen in a continuous manner (Soares et al., 2020). As H_2 concentrations increase, hydrogen production declines and metabolic pathways shift toward producing ethanol, acetone, butanol, lactate, or alanine. This shift is less pronounced at higher temperatures. To mitigate the impact of increased H_2 partial pressure, strategies such as nitrogen or carbon dioxide sparging or applying a vacuum have been explored (Soares et al., 2020). For example, reducing pressure in a continuous stirred tank reactor (CSTR) from 760 to 380 mmHg resulted in an 8% increase in hydrogen production (Lee et al., 2012).
6. **Fermentation End-Products**: Fermentative hydrogen production yields organic acids such as acetic, butyric, and propionic acids, which decrease pH and may inhibit hydrogen-producing bacteria (Fang et al., 2015). The maximum theoretical hydrogen yield that can be achieved by acetic acid is 4 mol H_2/mol of glucose. However, the yield is reduced to 2 mol H_2/mol of glucose when butyric acid is taken into consideration. By-products such as lactic acid and ethanol do not contribute to the biohydrogen balance, which results in a reduction in overall yields. Propionic acid, on the other hand, makes use of biohydrogen (Soares et al., 2020).
7. **Bioreactor Configuration**: The bioreactor type has a significant effect on biohydrogen production. CSTR is the most common reactor type, known for a brief start-up phase and efficient mass transfer capabilities (Abreu, 2011). Nevertheless, it requires careful management to prevent cell washout. Alternative methods, including anaerobic membrane bioreactors and immobilized designs such as fluidized bed reactors, are increasingly favored for their capacity to boost hydrogen generation efficiency in lower amounts (Bakonyi et al., 2014). The up-flow anaerobic sludge blanket reactor (UASBR) displays an extended start-up phase; however, it provides significant operational stability once the bacterial granules are formed. Granular systems in UASBRs can be swiftly transitioned from methane-producing to hydrogen-producing granules, thereby significantly decreasing the start-up duration (Bakonyi et al., 2014). Another option is the anaerobic sequencing batch reactor, which can also be used for continuous hydrogen fermentation. Each of these factors must be carefully optimized to maximize biohydrogen production through DF (Ntaikou et al., 2010).

DF is receiving attention as a promising approach for biohydrogen production; however, several challenges are preventing its large-scale application. One of the primary issues is its relatively low hydrogen yield compared to other processes, such as PF or thermochemical methods. Biohydrogen production depends mainly on the type of substrate and the microbial community used (Ntaikou et al., 2010). LCB, which is commonly used as a substrate, presents a challenge due to its complex composition of cellulose, hemicellulose, and lignin, which can limit microbial access to fermentable sugars. This complexity requires effective pretreatment methods, which increase costs and energy demands. On the other hand, the hydrolysis of LCB can release inhibitory compounds like phenolics and organic acids, which can negatively impact microbial activity and reduce hydrogen production (Soares et al., 2020). The metabolic pathways of hydrogen-producing bacteria are also intricate, making it crucial to optimize fermentation conditions—such as pH, temperature, and nutrient availability—to maximize hydrogen yields. In mixed anaerobic cultures, hydrogen-producing bacteria often face competition from methanogenic archaea, which consume hydrogen before it can be captured, further reducing yields. Overall, the economic viability of DF for hydrogen production is challenged by high costs related to feedstock pretreatment, substrate acquisition, and downstream processing (Soares et al., 2020).

13.5 BIOHYDROGEN PRODUCTION THROUGH PF

PF was initially discovered in 1949 with the use of photosynthetic bacteria to produce biohydrogen. PF is a biological process that involves the conversion of organic substrates into hydrogen and carbon dioxide by photosynthetic bacteria. Examples of these bacteria include purple non-sulfur (PNS) bacteria and green sulfur bacteria (Nirmala et al., 2023). PF has various advantages over DF, especially in hydrogen production. This process produces a greater yield of hydrogen per reaction than DF and prevents the buildup of acidic by-products. In addition, PF can occur using a variety of waste products, including agricultural waste, sugar beet molasses, and olive mill wastewater, making it a more adaptable option. Also, from an environmental perspective, it is considered more eco-friendly, with a lower global warming and acidification potential. Moreover, since PF relies on light as its energy source rather than sugars, it reduces competition for land with food crops. This method is also effective in producing high-purity hydrogen without oxygen (Guo et al., 2020). Despite these benefits, PF has some drawbacks, including the need for effective pretreatments to prevent toxicity and specific partial nitrogen conditions for optimum functioning. Photosynthetic or anaerobic bacteria, such as *Rhodobium*, *Rhodobacter*, *Rhodospirillum*, and *Rhodopseudomonas*, utilize light energy to degrade organic substances with the assistance of nitrogenase. This process is referred to as PF (Łukajtis et al., 2018). In PF, the photo-reactor is filled with a fermentation broth including substrate, inoculum, and medium. The photo-reactor of suitable volume is then incubated under a constant light intensity. Next, 5 M HCl or 5 M NaOH is used to bring the pH down to an ideal level. Hydrogen was produced with steady stirring for several days, with samples obtained every 12 hours. It is then collected, and its concentration was measured using a gas chromatograph. The inability of photosynthetic bacteria to efficiently capture sunlight is one of the main challenges in PF. This can lead to a low overall efficiency in the conversion of light energy into hydrogen (Tiang et al., 2020). These bacteria need sterile and controlled conditions to grow and produce hydrogen. Additionally, cell shadowing reduces light penetration in the photo-reactor and may lower hydrogen production. For large-scale use, a large land area is needed for an efficient anaerobic photo-bioreactor (PBR).

13.5.1 Microbes Involved in PF

Purple sulfur, green sulfur, and non-sulfur bacteria are examples of microbial communities involved in PF. These bacteria convert organic substrates into hydrogen and carbon dioxide by absorbing light energy without generating oxygen (V. Singh & Das, 2019). However, comprehensive research data is not available for PF. Various organisms utilize microbial electrolysis to convert organic substrates into hydrogen.

Microbial Electrolysis: Different types of microorganisms and organic substrates are used in these biological processes to generate hydrogen (Hitam & Jalil, 2023). Both monosaccharides and volatile fatty acids (VFAs) are examples of substrates that can be found in lignocellulosic hydrolysate or wastewater samples (Karimi, 2015). A microbial electrolysis cell (MEC) can turn these compounds into hydrogen. The key advantage of this method is that it overcomes obstacles seen in DF processes while producing a significant amount of hydrogen (Hitam & Jalil, 2023). Additionally, in comparison to other methods of producing biohydrogen, this approach not only consumes less energy but also has a lower impact on the environment. Through the utilization of a low additional voltage (below 1.23 V) and particular microorganisms, it is typically capable of converting organic compounds into hydrogen energy (Jayabalan et al., 2020). In MECs, thermophiles or extremophiles and appropriate electrodes, such as carbon cloth or graphite, are essential for hydrogen production. Anaerobic bacteria at the anode are responsible for the metabolism of organic matter, which ultimately leads to the creation of hydrogen gas through a process that is referred to as electron hydrogenesis. These anaerobic bacteria are generally referred to as anode-respiring bacteria (ARB) due to the fact that they are able to transfer electrons from a biodegradable substrate to a solid electrode (Ntaikou et al., 2010). The primary by-products of ARB metabolism in MEC are protons, electrons, and CO_2. These are carried to the anode and subsequently flow to the cathode via an external connection (Fang et al., 2015). At the cathode, electrons combine with protons to generate hydrogen gas. An external voltage of at least 0.2 V is required under conditions of 1 atm pressure, 30°C temperature, and pH 7 to achieve this. This method, while beneficial, has a limitation: hydrogen produced can be readily utilized by methanogens, homoacetogens, and exoelectrogens in a single-chamber configuration. This has resulted in reduced hydrogen production efficiency of MEC (Lin et al., 2020).

13.5.2 Biological Pathway for PF

Through the biological process of PF, photosynthetic bacteria—like the PNS bacterium *R. sphaeroides*—metabolize organic substrates in the presence of light to create hydrogen gas. In order to create a proton gradient across their cell membrane, these bacteria use light-absorbing pigments to absorb light energy (Figure 13.2). This excites electrons in their photosystem, which in turn drives an electron transport chain. Organic compounds such as lactate and acetate undergo oxidation, resulting in the release of electrons and protons. Electrons are transferred to hydrogenase enzymes, facilitating the reduction of protons to

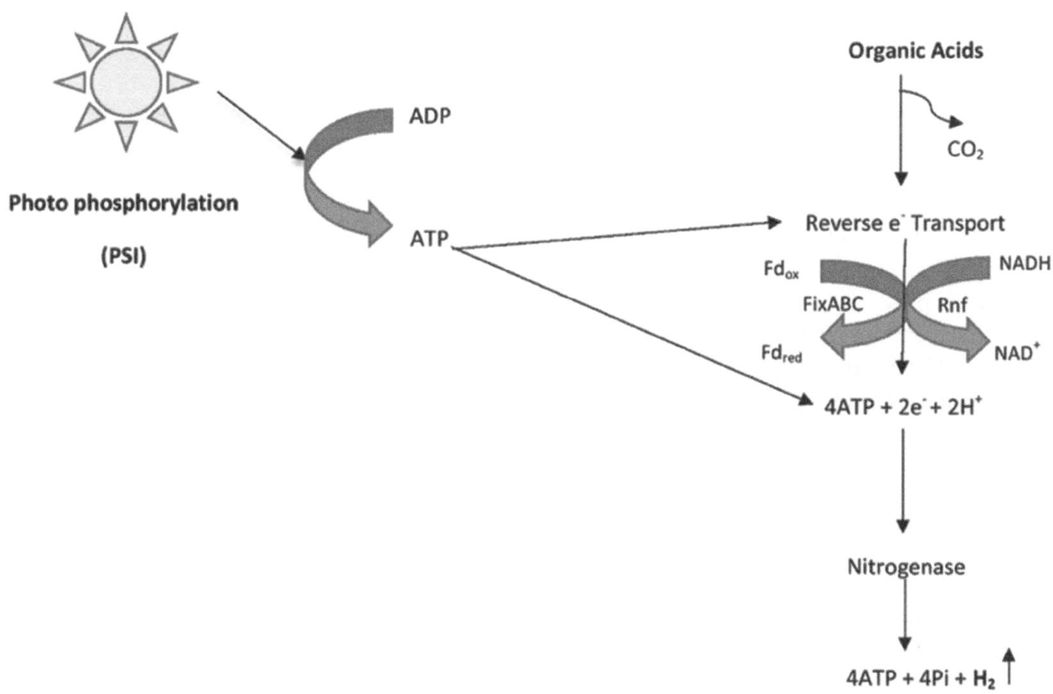

FIGURE 13.2 Biohydrogen production through photo-fermentation: The image represents hydrogen production via photo-fermentation in photosynthetic microorganisms, where light energy is utilized to generate ATP through photophosphorylation. This ATP powers reverse electron transport, which, along with reduced ferredoxin (Fd) and NADH, provides the necessary electrons and energy for the nitrogenase enzyme to convert protons into hydrogen gas (H_2). Organic acids serve as the carbon source, producing CO_2 as a by-product. Thus, the pathway demonstrates how photo-fermentation couples photosynthesis with hydrogen production in microorganisms such as purple non-sulfur bacteria.

produce hydrogen (Fang et al., 2015). The process takes place under anaerobic conditions, as oxygen inhibits hydrogenase activity and necessitates continuous light as an energy source. PF, although promising for renewable energy production, encounters challenges including low efficiency and sensitivity to oxygen.

In PF, Photosystem I (PSI) captures light energy and converts ADP+Pi into ATP. This ATP is essential for powering the subsequent steps, especially nitrogenase activity, which drives the production of hydrogen (H_2).

$$2H^+ + 2e^- + ATP \rightarrow H_2 \tag{13.7}$$

Organic acids (such as acetic acid, butyrate, or lactate) serve as the electron donors in the system (Rehm et al., 2008). These organic acids undergo oxidation, which releases electrons. These electrons pass through a reverse electron transport system involving proteins like FixABC and Rnf, leading to the reduction of NAD^+ into NADH. For example, the oxidation of acetic acid in a stoichiometric reaction is:

$$C_2H_4O_2 + 2H_2O \rightarrow 4H_2 + 2CO_2 \tag{13.8}$$

This equation shows how acetic acid is oxidized, producing electrons, hydrogen (H_2), and carbon dioxide (CO_2). These electrons are funneled into the reverse electron transport system, which eventually reduces ferredoxin. Different organic substrates yield varying stoichiometric outcomes.

Butyrate

$$C_4H_8O_2 + 6H_2O \rightarrow 10H_2 + 4CO_2 \tag{13.9}$$

Lactate

$$C_3H_8O_3 + 4H_2O \rightarrow 8H_2 + 3CO_2 \tag{13.10}$$

Each substrate provides electrons for the reverse electron transport, which eventually ends in the production of NADH, a reducing equivalent for the nitrogenase reaction.

Protons (H^+) are reduced to molecular hydrogen (H_2) by the enzyme nitrogenase, which is powered by the ATP generated by photophosphorylation. In this reaction, four ATP molecules are utilized for each hydrogen produced. Electrons and protons from the substrate oxidation process are used to generate hydrogen.

$$2H^+ + 2e^- + ATP \rightarrow H_2 \tag{13.11}$$

This reaction highlights how ATP, generated via photophosphorylation, enables nitrogenase to reduce protons (H^+) into molecular hydrogen (H_2). The protons come from the water molecules involved in substrate oxidation, such as in the acetic acid oxidation reaction. For example, with acetic acid as the substrate:

$$C_2H_4O_2 + 2H_2O \rightarrow 4H_2 + 2CO_2 \tag{13.12}$$

This process yields four hydrogen molecules via nitrogenase activity and 2 CO_2 molecules as a by-product of organic substrate oxidation. Photophosphorylation provides ATP for nitrogenase activity. Reverse electron transport supplies electrons from organic substrates, which are crucial for reducing protons to hydrogen. The organic substrate oxidation equations (like those of acetic acid, butyrate, and lactate) show how organic acids donate electrons and protons, which are then used by nitrogenase, powered by ATP, to produce molecular hydrogen. Thus, the reactions (photophosphorylation, reverse electron transport, and nitrogenase activity) and the stoichiometric equations are interconnected, creating a sustainable cycle of biohydrogen production from organic substrates under light-driven conditions. The equations illustrate the utilization of organic acids and light energy in PF pathways for hydrogen production, with carbon dioxide as a by-product. The process is significantly influenced by the bacterial strain, the organic substrate employed, and environmental factors such as light intensity and oxygen availability (H. Liu et al., 2020).

13.5.3 Factors Influencing PF

Several factors significantly affect biohydrogen production via PF, including pH, light intensity, stirring mode, substrate concentration, and pretreatment conditions (Ntaikou et al., 2010).

1. **pH**: A starch agricultural waste substrate was used to study the impact of initial pH on PF. This study employed the Gompertz model to monitor hydrogen yield, with initial pH levels ranging from 5 to 9. This model is frequently used in biology to explain how bacteria, cancer cells, plants, and animals grow (Ntaikou et al., 2010). Butyric and acetic acid pathways were the primary means of producing hydrogen during the PF of starch. At pH 7, the highest rates of hydrogen yield (~642 mL), sugar consumption (0.26 g/(L h)), and formation (77.78 mL/(L h)) were attained. Lower pH values produced less hydrogen and slower production rates, most likely due to decreased bacterial growth. Other research has looked at the effects of the initial pH of the buffer on the kinetics of biohydrogen production in maize stalk PF (Ntaikou et al., 2010). Numerous buffers, such as potassium phosphate, sodium sulfate, ferric citrate, and citric acid-sodium citrate, have been employed in PF. At a pH of 6.5, the utilization of a potassium phosphate buffer resulted in the maximum hydrogen production rate, which was around 23.96 mL/h (Fang et al., 2015). Rates were diminished at pH 5.0 and 7.5. Lower pH values enhanced hydrogen production, whereas higher pH values diminished it, likely due to the adverse effects of alkaline conditions on bacterial activity (Guo et al., 2020).
2. **Light Intensity**: A study examined the impact of mixing speed and light intensity on the PF of corn cob using HAU-M1 bacteria. Dynamic mixing and lighting (DM-DL) were found to achieve the highest hydrogen yield of 84.7 mL/g TS (Z. Zhang et al., 2020). They found that optimizing both mixing and lighting conditions enhances hydrogen production, especially with higher substrate concentrations (Goren et al., 2024). Rheological properties describe how materials respond to external forces, affecting their behavior, whether they are liquids, solids, or gases (S. Zhu et al., 2020).
3. **Type of Substrate**: The biodegradable carbon source glucose is derived from industrial effluents and agricultural wastes. The maximum hydrogen yield from glucose is between 2.0 and 2.4 mol/mol, while lower yields could result from partial biodegradation and butyrate synthesis. Compared to other simple sugars, sucrose has been thoroughly investigated for its ability to produce hydrogen gas in both batch and continuous forms with greater yields. Compared to starch, which only produces 140 mL of hydrogen overall, sucrose produces 300 mL of hydrogen. Of the carbon sources that were investigated, *Enterobacter cloacae* ITT-BY 08 yielded the most. All kinds of disaccharides, however, have almost the same yield per mole of hexose (Kapdan & Kargi, 2006).

Reducing carbon emissions is the primary driver of the growing demand for hydrogen as a clean fuel. Due to their reliance on fossil fuels, conventional hydrogen generation techniques including partial oxidation, auto-thermal reforming, steam reforming, and dry reforming raise greenhouse gas emissions. PF is being investigated as a sustainable alternative for biohydrogen production. High hydrogen yields under mild conditions with little energy use are possible using this approach. Improvements are necessary to increase its efficiency. PF necessitates light for the processes of electron transport and hydrogen production. The reaction site is separated into three zones—light development, light inhibition, and light restraining—as a result of the unequal dispersion of light (Li et al., 2020). The three light zones in PF do not use light equally well. One zone often lacks enough light to support efficient biohydrogen production. So, to boost hydrogen production, ongoing research is needed to improve light distribution in photo-reactors. The color of waste waters affects light penetration, for example, light brown indicates fresh wastewater, light to medium gray means it's been stored, and dark gray/black/brown shows it's highly decomposed. Colors in wastewater can block light in PBRs, decreasing the effectiveness of light conversion and limiting the amount of wastewater and residues that can be used for PF. Pretreating these substrates to increase light penetration and broaden the light growth zone is essential to boosting hydrogen production. Forest leftovers are a promising lignocellulosic resource for the production of biohydrogen and account for around 50% of forest biomass. The residual waste from forests is made up of wood chips, sawdust, and wood bark, and it is believed that it has the capacity to generate around 10,000 MW of electricity (Hitam & Jalil, 2023). One of the most defining features of forest biomass is the presence of woody ash, which, in contrast to crop leftovers, makes it stable and easy to handle while it is being transported. This is one of the most important qualities of forest biomass (Fatma et al., 2018). Forestry LCB is abundant, carbon-neutral, and doesn't compete with food supplies, making it ideal for developing cost-effective and eco-friendly biohydrogen production. Wastewater and food wastes often contain harmful compounds like furfural, ammonium, and polyphenols, which can inhibit hydrogen production. Furfural is toxic, ammonium can suppress hydrogen-producing enzymes, and polyphenols can darken the waste, reducing pretreatment efficiency. Removing these inhibitors during pretreatment is crucial. Currently, pretreatment methods are complex and energy-intensive, so simpler, cost-effective approaches are needed.

Challenges for DF, PF, and fermentative biohydrogen generation technologies include low yield and high costs for DF and low hydrogen production rates for PF. These problems might be resolved by integrating these processes with LCB as a substrate. Without requiring significant pH modifications, the overall hydrogen output can be increased by using VFAs from DF as substrates for PF. This integrated approach, operated either in a single or sequential two-stage system, has shown promise in increasing hydrogen production efficiency, as confirmed by various studies (Shanmugam et al., 2020).

13.6 INTEGRATED BIOHYDROGEN PRODUCTION

Challenges like low yield, unavailability of substrates, poor hydrogen production rate, and high cost of production in PF and DF could be overcome by using LCB as the substrate and integrating both these pathways (Cheng et al., 2011). Integrated fermentation refers to the combining of DF and PF, which involves the use of both the types of microbes and the respective optimum conditions to produce a better yield of biohydrogen. Integrated hydrogen production requires nitrogenases and hydrogenases. The effluent of DF, characterized by the production of substantial quantities of VFAs, can serve as an initial substrate for the PF process (Goren et al., 2024). This technique has the potential to produce increased hydrogen yields (Patel et al., 2021). Utilizing organic VFAs that have been left over from the PF process, such as acetic, butyric, lactate, formate, and propionic acids, is an excellent method that shows a great deal of promise for application in integrated fermentation technology (Z. Liu & Smith, 2021). This occurs in either a single-batch system or a two-step batch system. Microorganisms from both the dark fermentation and light fermentation methods are used in combined form in a single batch. The microorganisms involved in both processes may encounter mutual metabolite inhibitory effects, in addition to substrate and product inhibition factors. In single-batch integrated fermentation, pH control is challenging.

Table 13.1 presents a comparison of the advantages and disadvantages associated with various fermentation modes for biohydrogen production. Integrated fermentation synthesizes the advantages of DF and PF, leading to enhanced hydrogen yields and minimized by-product formation (Sivagurunathan et al., 2018).

TABLE 13.1 Different fermentation methods for biohydrogen production

PROCESS	*ADVANTAGES*	*DISADVANTAGES*	*REFERENCES*
Dark fermentation	High hydrogen yields, versatile substrates, lower light requirements	Lower hydrogen production rates, accumulation of inhibitory by-products, potential for methane production	Soares et al. (2020)
Photo-fermentation	Higher hydrogen yields, can utilize organic acids, potential for simultaneous production of other valuable compounds	Requires light energy, sensitive to environmental conditions, lower productivity	Hitam and Jalil (2023)
Integrated fermentation	Higher overall hydrogen yields, reduced by-product accumulation, enhanced substrate utilization	More complex process, potential for compatibility issues between microbial strains	T. Singh et al. (2021)

13.6.1 Methods of Integrated Biohydrogen Production

13.6.1.1 Sequential Biohydrogen Production

Sequential biohydrogen production in an integrated bioprocess refers to the staged generation of hydrogen using different microorganisms or metabolic pathways in a well-coordinated manner. It typically involves two main fermentation processes. Anaerobic bacteria, such as *Clostridium* sp., are responsible for the decomposition of organic substrates, which are typically LCB or organic waste. This process results in the production of hydrogen, organic acids, and additional by-products, including carbon dioxide. The first phase of fermentation is known as DF. Fermentation of glucose takes place during this step, which results in the creation of hydrogen, carbon dioxide, and acetic acid (also known as VFA) (Tawfik & El-Qelish, 2012). The acetic acid evolved is a by-product. In the DF stage, 4 mol hydrogen has evolved.

$$C_6H_{12}O_6 + 2H_2O \rightarrow 2CH_3COOH + 2CO_2 + 4H_2 \quad (13.13)$$

The fundamental constraint of DF is that it is unable to completely utilize all of the organic matter that is available. This leads to a limited hydrogen yield as a result of the accumulation of by-products such as VFAs (Maman et al., 2022). The second stage is PF, photosynthetic bacteria (such as *Rhodobacter* species) are used to further convert the residual organic acids and by-products from DF into additional hydrogen under light conditions. PF offers the advantage of higher hydrogen yields from the intermediate products left behind by DF. In the initial step, the acetic acid generated is utilized by photosynthetic bacteria to produce 8 mol of hydrogen (Mishra et al., 2016).

$$2CH_3COOH + 4H_2O \rightarrow 4CO_2 + 8H_2 \quad (13.14)$$

Combining the two stages would theoretically yield 12 mol of hydrogen. This illustrates the principle of two-step sequential integrated hydrogen production. This integrated approach optimizes hydrogen production by overcoming the limitations of each process. For example, DF is more efficient at breaking down complex substrates quickly but produces a limited amount of hydrogen, while PF can produce additional hydrogen by utilizing the by-products more efficiently. The sequential nature of the process ensures better substrate utilization and improves the overall biohydrogen yield.

Biohydrogen was produced from Palm Oil Mill Effluent (POME) using a two-stage integrated system (Wang & Yin, 2017a). In the first DF step, 0.784 mL H_2/mL POME was produced using *C. butyricum* LS2. After that, *Rhodopseudomonas palustris* was used as the inoculum in the PF stage using the effluent from the DF stage. This sequential process resulted in a notable enhancement of hydrogen yield, achieving 3.064 mL H_2/mL POME (Mishra et al., 2019). The combined DF and PF processes offer several advantages, including higher hydrogen yields and the potential for wastewater treatment through chemical oxygen demand (COD) reduction.This integrated approach demonstrates the synergistic benefits of combining different microbial processes for enhanced biohydrogen production (Figure 13.3).

13.6.1.2 Biohydrogen Production Using Mixed Co-culture Method

The mixed co-culture method involves cultivating two or more distinct microbial species within the same bioreactor, allowing them to work together to carry out a desired bioprocess. In biohydrogen production, this method takes advantage of the complementary metabolic activities of different microbes to enhance hydrogen yields. Typically, one set of bacteria, such as DF bacteria, breaks down complex organic matter to produce hydrogen and metabolic by-products like organic acids. Meanwhile, another group of microorganisms, such as homofermentative bacteria, simultaneously consumes these by-products to generate additional hydrogen. This process creates a synergistic relationship where each microorganism benefits from the activities of the others, leading to improved overall efficiency.

One of the main advantages of the mixed co-culture system is its ability to operate within a single bioreactor, simplifying the overall design and reducing operational costs. Unlike sequential biohydrogen production, which requires separate reactors for each stage (DF and PF), mixed co-culture integrates both processes into a single environment. This streamlined setup can be particularly beneficial in large-scale biohydrogen production, as it allows for better resource utilization and reduced infrastructure costs. Furthermore, microbial cooperation in co-culture systems helps reduce the accumulation of inhibitory compounds, such as VFAs, that can halt the hydrogen production process in a single-culture system.

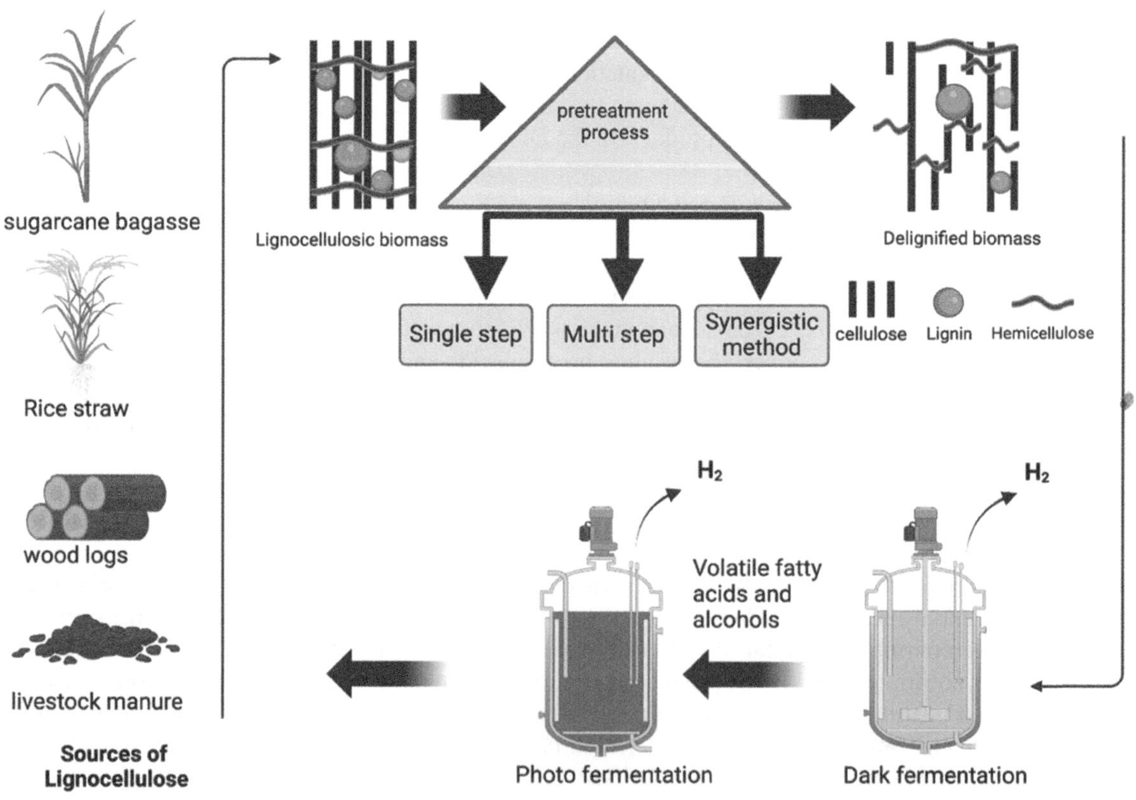

FIGURE 13.3 Biohydrogen production through integrated fermentation: The schematic diagram illustrates the strategies to improve biohydrogen production from LCB through integrated fermentation.

A mixed co-culture system reduces operational costs and minimizes contamination levels. Additionally, the synergistic interactions between different microbial species in mixed cultures can optimize metabolic pathways and enhance hydrogen yield. Odom and Wall (1983) conducted pioneering research on mixed co-culture hydrogen production, utilizing *Cellulomonas* strain and *Rhodopseudomonas capsulata*. Their findings demonstrated the feasibility of this approach and highlighted the potential for improved hydrogen yields. Subsequent studies by Asada et al. (2006) further explored the capabilities of mixed co-culture systems. The utilization of immobilized *Lactobacillus delbrueckii* and *Rhodopseudomonas sphaeroides* resulted in a hydrogen yield of 7.1 mol H_2/mol of glucose. This finding highlights the efficacy of mixed co-culture systems in biohydrogen production (Vertés et al., 2020). Genetic engineering techniques can also be employed to enhance hydrogen yield in mixed co-culture systems. For example, eliminating genes responsible for hydrogen consumption or modifying metabolic pathways can redirect carbon flux toward hydrogen production. Wu et al. (2010) successfully demonstrated this approach by inactivating the *adhE* gene in *Klebsiella oxytoca*, resulting in reduced alcohol production and increased hydrogen yield.

However, mixed co-culture systems come with their challenges. Balancing the growth and activity of different microorganisms can be difficult, as each species may have unique environmental needs, such as specific pH, temperature, and nutrient requirements (Rai & Singh, 2016). Several factors influence hydrogen yield in mixed co-culture systems, including substrate type and pretreatment, inoculum selection, light intensity, bioreactor design, and operational parameters. Optimizing these factors is crucial for maximizing hydrogen production. Additionally, the interactions between the different microbes must be carefully managed to avoid competition or dominance by one species over the others. Despite these challenges, mixed co-culture offers a promising approach to improving biohydrogen production by enabling more efficient substrate utilization and higher hydrogen yields. Further research is needed to optimize these systems and explore their potential for large-scale applications (Kumar et al., 2024). By understanding the factors influencing hydrogen yield and utilizing genetic engineering techniques, researchers can unlock the full potential of mixed co-culture systems for sustainable biohydrogen production.

13.6.3 Advantages of Integrated Biohydrogen Production

Higher Hydrogen Yield: In addition to producing hydrogen, DF creates organic acid by-products such as butyric and acetic acid. The total hydrogen yield can be further increased by using these by-products as efficient substrates in PF. In one investigation, the *E. cloacae* DM11 strain dark fermented glucose as a substrate, yielding 1.86 mol H_2/mol glucose of hydrogen (Wang & Yin, 2017a). A hydrogen yield of 1.72 mol H_2/glucose was obtained by photo-fermenting the waste medium from DF with *R. sphaeroides* O.U. 001 (Wang & Yin, 2017a). Interestingly, the integrated system outperformed single-stage devices in terms of overall hydrogen yield (Nath et al., 2005).

Efficient Waste Utilization: DF generates partially degraded substrates rich in VFAs. Instead of treating these by-products as waste, PF can effectively convert them into additional hydrogen, reducing waste output and improving the sustainability of the overall process. A study utilizing sucrose as a substrate illustrated the efficacy of this integrated approach. DF conducted by *C. pasteurianum* resulted in the production of 3.80 mol H_2/mol of sucrose, generating a by-product stream that is abundant in butyric and acetic acids. The by-products were subsequently employed in PF by *R. palustris*, leading to an enhanced hydrogen yield of 10.25 mol H_2/mol of sucrose (Ramprakash & Karuppan, 2019). This two-stage process demonstrates the combined advantages of DF and PF in improving biohydrogen production (Chen et al., 2010).

Improved Energy Efficiency: Integrating the two processes enhances the system's energy efficiency. DF transpires in the absence of light, whereas PF necessitates light for its process (Soares et al., 2020). Coupling these processes optimizes energy inputs according to the specific requirements of each process. Studies have demonstrated that enhanced energy efficiency can result from optimizing operating parameters such as pH, temperature, and retention duration in integrated biohydrogen production systems. *R. sphaeroides* and *Enterobacter aerogenes*, DF and PF bacteria, were utilized in a 1:7 (w/w) ratio for hydrogen production (Sivagurunathan et al., 2018). A decrease in pH was noted within the PBR during fermentation, attributed to the accumulation of VFAs. A total of 53% of the initial carbohydrate was converted to hydrogen, resulting in a production yield of 6.31 mol H_2/mol of carbohydrate (Das & Basak, 2024). The research showed that there is high energy efficiency in the integrated process.

Reduced Environmental Impact: Integrated fermentation systems yield reduced levels of inhibitory by-products, including organic acids and alcohols, due to the ability of the PF process to utilize these intermediates. This leads to reduced environmental pollution and improved waste management. A study looked into how well *Spirulina platensis* hydrolysate worked as a substrate for *Bacillus firmus*-mediated DF. The findings showed a hydrogen output of 1.92 ± 0.20 mmol H_2/g reduction in COD. Under ideal circumstances, the subsequent PF step with *R. sphaeroides* produced a hydrogen yield of $1{,}510 \pm 35$ mL/L H_2 (Pandey et al., 2021).

Cost-Efficiency Process: Integrating DF and PF reduces the cost of waste treatment as organic acids produced from DF are not discarded but used in PF, thus lowering operational costs (Rai & Singh, 2016). There is also a reduction in the need for external substrate inputs, which can decrease the overall process cost.

Possibility of Continuous Operation: The integration allows for continuous hydrogen production. DF can be operated continuously with organic wastes, and its by-products can feed directly into a PF system, leading to a more streamlined and efficient process. It is mainly done in sequential biohydrogen production. For instance, *C. pasteurianum* was used in the DF stage using sucrose as the substrate. This step yielded 3.85 mol H_2/mol sucrose. In the PF step, *R. palustris* WP3-5 yielded 10.25 mol H_2/mol sucrose (Chen et al., 2008). A comparison of hydrogen yields in different methods is illustrated in Table 13.2.

13.7 CONCLUSION

Integrated biohydrogen production combines the positive outlooks of DF and PF and overcomes both their shortcomings. Through several research studies conducted, it is evident that integrated process enhances biohydrogen production and substrate utilization. Many factors should be considered for the optimization of conditions. Substrate composition determines biohydrogen yield. Organic feedstocks are used, which contain lignocellulose. The amount of lignin in LCB affects the production and yield of biohydrogen. LCB is a potential substrate which reduces production costs and helps in bioremediation and waste reductions. By implementing nanomaterials as catalysts, the integrated process can be more effective and sustainable for improving process economy. Genetic engineering of microbes used as inoculum is the next stage for producing higher yields of biohydrogen, as seen in the research conducted by scientists who genetically manipulated bacteria to improve production. Light intensity, pH and hydraulic retention rate (HRT) should be optimized in two-stage systems. More research studies should be conducted on a large scale. Also, more studies should focus on removing inhibitory components formed after DF. There should be a system to properly neutralize the acidic effects of effluents so that photo bacteria can act on the effluent to generate hydrogen during PF. This process should be further conducted as continuous fermentation because it extends fermentation periods and enhances process yields. To conclude, integrated production will be the future, and this can be achieved soon by implementing proper substrate, nano-catalyst, and genetic manipulation techniques. A multidisciplinary expertise from chemical, biological, and material sciences can escalate the process and yields.

TABLE 13.2 Comparative hydrogen yields from lignocellulosic biomass via dark, photo-, and integrated fermentation

	SUBSTRATE	*PRETREATMENT METHOD*	*FERMENTATION TYPE*	*HYDROGEN YIELD (mL/g SUBSTRATE)*	*REFERENCE*
Dark fermentation	Cornstalk	Fungal hydrolysis (*Trichoderma viride*)	Dark (*Thermoanaerobacterium thermosaccharolyticum* W16)	73.5 (3.28 mmol/g)	Kucharska et al. (2019) Reginatto and Antônio (2015)
	Sugarcane bagasse	NaOH (0.1 mol/L)+enzymatic hydrolysis	Dark (preheated anaerobic sludge)	300 (13.4 mmol/g TVS)	Reginatto and Antônio (2015)
	Wheat straw	Ozone+enzymatic hydrolysis	Dark (preheated cow manure/sediment)	71.7 (3.2 mmol/g)	Reginatto and Antônio (2015)
Photo-fermentation	Sweet potato stem	None (raw biomass)	Photo (mixed culture)	65.28	H. Zhang et al. (2022)
	Sweet potato leaf	None (raw biomass)	Photo (mixed culture)	18.42	H. Zhang et al. (2022)
Integrated fermentation	Corn stover	Delignification (2% NaOH)+enzymatic hydrolysis	Dark (*Thermoanaerobacter mathranii* A3N)+Photo (*Rhodobacter sphaeroides*)	168 (7.5 mmol/g)	Reginatto and Antônio (2015) Chandran and Mohan (2023)

REFERENCES

Abraham, A., Mathew, A. K., Park, H., Choi, O., Sindhu, R., Parameswaran, B., Pandey, A., Park, J. H., & Sang, B. I. (2020). Retracted: Pretreatment strategies for enhanced biogas production from lignocellulosic biomass. *Bioresource Technology*, 301, 122725. https://doi.org/10.1016/J.BIORTECH.2019.122725

Abreu, A. A. (2011). *Optimization of Dark Fermentation Processes for Biohydrogen Production from Sugars*. https://repositorium.sdum.uminho.pt/handle/1822/14268

Alinia, R., Zabihi, S., Esmaeilzadeh, F., & Kalajahi, J. F. (2010). Pretreatment of wheat straw by supercritical CO_2 and its enzymatic hydrolysis for sugar production. *Biosystems Engineering*, 107(1), 61–66. https://doi.org/10.1016/J.BIOSYSTEMSENG.2010.07.002

Asada, Y., Tokumoto, M., Aihara, Y., Oku, M., Ishimi, K., Wakayama, T., Miyake, J., Tomiyama, M., & Kohno, H. (2006). Hydrogen production by co-cultures of Lactobacillus and a photosynthetic bacterium, Rhodobacter sphaeroides RV. *International Journal of Hydrogen Energy*, 31(11), 1509-1513. https://doi.org/10.1016/j.ijhydene.2006.06.017

Bak, J. S., Ko, J. K., Han, Y. H., Lee, B. C., Choi, I. G., & Kim, K. H. (2009). Improved enzymatic hydrolysis yield of rice straw using electron beam irradiation pretreatment. *Bioresource Technology*, 100(3), 1285–1290. https://doi.org/10.1016/J.BIORTECH.2008.09.010

Bakonyi, P., Nemestóthy, N., Simon, V., & Bélafi-Bakó, K. (2014). Review on the start-up experiences of continuous fermentative hydrogen producing bioreactors. *Renewable and Sustainable Energy Reviews*, 40, 806–813. https://doi.org/10.1016/j.rser.2014.08.014

Ballesteros, I., Oliva, J. M., Navarro, A. A., González, A., Carrasco, J., & Ballesteros, M. (2000). Effect of chip size on steam explosion pretreatment of softwood. *Applied Biochemistry and Biotechnology – Part A Enzyme Engineering and Biotechnology*, 84–86, 97–110. https://doi.org/10.1385/ABAB:84-86:1-9:97

Bastidas-Oyanedel, J. R., Bonk, F., Thomsen, M. H., & Schmidt, J. E. (2015). Dark fermentation biorefinery in the present and future (bio)chemical industry. *Reviews in Environmental Science and Biotechnology*, 14(3), 473–498. https://doi.org/10.1007/s11157-015-9369-3

Bhatia, S. K., Jagtap, S. S., Bedekar, A. A., Bhatia, R. K., Rajendran, K., Pugazhendhi, A., Rao, C. V., Atabani, A. E., Kumar, G., & Yang, Y. H. (2021). Renewable biohydrogen production from lignocellulosic biomass using fermentation and integration of systems with other energy generation technologies. *Science of the Total Environment*, 765, 144429. https://doi.org/10.1016/J.SCITOTENV.2020.144429

Bundhoo, M. A. Z., Mohee, R., & Hassan, M. A. (2015). Effects of pre-treatment technologies on dark fermentative biohydrogen production: A review. *Journal of Environmental Management*, 157, 20–48. https://doi.org/10.1016/j.jenvman.2015.04.006

Cabrol, L., Marone, A., Tapia-Venegas, E., Steyer, J. P., Ruiz-Filippi, G., & Trably, E. (2017). Microbial ecology of fermentative hydrogen producing bioprocesses: Useful insights for driving the ecosystem function. *FEMS Microbiology Reviews*, 41(2), 158–181. https://doi.org/10.1093/FEMSRE/FUW043

Carrère, H., Dumas, C., Battimelli, A., Batstone, D. J., Delgenès, J. P., Steyer, J. P., & Ferrer, I. (2010). Pretreatment methods to improve sludge anaerobic degradability: A review. *Journal of Hazardous Materials*, 183(1–3), 1–15. https://doi.org/10.1016/j.jhazmat.2010.06.129

Chandran, E. M., & Mohan, E. (2023). Sustainable biohydrogen production from lignocellulosic biomass sources - metabolic pathways, production enhancement, and challenges. *Environmental Science and Pollution Research International*, 30(46), 102129–102157. https://doi.org/10.1007/S11356-023-29617-Z

Chen, C. Y., Yang, M. H., Yeh, K. L., Liu, C. H., & Chang, J. S. (2008). Biohydrogen production using sequential two-stage dark and photo fermentation processes. *International Journal of Hydrogen Energy*, 33(18), 4755–4762. https://doi.org/10.1016/J.IJHYDENE.2008.06.055

Chen, C. Y., Yeh, K. L., Lo, Y. C., Wang, H. M., & Chang, J. S. (2010). Engineering strategies for the enhanced photo-H_2 production using effluents of dark fermentation processes as substrate. *International Journal of Hydrogen Energy*, 35(24), 13356–13364. https://doi.org/10.1016/J.IJHYDENE.2009.11.070

Cheng, C. L., Lo, Y. C., Lee, K. S., Lee, D. J., Lin, C. Y., & Chang, J. S. (2011). Biohydrogen production from lignocellulosic feedstock. *Bioresource Technology*, 102(18), 8514–8523. https://doi.org/10.1016/J.BIORTECH.2011.04.059

Das, S. R., & Basak, N. (2024). Optimization of process parameters for enhanced biohydrogen production using potato waste as substrate by combined dark and photo fermentation. *Biomass Conversion and Biorefinery*, 14(4), 4791–4811. https://doi.org/10.1007/s13399-022-02588-w

Fang, Z., Smith, R. L., & Qi, X. (Eds.). (2015). *Production of Hydrogen from Renewable Resources*. Springer Dordrecht. https://doi.org/10.1007/978-94-017-7330-0

Fatma, S., Hameed, A., Noman, M., Ahmed, T., Shahid, M., Tariq, M., Sohail, I., & Tabassum, R. (2018). Lignocellulosic biomass: A sustainable bioenergy source for the future. *Protein & Peptide Letters*, 25(2), 148–163. https://doi.org/10.2174/0929866525666180122144504

Ferdeş, M., Dincă, M. N., Moiceanu, G., Zabava, B. Ş., & Paraschiv, G. (2020). Microorganisms and enzymes used in the biological pretreatment of the substrate to enhance biogas production: A review. *Sustainability (Switzerland)*, 12(17), 7205. https://doi.org/10.3390/SU12177205

Ghimire, A., Frunzo, L., Pirozzi, F., Trably, E., Escudie, R., Lens, P. N. L., & Esposito, G. (2015). A review on dark fermentative biohydrogen production from organic biomass: Process parameters and use of by-products. *Applied Energy*, 144, 73–95. https://doi.org/10.1016/j.apenergy.2015.01.045

Goren, A. Y., Dincer, I., & Khalvati, A. (2024). Green biohydrogen production from renewable plant-based resources: A comparative evaluation. *Process Safety and Environmental Protection*, 185, 947–977. https://doi.org/10.1016/J.PSEP.2024.03.056

Guo, S., Lu, C., Wang, K., Wang, J., Zhang, Z., Jing, Y., & Zhang, Q. (2020). Enhancement of pH values stability and photo-fermentation biohydrogen production by phosphate buffer. *Bioengineered*, 11(1), 291–300. https://doi.org/10.1080/21655979.2020.1736239

Hawkes, F. R., Dinsdale, R., Hawkes, D. L., & Hussy, I. (2002). Sustainable fermentative hydrogen production: Challenges for process optimisation. *International Journal of Hydrogen Energy*, 27, 1339–1347.

Hay, J. X. W., Wu, T. Y., Juan, J. C., & Md. Jahim, J. (2013). Biohydrogen production through photo fermentation or dark fermentation using waste as a substrate: Overview, economics, and future prospects of hydrogen usage. *Biofuels, Bioproducts and Biorefining*, 7(3), 334–352. https://doi.org/10.1002/BBB.1403

Hitam, C. N. C., & Jalil, A. A. (2023). A review on biohydrogen production through photo-fermentation of lignocellulosic biomass. *Biomass Conversion and Biorefinery*, 13(10), 8465–8483. https://doi.org/10.1007/s13399-020-01140-y

I. Eroglu, P. Sevinc, U, G., & M. Yucel. (2010). *The Effect of Temperature and Light Intensity on Hydrogen Production by Rhodobacter Capsulatus*. Forschungszentrum Jülich.

Jayabalan, T., Matheswaran, M., Preethi, V., & Naina Mohamed, S. (2020). Enhancing biohydrogen production from sugar industry wastewater using metal oxide/graphene nanocomposite catalysts in microbial electrolysis cell. *International Journal of Hydrogen Energy*, 45(13), 7647–7655. https://doi.org/10.1016/J.IJHYDENE.2019.09.068

Jiang, D., Ge, X., Zhang, T., Chen, Z., Zhang, Z., He, C., Zhang, Q., & Li, Y. (2020). Effect of alkaline pretreatment on photo-fermentative hydrogen production from giant reed: Comparison of NaOH and $Ca(OH)_2$. *Bioresource Technology*, 304, 123001. https://doi.org/10.1016/J.BIORTECH.2020.123001

Kapdan, I. K., & Kargi, F. (2006). Bio-hydrogen production from waste materials. *Enzyme and Microbial Technology*, 38(5), 569–582. https://doi.org/10.1016/J.ENZMICTEC.2005.09.015

Karimi, K. (Ed.). (2015). *Lignocellulose-Based Bioproducts. 1*. https://doi.org/10.1007/978-3-319-14033-9

Kirli, B., & Karapinar, I. (2018). The effect of HRT on biohydrogen production from acid hydrolyzed waste wheat in a continuously operated packed bed reactor. *International Journal of Hydrogen Energy*, *43*(23), 10678–10685. https://doi.org/10.1016/J.IJHYDENE.2018.01.175

Kongjan, P. (2017). *General Rights Biohydrogen Production from Sugar Rich Substrates Using the Dark Fermentation Process*. Technical University of Denmark.

Koukios, E., & Panagiotopoulos, I. (John) A. (2018). Biohydrogen Production from Agricultural Residues. In *Encyclopedia of Sustainability Science and Technology*. Springer. https://doi.org/10.1007/978-1-4939-2493-6_949-1

Kucharska, K., Cieśliński, H., Rybarczyk, P., Słupek, E., Łukajtis, R., Wychodnik, K., & Kamiński, M. (2019). Fermentative conversion of two-step pre-treated lignocellulosic biomass to hydrogen. *Catalysts*, 9(10), 858. https://doi.org/10.3390/CATAL9100858

Kucharska, K., Rybarczyk, P., Hołowacz, I., Łukajtis, R., Glinka, M., & Kamiński, M. (2018). Pretreatment of lignocellulosic materials as substrates for fermentation processes. *Molecules*, 23(11), 2937. https://doi.org/10.3390/MOLECULES23112937

Kumar Sharma, A., Kumar Ghodke, P., Goyal, N., Nethaji, S., & Chen, W. H. (2022). Machine learning technology in biohydrogen production from agriculture waste: Recent advances and future perspectives. *Bioresource Technology*, 364, 128076. https://doi.org/10.1016/J.BIORTECH.2022.128076

Kumar, V., Kumar, S., Verma, P., & Bhat, S. A. (2024). Microbial nexus for sustainable wastewater treatment: Resources, efficiency, and reuse. In *Microbial Nexus for Sustainable Wastewater Treatment: Resources, Efficiency, and Reuse* (pp. 1–416). Routledge Taylor & Francis Group.

Lee, K. S., Tseng, T. S., Liu, Y. W., & Hsiao, Y. D. (2012). Enhancing the performance of dark fermentative hydrogen production using a reduced pressure fermentation strategy. *International Journal of Hydrogen Energy*, *37*(20), 15556–15562. https://doi.org/10.1016/J.IJHYDENE.2012.04.039

Li, Y., Zhang, Z., Xia, C., Jing, Y., Zhang, Q., Li, S., Zhu, S., & Jin, P. (2020). Photo-fermentation biohydrogen production and electrons distribution from dark fermentation effluents under batch, semi-continuous and continuous modes. *Bioresource Technology*, 311, 123549. https://doi.org/10.1016/J.BIORTECH.2020.123549

Lin, R., Deng, C., Cheng, J., & Murphy, J. D. (2020). Low concentrations of furfural facilitate biohydrogen production in dark fermentation using Enterobacter aerogenes. *Renewable Energy*, 150, 23–30. https://doi.org/10.1016/J.RENENE.2019.12.106

Liu, H., Zhang, Z., Zhang, H., Lee, D. J., Zhang, Q., Lu, C., & He, C. (2020). Evaluation of hydrogen yield potential from Chlorella by photo-fermentation under diverse substrate concentration and enzyme loading. *Bioresource Technology*, 303, 122956. https://doi.org/10.1016/J.BIORTECH.2020.122956

Liu, Z., & Smith, S. R. (2021). Enzyme recovery from biological wastewater treatment. *Waste and Biomass Valorization*, 12(8), 4185–4211. https://doi.org/10.1007/s12649-020-01251-7

Łukajtis, R., Hołowacz, I., Kucharska, K., Glinka, M., Rybarczyk, P., Przyjazny, A., & Kamiński, M. (2018). Hydrogen production from biomass using dark fermentation. *Renewable and Sustainable Energy Reviews*, 91, 665–694. https://doi.org/10.1016/J.RSER.2018.04.043

Maman, M., Sundaram, M., Sarathchandra, G., Kanagaraj, V., & Sundaram, S. M. (2022). Biohydrogen production – An overview on the factors affecting, microbes and enzymes involved. *Asian Journal of Microbiology, Biotechnology and Environmental Sciences*, 24, 472–479. https://doi.org/10.53550/ajmbes.2022.v24i03.005

Mishra, P., Thakur, S., Singh, L., Wahid, Z. A., Sakinah, M. (2016). Enhanced hydrogen production from palm oil mill effluent using two stage sequential dark and photo fermentation. *International Journal of Hydrogen Energy*, *41*(41), 18431–18440. https://doi.org/ 10.1016/j.ijhydene.2016.07.138

Mishra, P., Krishnan, S., Rana, S., Singh, L., Sakinah, M., & Ab Wahid, Z. (2019). Outlook of fermentative hydrogen production techniques: An overview of dark, photo and integrated dark-photo fermentative approach to biomass. *Energy Strategy Reviews*, *24*, 27–37. https://doi.org/10.1016/j.esr.2019.01.001

Monlau, F., Barakat, A., Trably, E., Dumas, C., Steyer, J. P., & Carrère, H. (2013). Lignocellulosic materials into biohydrogen and biomethane: Impact of structural features and pretreatment. *Critical Reviews in Environmental Science and Technology*, 43(3), 260–322. https://doi.org/10.1080/10643389.2011.604258

Nasr, Noha El-Sayed, "Investigation of Bio-hydrogen Production Optimization from Synthetic and Real Wastes using Pure and Mixed Cultures" (2017). Electronic Thesis and Dissertation Repository. 4537. https://ir.lib.uwo.ca/etd/4537

Nath, K., Kumar, A., & Das, D. (2005). Hydrogen production by *Rhodobacter sphaeroides* strain O.U.001 using spent media of *Enterobacter cloacae* strain DM11. *Applied Microbiology and Biotechnology*, 68(4), 533–541. https://doi.org/10.1007/s00253-005-1887-4

Nirmala, N., Praveen, G., AmitKumar, S., SundarRajan, P. S., Baskaran, A., Priyadharsini, P., SanjayKumar, S. P., Dawn, S. S., Pavithra, K. G., Arun, J., & Pugazhendhi, A. (2023). A review on biological biohydrogen production: Outlook on genetic strain enhancements, reactor model and techno-economics analysis. *Science of the Total Environment*, 896, 165143. https://doi.org/10.1016/J.SCITOTENV.2023.165143

Ntaikou, I., Antonopoulou, G., & Lyberatos, G. (2010). Biohydrogen production from biomass and wastes via dark fermentation: A review. *Waste and Biomass Valorization*, 1(1), 21–39. https://doi.org/10.1007/s12649-009-9001-2

Ong, V. Z., & Wu, T. Y. (2020). An application of ultrasonication in lignocellulosic biomass valorisation into bio-energy and bio-based products. *Renewable and Sustainable Energy Reviews*, 132. https://doi.org/10.1016/j.rser.2020.109924

Odom, J. M., & Wall, J. D. (1983). Photoproduction of H2 from cellulose by an anaerobic bacterial coculture. *Applied and Environmental Microbiology*, *45*(4), 1300–1305. https://doi.org/10.1128/aem.45.4.1300-1305.1983

Pachapur, V. L., Kutty, P., Pachapur, P., Brar, S. K., Le Bihan, Y., Galvez-Cloutier, R., & Buelna, G. (2019). Seed pretreatment for increased hydrogen production using mixed-culture systems with advantages over pure-culture systems. *Energies 2019, Vol.* 12*,* Page 530, 12(3), 530. https://doi.org/10.3390/EN12030530

Pandey, A., Sinha, P., & Pandey, A. (2021). Hydrogen production by sequential dark and photofermentation using wet biomass hydrolysate of Spirulina platensis: Response surface methodological approach. *International Journal of Hydrogen Energy*, 46(10), 7137–7146. https://doi.org/10.1016/J.IJHYDENE.2020.11.205

Patel, A., Sarkar, O., Rova, U., Christakopoulos, P., & Matsakas, L. (2021). Valorization of volatile fatty acids derived from low-cost organic waste for lipogenesis in oleaginous microorganisms-A review. *Bioresource Technology*, 321, 124457. https://doi.org/10.1016/J.BIORTECH.2020.124457

Pattra, S., Sangyoka, S., Boonmee, M., & Reungsang, A. (2008). Bio-hydrogen production from the fermentation of sugarcane bagasse hydrolysate by *Clostridium butyricum*. *International Journal of Hydrogen Energy*, 33(19), 5256–5265. https://doi.org/10.1016/J.IJHYDENE.2008.05.008

Rai, P. K., & Singh, S. P. (2016). Integrated dark- and photo-fermentation: Recent advances and provisions for improvement. *International Journal of Hydrogen Energy*, 41(44), 19957–19971. https://doi.org/10.1016/J.IJHYDENE.2016.08.084

Ramprakash, B., & Karuppan, M. (2019). Hydrogen production from wastewater by biochemical methods. In *Encyclopedia of Water* (pp. 1–17). https://doi.org/10.1002/9781119300762.WSTS0016

Reddy, K., Nasr, M., Kumari, S., Kumar, S., Gupta, S. K., Enitan, A. M., & Bux, F. (2017). Biohydrogen production from sugarcane bagasse hydrolysate: effects of pH, S/X, Fe^{2+}, and magnetite nanoparticles. *Environmental Science and Pollution Research*, 24(9), 8790–8804. https://doi.org/10.1007/S11356-017-8560-1

Reginatto, V., & Antônio, R. V. (2015). Fermentative hydrogen production from agroindustrial lignocellulosic substrates. *Brazilian Journal of Microbiology*, 46(2), 323. https://doi.org/10.1590/S1517-838246220140111

Rehm, H. J., Reed, G., Pühler, A., & Stadler, P. J. W. (2008). *Biotechnology: Second, Completely Revised Edition* (pp. 1–12). Wiley. https://doi.org/10.1002/9783527620999

Ren, N. Q., Zhao, L., Chen, C., Guo, W. Q., & Cao, G. L. (2016). A review on bioconversion of lignocellulosic biomass to H_2: Key challenges and new insights. *Bioresource Technology*, 215, 92–99. https://doi.org/10.1016/J.BIORTECH.2016.03.124

Richardson, Y., Drobek, M., Julbe, A., Blin, J., & Pinta, F. (2015). Biomass gasification to produce syngas. In *Recent Advances in Thermochemical Conversion of Biomass* (pp. 213–250). Elsevier Inc. https://doi.org/10.1016/B978-0-444-63289-0.00008-9

Roque, B. A. C., Cavalcanti, M. H. C., Brasileiro, P. P. F., Gama, P. H. R. P., dos Santos, V. A., Converti, A., Benachour, M., & Sarubbo, L. A. (2025). Hydrogen-powered future: Catalyzing energy transition, industry decarbonization and sustainable economic development: A review. *Gondwana Research*, 140, 159–180. https://doi.org/10.1016/J.GR.2025.01.012

Shanmugam, S., Hari, A., Pandey, A., Mathimani, T., Felix, L. O., & Pugazhendhi, A. (2020). Comprehensive review on the application of inorganic and organic nanoparticles for enhancing biohydrogen production. *Fuel*, 270, 117453. https://doi.org/10.1016/J.FUEL.2020.117453

Singh, T., Alhazmi, A., Mohammad, A., Srivastava, N., Haque, S., Sharma, S., Singh, R., Yoon, T., & Gupta, V. K. (2021). Integrated biohydrogen production via lignocellulosic waste: Opportunity, challenges & future prospects. *Bioresource Technology*, 338, 125511. https://doi.org/10.1016/J.BIORTECH.2021.125511

Singh, V., & Das, D. (2019). Potential of hydrogen production from biomass. In *Science and Engineering of Hydrogen-Based Energy Technologies: Hydrogen Production and Practical Applications in Energy Generation* (pp. 123–164). https://doi.org/10.1016/B978-0-12-814251-6.00003-4

Sivagurunathan, P., Kuppam, C., Mudhoo, A., Saratale, G. D., Kadier, A., Zhen, G., Chatellard, L., Trably, E., & Kumar, G. (2018). A comprehensive review on two-stage integrative schemes for the valorization of dark fermentative effluents. *Critical Reviews in Biotechnology*, 38(6), 868–882. https://doi.org/10.1080/07388551.2017.1416578

Soares, J. F., Confortin, T. C., Todero, I., Mayer, F. D., & Mazutti, M. A. (2020). Dark fermentative biohydrogen production from lignocellulosic biomass: Technological challenges and future prospects. *Renewable and Sustainable Energy Reviews*, 117, 109484. https://doi.org/10.1016/J.RSER.2019.109484

Sołowski, G., Konkol, I., & Cenian, A. (2020). Production of hydrogen and methane from lignocellulose waste by fermentation. A review of chemical pretreatment for enhancing the efficiency of the digestion process. *Journal of Cleaner Production*, 267, 121721. https://doi.org/10.1016/J.JCLEPRO.2020.121721

Tawfik, A., & El Qelish, M. (2012). Continuous hydrogen production from co-digestion of municipal food waste and kitchen wastewater in mesophilic anaerobic baffled reactor. *Bioresource Technology*, *114*, 270–274. https://doi.org/10.1016/j.biortech.2012.02.016

Tiang, M. F., Fitri Hanipa, M. A., Abdul, P. M., Jahim, J. M. d., Mahmod, S. S., Takriff, M. S., Lay, C. H., Reungsang, A., & Wu, S. Y. (2020). Recent advanced biotechnological strategies to enhance photo-fermentative biohydrogen production by purple non-sulphur bacteria: An overview. *International Journal of Hydrogen Energy*, 45(24), 13211–13230. https://doi.org/10.1016/J.IJHYDENE.2020.03.033

Tornyie, F., Ampomah-Benefo, K., Dogli, M., Apperkon, A. K., Sedzro, D. M., & Bayitse, R. (2024). Biofuels: Bioethanol, biogas, and biohydrogen production from cassava peels. In *Sustainable Cassava: Strategies from Production through Waste Management* (pp. 361–378). Academic Press. https://doi.org/10.1016/B978-0-443-21747-0.00003-5

Ummalyma, S. B., Supriya, R. D., Sindhu, R., Binod, P., Nair, R. B., Pandey, A., & Gnansounou, E. (2019). Biological pretreatment of lignocellulosic biomass—Current trends and future perspectives. In *Second and Third Generation of Feedstocks: The Evolution of Biofuels* (pp. 197–212). Elsevier. https://doi.org/10.1016/B978-0-12-815162-4.00007-0

Vertés, A. A., Qureshi, N., Blaschek, H. P., & Yukawa, H. (2020). Green energy to sustainability: Strategies for global industries. In *Green Energy to Sustainability: Strategies for Global Industries* (pp. 1–664). John Wiley & Sons Ltd. https://doi.org/10.1002/9781119152057

Wang, J., & Wan, W. (2008). Effect of temperature on fermentative hydrogen production by mixed cultures. *International Journal of Hydrogen Energy*, 33(20), 5392–5397. https://doi.org/10.1016/j.ijhydene.2008.07.010

Wang, J., & Yin, Y. (2017a). *Biohydrogen Production from Organic Wastes*. Springer. https://doi.org/10.1007/978-981-10-4675-9

Wang, J., & Yin, Y. (2017b). Influencing factors for biohydrogen production. In *Green Energy and Technology*. Springer. https://doi.org/10.1007/978-981-10-4675-9_5

Wu, X., Li, Q., Dieudonne, M., Cong, Y., Zhou, J., & Long, M. (2010). Enhanced H_2 gas production from bagasse using adhE inactivated Klebsiella oxytoca HP1 by sequential dark-photo fermentations. *Bioresource Technology*, 101(24), 9605–9611. https://doi.org/10.1016/j.biortech.2010.07.095

Yagmur Goren, A., Dincer, I., & Khalvati, A. (2023). A comprehensive review on environmental and economic impacts of hydrogen production from traditional and cleaner resources. *Journal of Environmental Chemical Engineering*, 11(6), 111187. https://doi.org/10.1016/J.JECE.2023.111187

Zhang, H., Lei, T., Lu, S., Zhu, S., Li, Y., Zhang, Q., & Zhang, Z. (2022). Study on comparisons of bio-hydrogen yield potential and energy conversion efficiency between stem and leaf of sweet potato by photo-fermentation. *Fermentation*, 8(4), 165. https://doi.org/10.3390/FERMENTATION8040165

Zhang, J., Kong, C., Yang, M., & Zang, L. (2020). Comparison of calcium oxide and calcium peroxide pretreatments of wheat straw for improving biohydrogen production. *ACS Omega*, 5(16), 9151–9161. https://doi.org/10.1021/ACSOMEGA.9B04368

Zhang, Q., Zhu, S., Zhang, Z., Zhang, H., & Xia, C. (2021). Enhancement strategies for photo-fermentative biohydrogen production: A review. *Bioresource Technology*, 340, 125601. https://doi.org/10.1016/J.BIORTECH.2021.125601

Zhang, Z., Zhang, H., Li, Y., Lu, C., Zhu, S., He, C., Ai, F., & Zhang, Q. (2020). Investigation of the interaction between lighting and mixing applied during the photo-fermentation biohydrogen production process from agricultural waste. *Bioresource Technology*, 312, 123570. https://doi.org/10.1016/J.BIORTECH.2020.123570

Zhao, L., Sun, Z. F., Zhang, C. C., Nan, J., Ren, N. Q., Lee, D. J., & Chen, C. (2022). Advances in pretreatment of lignocellulosic biomass for bioenergy production: Challenges and perspectives. *Bioresource Technology*, 343, 126123. https://doi.org/10.1016/J.BIORTECH.2021.126123

Zhu, S., Zhang, Z., Zhang, H., Jing, Y., Li, Y., & Zhang, Q. (2020). Rheological properties of corn stover hydrolysate and photo-fermentation bio-hydrogen producing capacity under intermittent stirring. *International Journal of Hydrogen Energy*, 45(6), 3721–3728. https://doi.org/10.1016/J.IJHYDENE.2019.08.069

Zhu, Z., Shi, J., Zhou, Z., Hu, F., & Bao, J. (2010). Photo-fermentation of *Rhodobacter sphaeroides* for hydrogen production using lignocellulose-derived organic acids. *Process Biochemistry*, 45(12), 1894–1898. https://doi.org/10.1016/J.PROCBIO.2010.08.017

Ziemiński, K., Romanowska, I., & Kowalska, M. (2012). Enzymatic pretreatment of lignocellulosic wastes to improve biogas production. *Waste Management*, 32(6), 1131–1137. https://doi.org/10.1016/J.WASMAN.2012.01.016

Bio-Hydrogen Production from Food Waste via Fungal and Yeast Species

14

Sakshi Kumari, Vishal Malhotra, Monalisa Ghara, and Debarati Paul

14.1 INTRODUCTION

The waste produced by the agri-food sector has great relevance as a critical worldwide issue that cuts over national boundaries and socioeconomic strata on the development of sustainable food systems and idea of the circular bio-economy. This organic waste, which includes residues and by-products from the agri-food sector as well as food loss and waste (FLW), is an intricate problem and an essential part of the larger conversation about environmental sustainability and the circular bio-economy. It is critical to comprehend and take care of these elements in order to advance sustainability in the agri-food industry (Tamaian, 2024). Food loss is the term used to describe the decrease in food quantity or quality that occurs in the initial phases of the food supply chain, from distribution to production. Food loss mostly happens prior to food reaching customers and is caused by supply chain logistics issues as well as inefficiencies in the agriculture industry. The Food and Agriculture Organization of the United Nations (FAO) reports that 828 million people suffered from hunger in 2021—a rise of roughly 46 million from 2020 and 150 million from 2019—while 3.1 billion people are thought to be unable to access a decent diet (FAO et al., 2022). These startling statistics not only make resource allocation and hunger problems worse, but they also greatly worsen environmental challenges, such as soil degradation and greenhouse gas emissions, since 8%–10% of global greenhouse gas emissions come from food waste (Kumar et al., 2022). Second, food processing leftovers and by-products are essential to the circular bio-economy paradigm's advancement. These organic materials include food product trimmings, peels, shells, marc and pomace, and other components that might not be suitable for human consumption. It is especially interesting that agri-food waste is being used as a substrate for the synthesis of bio-hydrogen as given in Table 14.1. Food waste is an economically appealing resource for the manufacture of biofuels because of its high starch and protein content (Cheng et al., 2020). But there are several obstacles in the way of realizing this promise. Converting macromolecules like protein and starch into usable carbon sources like glucose and

DOI: 10.1201/9781003568612-14

TABLE-14.1 Different organisms used in production of $BioH_2$ and their parameters

SOURCE OF WASTE	SPECIES	PRETREATMENT METHOD	PH	TEMPERATURE	YIELD (ML H_2/G VS)	REACTOR TYPE	REFERENCES
Fruit and vegetable waste	*Clostridium* sp.	Heat treatment	6.5	37°C	249	Continuous stirred tank reactor (CSTR)	Han et al. (2016)
Food processing waste	*Escherichia coli*	Enzymatic hydrolysis	5.5	30°C	245.7	Fed-batch reactor	Nam (2023)
Cheese whey	*Saccharomyces cerevisiae*	Acidic pretreatment	5.0	35°C	205.8	Batch reactor	Samrot et al. (2023)
Kitchen waste	Anaerobic digester sludge	Sonication	7.0	55°C	165	Anaerobic digestion reactor	Han et al. (2016)
Fruit juice wastewater	*Clostridium thermocellum*	Thermal hydrolysis	6.0	37°C	5850	Continuous stirred tank reactor (CSTR)	Nam (2023)
Banana peels	Mixed microbial cultures	Physical (Grinding)	7.0	37°C	97	Batch reactor	Tamaian (2024)
Bread waste	*Bacillus subtilis*	Chemical (NaOH)	6.5	40°C	77	Batch reactor	Dzulkarnain et al. (2022)
Potato peel waste	*Clostridium butyricum*	Microwave irradiation	6.5	35°C	60	Batch reactor	Ahmad et al. (2022)
Corn stover	*Clostridium beijerinckii*	Dilute acid hydrolysis	6.0	35°C	55	Continuous stirred tank reactor (CSTR)	Han et al. (2016)
Rice straw	*Enterobacter cloacae*	Alkali pretreatment	7.0	37°C	50	Batch reactor	Ahmad et al. (2022)
Wheat straw	*Clostridium acetobutylicum*	Steam explosion	6.5	37°C	40	Continuous stirred tank reactor (CSTR)	Han et al.(2016)
Sugarcane bagasse	*Thermoanaerobacterium*	Hydrothermal pretreatment	7.0	55°C	35	Batch reactor	Ahmad et al. (2022)
Apple pomace	*Clostridium pasteurianum*	Enzymatic hydrolysis	6.0	37°C	30	Batch reactor	Han et al.(2016)
Pineapple peel	*Enterobacter aerogenes*	Acid hydrolysis	6.5	37°C	25	Batch reactor	Ahmad et al. (2022)
Tomato waste	*Bacillus cereus*	Alkaline peroxide pretreatment	7.0	37°C	20	Batch reactor	Han et al.(2016)

(Continued)

TABLE-14.1 (*Continued*) Different organisms used in production of $BioH_2$ and their parameters

SOURCE OF WASTE	*SPECIES*	*PRETREATMENT METHOD*	*PH*	*TEMPERATURE*	*YIELD (ML H_2/G VS)*	*REACTOR TYPE*	*REFERENCES*
Onion waste	*Klebsiella pneumoniae*	Ultrasonic pretreatment	6.5	37°C	18	Batch reactor	Ahmad et al. (2022)
Cabbage waste	*Enterobacter aerogenes*	Acid and alkali pretreatment	7.0	37°C	15	Batch reactor	Han et al.(2016)
Broccoli waste	*Bacillus licheniformis*	Microwave-assisted hydrolysis	7.0	37°C	10	Batch reactor	Yun et al.(2018)
Lettuce waste	*Klebsiella oxytoca*	Acid and alkali pretreatment	6.5	37°C	8	Batch reactor	Ghimire (2022)
Orange peel waste	*Candida tropicalis*	Enzymatic hydrolysis	6.5	35°C	28	Batch reactor	Han et al.(2016)
Mango peel waste	*Saccharomyces boulardii*	Acid hydrolysis	5.0	30°C	32	Batch reactor	Nam (2023)
Pineapple core waste	*Trichoderma reesei*	Alkali pretreatment	6.8	37°C	40	Batch reactor	Samrot et al. (2023)
Grape pomace	*Aspergillus niger*	Heat treatment	6.0	35°C	22	Batch reactor	Yun et al.(2018)
Banana peel waste	*Rhizopus oryzae*	Steam explosion	6.5	40°C	26	Batch reactor	Ahmad et al. (2022)
Sugar beet waste	*Saccharomyces cerevisiae*	Microwave irradiation	7.0	37°C	38	Batch reactor	Yun et al.(2018)

free amino nitrogen (FAN), which are necessary for biotechnological operations, is the difficult part. In the majority of bioprocesses, this conversion process—known as hydrolysis—often turns out to be the rate-limiting one. This analysis looks at a sustainable strategy to overcome the hydrolysis constraint and boost the output of bio-hydrogen more effectively. This research explores the use of agri-food waste as a substrate, emphasizing its twofold benefit in reducing waste disposal issues and producing alternative energy (Tamaian, 2024). Furthermore, a detailed assessment is conducted on two well-known biological processes for the synthesis of bio-hydrogen: dark fermentation (DF) and photosynthesis.

14.2 USING AGRI-FOOD WASTE AS A RESOURCE

Agri-food waste's enormous scope and potential for resource recovery make it a serious environmental concern that has to be addressed. To fully appreciate the importance of using food waste as a valuable resource (low-cost feedstock) for bio-hydrogen production, one must have a clear understanding of the scope of this problem.

Agri-food waste's composition is one of the main factors contributing to its potential as a resource for the generation of bio-hydrogen. Carbohydrates, notably proteins and starch, are abundant in food waste. A polysaccharide made up of glucose units, starch is a common ingredient in a variety of foods, including pasta, bread, rice, and potatoes. Conversely, proteins are abundant in a variety of food sources such as meat, dairy, and legumes and are made up of amino acids. Since these proteins and carbohydrates may be transformed into bio-hydrogen by microorganisms, they are important feedstock for the synthesis of biofuel. Food waste is a barrier due to its complex nature, despite its potential as a resource. For the synthesis of bio-hydrogen, macromolecules like proteins and starch must be broken down into more manageable forms. For microbes that produce hydrogen to ferment glucose, starch must first undergo enzymatic hydrolysis. Proteins, rich in amino acids, require enzymatic or microbial degradation to yield FAN, which is a crucial nutrient for the growth and activity of hydrogen-producing microorganisms. The conversion of these complex substrates into simpler forms is often a rate-limiting step in bio-hydrogen production processes. The challenge lies in efficiently converting these complex substrates into glucose (or another accessible carbon sources) and FAN to facilitate bio-hydrogen production (Tamaian, 2024).

14.3 DISTINCT HYDROLYSIS AND FERMENTATION METHODOLOGY

One important concept in strategic bioprocessing is the separate hydrolysis and fermentation (SHF) strategy. This approach converts complicated substrates from agri-food waste more efficiently into valuable nutrient-rich solutions as given in Figure 14.1, which in turn increases the production of bio-hydrogen (Rafieenia et al., 2018). Several production process phases are involved in this method, each of which is optimized for a particular purpose. The SHF technique splits the entire process of producing bio-hydrogen into two distinct phases: fermentation and hydrolysis.

14.3.1 Grinding of Waste Particles

Grinding trash is an essential pretreatment step in the manufacture of bio-hydrogen. It lowers the particle size of the food waste, increasing its surface area and improving its accessibility for microbial digestion.

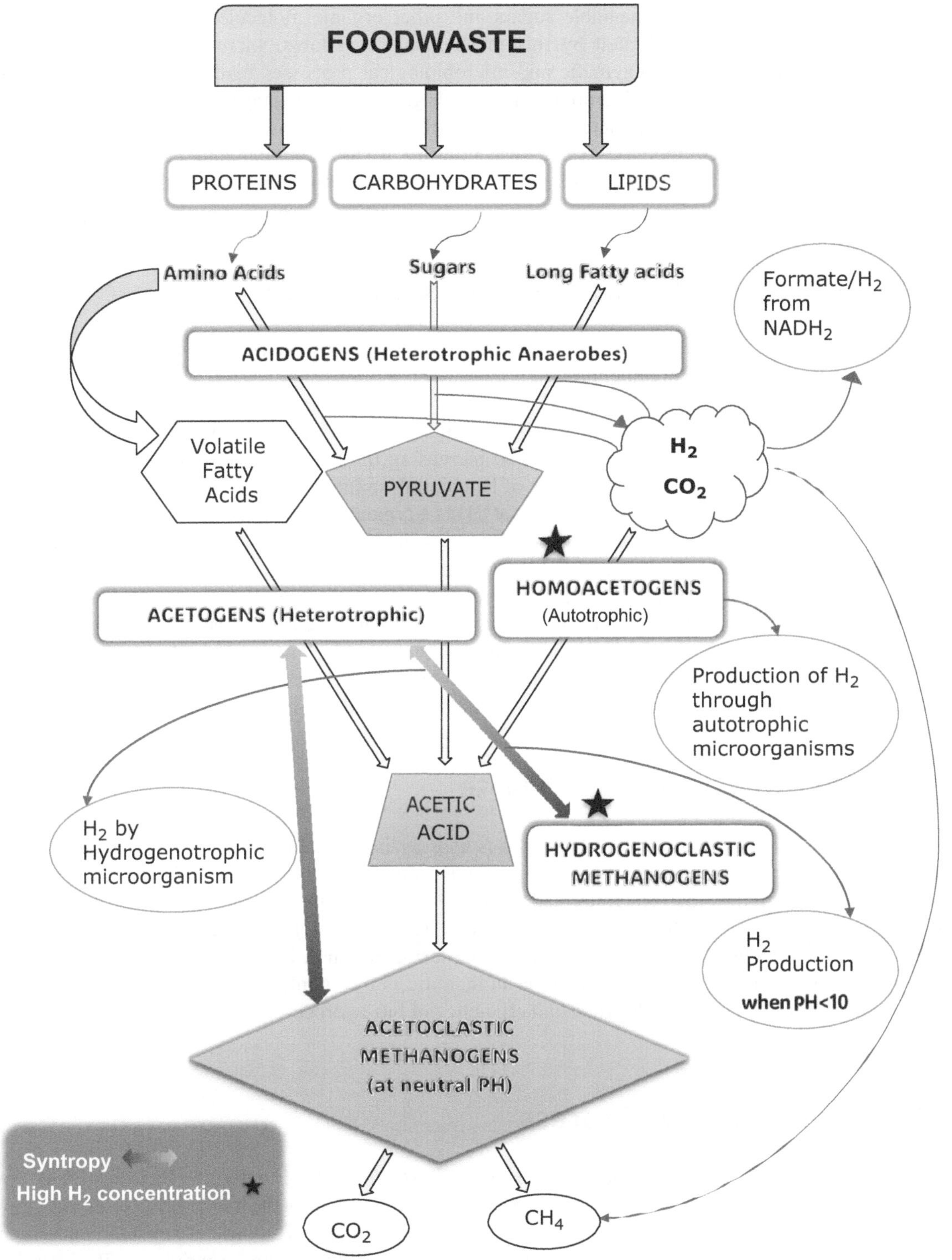

FIGURE 14.1 Bio-hydrogen (H_2) production through dark fermentation of food waste utilizes bacteria to generate clean energy without light or oxygen.

The following hydrolysis of fermentable sugars and other organic molecules, which is necessary for microbial fermentation, is facilitated by this mechanical breakdown. Increased bio-hydrogen yields result from the more effective enzymatic and microbiological processes made possible by the smaller and more homogeneous particle size obtained through grinding. This stage is especially crucial when working with the lignocellulosic materials found in food waste because it aids in the disruption of intricate structures, which increases their susceptibility to fermentation and enzymatic hydrolysis. Enhancing the overall effectiveness and financial sustainability of systems for producing bio-hydrogen requires taking this crucial step.

14.3.2 Chemical Pretreatment of Food Waste

In the pretreatment of food waste for bio-hydrogen production, acids, bases, and enzymes are commonly used to break down complex organic matter:

Acid Pretreatment: Food waste lignocellulosic materials are often hydrolyzed with sulfuric acid (H_2SO_4). By rupturing the cell walls and increasing the cellulose's accessibility to enzymes, this potent acid aids in the breakdown of hemicellulose into fermentable sugars.

Base Pretreatment: Sodium hydroxide (NaOH) is a typical alkali utilized in this process. It aids in the process of delignification, which increases the amount of cellulose available for microbial fermentation by removing lignin, a complex polymer found in plant cell walls.

Enzymatic Pretreatment: Cellulose and hemicellulose are broken down into simple sugars by enzymes such as cellulases and hemicellulases. The complex carbohydrates in food waste must be converted into fermentable sugars by these enzymes, which are made by bacteria and fungi. Fermentation uses these sugars to make hydrogen. The yield of bio-hydrogen in different pretreatment methods is depicted in Table 14.2. The pros and cons are depicted in Table 14.3.

14.3.3 Hydrolysis of Food Waste

A crucial stage in the synthesis of bio-hydrogen is hydrolysis, in which enzymes and water break down complex organic components in food waste into simpler molecules, mostly sugars. This process can happen biologically, with the aid of enzymes made by bacteria and fungus, or chemically, with the application of acids and alkalis. Polysaccharides like cellulose and hemicellulose hydrolyze to produce monosaccharides, which are then used as substrates for microbial fermentation in later phases. Hydrogen production during fermentation is enhanced when there is an increase in the supply of fermentable substrates due to effective hydrolysis. To increase the yield and efficiency of bio-hydrogen from organic waste sources, this step is crucial.

14.3.4 Fermentation Process

Through metabolic pathways like DF, microbes like *Clostridium* and *Enterobacter* species break down organic substrates during this process, generating carbon dioxide, hydrogen, and organic acids as by-products. The particular microbial population involved, temperature, and pH all affect how efficiently hydrogen is produced. In order to turn trash into clean energy, fermentation is an essential step. By optimizing these conditions, hydrogen output can be increased.

TABLE 14.2 Different pretreatments used in production of bio-hydrogen

PRETREATMENT METHOD	*CONCENTRATION*	*SPECIES*	*PH*	*TEMPERATURE*	*YIELD (ML H_2/G VS)*	*REFERENCES*
Physical (grinding)	–	Mixed microbial cultures	7.0	37°C	97	Habashy et al. (2021)
Acid hydrolysis	0.5%–2% H_2SO_4	*Clostridium acetobutylicum*	6.5	90°C–110°C	120	Smith et al. (2020)
Alkali pretreatment	2–2.5 M NaOH	*Enterobacter aerogenes*	7.0	50°C–80°C	80	Johnson et al. (2020)
Enzymatic hydrolysis	–	*Clostridium beijerinckii*	6.5	37°C	150	Brown et al. (2020)
Microwave irradiation	–	*Bacillus cereus*	7.0	55°C	90	Zhang et al. (2020)
Ultrasonic pretreatment	–	*Klebsiella pneumoniae*	6.5	37°C	75	Davis et al. (2020)
Thermal hydrolysis	–	*Clostridium thermocellum*	6.5	55°C	130	Kim et al. (2020)
Dilute acid hydrolysis	0.5% H_2SO_4	*Enterobacter cloacae*	6.0	121°C	100	Patel et al. (2020)
Steam explosion	–	*Clostridium butyricum*	6.5	55°C	110	Weldehans and Halefom (2015)
Hydrothermal pretreatment	–	*Thermoanaerobacterium*	7.0	55°C	90	Bauer et al. (2014)

TABLE 14.3 Pros and cons of pretreatment methods

PRETREATMENT METHOD	*PROS*	*CONS*
Physical (grinding)	• Increases surface area for microbial access • Simple and cost-effective	• May require additional energy for grinding • Limited impact on lignocellulosic materials
Acid hydrolysis	• Effective for breaking down hemicellulose and cellulose • High yield of fermentable sugars	• Corrosive chemicals required • Generation of inhibitors that can affect fermentation
Alkali pretreatment	• Enhances cellulose accessibility • Reduces lignin content	• High chemical costs • Requires neutralization post-treatment
Enzymatic hydrolysis	• Specific and efficient breakdown of complex carbohydrates • Produces fewer inhibitors	• High enzyme costs • Requires optimal conditions for enzyme activity
Microwave irradiation	• Rapid heating and effective biomass disruption • Can enhance enzymatic activity	• Equipment cost can be high • Potential degradation of sugars at high temperatures
Ultrasonic pretreatment	• Increases mass transfer and enhances enzyme accessibility • Short treatment time	• High energy consumption • Equipment maintenance can be costly
Thermal hydrolysis	• Effective sterilization of biomass • Increases solubilization of organic matter	• High energy requirements • Potential loss of volatile compounds
Dilute acid hydrolysis	• Effective for hemicellulose solubilization • Can be integrated into continuous processes	• Requires neutralization • Can produce inhibitory by-products
Steam explosion	• Effective for breaking down lignocellulosic materials • Increases accessibility for fermentation	• High operational costs • Requires specialized equipment
Hydrothermal pretreatment	• Utilizes water as a solvent • Can solubilize a wide range of biomass types	• High energy input required • Potential for sugar degradation at high temperatures

14.4 ENHANCEMENT OF OPERATIONAL ENVIRONMENT

Optimizing operating conditions, especially in the pretreatment phase, is critical to the effectiveness of the distinct hydrolysis and fermentation approaches in increasing bio-hydrogen generation from agri-food waste. These parameters can be adjusted to optimize conversion effectiveness and handle issues brought on by the intricacy of the substrates used to produce agri-food waste. Operating conditions are a group of variables that can be changed in order to attain the best conversion efficiency possible before treatment. Temperature, pH, residency duration, and the selection of microbes or enzymes are some of these variables (Mukherjee and Mohan, 2022).

Temperature: During pretreatment, alterations in temperature can have a major effect on microbial or enzymatic activity. While higher temperatures can speed up reactions, they must still be within the range that is appropriate for the particular enzymes or microbes being utilized.

pH Levels: Microorganisms and enzyme activity are impacted by varying pH values. There are pH ranges where certain enzymes work best, and you may maximize their efficiency by bringing the pH within these limits.

Residence Time: It is possible to maximize the amount of time that agri-food waste is exposed to pretreatment conditions. Although more comprehensive substrate breakdown may result from longer residence times, it is important to strike a balance to prevent using too much energy.

Microorganisms or Enzymes: The pretreatment's selection of microbial strains or enzymes can be adjusted to more successfully target particular substrates found in agri-food waste. Biotechnologically, bio-hydrogen can be created by a number of techniques, such as DF, dark-photo sequential fermentation (DF-PF), photo-fermentation (PF), direct photolysis, and indirect photolysis (Mukherjee and Mohan, 2022).

DF is unique in that it can generate hydrogen more quickly and effectively at room pressure than photosynthetic processes. It is adaptable and able to use many kinds of agri-food waste as feedstock because it functions under mild reaction conditions. DF is thought to be eco-friendly and has potential for producing hydrogen for commercial use. The ability of PF to transform lignocellulosic biomass into bio-hydrogen makes it noteworthy. Its ability to capture a broad range of light wavelengths improves its solar energy utilization efficiency. PF produces treatable and manageable effluent. When compared to alternative technologies, it has a better substrate conversion efficiency, lower pollution emissions, and the flexibility to use different carbon sources. DF-PF appears to be a technique that can produce a significant amount of bio-hydrogen while still being economically and practically viable (Rai and Singh, 2016; Lee, 2021). It has been determined to be the most effective method for converting substrates into hydrogen, making it a strong contender for both commercial bio-hydrogen production and sustainable resource management.

14.5 THE POTENTIAL CONTRIBUTION OF COMPUTATIONAL METHODS TO THE DEVELOPMENT OF BIO-HYDROGEN PRODUCTION

The Complementary Effects of Genomics Technologies and Computational Methods: Historically, the identification of hydrogen-producing bacteria required labor-intensive, time-consuming, expensive, and frequently narrowly focused wet-lab research. However, a paradigm shift has been brought about in this field by methods related to genomics and computational biology. Today, scientists may investigate the wide genetic diversity of microorganisms, from archaea to algae, using cutting-edge bioinformatics and genomic analysis in an effort to determine which have the greatest potential to produce bio-hydrogen (Ramakodi, 2019).

Through the examination of microorganisms' genetic composition, researchers can acquire knowledge about the metabolic processes in charge of producing hydrogen. It is possible to identify the important genetic markers and enzymes connected to the production of hydrogen. Furthermore, bioengineering is essential to enhancing microorganisms' capacity to produce hydrogen. Researchers can increase the efficiency and hydrogen gas output of these creatures by modifying their genomes. This strategy may hasten the formation of consortia with high hydrogen production efficiency and facilitate the generation of microorganisms specifically engineered for the bio-hydrogen production process.

Bio-Hydrogen Process Modeling Advances: Traditionally, experimentation and empirical models have been used in the making and enhancement of bio-hydrogen production systems. But a new age has begun with recent developments in computational approaches.

Process optimization has benefited greatly from the application of empirical models, which include statistical techniques and experimental design strategies (Chezeau and Vial, 2019). These models aid in determining the critical elements impacting the synthesis of bio-hydrogen and direct experimental endeavors. Furthermore, cutting-edge methods such as artificial neural networks have the potential to expand on the present modeling capabilities, enabling researchers to capture intricate interactions among process variables.

14.6 CONCLUSION

The low rate of substrate breakdown and $BioH_2$ generation present challenges for the process of manufacturing $BioH_2$ from biomass. There are advantages and disadvantages to both DF and PF, indicating that these processes should not be approached with a one-size-fits-all approach. For treating mixed organic waste, for instance, DF, a more reliable method, might be a good option in addition to or instead of bio-methane.

On the other hand, PF, which needs a cleaner feedstock and light, may be utilized to treat particular waste materials while also producing other high-value compounds or in nations where light variation is not a problem. It is significant to note that the storage techniques for $BioH_2$ generated by biological pathways are same to those used for $BioH_2$ created by other processes (e.g. electrolysis). As a result, consideration of $BioH_2$ separation and purification comes before storage, which poses a significant obstacle to the establishment of a viable and sustainable $BioH_2$ economy.

Enhancing and optimizing storage still presents a significant problem. Depending on the desired duration of the transfer, the amount of $BioH_2$ involved, geographical aspects, as well as technical and financial constraints, there are a variety of methods for carrying out $BioH_2$ transportation. The difficulties facing $BioH_2$ transport today include the energy required for compression, security concerns, and economic considerations.

REFERENCES

Ahmad, S., Singh, M., & Tiwari, A. (2022). Production of bio-hydrogen from banana waste by using anaerobic fermentation. *International Journal for Research in Applied Science and Engineering Technology*, *10*, 1202–1205. https://doi.org/10.22214/ijraset.2022.40831.

Bauer, A., Lizasoain, J., Theuretzbacher, F., Agger, J., Rincón, M., Menardo, S., Saylor, M., Enguidanos Requena, R., Nielsen, P., Potthast, A., Eichinger, T., Gronauer, A., & Horn, S. (2014). Steam explosion pretreatment for enhancing biogas production of late harvested hay. *Bioresource Technology*, *166C*, 403–410. https://doi.org/10.1016/j.biortech.2014.05.025.

Brown, M., et al. (2020). Alkali pretreatment of banana peels for enhanced biohydrogen production. *Bioresource Technology*, *302*, 122–130.

Cheng, S. Y., Tan, X., Show, P. L., Rambabu, K., Banat, F., Veeramuthu, A., Lau, B. F., Ng, E. P., & Ling, T. C. (2020). Incorporating biowaste into circular bioeconomy: a critical review of current trend and scaling up feasibility. *Environmental Technology & Innovation*, *19*, 101034.

Chezeau, B., & Vial, C. (2019). Modeling and simulation of the biohydrogen production processes. In Pandey, A., Mohan, S. V., Chang, J.-S., Hallenbeck, P. C., Larroche, C., Eds. *Biohydrogen, Biomass, Biofuels, Biochemicals* (pp. 445–483). Elsevier: Amsterdam, the Netherlands.

Davis, T., et al. (2020). Enzymatic hydrolysis of banana peels for biohydrogen production. *Biotechnology and Bioengineering*, *117*, 800–810.

Dzulkarnain, E. L. N., Audu, J. O., Dagang, W. R. Z. W., & Abdul-Wahab, M. F. (2022). Microbiomes of biohydrogen production from dark fermentation of industrial wastes: current trends, advanced tools and future outlook. *Bioresources and Bioprocessing*, *9*(1), 16.

FAO; IFAD; UNICEF; WFP; & WHO. (2022). *The State of Food Security and Nutrition in the World 2022: Repurposing Food and Agricultural Policies to Make Healthy Diets More Affordable; The State of Food Security and Nutrition in the World (SOFI)*. FAO; IFAD; UNICEF; WFP; WHO: Rome, Italy.

Ghimire, A. (2022). Microbiomes of biohydrogen production from dark fermentation of industrial wastes: current trends, advanced tools and future outlook. *Bioresources and Bioprocessing*, *9*(1), 35.

Habashy, M.M., Ong, E.S., Abdeldayem, O.M., Al-Sakkari, E.G., & Rene, E.R. (2021). Food waste: a promising source of sustainable biohydrogen fuel. *Trends in Biotechnology*, *39*(11), 1175–1190.

Han, W., Yan, Y., Shi, Y., Gu, J., Tang, J., & Zhao, H. (2016). Biohydrogen production from enzymatic hydrolysis of food waste in batch and continuous systems. *Scientific Reports*, *6*(1), 38395.

Johnson, K., et al. (2020). Biohydrogen production from banana peels using acid hydrolysis pretreatment. *International Journal of Hydrogen Energy*, *45*, 300–310.

Kim, J., et al. (2020). Ultrasonic pretreatment of banana peels for enhanced biohydrogen production. *Ultrasonics Sonochemistry*, *61*, 104812.

Kumar, L., Chhogyel, N., Gopalakrishnan, T., Hasan, M. K., Jayasinghe, S. L., Kariyawasam, C. S., Kogo, B. K., Ratnayake, S. (2022). Climate change and future of agri-food production. In Bhat, R., Ed., *Future Foods* (pp. 49–79). Academic Press: Cambridge, MA.

Lee, D. -H. (2021). Biohydrogen yield efficiency and the benefits of dark, photo and dark-photo fermentative production technology in circular Asian economies. *The International Journal of Hydrogen Energy*, *46*, 13908–13922.

Mukherjee, T., Mohan, S. V. (2022). Bio-Waste to Hydrogen Production Technologies. In Tuli, D., Kasture, S., Kuila, A., Eds. *Advanced Biofuel Technologies* (pp. 389–407). Elsevier: Amsterdam, the Netherlands.

Nam, J. (2023). Optimum conditions for enhanced biohydrogen production from a mixture of food waste and sewage sludge with alkali pretreatment. *Energies*, *16*(7), 3281.

Patel, S., et al. (2020). Thermal hydrolysis of banana peels for biohydrogen production. *Chemical Engineering Journal*, *385*, 123456.

Rafieenia, R., Lavagnolo, M. C., Pivato, A. (2018). Pre-treatment technologies for dark fermentative hydrogen production: current advances and future directions. *Waste Management*, *71*, 734–748.

Rai, P. K., Singh, S. P. (2016). Integrated dark- and photo-fermentation: recent advances and provisions for improvement. *International Journal of Hydrogen Energy*, *41*, 19957–19971.

Ramakodi, M. P. (2019). Computational biology and genomics tools for biohydrogen research. In Pandey, A., Mohan, S. V., Chang, J.-S., Hallenbeck, P. C., Larroche, C., Eds. *Biohydrogen: Biomass, Biofuels, Biochemicals* (2nd ed., pp. 435–444). Elsevier: Amsterdam, The Netherlands.

Samrot, A. V., Rajalakshmi, D., Sathiyasree, M., Saigeetha, S., Kasipandian, K., Smith, J., et al. (2020). Optimization of physical pretreatment for enhanced biohydrogen production from banana peels. *Journal of Environmental Science and Health, Part B*, *54*, 1–10.

Tamaian, R. (2024). Enhanced biohydrogen production from food waste via separate hydrolysis and fermentation: a sustainable approach. *Biology and Life Sciences Forum*, *31*, 14.

Weldehans, M. G. & Halefom, A. (2015). Optimization of banana peels hydrolysis for the production of bioethanol: response surface methodology. *International Letters of Natural Sciences*, *48*, 53–60.

Yun, Y.-M., Lee, M.-K., Im, S.-W., Marone, A., Trably, E., Shin, S.-R., Kim, M.-G., Cho, S.-K., Kim, D.-H. (2018). Biohydrogen production from food waste: current status, limitations, and future perspectives. *Bioresource Technology*, *248*, 60–70.

Zhang, Y., et al. (2020). Hydrothermal pretreatment of banana peels for biohydrogen production. *Green Chemistry*, *22*, 4000–4010.

15 Nanoparticle-Conjugated Enzyme-Mediated Enhanced Saccharification of Lignocellulosic Biomass for Biohydrogen Generation

Arghadeep Bhowmick, Sneha Banerjee, Biva Ghosh, Mainak Mukhopadhyay, and Debalina Bhattacharya

15.1 INTRODUCTION

Lignocellulosic biomasses (LCBs) are the most preferred feedstocks to produce a variety of valuable bio-based products in second-generation biorefineries (Kucharska et al., 2018). The primary components of plant biomass are cellulose, hemicellulose, and lignin, with lesser quantities of pectin, protein, extractives, and ash. The composition of these constituents can differ among different plant species. For instance, hardwoods contain higher concentrations of cellulose, whereas leaves, rice, and wheat straws contain a larger quantity of hemicellulose. The intricate structure of these biomasses requires their pre-processing to liberate the carbohydrate (hemicellulose and cellulose) and non-carbohydrate (lignin) polymers that are present. The pretreatment of LCB can break hemicellulose and cellulose, which are the primary sources of fermentable sugars. Many physical, chemical, and biological pretreatment methods have been proposed and are routinely employed in accordance with the type of LCB (Ingle et al., 2019).

DOI: 10.1201/9781003568612-15

Out of various bio-based products obtained from LCBs, biofuel production is mostly emerging due to vivid environmental safety reasons. Among other biofuels, biohydrogen production has attracted maximum attention because of independence of non-renewable resources, pollution-free production (water as product of combustion), and ability to waste degradation (Zhang et al., 2017a). Furthermore, the optimization of initial pH, substrate concentration, and cellulase loading during biohydrogen production will markedly enhance biohydrogen production (Zhang et al., 2017b, 2017c, 2017d). Hence, cellulase plays an important role in pretreatment processes. However, only the proper amount of cellulase loading can increase biohydrogen production. Therefore, the key to solve the problem is to integrate various enzymes in the saccharification process after pretreatment that will yield a higher amount of biohydrogen. Saccharification, also known as enzyme hydrolysis, is the process that uses enzymes to degrade the complex carbohydrates into simple monomers. The typical objective of saccharification is to effectively hydrolyse both the cellulose and hemicellulose constituents of LCB into fermentable sugar monomers, as LCB contains both (Vidhya, 2021).

It is a synergetic process that is done after the pretreatment and facilitated by a variety of hydrolytic enzymes, including cellulases and hemicellulases. Hemicellulases are primarily responsible for the hydrolysis of hemicellulose hydrolysate (pentosan) into C5 monomers, primarily xylose, while cellulases (which include both the endo-acting and exo-acting enzymes) are primarily responsible for the hydrolysis of cellulose hydrolysate (hexosan) into C6 monomers (Vidhya, 2021). Chemical treatments like concentrated and dilute acid hydrolysis processes are also frequently employed for biomass hydrolysis due to their simplicity and ease of operation in place of enzyme-based saccharification. Nevertheless, concentrated acid mandates specialized apparatus and is costly due to its high corrosivity. Additionally, it mandates extensive cleansing and acid removal, which elevates the process cost (Keskin et al., 2019). Dilute acid is often preferred over concentrated acid for biomass pretreatment; however, it has its own set of constraints. The efficacy of subsequent processes and the degradation of carbohydrates can be influenced by toxic inhibitors, including acetic acid, furfural, and hydroxymethylfurfural (HMF; De Jong & Gosselink, 2014). Extensive cleansing and detoxification are necessary to eliminate these inhibitors. These methodologies are antiquated, demanding the development of innovative alternatives to enhance efficiency and revolutionize systems (Bhatia et al., 2020). The concerns regarding acid hydrolysis methods needed the urgent research and development of innovative catalysts that can effectively pretreat LCBs. It has been established that enzymes employed in the hydrolysis of biomass are more effective than any other technique (Aarti et al., 2018). As per Palacios-Bereche et al. (2017), enzymes exhibit a higher degree of specificity than any other catalytic mechanism.

Through the past decade, nanobiotechnology has been applied in a wide range of scientific and technological fields, including biomedicine, drug delivery, bioremediation, biosensors, fabrics, cosmetics, and coating (Judy et al., 2021). Thus, a few nanotechnology-based approaches have been proposed, and some attempts have been made by Arora et al. (2020) for implementing various nanoparticles as catalysts for pretreatment and hydrolysis of LCB. The utilization of nanomaterials (NMs) in bioenergy sectors for the hydrolysis of LCB, which is followed by the production of biofuel, has garnered the attention of researchers worldwide (Aarti et al., 2021). Nanoparticles not only increase productivity but also improve the pretreatment and saccharification efficiency of biomass, enzymatic hydrolysis, and reaction rate (Cherian et al., 2015). Several kinds of nanoparticles including magnetic, nickel, iron, cobalt, silver, and silicon are being widely used to adsorb hydrolytic enzymes for the conversion of LCB into biofuel (Singhvi & Kim, 2020). Nanoparticles could improve the catalytic activity through improving the enzyme loading due to a higher surface area (Cherian et al., 2015).

Thus, based on the present and potential advantages of nanoparticle-mediated enzyme catalysis such as the improvement of biohydrogen yield and the production of simple sugar (Singhvi & Kim, 2020), and the existing literature gap regarding the range of nanoparticles used in biofuel production (Aarti et al., 2022) (Figure 15.1), this chapter is a summary of the process of saccharification enhanced by the use of various nanoparticles to produce biohydrogen. We will acquire knowledge regarding the efficacy of a variety of nanoparticles in the production of biohydrogen. This chapter also emphasizes the role of preliminary processes that allow for biohydrogen generation through light and dark fermentation, by using

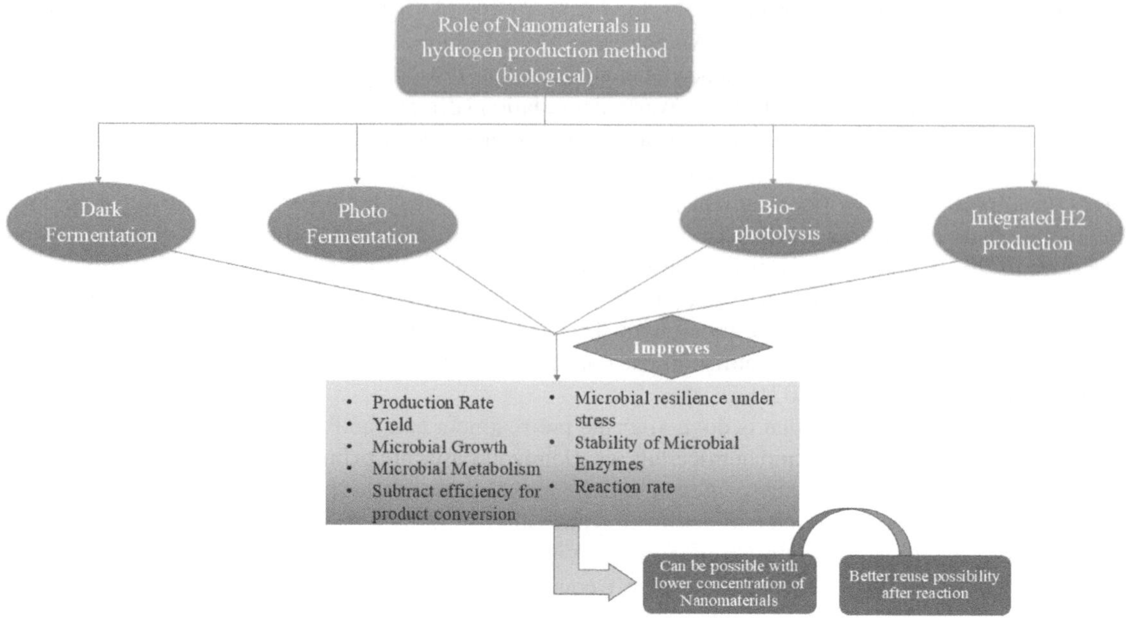

FIGURE 15.1 Role of nanomaterials in biohydrogen production.

different microorganisms. It also sheds light on the effect of advanced catalysis and new enzyme-immobilized nanoparticles in raising the sugar yield in the experimental saccharification process in order to improve the biohydrogen production rate.

15.2 LCB: THE ULTIMATE RESOURCE OF SUGAR PRODUCTION

The carbon-neutral raw material lignocellulose generates 200 billion metric tonnes per year from forestry and horticultural as well as agricultural waste products (Ufodike et al., 2020). The three principal chemical compositions of lignocellulose include cellulose and hemicellulose polysaccharides in addition to lignin polyphenols, and Mood et al. (2013) documented secondary compounds. A representation of the standard LCB content appears in Figure 15.2. The production of biohydrogen along with biochar and biodiesel and biomethane benefits from lignocellulose since this material is plentiful and produces minimal emissions while remaining financially viable (Saravanan et al., 2021). The fourth-ranking source of energy following fossil fuels provides LCB with extensive scope for renewable fuel production according to Wang and Yin (2021) so researchers must study its resource base and extraction approaches.

15.2.1 Resources of LCBs

The renewable source of LCB consists of cellulose and hemicellulose and lignin together with organic and inorganic compounds that supplies both thermal energy and biofuels. The sustainable production of biofuels with agricultural and forest-derived waste products uses abundant and cost-effective and efficient components that show potential as a promising biofuel solution according to Cai et al. (2017).

Table 15.1 represents different resources of LCB with multiple industrial products.

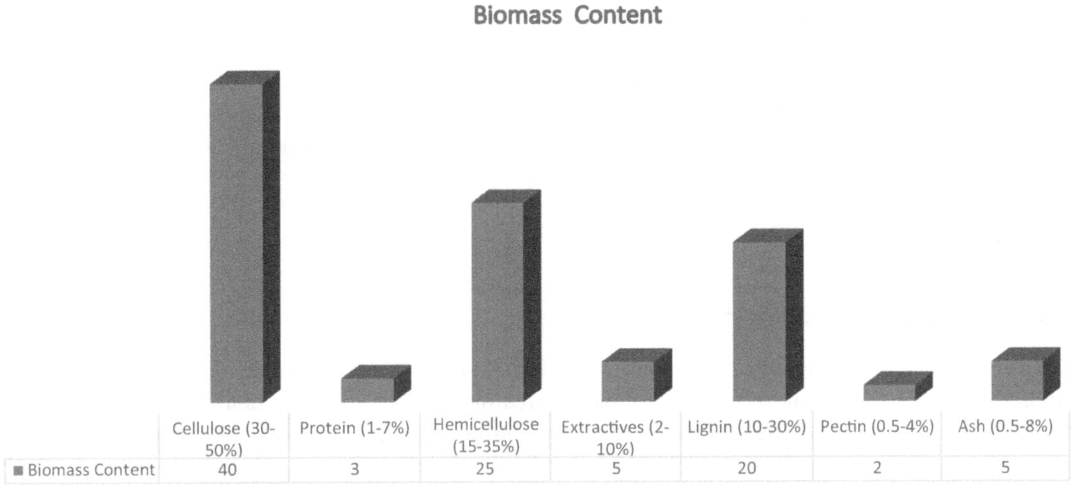

FIGURE 15.2 Various composition and its percentage of lignocellulosic biomass.

TABLE 15.1 Different resources of lignocellulosic biomass with multiple industrial products

RESOURCES OF LCB	*PRETREATMENT*	*INDUSTRIAL PRODUCT*	*REFERENCE*
Wheat straw	• Stream explosion. • Alkali pretreated; enzyme hydrolysed. • Acid, stream explosion, and enzyme. • NaOH pretreated.	• Bioethanol (56.9%). • Acetone butanol ethanol (ABE) (17.7 g/L). • Biohydrogen (2.62 mmol/g sugar). • Biogas (87.5% increase).	Alvira et al. (2013), Marchal et al. (1992), Bhatia et al. (2020), Voloshin et al. (2015)
Rice straw	• Dilute acid-dilute alkali. • Alkali pretreated+enzyme hydrolysed. • Acid. • Hydrothermal+5% NaOH.	• Bioethanol (73.4%). • Acetone butanol ethanol (ABE) (13.0 g/L). • Biohydrogen (0.44 mmol/g sugar). • Biogas (225.6%).	Soni et al. (1982), Liu et al. (2013), Razzak et al. (2013)
Bagasse	• Hydrothermal, Formiline. • Alkali pretreated+enzyme hydrolysed.	• Bioethanol (70.4%, 82.7%). • Acetone butanol ethanol (ABE) (18.1 g/L).	Katsimpouras et al. (2017), Zhao et al. (2013), Soni et al. (1982)
Corn stover	• Acetic acid catalysed hydrothermal. • Acid.	• Bioethanol (51.3%). • Biohydrogen (4.17 mmol/g sugar).	Zhang et al. (2014), Zhang et al. (2014)
Corncob	• Dilute acid hydrolysis alkaline extraction. • High-pressure autohydrolysis.	• Bioethanol (89.4%). • Biohydrogen (31 mmol/g sugar).	Lei et al. (2014)
Cornstalk	• Wet stream explosion.	• Biohydrogen (12 mmol/g sugar).	Liu et al. (2014), Liu et al. (2014)
Wood chopsticks	• Alkaline and enzyme.	• Biohydrogen (195 mmol/g sugar).	Phummala et al. (2014)

15.2.2 Pretreatment of LCB

The first process involved in the disintegration of LCB and its conversion into bioenergy with a high yield entails the pretreatment of the three components of it (Figure 15.3). To ensure the effectiveness of an LCB deconstruction that involves the various components in its construction, it is essential to ensure the effectiveness of a pretreatment process (Mankar et al., 2021). Hence on the basis of the efficacy of a pretreatment method some of the physical, chemical, physicochemical, and biological approaches are discussed below:

15.2.2.1 Physical Treatment

The general purpose of physical pretreatment methods is to decrease the particle size of LCB. Efficient LCB to biohydrogen conversion requires multiple pretreatments, including physical, chemical, and thermal methods, to improve feedstock quality before processing (Singh et al., 2022). Mechanical tools and techniques, including millers, grinders, fasteners, and ultraviolet or microwave radiations, are usually implemented during physical pretreatments. In addition, the process is facilitated using secondary or tertiary sources, including compounds, solvents, or enzymes. Some major physical pretreatment methods are discussed below:

15.2.2.1.1 Milling Pretreatment

Milling is a pretreatment method used to reduce the size of biomass particles, which can also be achieved through grinding, milling, pulverizing, or flaking. The ball milling method is widely used for pretreatment of various biomasses (Veluchamy et al., 2018).

In a study conducted by Zakaria et al. (2014), the crystallinity index of oil palm empty fruit bunch (OPEFB) was reduced from 56.1% (untreated) to 9.3% (ball-milled) because of the pretreatment process using balling milling (120 min). Consequently, the ball-milled sample exhibited higher glucose/xylose yields (67.5%/80.1%) following enzymatic hydrolysis (50°C, 72 h) than the untreated sample (15.9%/5.4%).

15.2.2.1.2 Extrusion-Based Pretreatment

High shear forces delivered by single or twin extruder screws operating inside a compact barrel both heat and pressurize biomass as a pretreatment method to enhance enzyme accessibility (Duque et al., 2017; Kumar & Sharma, 2017). The increase of screw speed from 280 to 350 rpm results in better glucose production rates through shorter fibre lengths which create increased surface contact. Maximum heat within the barrel reduces the yield of glucose while keeping the combustion rate at a high level. The optimal condition for sugar recovery generated 49.2% for PCG along with 66.2% for BB and 28.2% for SG. Sucrose recovery suffers a reduction of −23% with increased moisture as this decreases friction through lubrication according to Karunanithy et al. (2012).

15.2.2.1.3 Ultrasound Pretreatment Method

Ultrasound, acoustic energy with frequencies exceeding 20 kHz, is a leading method for pretreatment of LCB. Acoustic cavitation occurs when ultrasound waves propagate in a medium, triggering the formation of microbubbles. These microbubbles release shockwaves and local hotspots, causing the solid particles in the medium to disintegrate (Ashokkumar, 2011). The high frequency of acoustic oscillations also results in the formation of oxidizing radicals, which rupture glycosidic bonds within the complex network of LCB (Luo et al., 2014; Velmurugan & Muthukumar, 2012). Ultrasound has been employed in enhancing numerous reactions for biomass valorization due to advantages such as short residence time, high activation energy, and the desired mass transfer.

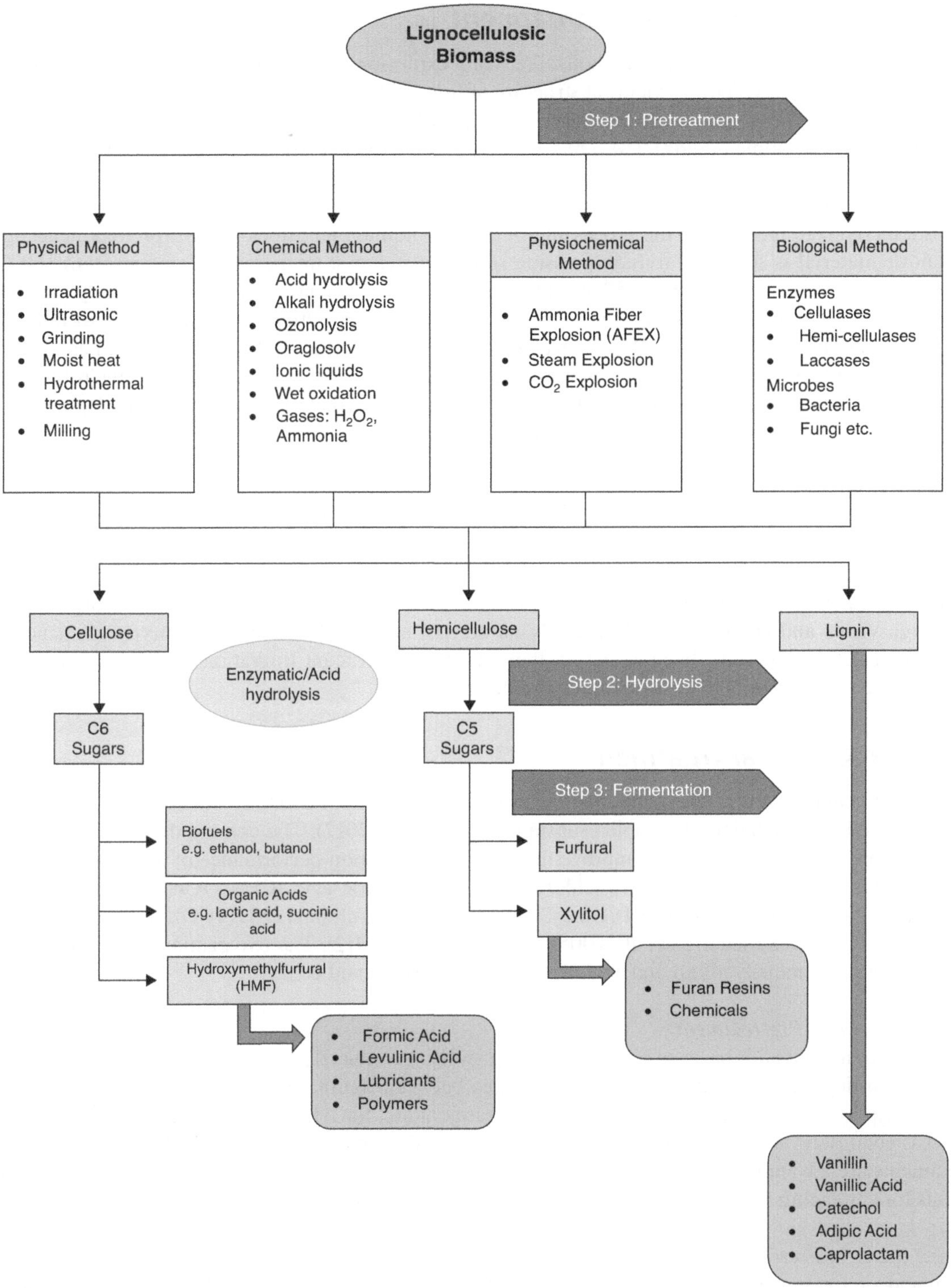

FIGURE 15.3 Lignocellulosic biomass pretreatment for the production of various valuable industrial products.

15.2.2.2 Physicochemical Pretreatment

The physicochemical pretreatment methods like steam explosion (SE) and microwave pretreatment methods are discussed below. The recalcitrant structure of lignocellulose materials is altered through chemical reactions in this type of pretreatment. The most frequently employed techniques involve the application of oxidizing agents, ionic liquids (ILs), acids, alkalines, or organosolv treatment (Jędrzejczyk et al., 2019).

15.2.2.2.1 Steam Explosion

SE is a recognized method for the processing of diverse biomass feedstocks. In SE pretreatment, lignocellulosic material is subjected to high-pressure saturated steam at temperatures ranging from 160°C to 260°C and pressures between 5 and 50 atm for several minutes. The pressure is incrementally alleviated, allowing the steam to expand within the lignocellulosic matrix, resulting in the separation of individual fibres and the disruption of the cell wall structure (Kumar et al., 2009). Acid may be used as a catalyst during SE, while its inclusion is not obligatory. Steam pretreatment is referred to as auto-hydrolysis when no external acid catalyst is introduced to the plant biomass. Nevertheless, enhanced lignin depolymerization can be accomplished with 1% acid treatment.

15.2.2.2.2 Microwave Pretreatment

Non-ionizing radiation-based microwave pretreatment effectively destroys the structures of LCB while adopting principles of green chemistry (Shi et al., 2011). The process of crystalline rearrangement under microwaves develops both interior pressure and heat according to Palav and Seetharaman (2007). The benefits of microwave pretreatment include low activation energy together with rapid heating capability and high yields and environmentally friendly nature (Li et al., 2016). Although energy efficiency and technological progress have been achieved, there are two significant limitations in biomass processing that involve high operational costs and pressure demands.

15.2.2.3 Chemical Treatment

Using chemical methods of pretreatment prevails over other methods due to its effectiveness and ability to enhance the biodegradation of complex substances (Liu et al., 2012). The commonly used chemicals for chemical pretreatment methods to improve the anaerobic digestion performance of agricultural residues include: sulphuric acid (H_2SO_4), hydrochloric acid (HCl), acetic acid (CH_3COOH), sodium hydroxide (NaOH), potassium hydroxide (KOH), lime $Ca(OH)_2$, aqueous ammonia ($NH_3{\cdot}H_2O$), and hydrogen peroxide (H_2O_2) (González Medeiros et al., 2005; Us & Perendeci, 2012). The two primary chemical pretreatment procedures, namely alkali and acid pretreatment, are outlined below:

15.2.2.3.1 Alkali Pretreatment

Alkali pretreatment adds bases to biomass, causing swelling, reduced polymerization, crystallinity reduction, lignin degradation, and disruption of linkages between lignin and other polymers. The most frequently reported chemicals used in alkaline pretreatment are NaOH, KOH, and $Ca(OH)_2$. Although the process conditions are comparatively mild, reaction periods can be lengthy. In some capacity, these pretreatments are advantageous in the partial hydrolysis of LCBs. Currently, the most effective alkali treatments for enhancing the digestibility of biomass are NaOH and KOH (Badiei et al., 2014).

15.2.2.3.2 Acid Pretreatment

The acid pretreatment of LCB primarily serves to eliminate hemicellulose. The acid pretreatment of biomass enhances enzyme accessibility to cellulose (Veluchamy et al., 2018). The predominant acids employed for pretreatment are sulphuric acid, acetic acid, and phosphoric acid (Rezania et al., 2020). Pretreatment with acids can be conducted in two distinct methods: dilute acid (0.1%) pretreatment at elevated temperatures (>200°C) and concentrated acid (30%–70%) pretreatment at comparatively lower temperatures (<50°C) (Den et al., 2018).

15.2.2.4 *Biological Treatment*

Early hydrolysis involves the biological method of pretreatment that uses microorganisms or enzymes to break down lignin structures. Cellulase enzymes catalyse enzymatic saccharification, transforming cellulose into its basic sugar units known as monomers, according to Tanjore and Richard (2015). The decomposition method of *Actinomycetes* bacteria targets grasses yet white-rot fungi particularly remove lignin with hemicellulose using *Phanerochaete chrysosporium*. The fungi create hydrolytic enzyme enzymes which activate bacterial growth using solid-state fermentation on biomass (Saritha, 2012).

15.2.3 Extraction Process of LCB

The first important stage of physical pretreatment serves to make LCB accessible for cellulose and hemicellulose extraction while breaking down the lignin protective layer (Ruiz et al., 2019). The focus has evolved from releasing sugars to obtaining high-value compounds because researchers now seek 5-HMF and furfural—previously considered inhibitory substances. The new approach gives priority to extractable value-added chemicals together with better accessibility to lignin compared to extracting polysaccharides from biomass. Research efforts have started developing economical extraction procedures which demonstrate promise for biochemical value creation. This document reviews multiple extraction methods throughout the next sections.

15.2.3.1 *Solvent Extraction*

Maceration extraction (ME) stands as the main simple solvent extraction method that extracts compounds at low temperatures through basic solvent treatment (Yan et al., 2021; Zuorro et al., 2019). An extraction process begins by soaking biomass while applying minimal heat in a solvent that usually requires pretreatment such as milling or pulverization for improving cellular penetration. The application of ME occurs before, during, or after fermentation time with wine production being an example since sugar transformation into ethanol results in modified solvent polarity and surface tension changes (Chong et al., 2021). Organic solvents' miscibility with water eases these variations (Silva et al., 2021).

15.2.3.2 *Subcritical Water Extraction*

Subcritical water extraction (SWE) process operates by heating water between 100°C and 374°C under 0.1 and 22.1 MPa pressure to avoid vaporization, thus adjusting substance viscosity and hydrogen proton count without requiring toxic solvents (Zhang et al., 2019). The water characteristics in subcritical conditions enable the substance to perform like acidic solutions and organic solvents. The optimal conditions for phenolic compound extraction by SWE according to Zhang et al. (2019) involved temperatures between 160°C and 220°C and pressures ranging from 2 to 6 MPa during 16–45 min of operation. Factors of temperature control between 100°C and 140°C and increased pressure from 10 to 15 MPa made extraction process efficient through parameter optimization.

15.2.3.3 *Extraction at Varying pH*

The product can be purified from other compounds that will dissolve in acidic or alkaline solutions by altering the pH during the extraction process, thereby extracting unwanted compounds. Furfural resins, 2,5-furandicarboxylic acid polymers, and polymers and resin chitosan are among the examples.

Alkalis are frequently employed to pretreat or extract specific compounds from biomass (Xu et al., 2016). For preventing the irreversible degradation of numerous compounds with low pK_a values, such as organic acids, it is necessary to evaluate the stabilities of compounds of interest.

15.2.3.4 Enzymatic Extraction

Zikeli et al., in 2016 demonstrated the presence of arabinoxylan–lignin and glucan–lignin linkage with the hydroxycinnamic acids (HCAs) ferulic acid and *p*-coumaric acid and the flavonoid tricin in wheat straw. The ferulic acid linkage is the most prevalent of the carbohydrate–lignin linkages in wheat straw. Additionally, significant quantities of tricin linkages are observed between glucan and lignin. Enzymatic hydrolysis could be examined as a component of an extraction cascade, as various extraction methods and solvents have been demonstrated to target the extraction of distinct phenolics (Zhu et al., 2015; Galanakis et al., 2013). This is because these intermolecular bonds are too robust to be disrupted by solvents.

15.2.4 Efficacy of Sugar Production

Sugarcane and sugar beetroot together with cereal starch serve as major sources to produce fuels and chemicals. Brazil relies on sugarcane sucrose for ethanol creation at the same time the U.S. uses maize starch to produce ethanol. The widespread use of carbohydrates specifically raises ethical uncertainties because these resources function as essential food along with feed resources. Sugarcane together with maize needs significantly higher amounts of water as well as fertilizers and energy compared to natural plant-derived lignocellulosic materials. The future of sustainable biorefinery operations requires the use of inedible LCB extracted from natural plants and energy farms to replace sugar raw materials but current extraction techniques face technical and economic limitations.

The entire sugar cane production process goes from milling to diffusion followed by clarification and evaporation and then boiling of products. Milling produces bagasse for energy application whereas molasses from boiling lead to bioethanol fermentation resulting in vinasse output. Nighttime fermentation of biohydrogen becomes feasible through exploiting these residues as potential fermentation sources (Kumari & Das, 2015, 2016).

The total global maize production reaches 1.06 billion tonnes mainly through food consumption but corn starch leads to ethanol production. Research shows that untreated corn stover contains biohydrogen reserves of 297.8 PJ/year yet they can be increased up to 621.0 PJ/year through pretreatment of the material.

Wheat is harvested for its 749 million tonnes annual production which generates 1.5 tonnes of straw as byproduct per each tonne of wheat. Biohydrogen generation utilizes wheat straw as a suitable biofuel source because of its 35% cellulose content (Li et al., 2014).

The production of rice during 2016 reached 741 million tonnes which led to the generation of cellulose-rich straw (31.9%) and husk (38.6%). The pretreatment procedure increases the hydrogen yield of rice straw as reported by Chang et al. (2011), and Gonzales and Kim (2017) demonstrated enhanced hydrogen production following strain treatment. The following segment will discuss how nanoparticles achieve additional improvements for sugar output levels.

15.3 NANOPARTICLES: SHORT BRIEFING ABOUT DIFFERENT SYNTHESIS METHODS

NMs exist as materials with at least one size dimension within the nanometre bandwidth from 1 to 100 nm (Jeevanandam et al., 2018). NMs exhibit different classifications through properties such as dimensionality, morphology, porosity, composition, and phase and origin and they appear in single-phase and dispersed/aggregated as well as multiphase forms (Meftahi et al., 2022). Various methods enable the synthesis of NMs including biological/biochemical processes in solution (liquid phase synthesis), nucleation and deposition under vacuum or plasma (gas phase synthesis) and grain formation and alloying (solid phase synthesis)

(Ijaz et al., 2020). The synthesis of NMs can be divided into physical and chemical and biological techniques but modern approaches now encompass physiochemical and mechanochemical methods. The manufacturing processes extensively determine both the dimensions of NMs and their structural characteristics and outer properties. The production of nanoparticles happens either by bottom-up procedures that assemble atoms and molecules from solid, liquid, and gas elements *or* by top-down approaches that break bulk materials into nanoscale particles.

Humans have been unintentionally utilizing NMs together with nanocomposites since ancient times according to Harish et al. (2022). Archaeologists have discovered through studies that early humans produced carbon-based NMs by using fire. The analysis using carbon dating techniques showed the existence of fat together with charcoal and paint-derived carbon NMs in 40,000 BCE cave art discovered in Sulawesi, Indonesia (Barhoum et al., 2019). The ancient production of PbS nanoparticles with sizes ranging from 5 nm for Egyptian hair colouring occurred at around 4,000 years ago and simultaneously during the 14th–15th centuries BCE when Egypt and Mesopotamia made glassware with inorganic NMs (Barhoum et al., 2022). During his 1959 presentation entitled "There's Plenty of Room at the Bottom" Feynman first presented nanotechnology to the world (Gaur et al., 2021). Scientists at present investigate materials under 100 nm dimensions in nanoscience research for water purification and hydrogen generation and fuel cells and sensors and diagnostics and batteries and drug delivery systems (Barhoum et al., 2021).

15.3.1 Synthesis

Nanoparticles can be synthesized based on three different factors—function of starting material, function of the deriving forces, and function of the reaction phase. In this context the short overview on first two factors-associated synthesis methods have been discussed (Harish et al., 2022).

15.3.1.1 Based on Function of Starting Material

The synthesis method is categorized into three approaches based on the starting material: Top-Down Approaches; Bottom-Up Approaches; Hybrid Approaches (Harish et al., 2022).

15.3.1.1.1 Top-Down Approaches

Large-scale NMs are reduced into smaller sizes by mechanical and chemical methods including attrition, milling and etching (Madkour, 2019). This method produces thin films and NMs exceeding 100 nm that include integrated circuits with micro-components among other products. These techniques face persistent resolution barriers according to the findings from Isaacoff and Brown (2017).

15.3.1.1.2 Bottom-Up Approaches

Bottom-up techniques are used to develop nanostructures with fewer faults, using atoms, molecules, and nanoparticles as foundation materials. Examples include ionic and molecular self-assembly, which assembles individual building blocks or molecules into larger entities using chemical or physical forces. These strategies allow spatial control over the characteristics and composition of individual components, with the size of the building block based on properties searched after. Wet chemical techniques like sol-gel, microemulsion, and co-precipitation also represent bottom-up approaches (Isaacoff & Brown, 2017).

15.3.1.1.3 Hybrid Approaches

Hybrid approaches integrate top-down and bottom-up fabrication techniques to create nanostructured platforms. Fabricating these hierarchically organized structures is generally challenging or unfeasible when relying solely on top-down or bottom-up techniques.

Photolithography illustrates a hybrid approach, where etching serves as the top-down technique, while layer formation occurs through growth with ions as the bottom-up method (Marcelo et al., 2020).

15.3.1.2 *Based on Function of the Deriving Forces*

The synthesis method is further classified based on function of the deriving forces: mechanical methods; physical synthesis methods; chemical and physicochemical synthesis methods; biological or green synthesis methods (Harish et al., 2022).

15.3.1.2.1 Mechanical Methods

The mechanical techniques of grinding and high-energy ball milling and cutting produce nanoparticles from solids yet they need large amounts of energy while sometimes creating contamination (Barhoum et al., 2020). Liu et al. (2021) demonstrate that mechanical alloying along with compaction and deformation and scrubbing techniques are applied yet particle sizes remain inconsistent. The use of ultrasonic devices leads to the creation of nano-droplets according to Singh et al. (2020).

15.3.1.2.2 Physical Synthesis Methods

The combination of evaporation, condensation, and laser ablation allows uniform nanoparticle distribution for optical and optoelectronic and microelectronic and magnetic devices yet research requires enormous space usage alongside high-energy input (Chavali & Nikolova, 2019). The single-step synthesis of different materials through physical vapour deposition (PVD) enables utilization of methods like spray drying, freeze drying, and plasma spraying (Sriram et al., 2022). The freeze-drying method evacuates solvents through sublimation under the protective management of agents for ice crystal control (Su et al., 2021). During electrospinning nanofibre production occurs through electrohydrodynamic forces which rely on solution properties and environmental conditions (Rasouli et al., 2018; Zhou et al., 2019).

15.3.1.2.3 Chemical and Physicochemical Synthesis Methods

The synthesis of nanoparticles through chemical methods uses four main techniques which include chemical vapour deposition (CVD), sol-gel, solvothermal, and polymerization and precipitation techniques (Tudose et al., 2019). The fusion of electrochemical with laser deposition techniques unites physical and chemical production methods although they pose considerable expenses together with environmental challenges. The wet chemical synthesis approach produces various nanoparticles yet it generates hazardous waste products. Diverse thin films such as ZnS or CuSe or InS or CdS are fabricated through CVD and electrochemical deposition together with chemical bath deposition (Jalali-Moghadam & Shariatinia, 2018). The synthesis of nanostructures with high purity occurs through electrochemical methods and ionizing radiation serves as a synthesis aid for NMs (Lebedeva et al., 2021; Guo et al., 2020).

15.3.1.2.4 Biological or Green Synthesis Methods

The synthesis of green nanoparticles requires three different approaches including biogenesis with plant extracts or microorganisms or bio templates (Jeevanandam et al., 2016). Leaves from different plants and microorganisms can combine with metallic elements to generate nanoparticles through their biological extracts. The low-energy method with its environmentally friendly approach makes it suitable for industrial applications (Mittal et al., 2013). Antibacterial drugs that are safe for production emerge from nano factories operated by microbes (Iravani & Zolfaghari, 2021).

15.4 DIVERSITY OF NANOPARTICLE CONJUGATED CATALYST

Several nanoparticles had been used for immobilization of different enzymes for highly efficient production of sugars through saccharification from LCB. Some of the nanoparticles such as magnetic nanoparticles (MNPs), carbon nanotube, metallic nanoparticles, and other nanoparticles are listed with mechanisms and properties in Table 15.2.

TABLE 15.2 Diversity of nanoparticle-conjugated catalyst and their mechanism and properties

NANOPARTICLE CONJUGATE	*CATALYSIS MECHANISM*	*PROPERTIES*	*REFERENCE*
Gold nanoparticles	Cellulase was immobilized on gold nanoparticles for the saccharification of alkali (4% w/v NaOH) pretreated aquatic weed biomass, specifically *Alternanthera philoxeroides* and *Brachiaria mutica*.	• The pretreated biomass of *A. philoxeroides* demonstrated a maximum total reducing sugar (TRS) yield of 18.95±0.18 mg/g after 72 hours and saccharification degree of 38.25%±0.8%, 49.05%±0.7%, 67.05%±0.7%, 85.27%±0.8%, and 67.99%±0.8% over the period from 12 to 96 hours. • The pretreated biomass of *B. mutica* demonstrated a maximum TRS yield of 20.98±0.17 mg/g after 72 hours and saccharification degrees of 44.46%±0.7%, 55.53%±0.8%, 73.26%±0.7%, 94.41%±0.8%, and 73.3%±0.7% over the time intervals from 12 to 96 hours.	Aarti et al. (2022)
Silver nanoparticles	Improved saccharification through fabricated silver nanoparticles using sodium borohydride along with immobilized alpha amylase enzyme for starch degradation.	The purified bacterial amylase enzyme exhibited a 40-fold increase in the formation of reducing sugars, while the crude bacterial amylase enzyme showed a 9-fold increase. This confirms the catalytic activity of the purified bacterial amylase enzyme when combined with Ag-NPs as a nanocatalyst, facilitating the saccharification of complex starch molecules into simple monosaccharides.	Krishnakumar et al. (2018)
Fe_3O_4 magnetic nanoparticles	1. Improved saccharification by immobilizing xylanases, cellulases, and amylases with glutaraldehyde-activated Fe_3O_4 nanoparticles. 2. Another approach also introduced by functionalized alkyl sulfonic acid and butyl carboxylic acid with silicon-fabricated Fe_3O_4 magnetic nanoparticle.	1.1. Immobilization efficiency of 92%, 45%, and 93% respectively. Immobilized enzyme activation at 50°C. 1.2. Exhibits high stability during storage, operational stability, and reusability, maintaining 69%, 48%, and 50% residual activity after 13 uses respectively. 2.1. The maximum concentration of 17.06 g/L for Fe_3O_4-MNPs-Si-AS (alkyl sulfonic acid) and 15.40 g/L for Fe_3O_4-MNPs-Si-BCOOH (butyl carboxylic acid) sugar at 500 mg was greater than normal acid (H_2SO_4) (14.63 g/L) and non-treated sugarcane straw (0.24 g/L).	Kumari et al. (2018), Ingle et al. (2020)

(Continued)

TABLE 15.2 (*Continued*) Diversity of nanoparticle-conjugated catalyst and their mechanism and properties

NANOPARTICLE CONJUGATE	*CATALYSIS MECHANISM*	*PROPERTIES*	*REFERENCE*
Silica nanoparticles	Covalent co-immobilization of SiO_2 with the amalgamation of the spontaneous synthesis of covalent bonds between the proteins Spy Catcher and Spy Tag (SC/ST) and elastin-like polypeptide (ELP)-mediated green biomimetic silicification. This approach involved genetically fusing the SC with the ELP, and then using the inverse transition cycling (ITC) method to purify the produced SC-ELP (SCE). Through a single-step biomimetic silicification process, the nanocarrier (SCE@SiO_2) was rapidly generated at room temperature, neutral pH, and ambient pressure. The ST-tagged recombinant bi-enzyme xylanase-ST-lichenase (XSTL) was covalently self-conjugated onto SCE@ SiO_2 via the SC and ST reaction, thereby eliminating the need for cross-linking reagents.	• Target-specific immobilization from cell crude extracts achieved excellent activity recovery (over 90%). • Co-immobilized XSTL illustrated excellent reusability, maintaining 83.9% (lichenase) and 81.4% (xylanase) of the initial activities by the seventh cycle, respectively. • XSTL has the potential to enhance the efficiency of producing fermentable sugars from renewable biomass polysaccharides.	Lin et al. (2021)
Nickel ferrite	Conjugation of nickel ferrite nanoparticles with cellulase enzymes utilizing algal biomass of *Lyngbya limnetica* (cyanobacteria) as a substrate, resulting in increased saccharification efficiency.	• 2.5-fold increased saccharification efficiency after 72 hours at specific pH and temperature in comparison with control. • $NiFe_2O_4$ NPs-mediated crude cellulase enzyme-hydrolysed rice straw to produce sugar hydrolysate.	Srivastava et al. (2021)

15.5 SACCHARIFICATION: MANUFACTURE OF FERMENTABLE SUGAR

The saccharification stage is the primary limitation for all biofuel generation methods, such as biohydrogen and bioethanol. It effectively facilitates the biodegradation of cellulosic constituents into sugar monomers. LCB has both cellulose and hemicellulose, and the primary objective of saccharification is to efficiently hydrolyse both components into fermentable sugar monomers (Zabed et al., 2016). Lignocellulose hydrolysis involves cellulases and hemicellulases, which break down cellulose hydrolysate into C6 and C5 monomers, respectively (Binod et al., 2019). In addition to lignocellulolytic enzymes, numerous additional proteins appear to be involved in lignocellulose breakdown; however, their specific contributions remain inadequately elucidated. Saccharification can be executed by two approaches, nanocomposite-free enzyme-based saccharification and nanocomposite-conjugated enzyme-immobilized saccharification.

15.5.1 Nanocomposite-Free Enzyme-Based Saccharification

The primary challenge of cellulosic biofuels, such as biohydrogen and bioethanol that generate at a commercial scale, is the breakdown of lignocellulosic biopolymers into reducing sugars. Enzymatic hydrolysis is a critical and expensive factor in the bioconversion of lignocellulose. Consequently, the expense and efficacy of the enzymes, or biocatalysts, employed for lignocellulose hydrolysis are significant issues (Chandel et al., 2019). Existing lignocellulose bioconversion strategies are categorized into off-site and on-site saccharification methods, based on production methods and biocatalysts used for enzymatic hydrolysis. On-site saccharification involves simultaneous enzyme production and hydrolysis, while off-site saccharification involves controlled enzyme production before saccharification.

15.5.1.1 Off-Site Saccharification

Both separate hydrolysis and fermentation (SHF) and simultaneous saccharification fermentation (SHF) primarily utilize free cellulases sourced from fungi as biocatalysts to carry out off-site saccharification (Liu et al., 2019a). SHF is structured to carry out enzymatic hydrolysis followed by fermentation in a sequential manner (Akhtar et al., 2013). This approach is regarded as the simplest strategy that integrates existing techniques to meet the requirements of multi-step lignocellulose bioconversion. The combination and simultaneous execution of the two process steps is known as simultaneous saccharification and fermentation (SSF). According to reports, SSF has a competitive edge over SHF regarding lower capital costs and reduced contamination risks (Öhgren et al., 2007). To enhance the conversion of fermentable sugars, simultaneous saccharification and cofermentation (SSCF) was modified for the simultaneous saccharification and co-fermentation of both pentose and hexose sugars, relying on the principles of SSF. *Trichoderma reesei* is recognized as a leading producer of cellulase whereas Novozymes, Inc. stands out as the leading supplier of commercial enzyme cocktails to date (Chandel et al., 2019). The essential expense for producing fungal cellulase is estimated to be approximately $10 per kg of protein. The sugar yield stands at 80%, and with a loading of 10 FPU/g cellulose (approximately 20 mg cellulase/g cellulose), the cost of cellulase amounts to roughly $250 per tonne of sugar. Additional enzymes are necessary for the saccharification of substrates with high solid loading (Hodge et al., 2008), which will consequently raise the cost. Conversely, the production of cellulosic biofuels such as ethanol and biohydrogen becomes unviable when the cost of sugar production exceeds $100 per tonne (Chandel et al., 2019). Multiple investigations have been launched to reduce off-site saccharification enzyme expenses because researchers work on enhancing enzyme manufacturing methods and developing new enzymes with superior heat-stable properties and resistance to reaction products while maintaining their catalytic capacity (Chandel et al., 2019). Fungal enzymes embody the major barrier to the advancement of lignocellulose bioconversion because they currently represent the most significant financial challenge (Lynd et al., 2017). Off-site saccharification involves lengthy preparation steps, lengthy transport requirements, excessive carbon and nitrogen source usage for fermentation, and the potential risk of contamination according to Chandel et al. (2019). The use of enzymes and nanoparticles represents essential research areas which must be improved for successful deployment of this technology.

15.5.1.2 On-Site Saccharification

The needs of industrial production call for simpler on-site saccharification techniques such as consolidated bioprocessing (CBP) and its derivatives because Lynd et al. (2017) introduced them. The concept of CBP emerged through Liu et al. (2020) work in the first part of the 21st century. This process distinguishes itself from other traditional saccharification approaches. Production of cellulases as well as target products takes place in one integrated step inside one reactor using CBP to diminish enzyme manufacturing expenses. Co-treatment (CT) was subsequently proposed as an advanced CBP technology, demonstrating improved degradation efficiency of lignocellulosic substrates (Balch et al., 2017). Both CBP and CT primarily focus on the production of cellulosic ethanol, and they are theoretically applicable to other fermentation products as well like biohydrogen. The CBS strategy has recently evolved by decoupling the fermentation step from the integrated CBP process (Liu et al., 2019b; Zhang et al., 2017), emphasizing the cost-effective production of lignocellulosic fermentable sugars instead of the final metabolites.

15.5.1.2.1 Consolidated Bioprocessing (CBP)

In contrast to SSF, CBP enhances enzyme production to facilitate on-site saccharification. It depends on effective biocatalysts that are capable of concurrently executing lignocellulose degradation and sugar conversion into specified products (Lin et al., 2015). Suitable CBP microorganisms should possess distinct characteristics, such as effective cellulase synthesis and secretion for rapid lignocellulose solubilization. Additionally at the same time absorption of C5 and C6 sugars takes place, and there is resilience to lignin-derived toxins and end metabolites. Various CBP biocatalysts have been engineered and developed using natural or recombinant methodologies (Olson et al., 2012). The native technique entails the augmentation of yield and titre of certain products through the genetic alteration of cellulolytic microbes (Chung et al., 2014). The recombinant approach includes the development of foreign cellulases and hemicellulases in host organisms, such as *Saccharomyces cerevisiae*, capable of generating desired products (Guo et al., 2018b; Tabañag et al., 2018). Certain model microorganisms, such as *Escherichia coli* and *Bacillus subtilis*, are incapable of degrading cellulose or synthesizing specific products. It has been employed to develop CBP biocatalysts due to their ability to co-utilize hexose and pentose, as well as the relative ease of their genetic manipulation (Zhang & Zhang, 2010). Moreover, the documented recombinant cellulase-producing bacteria are incapable of utilizing cellulose as the exclusive carbon source without supplementary nutrients like yeast extract. Therefore, the traditional method demonstrates greater viability for practical implementation. The hydrolysis rates of cellulases typically rise with elevated temperatures; nevertheless, solventogenic organisms normally are unable to resist high temperatures. Hence, different clostridial strains, such as *Clostridium cellulolyticum* (Xu et al., 2015), *C. papyrosolvens* (Zou et al., 2018), *C. cellulovorans* (Bao et al., 2019), and *C. josui* (Jindou et al., 2002), are recognized as candidates for CBP. The thermophilic cellulolytic bacteria *Clostridium thermocellum*, also known as *Hungateiclostridium thermocellum* or *Ruminiclostridium thermocellum*, is the optimal selection for CBP because of its capacity to digest cellulose and produce ethanol. *C. thermocellum* is reported to produce 22.4 g/L of ethanol from 60 g/L of microcrystalline cellulose (Avicel) as the substrate. This constitutes 75% of the maximum theoretical yield, and the potential sugar conversion ratio was suggested to reach 85% with additional genetic engineering (Tian et al., 2016). The co-cultivation of cellulolytic and solventogenic bacteria has been documented to yield ethanol with an elevated titre. A tailored coculture of *C. thermocellum* and *Thermoanaerobacterium saccharolyticum* produced 38 g/L of ethanol from 92 g/L of Avicel (Park et al., 2012). Productivity may be constrained by the non-solventogenic microbe in the coculture due to product tolerance. Therefore, anaerobic CBP conditions will limit the technology's use and the variety of viable products. CT has been suggested as a technique evolved from CBP, integrating the CBP process with intensive mechanical milling under thermophilic conditions, utilizing the cellulosome-producing *C. thermocellum* as the biocatalyst. *C. thermocellum* has been utilized to generate CBP isobutanol (5.4 g/L) and butanol (0.36 g/L) by augmenting the natural route and including heterologous enzymes into the organism (Lin et al., 2015; Tian et al., 2019).

15.5.1.2.2 Consolidated Bio-saccharification (CBS)

CBP has demonstrated significant benefits in lowering enzyme production costs and optimizing operational processes; nevertheless, this approach is also limited by its very narrow range of applications and products generated. Due to the constraints of anaerobic cultivation conditions for cellulolytic microbes, contemporary CBP technology is primarily designed for biofuel production, making it challenging to integrate with aerobic fermentations. Furthermore, for achieving versatile downstream applications for lignocellulose, it is essential to detach fermentation from the anaerobic and high-temperature hydrolysis phase. A CBS technique has been presented that separates fermentation from the integrated process. CBS generates fermentable sugars as the primary product instead of end metabolites like ethanol (Liu et al., 2019b;

Zhang et al., 2017). As a result, fermentation will not be constrained by hydrolysis conditions, allowing for the optimization of cellulolytic capacity. Consequently, CBS offers novel perspectives on lignocellulose bioconversion. CBS utilizes cellulolytic bacteria as biocatalysts for on-site saccharification, like CBP. Cellulosome-producing bacteria are favoured among natural cellulolytic microorganisms due to their superior cellulase-degrading activity compared to free enzymes produced by fungi, particularly regarding crystalline cellulosic substrates (Thomas et al., 2017b). *Clostridium thermocellum* is regarded as a premier option for CBS, owing to its extremely effective cellulosomal system (Hong et al., 2014); however, specific genetic modification is necessary. It is essential to ensure that CBS biocatalysts can endure elevated concentrations of reducing sugars, as high saccharification efficiency and sugar output are vital for CBS to meet the demands of diverse fermentation processes. Prior research indicates that the utilization of *C. thermocellum* and its cellulosome in lignocellulose bioconversion is predominantly constrained by the significant inhibitory effect of the hydrolysate product cellobiose, and this feedback inhibition can be substantially mitigated through the incorporation of β-glucosidase (BGL) (Qu et al., 2017). Consequently, the expression of BGL is pivotal to the development of CBS biocatalysts. C. *thermocellum* presents higher complexity when compared to model microorganisms leading to extensive research for designing adequate genetic engineering methods which began in 2010 and continues to this day (Olson, 2010; Olson et al., 2015; Tripathi et al., 2010; Zhang et al., 2015, 2017). Two generations of CBS biocatalysts were developed through the fused expression of BGL with cellulosomal exoglucanases according to Zhang et al. (2017) and Liu et al. (2019b). Strong glucose solutions above 88 g/L can be gathered through fermentation processes (Zhang et al., 2017). Improving the CBS process requires thorough optimization both to generate higher sugar outputs and reduce operational expenses as described later. The significance of CBS increases because it generates fermentable sugars at affordable prices which meets requirements for the sugar platform in lignocellulosic biorefineries. A biological simultaneous enzyme production and saccharification (BSES) method has been suggested for lignocellulose saccharification utilizing *C. thermocellum* (Ichikawa et al., 2019). In contrast to CBS, BSES necessitates the incorporation of BGLs in the form of either free enzymes or bacterial cells (Ichikawa et al., 2019) and can be regarded as a hybrid approach that integrates on-site and off-site techniques.

Due to high recalcitrance to biodegradability of LCB it became difficult to execute the saccharification process with different enzymes even after the pretreatment. Due to the high expense of enzymes, it is getting difficult to process the saccharified product (sugar) for further development in green biofuel (biohydrogen, bioethanol) generation, in a feasible manner. That's why, scientists have developed an innovative approach of immobilization enzymes like cellulase on the nanoparticle matrices for higher lignocellulosic biodegradability along with higher sugar yield at a low cost (Abraham et al., 2014; Zhang et al., 2014).

15.5.2 Nanocomposite-Conjugated Enzyme-Immobilized Saccharification

Enzymes can be physically confined or localized in a specific area while maintaining their catalytic activities, a process referred to as enzyme immobilization. In one of the studies, Brena et al. (2013) state that immobilized enzymes are better than soluble enzymes because they enable repetitive usage. Immobilization proves to be an effective solution when handling selectivity and stability along with recyclability issues that occur during enzymatic catalysis. The tool successfully resolves problems that occur because of enzyme inhibitors which exist naturally in reaction medium (Qi et al., 2018). Using immobilized cellulases presents a pathway to decrease the overall production expenses of these enzymes along with improving their industrial usage possibilities. The specific activity increases after immobilizing cellulases because of the stable microenvironment created during immobilization. The enzyme-crosslinking agent covalent bond changes enzyme functional group ionization states to affect enzyme stability levels (Qi et al., 2018). The residual properties of porous organic composites were utilized to attach cellulase through a glutaraldehyde linkage according to Mo et al. (2020a, 2020b). Activity recovery exceeded 67% while reusability increased in 10 successive cycles for this immobilized cellulase. The immobilization support system created both plentiful sites for enzyme attachment while simultaneously designing an ideal biological interface that advanced substrate and enzyme connectivity. The free availability of cellulase encounters multiple limitations during reuse because it displays poor stability when exposed to varying conditions like pH and temperature and ionic strength changes (Zang et al., 2014). The enzyme can maintain stability once immobilized to the solid support which enables its repeated use at affordable costs.

15.5.2.1 Immobilization Methodologies

The current enzyme research interest revolves around cellulase because this enzyme shows restricted thermal and pH stability affecting its recyclable potential (Qamar et al., 2021). Research seeks to raise immobilization levels of cellulase and boost its reusability because a thorough understanding of the basic methods used to achieve this goal is essential. Four standard techniques to immobilize enzymes exist which include physical adsorption as well as covalent attachment and cross-linking followed by entrapment.

15.5.2.1.1 Physical Adsorption

Physical adsorption relies on establishing enzyme-solid material or support connections through bonds such as hydrogen bonding and hydrophobic contacts alongside ionic and electrostatic binding and Van der Waals forces. Physical adsorption maintains the native enzyme structure without modification because its approach does not result in structure changes (Liu et al., 2021). Enzyme immobilization undergoes drastic modification by the features of adsorbent surface including pore dimensions and available attachment sites and total available surface area (Mo et al., 2020a, 2020b). Mo et al. (2020a, 2020b) generated magnetic polyporous biochar through calcination before physically adsorbing cellulase onto the substrate which led to an immobilization efficiency of 73.6%. The procedure is both budget-friendly and easy to execute yet produces nonspecific adsorption as its main weakness (Liang et al., 2019).

15.5.2.1.2 Covalent Attachment

Covalent binding, conversely, exhibits enhanced enzyme thermostability due to the robust interactions between the enzyme's functional moieties and the corresponding moieties on the surface of the support material. Numerous linkages between the nanopore surface of the binding material and the active site of the catalyst are established during the reaction (Misson et al., 2015). It inhibits enzyme leakage and demonstrates significant potential for stabilizing enzyme molecules to facilitate reactions with substrate entities. The covalent immobilization of commercial cellulases on amino-functionalized magnetic beads results in the formation of advanced biocatalysts for continuous cellulose hydrolysis, facilitated by a Schiff base linkage between the support particles and the covalently bound cellulose complex (Mariño et al., 2021). The enzyme leakage together with active structural deformations that occur through covalent bonds leads to severe decreases in enzymatic activity (Kim et al., 2018).

15.5.2.1.3 Crosslinking

Enzyme stabilization requires the cross-linking approach since it connects enzyme molecules by using functional reactants. The addition of cross-linking agents in binding supports generates enlarged surface areas which boosts both active enzyme capacity and solution recovery potential (Zhou et al., 2019). The use of cross-linking faces particular challenges because structural enzyme alteration leads to function diminution and obstructed mass transport functions (Sassolas et al., 2012).

15.5.2.1.4 Entrapment

Enzyme membrane entrapment stands as a proven efficient method because it enables continuous downstream processing while enabling longer operations. Using this method enzymes can access support

TABLE 15.3 Advantages and disadvantages of different immobilization strategies

IMMOBILIZATION METHODOLOGIES	*ADVANTAGES*	*DISADVANTAGES*	*REFERENCE*
Adsorption immobilization	• Reversible, simple, and low cost • Low chance of damage to biocatalyst • Additional coupling or enzyme modification not required • Unchanged k_{cat} and k_m values, resulting in adequate catalytic activity of immobilized enzyme	• High chance of enzyme leaching and desorption due to weak and reversible interactions between carrier and enzymes. • Uncontrolled packing density of the immobilized enzymes, leads to loss of enzyme activity during time.	Ahmad (2015), Datta et al. (2013), Mohamad et al. (2015)
Covalent binding immobilization	Improved thermostability with strong and stable binding, enzyme leaching can be prevented.	For decreased substrate affinity of immobilized enzymes, the conformational restriction deactivates the enzymatic function.	Ahmad (2015), Datta et al. (2013) Mohamad et al. (2015)
Entrapment immobilization	• Protecting enzyme from mechanical shear, hydrophobic solvents, and gas bubbles • Simple downstream processing and continuous operation. • Protein integrity and enzyme efficacy retained.	Limitations include lesser enzyme loading and limiting mass transfer.	Ahmad (2015), Datta et al. (2013), Mohamad et al. (2015)
Cross-linking immobilization	No requirement of support matrix leads to high enzyme stability, desorption decreasing and easily recyclable and reusable.	The conformational change leads to decrease in diffusion rate and loss of the activity of enzyme.	Ahmad (2015), Datta et al. (2013), Mohamad et al. (2015)

matrix channels and pores to achieve high levels of loading. The membrane entrapment method enhances enzyme stability while blocking enzyme denaturation according to Datta et al. (2013).

The selected method for immobilization is contingent upon the individual enzyme type and the support matrix utilized. A synthesis of the aforementioned methods has been utilized to augment the efficiency of immobilization and to improve enzyme utilization (Qamar et al., 2021).

The advantages and disadvantages of the different immobilization techniques are tabulated in Table 15.3.

15.5.2.2 Cellulase Immobilization on Nanostructured Film

Cellulase is predominantly an essential enzyme utilized in large-scale enzymatic biofuel synthesis. Cellulases have been derived from several bacteria and fungi. Specifically, some fungus from the genera *Penicillium*, *Trichoderma*, *Aspergillus*, and *Schizophyllum* exhibited significant cellulase production potential (Srivastava et al., 2017). Cellulase catalyses the hydrolysis of LCB into fermentable sugars for the subsequent generation of biofuels, including bioethanol, biohydrogen, etc. Cellulases are classified into three primary categories based on their mechanism of action: (1) endoglucanases, which generate free chain ends; (2) exoglucanases or cellobiohydrolases, which hydrolyse the 1,4-glycosidic linkages to produce cellobiose; and (3) β-glucosidase, which transforms cello-oligosaccharides and the disaccharide cellobiose into glucose.

The primary challenge in cellulase-catalysed biofuel generation is the elevated cost of enzyme. Therefore, enzyme reusability and recovery are critical variables for industrial applications. Enzyme immobilization is an effective method to achieve these characteristics. The utilization of nano-immobilized cellulase in LCB hydrolysis presents significant potential for transformative outcomes (Abraham et al., 2014; Zhang et al., 2015). As previously discussed, the immobilization of enzymes on carriers such as NMs can enhance their thermal stability, efficiency, and reusability. Recently, numerous NMs have been employed as carriers for the immobilization of cellulase, including MNPs, silica nanoparticles, nickel nanoparticles, metal oxide nanoparticles, carbon nanotubes, and graphene-based nano catalysts.

15.5.2.2.1 MNPs Conjugated Immobilization

MNP-based carriers have been utilized for diverse biological applications due to their capacity to elevate enzyme reusability, activity, and stability, along with facilitating enhanced enzyme recovery (Sánchez-Ramírez et al., 2017). The nanoscale MNPs display superparamagnetic properties as well as stable dispersion within reaction fluids. The surface of MNPs must receive functionalization treatment to stop oxidation because they naturally agglomerate and oxidize when exposed to air (Zhang et al., 2015; Sánchez-Ramírez et al., 2017). MNP functionals for cellulase immobilization involve five different methods such as silica-functionalization, amino-functionalization, composite functionalization, chitosan-functionalization, and carrier-free functionalization (Zhang et al., 2015; Sánchez-Ramírez et al., 2017). The available reports demonstrated that making cellulase active on MNPs serves as a critical method for biofuel production. The immobilization process involving *T. reesei* cellulase on active magnetic support for covalent binding was examined by Abraham et al. (2014). The hydrolysis procedure of pretreated hemp hurd biomass (HHB) under 80°C showed 50% preserved activity throughout five repeated operations according to Abraham et al. (2014). The synthesis of magnetic-chitosan nanoparticles for cellulase immobilization through covalent binding used glutaraldehyde as a coupling agent according to Zang et al. (2014). A study revealed that the co-immobilization of β-glucosidase A and cellobiohydrolase D on superparamagnetic nanoparticles acquired 100% immobilization efficiency and enhanced enzyme stability for the conversion of cellulosic biomass into sugars for bioethanol production. A recent study demonstrated that immobilizing cellulase on zinc ferrite nanoparticles considerably enhances the enzymatic hydrolysis of pretreated *Crotalaria juncea* biomass. Furthermore, it was noted that immobilization confers thermal stability to the enzyme, allowing it to maintain its activity for up to three cycles.

The researchers showed that cellulase-based attachment to MNPs could cut down enzyme requirements during bioethanol fermentation from lignocellulosic materials. Laboratory tests indicated that cellulase obtained from *T. reesei* which was immobilized onto chitosan-coated MNPs through glutaraldehyde bonding showed 80% activity retention after 15 cycles of carboxymethylcellulose hydrolysis (Sánchez-Ramírez et al., 2017). Research showed that covalently bonding cellulase from *T. reesei* onto chitosan-coated MNPs with glutaraldehyde as the connector produced nanoparticles with enhanced catalytic function. The application of a magnetic field allows researchers to recover 80% of immobilized enzyme effectively. Data from extra study confirmed that the active period of the immobilized cellulase extended across 15 cycles without any significant performance deterioration. A research team investigated cellulase-immobilized MNPs under ultrasonic conditions which generated superior catalytic performance compared to the control (Ladole et al., 2017). Holocellulase isolated from *Aspergillus niger* SH3 underwent successful immobilization onto iron oxide MNPs through two methods which included physical adsorption and covalent coupling. Immobilized enzyme showed better alkali-treated paddy straw saccharification while keeping its stability across two saccharification cycles with a 55% enzyme preservation rate reported by Kumar et al. (2016).

T. reesei cellulase received separate investigation for immobilization onto magnetic and silica nanoparticles which yielded immobilization efficiencies of 85% and 76% respectively to strengthen its catalytic performance (Grewal et al., 2017). The scientists used immobilized cellulase to saccharify

1-ethyl-3-methylimidazoliumacetate [EMIM][Ac] treated sugarcane bagasse and wheat straw through in situ saccharification to obtain 89% saccharification yield (Grewal et al., 2017). Jia et al. (2017) developed MNPs attached to cellulase aggregates that advanced biomass bioconversion effectiveness (Jia et al., 2017). The combination of cellulases with amylases and xylenes from bacterial sources was immobilized onto MNPs for effective stability and recovery (Kumari et al., 2018). Researchers exploited glutaraldehyde cross-linking to immobilize cellulase, pectinase, and xylanase mixed enzymes obtained from *A. niger* culture onto MNPs in their most recent investigation (Muley et al., 2018). The researchers improved thermal stability and the tri-enzyme magnetic complex maintained 90% activity through four cycles while being stable for 36 days at 5°C.

Research on cellulase immobilization demonstrates that gold magnetic silica nanoparticles stand as a promising platform alongside various MNPs for this purpose. Poorakbar et al. (2018) created mAu@PSNs magnetic gold mesoporous silica nanoparticle core shells to serve as a carrier system for cellulase immobilization. The immobilized enzyme demonstrated a binding efficiency of 76%, exhibiting enhanced thermal stability and the capability to be reused for 5 cycles. All these studies indicated that enzyme-immobilized MNPs may serve as a promising source for biomass-derived biofuel production.

15.5.2.2.2 Carbon Nanotube Conjugated Immobilization

In addition to MNPs, carbon nanotubes have garnered significant interest for enzyme immobilization owing to their distinctive structure and the potential to provide remarkable properties, including high mechanical strength, as well as exceptional electrical and thermal characteristics (Ahmad & Khare, 2017). Carbon nanotubes consist of graphite sheets organized in a cylindrical formation. The classification into single-walled or multi-walled carbon nanotubes depends on the number of layers present.

Among these, multi-walled carbon nanotubes appeared to be more advantageous for cellulase immobilization due to their improved electronic properties, superior physical and chemical stability, cost-effectiveness, simpler preparation, and reduced toxicity compared to single-walled carbon nanotubes. The enzymes bound to carbon nanotubes exhibit stability and activity at elevated temperatures, possess a significant surface area, high enzyme content, and excellent dispersibility, along with additional beneficial characteristics like low mass transfer resistance and the ability to recycle the catalyst. The various characteristics may enhance the utilization of carbon nanotubes in enzyme immobilization and the production of platform chemicals (Mohamad et al., 2015).

15.5.2.2.3 Metallic Nanoparticles Conjugated Immobilization

Along with magnetic and carbon-based nanoparticles, lignocellulolytic enzymes like cellulase have also been successfully immobilized using metallic nanoparticle carriers. Chang et al. (2011) utilized chemical binding and physical adsorption to immobilize cellulase on two distinct mesoporous silica nanoparticles with variable surface areas, pore sizes, and particle sizes (Chang et al., 2011). Effective cellulose-to-glucose conversion surpassing 80% yield and good stability were demonstrated by the immobilized cellulase on mesoporous silica nanoparticles via chemical binding (Chang et al., 2011). Cellulase was shown to be immobilized by adsorption onto 40 nm silica nanoparticles by Lupoi and Smith (2011). During SSF reactions, the nano biocatalyst was employed to enhance the synthesis of ethanol through the hydrolysis of lignocellulosic materials. With a comparable increase in ethanol output of 2.1 to 2.3 times, immobilized cellulase was able to produce 1.6 times more glucose than cellulase (Lupoi & Smith, 2011).

In several biofuel production processes, such as the synthesis of biohydrogen from LCB and the processing of LCB to syngas (a mixture of hydrogen and carbon monoxide), other metallic nanoparticles, such as crystalline nickel nanoparticles, can function as catalysts. And it has been observed that gold and silver nanoparticles serve as a carrier for the immobilization of cellulase. The fabrication of silver and gold nanoparticles facilitated by cellulase was reported by Mishra and Sardar (2015).

15.5.2.2.4 Other Nanoparticles Conjugated Immobilization

Apart from the previously discussed nanoparticles, there exist additional NMs that can be employed for the immobilization of metal oxide NMs that resemble cellulases. In this regard, it has been shown that immobilizing

Aspergillus fumigatus cellulase on zinc oxide (ZnO) nanoparticles can result in increased catalytic efficiency and enzyme stability (Srivastava et al., 2016). High biocompatibility, a high isoelectric point (IEP ~ 9.5), and electron mobility—properties that give the enzyme long-term stability—were made possible by using ZnO as a nanocarrier. The investigation of enzyme catalytic efficiency enhancement requires ZnO nanoparticles to function as the optimal testing substance. The immobilized cellulase retained 53% relative activity together with thermal stability for up to 10 hours at a pH of 10.5 while operating at temperatures up to 65°C.

Investigation by Cherian et al. (2015) showed how MnO_2 nanoparticles improved both the activity and stability of *A. fumigatus* JCF-produced cellulase. The synthesized TiO_2-lignin hybrid substrate enabled physical contact immobilization of *A. niger* cellulase which demonstrated promising catalytic activity during the tenth cycle of reuse.

15.5.3 Comparison of Both Cases for Enhancing Saccharification

Ahmed et al. (2018) indicated that the cellulase enzyme in solution form demonstrated instability, having lost nearly all its activity after 30 days of storage. Conversely, the cellulase immobilized on a metal-organic framework exhibited remarkable storage stability, retaining nearly all its activity after being stored for an equivalent duration as free cellulase. This demonstrated that immobilization significantly restricted enzyme denaturation, therefore displaying enhanced catalytic activity.

The immobilized cellulase exhibited improved performance compared to the free enzyme across wider temperature and pH ranges, along with commendable reusability following recovery via magnetic separation. Similarly, the immobilization of cellulase extracted from *A. niger* fabricated onto b-cyclodextrin-Fe_3O_4 MNPs can be efficiently employed for the hydrolysis of rice straw (Xu et al., 2015). The immobilized cellulase acquired supermagnetism and was recycled and utilized a total of 16 times in this investigation. Furthermore, the immobilized cellulase generated a glucose concentration 20 times greater than that of free cellulase (Xu et al., 2015). Zhang et al. (2015) documented the immobilization of cellulase on three functionalized magnetic nanospheres. The results indicated that magnetized cellulase exhibited greater thermal stability and a wider pH stability range compared to free cellulase (Zhang et al., 2015).

Srivastava et al. (2017) assessed the effects of immobilization on the production and thermostability of the cellulase enzyme, which was isolated from thermotolerant *A. fumigatus* and immobilized on nickel-cobaltite nanoparticles. The investigation revealed that, when compared to the absence of nanoparticles, nickel-cobaltite nanoparticles at a concentration of 1 mM had an impact on the thermostability of the cellulase enzyme, resulting in increased enzyme stability at 80°C for 8 hours (Srivastava et al., 2017).

Compared to the free enzyme, the immobilized cellulase demonstrated enhanced heat stability and maintained 77%–80% activity. Additionally, there was no loss of enzyme activity when the immobilized cellulase was reused for up to six cycles (Mishra & Sardar, 2015).

15.6 FUTURE PROSPECTS

Bioenergy production depends on pretreatment methods which nanobiotechnology enhances through significant improvements. Complex biomass structures get penetrated by NMs which interact with cellulose along with hemicellulose and lignin to increase biochemical conversion performance while generating higher amounts of hydrolysis-ready carbohydrates. MNPs represent an excellent solution because they allow enzyme reuse through immobilization which decreases industrial expenses related to enzyme development and processing (Melo et al., 2023). Benatti and De Lourdes Teixeira De Moraes Polizeli (2024) explain that enzyme immobilization on NMs improves conversion efficiency through enhanced stability and longevity of reuse. Arend et al. (2023) show that enzymes immobilized on supports function for longer periods with fewer added enzyme doses. Such financial viability of systems depends on support matrix quality as well as enzyme loading capacity and recyclability according to Basso and Serban (2019).

For future progress to occur the research world needs to investigate enzyme stability and synthesis complexity while coping with high production costs (Tomke & Rathod, 2020). Surface modifications of NMs with improved techniques should enhance enzyme binding while extending operational duration. NMs must undergo an environmental safety check for successful commercial applications. The most important attributes for materials selection in nanotechnology applications include compatibility with biological systems as well as affordable manufacturing costs together with recyclability capabilities. Nanotechnology acts as a base for future biofuel and biorefinery advancement because continued research will remove existing barriers within technology and economics (Sheldon, 2019).

15.7 CONCLUSION

Researches examine modern approaches to pretreating LCB together with enzyme immobilization techniques which employ NMs. MNPs demonstrate great effectiveness as they enhance biomass hydrolysis and extend enzyme operational lifespan. The process of enzyme immobilization provides enhanced stability in addition to improved activity with cost benefits. Acid-functionalized nanocatalysts and cerium-doped iron oxide offer low-cost, efficient pretreatment with minimal inhibitor formation. NMs face ongoing challenges because researchers remain uncertain about their dangerous potential to both humans and the environment. Additional studies have to resolve safety concerns while optimizing nanocatalysts and immobilization methods for industrial applications because they are essential to develop environmentally sustainable bioenergy conversion processes leading to sustainable biomass bioconversion commercialization. The following chapter will explore these improvements elaborately, and will delve into newer possibilities and explore possible advancements.

REFERENCES

Aarti, C., Khusro, A., & Agastian, P. (2018). Carboxymethyl cellulase production optimization from *Glutamicibacter arilaitensis* strain ALA4 and its application in lignocellulosic waste biomass saccharification. *Preparative Biochemistry & Biotechnology*, *48*(9), 853–866. https://doi.org/10.1080/10826068.2018.1514513

Aarti, C., Khusro, A., & Agastian, P. (2021). Lignocellulosic biomass as potent feedstock resource for bioethanol production: recent updates. *World News of Natural Sciences*, *37*, 164–181.

Aarti, C., Khusro, A., Agastian, P., Kuppusamy, P., & Farraj, D. a. A. (2022). Synthesis of gold nanoparticles using bacterial cellulase and its role in saccharification and bioethanol production from aquatic weeds. *Journal of King Saud University – Science*, *34*(4), 101974. https://doi.org/10.1016/j.jksus.2022.101974.

Abraham, R. E., Verma, M. L., Barrow, C. J., & Puri, M. (2014). Suitability of magnetic nanoparticle immobilised cellulases in enhancing enzymatic saccharification of pretreated hemp biomass. *Biotechnology for Biofuels*, *7*(1), 90. https://doi.org/10.1186/1754-6834-7-90

Ahmad, R. (2015). Welcome foreword from the conference chair. In *International Conference on Smart Sensors and Application (ICSSA)*, Kuala Lumpur, Malaysia, 2015, pp. 1–1, doi: 10.1109/ICSSA.2015.7322489.

Ahmad, R., & Khare, S. K. (2017). Immobilization of *Aspergillus niger* cellulase on multiwall carbon nanotubes for cellulose hydrolysis. *Bioresource Technology*, *252*, 72–75. https://doi.org/10.1016/j.biortech.2017.12.082

Ahmed, I. N., Yang, X. L., Dubale, A. A., Li, R. F., Ma, Y. M., Wang, L. M., & Xie, M. H. (2018). Hydrolysis of cellulose using cellulase physically immobilized on highly stable zirconium-based metal-organic frameworks. *Bioresource Technology*, *270*, 377–382. https://doi.org/10.1016/j.biortech.2018.09.077

Akhtar, J., Idris, A., & Abd Aziz, R. (2013). Recent advances in production of succinic acid from lignocellulosic biomass. *Applied Microbiology and Biotechnology*, *98*(3), 987–1000.

Alvira, P., et al. (2013). Improving the fermentation performance of Saccharomyces cerevisiae by laccase during ethanol production from steam-exploded wheat straw at high-substrate loadings. *Biotechnology Progress*, *29*(1), 74–82.

Arend, M., Zimmer, D., Xu, R., Sommer, F., Mühlhaus, T., & Nikoloski, Z. (2023). Proteomics and constraint-based modelling reveal enzyme kinetic properties of Chlamydomonas reinhardtii on a genome scale. *Nature Communications*, *14*(1), 4781. https://doi.org/10.1038/s41467-023-40498-1

Arora, A., Nandal, P., Singh, J., & Verma, M. L. (2020). Nanobiotechnological advancements in lignocellulosic biomass pretreatment. *Materials Science for Energy Technologies*, *3*, 308–318. https://doi.org/10.1016/j.mset.2019.12.003Balch,

M. L., Holwerda, E. K., Davis, M. F., Sykes, R. W., Happs, R. M., Kumar, R., Wyman, C. E., & Lynd, L. R., (2017). Lignocellulose fermentation and residual solids characterization for senescent switchgrass fermentation by Clostridium thermocellum in the presence and absence of continuous in situ ball-milling. *Energy & Environmental Science*, *10*(5), 1252–1261.

Bao, T., Zhao, J., Li, J., Liu, X., & Yang, S.T., (2019). n-Butanol and ethanol production from cellulose by *Clostridium cellulovorans* overexpressing heterologous aldehyde/alcohol dehydrogenases. *Bioresource Technology*, *285*, 121316.

Barhoum, A., El-Maghrabi, H. H., Nada, A. A., Sayegh, S., Roualdès, S., Renard, A., Iatsunskyi, I., Coy, E., & Bechelany, M. (2021). Simultaneous hydrogen and oxygen evolution reactions using free-standing nitrogen-doped-carbon–Co/CoOx nanofiber electrodes decorated with palladium nanoparticles. *Journal of Materials Chemistry A*, *9*(33), 17724–17739. https://doi.org/10.1039/d1ta03704h

Barhoum, A., García-Betancourt, M. L., Jeevanandam, J., Hussien, E. A., Mekkawy, S. A., Mostafa, M., Omran, M. M., Abdalla, M. S., & Bechelany, M. (2022). Review on natural, incidental, bioinspired, and engineered nanomaterials: history, definitions, classifications, synthesis, properties, market, toxicities, risks, and regulations. *Nanomaterials*, *12*(2), 177. https://doi.org/10.3390/nano12020177

Barhoum, A., Jeevanandam, J., Rastogi, A., Samyn, P., Boluk, Y., Dufresne, A., Danquah, M. K., & Bechelany, M. (2020). Plant celluloses, hemicelluloses, lignins, and volatile oils for the synthesis of nanoparticles and nanostructured materials. *Nanoscale*, *12*(45), 22845–22890. https://doi.org/10.1039/d0nr04795c

Barhoum, A., Bechelany, M., & Makhlouf, A. S. H. (2019). *Handbook of Nanofibers*. Springer International Publishing: Cham, Switzerland.

Basso, A., & Serban, S. (2019). Industrial applications of immobilized enzymes—a review. *Molecular Catalysis*, *479*, 110607. https://doi.org/10.1016/j.mcat.2019.110607

Benatti, A. L. T., & De Lourdes Teixeira De Moraes Polizeli, M. (2023). Lignocellulolytic biocatalysts: the main players involved in multiple biotechnological processes for biomass valorization. *Microorganisms*, *11*(1), 162. https://doi.org/10.3390/microorganisms11010162

Bhatia, S. K., Jagtap, S. S., Bedekar, A. A., Bhatia, R. K., Patel, A. K., Pant, D., Banu, J. R., Rao, C. V., Kim, Y., & Yang, Y. (2020). Recent developments in pretreatment technologies on lignocellulosic biomass: effect of key parameters, technological improvements, and challenges. *Bioresource Technology*, *300*, 122724. https://doi.org/10.1016/j.biortech.2019.122724

Badiei, M., Asim, N., Jahim, J. M., & Sopian, K. (2014). Comparison of chemical pretreatment methods for cellulosic biomass. *APCBEE procedia*, *9*, 170–174. https://doi.org/10.1016/j.apcbee.2014.01.030

Binod, P., Sindhu, R., Janu, K. U., & Pandey, A. (2019). Hydrolysis of cellulosic and hemicellulosic biomass. In *Elsevier eBooks* (pp. 447–460). Academic Press. https://doi.org/10.1016/b978-0-12-816856-1.00019-1

Brena, B., González-Pombo, P., & Batista-Viera, F. (2013). Immobilization of enzymes: a literature survey. *Methods in Molecular Biology*, *1051*, 15–31. https:// doi. org/10.1007/978-1-62703-550-7_2

Cai, J., He, Y., Yu, X., Banks, S. W., Yang, Y., Zhang, X., Yu, Y., Liu, R., & Bridgwater, A. V. (2017). Review of physicochemical properties and analytical characterization of lignocellulosic biomass. *Renewable and Sustainable Energy Reviews*, *76*, 309–322. https://doi.org/10.1016/j.rser.2017.03.072

Chandel, A.K., Albarelli, J.Q., Santos, D.T., Chundawat, S.P., Puri, M., Meireles, M.A.A., (2019). Comparative analysis of key technologies for cellulosic ethanol production from Brazilian sugarcane bagasse at a commercial scale. *Biofuels, Bioproducts and Biorefining*, *13*(4), 994–1014.

Chang, R. H., Jang, J., & Wu, K. C. (2011). Cellulase immobilized mesoporous silica nanocatalysts for efficient cellulose-to-glucose conversion. *Green Chemistry*, *13*(10), 2844. https://doi.org/10.1039/c1gc15563f.

Chavali, M. S., & Nikolova, M. P. (2019). Metal oxide nanoparticles and their applications in nanotechnology. *SN Applied Sciences*, *1*(6), 607. https://doi.org/10.1007/s42452-019-0592-3

Cherian, E., Dharmendirakumar, M., & Baskar, G. (2015). Immobilization of cellulase onto MnO_2 nanoparticles for bioethanol production by enhanced hydrolysis of agricultural waste. *Chinese Journal of Catalysis*, *36*(8), 1223–1229. https://doi.org/10.1016/s1872-2067(15)60906-8

Chong, K. Y., Yuryev, Y., Jain, A., Mason, B., & Brooks, M. S. L. (2021). Development of pea protein films with haskap (Lonicera caerulea) leaf extracts from aqueous two-phase systems. *Food and Bioprocess Technology*, *14*(9), 1733–1750. https://doi.org/10.1007/s11947-021-02671-6

Chung, D., Cha, M., Guss, A.M., & Westpheling, J. (2014). Direct conversion of plant biomass to ethanol by engineered *Caldicellulosiruptor bescii. Proceedings of the National Academy of Sciences of the United States of America*, *111*(24), 8931–8936.

Datta, S., Christena, L. R., & Rajaram, Y. R. S. (2013). Enzyme immobilization: an overview on techniques and support materials. *3 Biotech*, *3*, 1–9. https://doi.org/10.1007/s13205-012-0071-7

De Jong, E., & Gosselink, R. J. (2014). Lignocellulose-based chemical products. In *Elsevier eBooks* (pp. 277–313). Academic Press. https://doi.org/10.1016/b978-0-444-59561-4.00017-6

Den, W., Sharma, V. K., Lee, M., Nadadur, G., & Varma, R. S. (2018). Lignocellulosic biomass transformations via greener oxidative pretreatment processes: access to energy and value-added chemicals. *Frontiers in chemistry*, *6*, 141. https://doi.org/10.3389/fchem.2018.00141

Duque, A., Manzanares, P., & Ballesteros, M. (2017). Extrusion as a pretreatment for lignocellulosic biomass: Fundamentals and applications. *Renewable Energy*, *114*, 1427–1441. https://doi.org/10.1016/j.renene.2017.06.050

Galanakis, C. M., Markouli, E., & Gekas, V. (2013). Recovery and fractionation of different phenolic classes from winery sludge using ultrafiltration. *Separation and Purification Technology*, *107*, 245–251. https://doi.org/10.1016/j.seppur.2013.01.034

Gaur, M., Misra, C., Yadav, A. B., Swaroop, S., Maolmhuaidh, F. Ó., Bechelany, M., & Barhoum, A. (2021). Biomedical applications of carbon nanomaterials: fullerenes, quantum dots, nanotubes, nanofibers, and graphene. *Materials*, *14*(20), 5978. https://doi.org/10.3390/ma14205978

González Medeiros, J. J., Pérez Cid, B., & Fernández Gómez, E. (2005). Analytical phosphorus fractionation in sewage sludge and sediment samples. *Analytical and Bioanalytical Chemistry*, *381*(4), 873-878. https://doi.org/10.1007/s00216-004-2989-z

Gonzales, R. R., & Kim, S. (2017). Dark fermentative hydrogen production following the sequential dilute acid pretreatment and enzymatic saccharification of rice husk. *International Journal of Hydrogen Energy*, *42*(45), 27577–27583. https://doi.org/10.1016/j.ijhydene.2017.08.185

Grewal, J., Ahmad, R., & Khare, S. K. (2017). Development of cellulase-nanoconjugates with enhanced ionic liquid and thermal stability for in situ lignocellulose saccharification. *Bioresource Technology*, *242*, 236–243.

Guo, K., Baidak, A., & Yu, Z. (2020). Recent advances in green synthesis and modification of inorganic nanomaterials by ionizing and non-ionizing radiation. *Journal of Materials Chemistry A*, *8*(44), 23029–23058. https://doi.org/10.1039/d0ta06742c

Guo, Z.P., Robin, J., Duquesne, S., O'Donohue, M.J., Marty, A., & Bordes, F. (2018). Developing cellulolytic Yarrowia lipolytica as a platform for the production of valuable products in consolidated bioprocessing of cellulose. *Biotechnology for Biofuels and Bioproducts*, *11*, 141.

Harish, V., Tewari, D., Gaur, M., Yadav, A. B., Swaroop, S., Bechelany, M., & Barhoum, A. (2022). Review on nanoparticles and nanostructured materials: bioimaging, biosensing, drug delivery, tissue engineering, antimicrobial, and agro-food applications. *Nanomaterials*, *12*(3), 457. https://doi.org/10.3390/nano12030457

Hodge, D. B., Karim, M. N., Schell, D. J., & McMillan, J. D., (2008). Soluble and insoluble solids contributions to high-solids enzymatic hydrolysis of lignocellulose. *Bioresource Technology*, *99* (18), 8940–8948.

Hong, W., Zhang, J., Feng, Y., Mohr, G., Lambowitz, A. M., Cui, G. -Z., Liu, Y. -J., & Cui, Q. (2014). The contribution of cellulosomal scaffoldins to cellulose hydrolysis by Clostridium thermocellum analyzed by using thermotargetrons. *Biotechnology for Biofuels and Bioproducts*, *7*, 80.

Ichikawa, S., Ichihara, M., Ito, T., Isozaki, K., Kosugi, A., & Karita, S. (2019). Glucose pro duction from cellulose through biological simultaneous enzyme production and saccharification using recombinant bacteria expressing the β-glucosidase gene. *Journal of Biosciences and Bioengineering*, *127*(3), 340–344.

Ijaz, I., Gilani, E., Nazir, A., & Bukhari, A. (2020). Detail review on chemical, physical and green synthesis, classification, characterizations and applications of nanoparticles. Taylor Fr, *13*, 59–81.

Ingle, A. P., Philippini, R. R., De Souza Melo, Y. C., & Da Silva, S. S. (2020). Acid-functionalized magnetic nanocatalysts mediated pretreatment of sugarcane straw: an eco-friendly and cost-effective approach. *Cellulose*, *27*(12), 7067–7078. https://doi.org/10.1007/s10570-020-03262-y.

Ingle, A. P., Philippini, R. R., Rai, M., & Da Silva, S. S. (2019). Catalytic hydrolysis of cellobiose using different acid-functionalised $Fe_3 O_4$ magnetic nanoparticles. *IET Nanobiotechnology*, *14*(1), 40–46. https://doi.org/10.1049/iet-nbt.2019.0181

Iravani, S., & Zolfaghari, B. (2021). Plant viruses and bacteriophages for eco-friendly synthesis of nanoparticles: recent trends and important challenges. *Comments on Inorganic Chemistry*, *42*(4), 226–248. https://doi.org/10.1080/02603594.2021.1993837.

Isaacoff, B. P., & Brown, K. A. (2017). Progress in top-down control of bottom-up assembly. *Nano Letters*, *17*(11), 6508–6510. https://doi.org/10.1021/acs.nanolett.7b04479

Jalali-Moghadam, E., & Shariatinia, Z. (2018). Quantum dot sensitized solar cells fabricated by means of a novel inorganic spinel nanoparticle. *Applied Surface Science*, *441*, 1–11. https://doi.org/10.1016/j.apsusc.2018.02.022

Jędrzejczyk, M. et al. (2019) Physical and chemical pretreatment of lignocellulosic biomass. In: *Elsevier eBooks*. pp. 143–196. https://doi.org/10.1016/B978-0-12-815162-4.00006-9

Jeevanandam, J., Chan, Y. S., & Danquah, M. K. (2016). Biosynthesis of metal and metal oxide nanoparticles. *ChemBioEng Reviews*, *3*(2), 55–67. https://doi.org/10.1002/cben.201500018

Jeevanandam, J., Barhoum, A., Chan, Y. S., Dufresne, A., & Danquah, M. K. (2018). Review on nanoparticles and nanostructured materials: history, sources, toxicity and regulations. *Beilstein Journal of Nanotechnology*, *9*, 1050–1074.

Jia, J., Zhang, W., Yang, Z., Yang, X., Wang, N., & Yu, X. (2017). Novel magnetic cross-linked cellulase aggregates with a potential application in lignocellulosic biomass bioconversion. *Molecules (Basel Switzerland)*, *22*(2), 269.

Jindou, S., Karita, S., Fujino, E., Fujino, T., Hayashi, H., Kimura, T., Sakka, K., & Ohmiya, K. (2002). α-Galactosidase Aga27A, an enzymatic component of the *Clostridium josui* cellulosome. *Journal of Bacteriology*, 184 (2), 600–604.

Judy, D. A., Sherlin, Y. S., Arasu, M. V., Al-Dhabi, N. A., Choi, K. C., & Bindhu. (2021). Environmental photochemistry in *Solanum trilobatum* mediated plasmonic nanoparticles as a probe for the detection of Cd^{2+} ions in water. *Environmental Research*, *202*, 111918. https://doi.org/10.1016/j.envres.2021.111918

Karunanithy, C., Muthukumarappan, K., & Gibbons, W. R. (2012). Effect of extruder screw speed, temperature, and enzyme levels on sugar recovery from different biomasses. *ISRN Biotechnology*, 1–13. https://doi.org/10.5402/2013/942810

Katsimpouras, C., Zacharopoulou, M., Matsakas, L., Rova, U., Christakopoulos, P., & Topakas, E. (2017). Sequential high gravity ethanol fermentation and anaerobic digestion of steam explosion and organosolv pretreated corn stover. *Bioresource Technology*, *244*, 1129–1136. https://doi.org/10.1016/j.biortech.2017.08.112

Keskin, T., Abubackar, H. N., Arslan, K., & Azbar, N. (2019). Biohydrogen production from solid wastes. In *Elsevier eBooks* (pp. 321–346). Academic Press. https://doi.org/10.1016/b978-0-444-64203-5.00012-5

Kim, H. S., Hong, S. G., Woo, K. M., Teijeiro Seijas, V., Kim, S., Lee, J. & Kim, J. (2018). Precipitation-based nanoscale enzyme reactor with improved loading, stability, and mass transfer for enzymatic CO_2 conversion and utilization. *ACS Catalysis*, *8*, 6526–6536. https://doi.org/10.1021/acscatal.8b00606

Krishnakumar, S., Janani, P., Mugilarasi, S., Kumari, G., & Janney, J. B. (2018). Chemical induced fabrication of silver nanoparticles (Ag-NPs) as nanocatalyst with alpha amylase enzyme for enhanced breakdown of starch. *Biocatalysis and Agricultural Biotechnology*, *15*, 377–383. https://doi.org/10.1016/j.bcab.2018.06.016.

Kucharska, K., Rybarczyk, P., Hołowacz, I., Łukajtis, R., Glinka, M., & Kamiński, M. (2018). Pretreatment of lignocellulosic materials as substrates for fermentation processes. *Molecules*, *23*(11), 2937. https://doi.org/10.3390/molecules23112937

Kumar, A., Singh, S., Tiwari, R., Goel, R., & Nain, L. (2016). Immobilization of indigenous holocellulase on iron oxide (Fe_2O_3) nanoparticles enhanced hydrolysis of alkali pretreated paddy straw. *International Journal of Biological Macromolecules*, *96*, 538–549. https://doi.org/10.1016/j.ijbiomac.2016.11.109

Kumar, P., Barrett, D. M., Delwiche, M. J., & Stroeve, P. (2009). Methods for pretreatment of lignocellulosic biomass for efficient hydrolysis and biofuel production. *Industrial & engineering chemistry research*, *48*(8), 3713–3729.

Kumar, A. K., & Sharma, S. (2017). Recent updates on different methods of pretreatment of lignocellulosic feedstocks: a review. *Bioresources and Bioprocessing*, *4*(1). https://doi.org/10.1186/s40643-017-0137-9

Kumari, S., & Das, D. (2016). Biologically pretreated sugarcane top as a potential raw material for the enhancement of gaseous energy recovery by two stage biohythane process. *Bioresource Technology*, *218*, 1090–1097. https://doi.org/10.1016/j.biortech.2016.07.070

Kumari, A., Kaila, P., Tiwari, P., Singh, V., Kaul, S., Singhal, N., & Guptasarma, P. (2018). Multiple thermostable enzyme hydrolases on magnetic nanoparticles: an immobilized enzyme-mediated approach to saccharification through simultaneous xylanase, cellulase and amylolytic glucanotransferase action. *International Journal of Biological Macromolecules*, *120*, 1650–1658. https://doi.org/10.1016/j.ijbiomac.2018.09.106

Kumari, S., & Das, D. (2015). Improvement of gaseous energy recovery from sugarcane bagasse by dark fermentation followed by biomethanation process. *Bioresource Technology*, *194*, 354–363. https://doi.org/10.1016/j.biortech.2015.07.038

Ladole, M. R., Mevada, J. S., & Pandit, A. B. (2017). Ultrasonic hyperactivation of cellulase immobilized on magnetic nanoparticles. *Bioresource Technology*, *239*, 117–126. https://doi.org/10.1016/j.biortech.2017.04.096

Lebedeva, O., Kultin, D., & Kustov, L. (2021). Electrochemical synthesis of unique nanomaterials in ionic liquids. *Nanomaterials*, *11*(12), 3270. https://doi.org/10.3390/nano11123270

Lei, C., et al. (2014). An alternative feedstock of corn meal for industrial fuel ethanol production: delignified corncob residue. *Bioresource Technology*, 167, 555–559.

Li, Q., Song, J., Peng, S., Wang, J. P., Qu, G. Z., Sederoff, R. R., & Chiang, V. L. (2014). Plant biotechnology for lignocellulosic biofuel production. *Plant Biotechnology Journal*, *12*(9), 1174–1192. https://doi.org/10.1111/pbi.12273

Li, H., Qu, Y., Yang, Y., Chang, S., & Xu, J. (2016). Microwave irradiation–A green and efficient way to pretreat biomass. *Bioresource Technology*, *199*, 34–41. https://doi.org/10.1016/j.biortech.2015.08.099

Liang, W., Xu, H., Carraro, F., Maddigan, N. K., Li, Q., Bell, S. G., & Doonan, C. J. (2019). Enhanced activity of enzymes encapsulated in hydro philic metal–organic frameworks. *Journal of the American Chemical Society*, *141*, 2348–2355. https://doi.org/10.1021/jacs.8b10302

Lin, P. P., Mi, L., Morioka, A. H., Yoshino, K. M., Konishi, S., Xu, S. C., Papanek, B. A., Riley, L. A., Guss, A. M., & Liao, J. C. (2015). Consolidated bioprocessing of cellulose to iso butanol using *Clostridium thermocellum*. *Metabolic Engineering*, *31*, 44–52.

Lin, Y., Jin, W., Cai, L., Liu, X., Qiu, Y., & Zhang, G. (2021). Green preparation of covalently co-immobilized multienzymes on silica nanoparticles for clean production of reducing sugar from lignocellulosic biomass. *Journal of Cleaner Production*, *314*, 127994. https://doi.org/10.1016/j.jclepro.2021.127994.

Liu, X., Wang, W., Gao, X., Zhou, Y., & Shen, R. (2012). Effect of thermal pretreatment on the physical and chemical properties of municipal biomass waste. *Waste Management*, *32*(2), 249–255.

Liu, S., Bilal, M., Rizwan, K., Gul, I., Rasheed, T., & Iqbal, H. M. (2021). Smart chemistry of enzyme immobilization using various support matrices–a review. *International Journal of Biological Macromolecules*, *190*, 396–408. https://doi.org/10.1016/j.ijbiomac.2021.09.006

Liu, C. G., Xiao, Y., Xia, X. X., Zhao, X. Q., Peng, L., Srinophakun, P., Bai, F. W., (2019a). Cellulosic ethanol production: progress, challenges and strategies for solutions. *Biotechnology Advances*, *37*(3), 491–504.

Liu, C.-M., et al., (2013). Biohydrogen production evaluation from rice straw hydro lysate by concentrated acid pre-treatment in both batch and continuous systems. *International Journal of Hydrogen Energy*, *38*(35), 15823–15829.

Liu, S., Liu, Y.-J., Feng, Y., Li, B., & Cui, Q., (2019b). Construction of consolidated bio-sac charification biocatalyst and process optimization for highly efficient lignocellulose solubilization. *Biotechnology for Biofuels and Bioproducts*, *12*(1), 35.

Liu, Y., Li, B., Feng, Y., & Cui, Q. (2020). Consolidated bio-saccharification: Leading lignocellulose bioconversion into the real world. *Biotechnology Advances*, *40*, 107535. https://doi.org/10.1016/j.biotechadv.2020.107535

Liu, Z., et al., (2014). Effects of operating parameters on hydrogen production from raw wet steam-exploded cornstalk and two-stage fermentation potential for biohythane production. *Biochemical Engineering Journal*, *90*, 234–238.

Lupoi, J. S., & Smith, E. A. (2011). Evaluation of nanoparticle-immobilized cellulase for improved ethanol yield in simultaneous saccharification and fermentation reactions. *Biotechnology and Bioengineering*, *108*(12), 2835–2843. https://doi.org/10.1002/bit.23246

Luo, J., Fang, Z., & Smith Jr, R. L. (2014). Ultrasound-enhanced conversion of biomass to biofuels. *Progress in Energy and Combustion Science*, *41*, 56–93. https://doi.org/10.1016/j.pecs.2013.11.001

Lynd, L.R., Liang, X., Biddy, M.J., Allee, A., Cai, H., Foust, T., Himmel, M.E., Laser, M.S., Wang, M., & Wyman, C. E., 2017. Cellulosic ethanol: status and innovation. *Current Opinion in Biotechnology*, *45*, 202–211.

Madkour, L. H. (2019). Introduction to nanotechnology (NT) and nanomaterials (NMs). In *Advanced Structured Materials* (pp. 1–47). https://doi.org/10.1007/978-3-030-21621-4_1

Mankar, A. R., Pandey, A., Modak, A., & Pant, K. (2021). Pretreatment of lignocellulosic biomass: a review on recent advances. *Bioresource Technology*, *334*, 125235. https://doi.org/10.1016/j.biortech.2021.125235

Marcelo, G. A., Lodeiro, C., Capelo, J. L., Lorenzo, J., & Oliveira, E. (2020). *Magnetic, Fluorescent and Hybrid Nanoparticles: From Synthesis to Application in Biosystems*. Elsevier: Amsterdam, the Netherlands.

Marchal, R., Ropars, M., Pourquié, J., Fayolle, F., & Vandecasteele, J. (1992). Large-scale enzymatic hydrolysis of agricultural lignocellulosic biomass. Part 2: Conversion into acetone-butanol. *Bioresource Technology*, *42*(3), 205–217. https://doi.org/10.1016/0960-8524(92)90024-r

Mariño, M. A., Moretti, P., & Tasic, L. (2021). Immobilized commercial cellulases onto amino-functionalized magnetic beads for biomass hydrolysis: enhanced stability by non-polar silanization. *Biomass Conversion and Biorefinery*, *13*(10), 1–11. https://doi.org/10.1007/s13399-021-01798-y.

Meftahi, A., Momeni Heravi, M. E., Barhoum, A., Samyn, P., Najarzadeh, H., & Alibakhshi, S. (2022). Cellulose nanofibers. In *Handbook of Nanocelluloses* (pp. 233–262). Springer.

Melo, R. L. F., Sales, M. B., De Castro Bizerra, V., De Sousa, P. G., Junior, Cavalcante, A. L. G., Freire, T. M., Neto, F. S., Bilal, M., Jesionowski, T., Soares, J. M., Fechine, P. B. A., & Santos, J. C. S. D. (2023). Recent applications and future prospects of magnetic biocatalysts. *International Journal of Biological Macromolecules*, *253*, 126709. https://doi.org/10.1016/j.ijbiomac.2023.126709

Mishra, A. and Sardar, M. (2015) Cellulase assisted synthesis of nano-silver and gold: application as immobilization matrix for biocatalysis. *International Journal of Biological Macromolecules*, *77*, pp. 105–113.

Misson M, Zhang H, Jin B (2015) Nanobiocatalyst advancements and bioprocessing applications. *Journal of the Royal Society Interface*, *12*, 20140891. https://doi.org/10.1098/rsif.2014.0891

Mittal, A. K., Chisti, Y., & Banerjee, U. C. (2013). Synthesis of metallic nanoparticles using plant extracts. *Biotechnology Advances*, *31*(2), 346–356. https://doi.org/10.1016/j.biotechadv.2013.01.003

Mo, Hv, Qiu, J., Yang, C., Zang, L., & Sakai, E. (2020a). Preparation and characterization of magnetic polyporous biochar for cellulase immobilization by physical adsorption. *Cellulose*, *27*, 4963–4973. https://doi.org/10.1007/s10570-020-03125-6

Mo, H., Qiu, J., Yang, C., Zang, L., Sakai, E., & Chen, J. (2020b). Porous biochar/chitosan composites for high performance cellulase immobilization by glutaraldehyde. *Enzyme and Microbial Technology*, *138*, 109561. https://doi.org/10.1016/j.enzmictec.2020.109561

Mohamad, N. R., Marzuki, N. H. C., Buang, N. A., Huyop, F., & Wahab, R. A. (2015). An overview of technologies for immobilization of enzymes and surface analysis techniques for immobilized enzymes. *Biotechnology & Biotechnological Equipment*, *29*(2), 205–220. https://doi.org/10.1080/13102818.2015.1008192.

Mood, S. H., Golfeshan, A. H., Tabatabaei, M., Jouzani, G. S., Najafi, G. H., Gholami, M., & Ardjmand, M. (2013). Lignocellulosic biomass to bioethanol, a comprehensive review with a focus on pretreatment. *Renewable and Sustainable Energy Reviews*, *27*, 77–93. https://doi.org/10.1016/j.rser.2013.06.033

Muley, A. B., Thorat, A. S., Singhal, R. S., & Babu, K. H. (2018). A tri-enzyme co-immobilized magnetic complex: process details, kinetics, thermodynamics and applications. *International Journal of Biological Macromolecules*, *118*, 1781–1795. https://doi.org/10.1016/j.ijbiomac.2018.07.022

Öhgren, K., Bura, R., Lesnicki, G., Saddler, J., Zacchi, G., 2007. A comparison between simultaneous saccharification and fermentation and separate hydrolysis and fer mentation using steam-pretreated corn stover. *Process Biochemistry*, *42*(5), 834–839.

Olson, D. G. (2010). *Electrotransformation of Gram-positive, Anaerobic, Thermophilic Bacteria*. Mascoma Corporation.

Olson, D. G., Maloney, M., Lanahan, A. A., Hon, S., Hauser, L. J., & Lynd, L. R., (2015). Identifying promoters for gene expression in Clostridium thermocellum. *Metabolic Engineering Communications*, *2*, 23–29.

Olson, D. G., McBride, J. E., Shaw, A. J., & Lynd, L. R., (2012). Recent progress in consolidated bioprocessing. *Current Opinion in Biotechnology*, *23*(3), 396–405.

Palacios-Bereche, R., Ensinas, A., Modesto, M., & Nebra, S. (2017). Enzymatic hydrolysis of sugarcane biomass and heat integration as enhancers of ethanol production. *Journal of Renewable Materials*, *6*(2), 183–194. https://doi.org/10.7569/jrm.2017.634175

Palav, T., & Seetharaman, K. (2007). Impact of microwave heating on the physico-chemical properties of a starch–water model system. *Carbohydrate Polymers*, *67*(4), 596–604. https://doi.org/10.1016/j.carbpol.2006.07.006

Park, E.Y., Naruse, K., & Kato, T., (2012). One-pot bioethanol production from cellulose by co-culture of Acremonium cellulolyticus and Saccharomyces cerevisiae. *Biotechnology for Biofuels and Bioproducts*, *5*, 64.

Phummala, K., et al. (2014). Delignification of disposable wooden chopsticks waste for fermentative hydrogen production by an enriched culture from a hot spring. *Journal of Environmental Sciences*, *26*(6), 1361–1368.

Poorakbar, E., Shafiee, A., Saboury, A. A., Rad, B. L., Khoshnevisan, K., Ma'mani, L., Derakhshankhah, H., Ganjali, M. R., & Hosseini, M. (2018). Synthesis of magnetic gold mesoporous silica nanoparticles core shell for cellulase enzyme immobilization: improvement of enzymatic activity and thermal stability. *Process Biochemistry*, *71*, 92–100. https://doi.org/10.1016/j.procbio.2018.05.012

Qamar, S. A., Qamar, M., Bilal, M., Bharagava, R. N., Ferreira, L. F. R., Sher, F., & Iqbal, H. M. (2021) Cellulose-deconstruction potential of nano biocatalytic systems: a strategic drive from designing to sustain able applications of immobilized cellulases. *International Journal of Biological Macromolecules*, *185*, 1–19. https://doi.org/10.1016/j.ijbiomac.2021.06.079

Qi, B., Luo, J., & Wan, Y. (2018). Immobilization of cellulase on a core-shell structured metal-organic framework composites: better inhibitors tolerance and easier recycling. *Bioresource Technology*, *268*, 577–582. https://doi.org/10.1016/j.biortech.2018.07.115

Qu, X.-S., Hu, B.-B., & Zhu, M.-J. (2017). Enhanced saccharification of cellulose and sugarcane bagasse by *Clostridium thermocellum* cultures with Triton X-100 and β-glucosidase/Cellic® CTec2 supplementation. *RSC Advances*, *7*(35), 21360–21365.

Rasouli, R., Barhoum, A., Bechelany, M., & Dufresne, A. (2018). Nanofibers for biomedical and healthcare applications. *Macromolecular Bioscience*, *19*(2), e1800256. https://doi.org/10.1002/mabi.201800256

Rezania, S., Oryani, B., Cho, J., Sabbagh, F., Rupani, P. F., Talaiekhozani, A., ... & Lotfi Ghahroud, M. (2020). Technical aspects of biofuel production from different sources in Malaysia—a review. *Processes*, *8*(8), 993. https://doi.org/10.3390/pr8080993

Ruiz, H. A., Conrad, M., Sun, S., Sanchez, A., Rocha, G. J., Romaní, A., Castro, E., Torres, A., Rodríguez-Jasso, R. M., Andrade, L. P., Smirnova, I., Sun, R., & Meyer, A. S. (2019). Engineering aspects of hydrothermal pretreatment: from batch to continuous operation, scale-up and pilot reactor under biorefinery concept. *Bioresource Technology*, *299*, 122685. https://doi.org/10.1016/j.biortech.2019.122685

Sánchez-Ramírez, J., Martínez-Hernández, J.L., Segura-Ceniceros, P. et al. Cellulases immobilization on chitosan-coated magnetic nanoparticles: application for *Agave Atrovirens* lignocellulosic biomass hydrolysis. *Bioprocess and Biosystems Engineering*, *40*, 9–22 (2017). https://doi.org/10.1007/s00449-016-1670-1

Saritha, M., Arora, A., & Lata. (2012). Biological pretreatment of lignocellulosic substrates for enhanced delignification and enzymatic digestibility. *Indian journal of microbiology*, *52*(2), 122-130. https://doi.org/10.1007/s12088-011-0199-x

Saravanan, A., Kumar, P. S., Jeevanantham, S., Karishma, S., & Vo, D. N. (2021). Recent advances and sustainable development of biofuels production from lignocellulosic biomass. *Bioresource Technology*, *344*, 126203. https://doi.org/10.1016/j.biortech.2021.126203

Sassolas, A., Blum, L. J., Leca-Bouvier, B. D. (2012). Immobilization strategies to develop enzymatic biosensors. *Biotechnology Advances*, *30*, 489–511. https://doi.org/10.1016/j.biotechadv.2011.09.003

Sheldon, R. A. (2019). CLEAs, combi-CLEAs and 'smart'magnetic CLEAs: Biocatalysis in a bio-based economy. *Catalysts*, *9*(3), 261. https://doi.org/10.3390/catal9030261

Shi, J., Pu, Y., Yang, B., Ragauskas, A., & Wyman, C. E. (2011). Comparison of microwaves to fluidized sand baths for heating tubular reactors for hydrothermal and dilute acid batch pretreatment of corn stover. *Bioresource Technology*, *102*(10), 5952–5961. https://doi.org/10.1016/j.biortech.2011.03.027

Singh, A. K., Kumar, A., Sharma, V., & Kala, P. (2020). Sustainable techniques in grinding: state of the art review. *Journal of Cleaner Production*, *269*, 121876. https://doi.org/10.1016/j.jclepro.2020.121876

Singh, R., Kumar, R., Sarangi, P. K., Kovalev, A. A., & Vivekanand, V. (2022). Effect of physical and thermal pretreatment of lignocellulosic biomass on biohydrogen production by thermochemical route: a critical review. *Bioresource Technology*, *369*, 128458. https://doi.org/10.1016/j.biortech.2022.128458

Singhvi, M., & Kim, B. S. (2020). Current developments in lignocellulosic biomass conversion into biofuels using nanobiotechnology approach. *Energies*, *13*(20), 5300. https://doi.org/10.3390/en13205300

Soni, B.K., Das, K., & Ghose, T. K. (1982). Bioconversion of agro-wastes into ace tone butanol. *Biotechnology Letters*, *4*(1), 19–22.

Sriram, S. R., Parne, S. R., Pothukanuri, N., & Edla, D. R. (2022). Prospects of spray pyrolysis technique for gas sensor applications – a comprehensive review. *Journal of Analytical and Applied Pyrolysis*, *164*, 105527. https://doi.org/10.1016/j.jaap.2022.105527

Srivastava, N., Hussain, A., Kushwaha, D., Haque, S., Mishra, P., Gupta, V. K., & Srivastava, M. (2021). Nickel ferrite nanoparticles induced improved fungal cellulase production using residual algal biomass and subsequent hydrogen production following dark fermentation. *Fuel*, *304*, 121391. https://doi.org/10.1016/j.fuel.2021.121391

Srivastava, N., Srivastava, M. Mishra, P. K., & Ramteke, P. W. (2016). Application of ZnO nanoparticles for improving the thermal and pH stability of crude cellulase obtained from *Aspergillus fumigatus* AA001. *Frontiers in Microbiology*, *7*, 514.

Srivastava, N., Srivastava, M., Mishra, P., Gupta, V. K., Molina, G., Rodriguez-Couto, S., Manikanta, A., & Ramteke, P. (2017). Applications of fungal cellulases in biofuel production: advances and limitations. *Renewable and Sustainable Energy Reviews*, *82*, 2379–2386. https://doi.org/10.1016/j.rser.2017.08.074

Su, W., Zhang, H., Xing, Y., Li, X., Wang, J., & Cai, C. (2021). A bibliometric analysis and review of supercritical fluids for the synthesis of nanomaterials. *Nanomaterials*, *11*(2), 336. https://doi.org/10.3390/nano11020336

Tabañag, I. D. F., Chu, I., Wei, Y., & Tsai, S. (2018). The role of Yeast-Surface-Display techniques in creating biocatalysts for Consolidated BioProcessing. *Catalysts*, *8*(3), 94. https://doi.org/10.3390/catal8030094

Tanjore, D., & Richard, T. L. (2015). A systems view of lignocellulose hydrolysis. *Advances in bioprocess technology*, 387-419. https://doi.org/10.1007/978-3-319-17915-5

Thomas, V. A., Kothari, N., Bhagia, S., Akinosho, H., Li, M., Pu, Y., Yoo, C. G., Pattathil, S., Hahn, M. G., Raguaskas, A. J., Wyman, C. E., & Kumar, R. (2017b). Comparative evaluation of *Populus* variants total sugar release and structural features following pretreatment and digestion by two distinct biological systems. *Biotechnology for Biofuels and Bioproducts*, *10*, 292.

Tian, L., Conway, P.M., Cervenka, N.D., Cui, J., Maloney, M., Olson, D.G., & Lynd, L.R. (2019). Metabolic engineering of *Clostridium thermocellum* for n-butanol production from cellulose. *Biotechnology for Biofuels and Bioproducts*, *12*, 186.

Tian, L., Papanek, B., Olson, D. G., Rydzak, T., Holwerda, E. K., Zheng, T., Zhou, J., Maloney, M., Jiang, N., Giannone, R. J., Hettich, R. L., Guss, A. M., & Lynd, L. R. (2016). Simultaneous achievement of high ethanol yield and titer in *Clostridium thermocellum*. *Biotechnology for Biofuels and Bioproducts*, *9*, 116.

Tomke, P. D., & Rathod, V. K. (2020). Facile fabrication of silver on magnetic nanocomposite (Fe_3O_4@Chitosan–AgNP nanocomposite) for catalytic reduction of anthropogenic pollutant and agricultural pathogens. *International Journal of Biological Macromolecules*, *149*, 989–999. https://doi.org/10.1016/j.ijbiomac.2020.01.183

Tripathi, S. A., Olson, D. G., Argyros, D. A., Miller, B. B., Barrett, T. F., Murphy, D. M., McCool, J. D., Warner, A. K., Rajgarhia, V. B., Lynd, L. R., Hogsett, D. A., & Caiazza, N. C. (2010). Development of pyrF-based genetic system for targeted gene deletion in *Clostridium thermocellum* and creation of a pta mutant. *Applied and Environmental Microbiology*, *76*(19), 6591–6599.

Tudose, I. V., Comanescu, F., Pascariu, P., Bucur, S., Rusen, L., Iacomi, F., Koudoumas, E., & Suchea, M. P. (2019). Chemical and physical methods for multifunctional nanostructured interface fabrication. In *Elsevier eBooks* (pp. 15–26). Academic Press. https://doi.org/10.1016/b978-0-12-814401-5.00002-5

Ufodike, C. O., Eze, V. O., Ahmed, M. F., Oluwalowo, A., Park, J. G., Okoli, O. I., & Wang, H. (2020). Evaluation of the inter-particle interference of cellulose and lignin in lignocellulosic materials. *International Journal of Biological Macromolecules*, *147*, 762–767. https://doi.org/10.1016/j.ijbiomac.2020.01.234

Us, E., & Perendeci, N. A. (2012). Improvement of methane production from greenhouse residues: optimization of thermal and H2SO4 pretreatment process by experimental design. *Chemical Engineering Journal*, *181*, 120–131. https://doi.org/10.1016/j.cej.2011.11.038

Veluchamy, C., Kalamdhad, A. S., & Gilroyed, B. H. (2018). Advanced pretreatment strategies for bioenergy production from biomass and biowaste. In *Springer eBooks* (pp. 1–19). Springer. https://doi.org/10.1007/978-3-319-58538-3_45-1

Velmurugan, R., & Muthukumar, K. (2012). Ultrasound-assisted alkaline pretreatment of sugarcane bagasse for fermentable sugar production: Optimization through response surface methodology. *Bioresource technology*, *112*, 293–299. https://doi.org/10.1016/j.biortech.2012.01.168

Vidhya, C. (2021). Microalgae—The ideal source of biofuel. In *Elsevier eBooks* (pp. 389–405). Academic Press. https://doi.org/10.1016/b978-0-323-85269-2.00007-1

Voloshin, R.A., et al. (2015). Photoelectrochemical cells based on photosynthetic systems: a review. *Biofuel Research Journal*, *2*(2), 227–235.

Wang, J., & Yin, Y. (2021). Clostridium species for fermentative hydrogen production: an overview. *International Journal of Hydrogen Energy*, *46*(70), 34599–34625. https://doi.org/10.1016/j.ijhydene.2021.08.052

Xu, C., Huang, R., Teng, L., Jing, X., Hu, J., Cui, G., Wang, Y., Cui, Q., & Xu, J. (2015). Cellulosome stoichiometry in *Clostridium cellulolyticum* is regulated by selective RNA processing and stabilization. *Nature Communications*, *6*, 6900.

Xu, H., Li, B., & Mu, X. (2016). Review of alkali-based pretreatment to enhance enzymatic saccharification for lignocellulosic biomass conversion. *Industrial & Engineering Chemistry Research*, *55*(32), 8691–8705. https://doi.org/10.1021/acs.iecr.6b01907

Xu, T., Chen, D., & Hu, X. (2015). Hydrogen-activating models of hydrogenases. *Coordination Chemistry Reviews*, *303*, 32–41.

Yan, D., Ji, Q., Yu, X., Li, M., Fakayode, O. A., Yagoub, A. E. A., ... & Zhou, C. (2021). Multimode-ultrasound and microwave assisted natural ternary deep eutectic solvent sequential pretreatments for corn straw biomass deconstruction under mild conditions. *Ultrasonics Sonochemistry*, *72*, 105414. https://doi.org/10.1016/j.ultsonch.2020.105414

Zabed, H., Sahu, J., Boyce, A., & Faruq, G. (2016). Fuel ethanol production from lignocellulosic biomass: an overview on feedstocks and technological approaches. *Renewable and Sustainable Energy Reviews*, *66*, 751–774. https://doi.org/10.1016/j.rser.2016.08.038

Zakaria, M. R., Fujimoto, S., Hirata, S., & Hassan, M. A. (2014). Ball milling pretreatment of oil palm biomass for enhancing enzymatic hydrolysis. *Applied Biochemistry and Biotechnology*, *173*(7), 1778–1789. https://doi.org/10.1007/s12010-014-0964-5

Zang, L., Qiu, J., Wu, X., Zhang, W., Sakai, E., & Wei, Y. (2014). Preparation of magnetic chitosan nanoparticles as support for cellulase immobilization. *Industrial & Engineering Chemistry Research*, *53*(9), 3448–3454. https://doi.org/10.1021/ie404072s

Zhang, J., Liu, S., Li, R., Hong, W., Xiao, Y., Feng, Y., Cui, Q., & Liu, Y.-J. (2017a). Efficient whole-cell-catalyzing cellulose saccharification using engineered *Clostridium thermocellum*. *Biotechnology for Biofuels and Bioproducts*, *10*, 124.

Zhang, J., Liu, Y.-J., Cui, G.-Z., Cui, Q., 2015. A novel arabinose-inducible genetic operation system developed for *Clostridium cellulolyticum*. *Biotechnology for Biofuels and Bioproducts*, *8*, 36.

Zhang, J., Wen, C., Zhang, H., Duan, Y., & Ma, H. (2019). Recent advances in the extraction of bioactive compounds with subcritical water: a review. *Trends in Food Science & Technology*, *95*, 183–195. https://doi.org/10.1016/j.tifs.2019.11.018

Zhang, K., Ren, N.-Q., & Wang, A.-J. (2014). Enhanced biohydrogen production from corn stover hydrolyzate by pretreatment of two typical seed sludges. *International Journal of Hydrogen Energy*, *39*(27), 14653–14662.

Zhang, Q., Hu, J., Lee, D., Chang, Y., & Lee, Y. (2017b). Sludge treatment: current research trends. *Bioresource Technology*, *243*, 1159–1172. https://doi.org/10.1016/j.biortech.2017.07.070

Zhang, Q., Wang, Y., Zhang, Z., Lee, D., Zhou, X., Jing, Y., Ge, X., Jiang, D., Hu, J., & He, C. (2017c). Photo-fermentative hydrogen production from crop residue: a mini review. *Bioresource Technology*, *229*, 222–230. https://doi.org/10.1016/j.biortech.2017.01.008

Zhang, W., Qiu, J., Feng, H., Zang, L., & Sakai, E. (2014). Increase in stability of cellulase immobilized on functionalized magnetic nanospheres. *Journal of Magnetism and Magnetic Materials*, *375*, 117–123. https://doi.org/10.1016/j.jmmm.2014.09.067

Zhang, X.-Z., & Zhang, Y.-H. P. (2010). One-step production of biocommodities from lignocellulosic biomass by recombinant cellulolytic *Bacillus subtilis*: opportunities and challenges. *Engineering in Life Sciences*, *10*(5), 398–406.

Zhang, Z., et al. (2014). Photo-fermentative bio-hydrogen production from agricultural residue enzymatic hydrolyzate and the enzyme reuse. *BioRes*, *9*, 2299–2310.

Zhang, Z., Li, Y., Zhang, H., He, C., & Zhang, Q. (2017d). Potential use and the energy conversion efficiency analysis of fermentation effluents from photo and dark fermentative bio-hydrogen production. *Bioresource Technology*, *245*, 884–889. https://doi.org/10.1016/j.biortech.2017.09.037

Zhang, Z., Zhou, X., Hu, J., Zhang, T., Zhu, S., & Zhang, Q. (2017e). Photo-bioreactor structure and light-heat-mass transfer properties in photo-fermentative bio-hydrogen production system: a mini review. *International Journal of Hydrogen Energy*, *42*(17), 12143–12152. https://doi.org/10.1016/j.ijhydene.2017.03.111

Zhao, X., et al., (2013). Batch and multi-step fed-batch enzymatic saccharification of Formiline-pretreated sugarcane bagasse at high solid loadings for high sugar and ethanol titers. *Bioresource Technology*, *135*, 350–356.

Zhou, M., Ju, X., Li, L., Yan, L., Xu, X., & Chen, J. (2019) Immobilization of cellulase in the non-natural ionic liquid environments to enhance cellulase activity and functional stability. *Applied Microbiology and Biotechnology*, *103*, 2483–2492. https://doi.org/10.1007/s00253-019-09647-9

Zhou, H., Shi, Z., Wan, X., Fang, H., Yu, D., Chen, X., & Liu, P. (2019). The relationships between process parameters and polymeric nanofibers fabricated using a modified coaxial electrospinning. *Nanomaterials*, *9*(6), 843. https://doi.org/10.3390/nano9060843

Zhu, Y., Li, T., Fu, X., Abbasi, A. M., Zheng, B., & Liu, R. H. (2015). Phenolics content, antioxidant and antiproliferative activities of dehulled highland barley (*Hordeum vulgare* L.). *Journal of Functional Foods*, *19*, 439–450. https://doi.org/10.1016/j.jff.2015.09.053

Zikeli, F., Ters, T., Fackler, K., Srebotnik, E., & Li, J. (2016). Wheat straw lignin fractionation and characterization as lignin-carbohydrate complexes. *Industrial Crops and Products*, *85*, 309–317. https://doi.org/10.1016/j.indcrop.2016.03.012

Zou, X., Ren, Z., Wang, N., Cheng, Y., Jiang, Y., Wang, Y., Xu, C., 2018. Function analysis of 5′-UTR of the cellulosomal xyl-doc cluster in *Clostridium papyrosolvens*. *Biotechnology for Biofuels and Bioproducts*, *11*, 43.

Zuorro, A., Lavecchia, R., González-Delgado, Á. D., García-Martinez, J. B., & L'Abbate, P. (2019). Optimization of enzyme-assisted extraction of flavonoids from corn husks. *Processes*, *7*(11), 804. https://doi.org/10.3390/pr7110804

Techniques for Enhanced Production of Biohydrogen from Microalgal Biomass

16

Monika Prakash Rai and Shrasti Vasistha

16.1 INTRODUCTION

The rise in ecological damage and a shortage of energy brought on by the excessive use of carbon-based fuels is a serious issue that will soon become urgent. Petroleum and other fossil fuels can be replaced by alternative energies involving sun, wind, and bioenergy from carbon and hydrogen. It's concerning how much energy the globe requires and uses. By 2040, it's expected to have increased by 48% to 815 quadrillion Btu (Nazir et al., 2019). In order to reduce dependency on fossil fuels, current research is concentrated on developing sustainable and environmentally friendly fuels. The search for alternative energy sources has become a critical issue amidst environmental degradation and the dwindling supply of non-renewable energy resources. Innovative, sustainable, and environmentally friendly alternatives are urgently required to meet the escalating energy demands. Hydrogen gas has emerged as a significant carbon-free energy source and a viable alternative to conventional fuels (Wadjeam et al., 2019). Projections suggest that by 2050, hydrogen gas could fulfil 18% of global energy demands. Researchers are diligently crafting innovative methods and technologies for sustainable hydrogen production. The combustion of hydrogen produces only water as a byproduct, rendering it an environmentally friendly option (Raheem et al., 2018).

Hydrogen is industrially produced through steam reforming of natural resources, specifically natural gas. However, biohydrogen generated from biomass via dark fermentation (DF) and photofermentation (PF) has demonstrated significant potential as a sustainable alternative to fossil-based hydrogen (Patel et al., 2020) (Figure 16.1). Furthermore, the biological agents involved, particularly hydrogen-producing bacteria (HPB), offer the benefits of self-reproduction, self-repair, and cost-effectiveness compared to the physical or chemical agents used in steam reforming processes. Biohydrogen was initially produced from first-generation feedstock, such as food and feed crop. The utilisation of first-generation biohydrogen has come under scrutiny due to its potential threat to food and feed supplies, particularly in developing nations (Limongi et al., 2021). Second-generation biohydrogen, derived from lignocellulosic crops, can alleviate this issue. Nonetheless, the production of lignocellulosic biohydrogen is hindered by several limitations,

 DOI: 10.1201/9781003568612-16

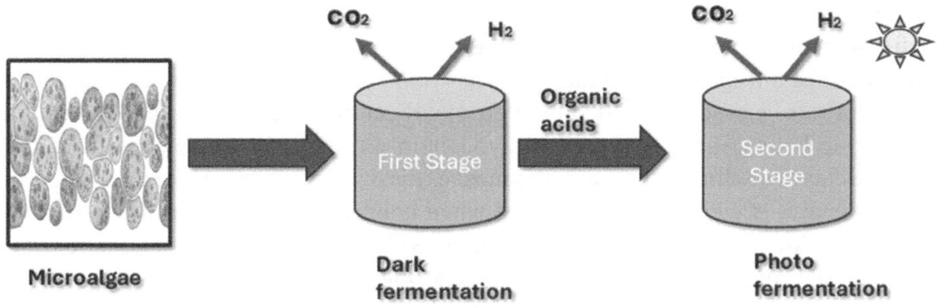

FIGURE 16.1 Integrated dark-photo biohydrogen production.

such as low conversion efficiency, the requirement for biomass pretreatment, and the demand for substantial quantities of water and plantation land (Ahmed et al., 2021). Third-generation biomass, specifically microalgal biomass, is now emerging as a replacement for first- and second-generation biomass.

Considering a great deal of study and development, there is still little use of microalgal biomass. Low rates of biomass bioconversion to hydrogen, inadequate cultivation and harvesting procedures, inefficient and expensive biomass pretreatment techniques, and a lack of preliminary and ecological analyses of the production process are some of the bottlenecks that are responsible for this (Mandotra et al., 2021).

It is crucial to improve process performance, carry out in-depth microbial community study, and expand on basic algae-based understanding in order to comprehend the synthesis of biohydrogen from microalgal biomass (Chandu, 2020). Examining the varied benefits of microalgae as an environmentally friendly fuel for the production of biohydrogen is the main objective of this thorough analysis. A thorough overview of the value-addition of microalgae for the production of biohydrogen is given in this review. Additionally the chapter throws light on the molecular techniques used in the process.

16.2 MECHANISM OF BIOHYDROGEN PRODUCTION FROM MICROALGAE

Nitrogenases and hydrogenases are essential enzymes found within cells involved in the synthesis of biohydrogen in microalgae. The majority of algal species typically produce hydrogenases as their primary enzymes.

Different algae types yield various hydrogenases. Specifically, microalgae synthesise three distinct types of enzymes: NiFe hydrogenase, Fe hydrogenase, and FeFe hydrogenase (Karstens et al., 2015). Cyanobacterial species that produce the NiFe hydrogenase enzyme are widely distributed across biological varieties. The enzyme's large subunit contains a Ni-Fe catalytic site, while the small subunit features highly conserved Fe-S clusters. The four identified groups of NiFe hydrogenases are ferredoxin-dependent hydrogenases, multimeric bidirectional cytoplasmic hydrogenases, membrane-bound hydrogenases, and hydrogenases with bidirectional uptake. The FeFe hydrogenase enzyme is most commonly associated with hydrogen generation. Its active centre, known as the H-cluster, consists of molecules linked to Fe–S and catalyses the production of protons and hydrogen. This enzyme is monomeric and encoded by the *HydA* gene. The FeFe hydrogenase produced by *Chlamydomonas reinhardtii* has been briefly utilised to study the activation process of FeFe hydrogen. According to the findings, the reduction of Fe–S leads to activated catalytic hydrogen generation in the reduced oxidised states of +1 and +2 (Sun et al., 2019). Hydrogen is produced when hydrogenases, electrons, and protons are oxidised. The process's reversibility simplifies hydrogen production from the proton-electron pair. Anaerobic fermentation conditions with active enzymes are crucial for hydrogen production. Nitrogenase is a crucial enzyme for hydrogen synthesis, working alongside hydrogenase. Most nitrogenase enzymes are found in cyanobacterial species, facilitating nitrogen fixation and biohydrogen production as a byproduct (Mandotra et al., 2021). Common cyanobacteria that produce nitrogenase enzymes include *Anabaena*, *Chlorococcus* sp., and *Microcystis*. The three main types of nitrogenases, classified by their metal centres, are iron, molybdenum, and vanadium nitrogenases, akin to hydrogenase enzymes. Molybdenum nitrogenase, which consists of two enzymes, dinitrogenase reductase and dinitrogenase, is the most extensively studied nitrogenase. The MoFe protein, known as dinitrogenase, is where molecules bind when converted to hydrogen. It features two types of prosthetic groups: the MoFe cofactor and the phosphorous cluster. During nitrogen fixation, a low electron flux facilitates proton reduction to hydrogen. Enzymes that produce nitrogenase generate less H_2 compared to hydrogenases (Harwood, 2020). However, in a nitrogen-deficient environment and without reducing substrates, nitrogenase enzymes undergo turnover, resulting in hydrogen production through proton reduction processes that require ATP. Nitrogenases in cyanobacteria are generally sensitive to oxygen and can be irreversibly impaired by reactive oxygen species. Consequently, hydrogenases are employed as the preferred and primary enzyme in advanced commercial methods for algal-based hydrogen production (Karishma et al., 2022).

16.3 CULTIVATION TECHNIQUES FOR MICROALGAE

The cultivation environment is a key factor, with options such as open ponds and enclosed photobioreactors (PBRs) requiring consideration. The comprehensive comparison of these systems shows that open ponds are simple and scalable but come with challenges such as contamination issue, lack of culture guard (Kumar et al., 2021). In contrast, PBRs allow for precise control on various consideration, enabling the cultivation of specific microalgae strains under optimal conditions for desired products. However, the initial capital investment significantly affects the feasibility of each method.

16.3.1 Open Pond

The raceway ponds are popular method for extensive microalgal cultivation due to their particular advantages including operation and maintenance costs (Ruiz-Marin et al., 2020; Sharma et al., 2021). In an open pond with domestic wastewater, various microalgae were cultivated. The author discovered that 45.50 ± 0.32 mL H_2/g VS, 38.43 ± 0.42 mL/H_2/g VS, and 34.83 ± 1.82 mL/H_2/g VS was the maximum amount of hydrogen produced (Satheesh et al., 2023). Pilot-scale research into integrated open pond and PBR systems may potentially be worthwhile in order to maximise biohydrogen production using waste materials (Kumar et al., 2021). Without a doubt, more scientific advancements could improve open pond cultivation's sustainability and economic viability.

16.3.2 Photobioreactor

Solving typical problems including pollution, water evaporation, and fluctuating productivity, PBRs provide considerable advantages over open ponds for microalgal cultivation. Research indicates that, in comparison to other systems, PBRs significantly reduce the risks of microbial invasion in heterotrophic or mixotrophic environments (Yaashikaa et al., 2023). In order to maximise microalgal biohydrogen outputs using DF/PF routes, researchers have created innovative PBR setups. Tubular PBRs, composed of tubes organised in vertical arrays, are still in demand owing to their modularity and efficient biohydrogen generation. One notable tubular PBR testing using a vertical tubular PBR generated 125 mL/L (Batista et al., 2015).

The main reasons flat-plate PBRs are preferred for producing biohydrogen are their broad surface, which permits sufficient light exposure. They are especially beneficial in extensive growth and maintaining dissolved oxygen level management. The biomass production of 14.8 g/m^2/day was recorded for *Nannochloropsis gaditana* in flat-plate gas-lift PBRs (Pfaffinger et al., 2016).

A further investigation demonstrated how to increase the production of cyanobacterial feedstock and biohydrogen from experimental to commercial levels using flat-plate PBRs. They are an intriguing option for producing sustained biohydrogen, as evidenced by the several studies that have demonstrated their efficacy (Oncel et al., 2014).

Biofilm-based PBRs are gaining popularity in the production of biohydrogen owing to minimal resource requirements. Simultaneous harvesting of the resultant microalgae biomass paste from these surfaces eliminates the need for time-consuming dewatering procedures. The primary advantages of biofilm-based PBRs are elevated microalgae production, hence enabled by the efficient delivery of dissolved inorganic carbon (Dai et al., 2017).

Investigation on biofilm-based PBR is still in its early stages, but the approach appears promising as a potential improvement over traditional growth methods.

According to a study by Wu et al. (2019), this approach provides a lot of benefits, including substantial biomass productivity and water utilisation efficiency.

The bacterial-algal symbiotic mixotrophic biofilms were investigated for their ability to remove nutrients. Consequently, biofilm-based PBRs offer a viable route for the generation of biohydrogen due to their efficiency, minimal demand for resources, and substantial production.

An essential stage in the creation of microalgal biomass is harvesting. These procedures aim to gather the microalgal biomass which will facilitate later procedures like the synthesis of biofuel and the separation of compounds that are bioactive (Barros et al., 2015). The effectiveness of a harvesting method is influenced by various factors, including the microalgal strain, cell size, and the products desired. Common harvesting methods are filtration, centrifugation, and flocculation (Tan et al., 2020). Filtration, for instance, can be done using membranes with different pore sizes to separate different microalgal strains. Although this method's main disadvantages are membrane clogging and shearing, centrifugation is more energy-intensive than other techniques, it can yield higher harvesting efficiency (Soomro et al., 2016).

16.4 PRETREATMENT OF MICROALGAE BIOMASS

16.4.1 Physical Treatment

Microwaves, ultrasonography, and high temperatures are examples of physical or mechanical techniques that transform microalgal biomass into nutrients that can be absorbed (Figure 16.2). A microwave is a type of electromagnetic radiation that is used to increase molecular collisions and heat production by stimulating ion movement and molecule vibration in the substrate. Additionally, the electromagnetic field raises pressure, which aids in breaking down the material's crystalline structures and speeds up hydrolysis. Microwave heating offers the benefits of a brief duration and the capacity for immediate initiation and cessation of the process (Nagarajan et al., 2020). However, this method requires a significant amount of energy and requires ongoing maintenance.

Ultrasonic pretreatment is an approach that produces significant shear stresses that can shatter microalgal cells at frequencies ranging from 10 to 20 kHz. This process involves the cavitation causing microturbulence in the liquid and shock waves that can disrupt the walls of microalgal cells (Naveena et al., 2015). In accordance with this approach, prolonged treatments and greater biomass destruction are

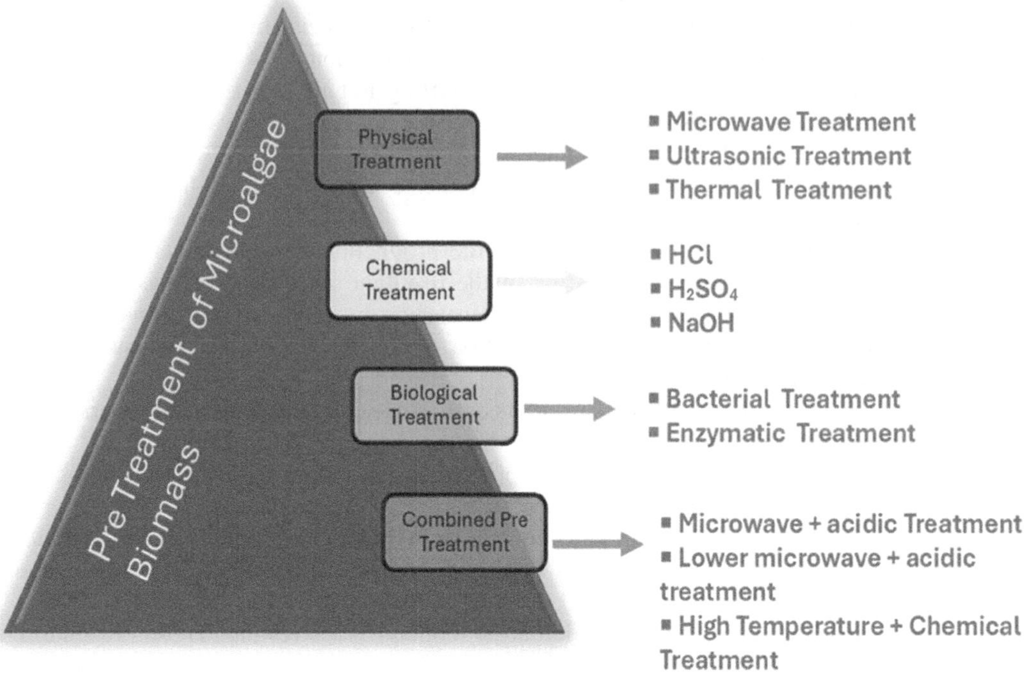

FIGURE 16.2 Schematic representation of microalgae biomass pretreatment.

correlated with higher energy consumption (Yun et al., 2013). However, this method may not be practical for big-scale operations and can be highly energy-intensive.

Another treatment is the thermal pretreatment where heat to break the hydrogen bonds in carbohydrates, a process known as thermal pretreatment, causes microalgal cells to swell and its components to become soluble. Temperature and time assess the hydrolysis process of the pretreatment. The breakdown of microalgae cells and cell organelles occurs at the range of 70°C to 121°C, raising the amounts of volatile fatty acids (Carrere et al., 2016). Several studies have demonstrated that because physical pretreatment processes use a lot of energy, they require a significant energy input (Wang et al., 2017).

16.4.2 Chemical Pretreatment

The chemical treatment provides effective percentage yield of biomass via using acid or alkali. Strong acids like HCl, H_2SO_4 are more effective in breaking the cell wall of the microalgae cells, hence resulting in conversion of sugars into hydrolysate (Phwan et al., 2018). The major impediment of acid pretreatment is the correction of the resultant pH hydrolysate formation which increases in the large volume, hence hampering the fermentation process via HPB activity (Giang et al., 2019).

One popular technique for removing lignin from lignocellulosic biomass is alkaline pretreatment. However, this technique is employed to break down and disturb the microalgal cell wall and solubilise organic molecules, especially protein, as microalgal biomass lacks lignin (Lunprom et al., 2019). However, it should be mindful that employing an alkali concentration (pH levels) that is too high in the pretreatment may result in reduced biogas production since this may harm the inoculum. Because heat was not applied during the pretreatment, it was also discovered that the usage of alkaline (NaOH) pretreatment required very low energy input, such as 1×10–4 kJ/g VS at pH 9, resulting in a very low energy ratio of 7.8×10^{-6} (Cho et al., 2013).

16.4.3 Biological Treatment

Biological pretreatment is a technique that breaks down the cell wall of microalgae and creates assimilable substrates by using enzymes or microorganisms. Usually, this technique is used at low temperatures and air pressure. Additionally, it is substrate-specific and requires little energy input, therefore no inhibitory chemicals are produced (Mahdy et al., 2016). In order to maximise enzyme activity during enzymatic pretreatment, temperature and pH must be regulated within ideal ranges. Microalgal biomass can also be pretreated using in situ enzymes produced by fungus and bacteria. Despite the limited usage of this approach, a recent study found that *Bacillus* sp. was effective in breaking down *Chlorella* sp. biomass for the subsequent synthesis of methane (Roy et al., 2014).

A significant increase in methane output was noted with the bacterial treatments. Furthermore, a process energy assessment showed that the application of the biological pretreatment could reduce the energy input to energy output ratio from 1.47 to 0.96, suggesting that this approach was successful in enhancing biomass energy production. Additionally, rumen fungi were effectively used to pretreat *Haematococcus pluvialis* biomass in order to produce biogas (Aydin et al., 2017).

16.4.4 Combined Pretreatment

To increase the effectiveness of the hydrolysis process, two pretreatment techniques were frequently combined or carried out one after the other. The most popular approach among all is most likely the mix of chemical and heat procedures. It has been observed that using acid at a high temperature (acid-thermal pretreatment) can efficiently break microalgal cells and break down the biomass's carbohydrates, which will make the hydrolysis process easier. Strong acids that are diluted, such as HCl and H_2SO_4, are typically employed in the acid-thermal process at temperatures higher than 80°C.

The efficacy of the acid-thermal pretreatment: The findings demonstrated that treating *Chlorella pyrenoidosa* biomass with steam heating and diluted acid (135°C, 15 min, 1% (v/v) acid) and microwave heating and diluted acid (140°C, 15 min, 1% (v/v) acid) enhanced reducing sugar yields to 0.30–0.31 g/g-TVS, which were more than 42 times what was achieved with stream heating, microwave heating, and ultrasonication (Xia et al., 2013). Following DF of the biomass processed with steam heating and diluted acid and microwave heating and diluted acid, the hydrogen yields (HY) increased to 75.6–83.3 mL/g-TVS, which were 8.6–9.5 times more than those produced using the raw biomass. An acid-thermal approach was also employed by Passos et al. (2016), however, at a lower temperature of 80°C. When 0.5% (w/w) HCl was used at 80°C for 2 hours, it was demonstrated that the acid mostly dissolved hemicellulose in microalgal biomass, increasing biogas production by 86% when compared to a control.

In a different investigation, *Chlorella sorokiniana* was pretreated for hydrogen generation using 2% (v/v) HCl for 20 min at 121°C. *Enterobacter cloacae* fermented the hydrolysate in this investigation, yielding 9 mol-H_2/kg-COD-reduced (Kumar et al., 2013). The initial pretreatment stage prior to an enzymatic hydrolysis might alternatively be acid-thermal. Total sugar (10.8 g/L) and glucose (5.7 g/L) were successfully produced from *Scenedesmus abundans* biomass using 3% H_2SO_4 at 110°C for 30 min and cellulase at 0.1 unit/g-algae (Guo et al., 2013). It is crucial to remember that using acid at very high temperatures (120°C–220°C) would cause hexoses and pentoses to get dehydrated, which would then result in the development of microbial inhibitors, such as 5-hydroxymethylfurfural (HMF) and furfural, respectively.

Microalgal biomass can also be dissolved by alkaline-thermal pretreatment; however, protein is more easily dissolved using this technique than carbohydrates. While HCl only slightly increased the concentration of soluble protein, NaOH was shown to be more successful at solubilising protein in *Chlorella vulgaris* biomass at the same temperature, increasing protein solubilisation by 1.9 times. In one study, Sole-Bundó et al. (2017) pretreated biomass of *Chlorella* sp. and *Scenedesmus* sp. using varying lime (CaO) concentrations (4% and 10%) and pretreatment temperatures (25°C, 55°C, and

72°C). The solubilisation of organic materials increased with temperature and chemical dose, according to the results. The study found that using 10% CaO at 72°C increased the COD concentration by up to 25% while also achieving 32.4% protein solubilisation and 31.4% carbohydrate solubilisation. In another investigation, lipid-extract *Scenedesmus* sp. biomass was treated with 8 g/L NaOH for 4 hours at 100°C. This resulted in 28%–58% protein solubilisation and 88%–90% COD solubilisation. When pretreated biomass was used to produce hydrogen, the result was 45.54 mL-H_2/g-VS, 168% more than the control (Yang et al., 2010).

16.5 TECHNIQUES FOR ENHANCED BIOHYDROGEN PRODUCTION

16.5.1 Microalgae Immobilisation

Immobilisation can happen spontaneously, as in the case of biofilm formation, or intentionally, for example, when cells adhere to solid surfaces or become trapped in matrices. These immobilisation techniques present chances to boost resource efficiency, increase process efficiency, and provide resilience to difficult environmental conditions. Encapsulating microalgal cells is a popular method for artificially immobilising them. Because of its many advantages, microbial immobilisation in solid media or on surfaces has attracted a lot of attention lately (Han et al., 2023).

These benefits encompass a high cell density in a small area and protecting cells from unfavourable factors like high salinity, harmful heavy metals, or abrupt pH changes (Lapponi et al., 2022). The above discussed process has the major disadvantage including slow rate of nutrient infusion followed by outflow of the product from the biomass which immobilises changes in solar gradients brought on by the high cell density of the immobilised biomass (Mishra et al., 2023).

16.5.2 Integrated Hybrid System

Effort is made to solve the shortcomings of conventional production methods while simultaneously optimising the distinctive features of each strategy in order to increase overall biohydrogen yields. The processes are genetic alteration, fermentation using MECs, PF and DF (Escapa et al., 2015). In addition, the wastewater as the nutrient media can be used to make process cost-effective providing the dual benefit of power generation and wastewater treatment (Olabi et al., 2024).

In an advanced study, the researchers have used the substrate from DF for the PF of miroalgae (Melitos et al., 2021). The researchers aimed to produce both biomethane and biohydrogen using combined technique of PF and DF proving to be more beneficial and effective. In an experiment by Nagarajan et al. (2020), three-stage process of PF, DF, and methanogenesis was applied on microalgae *C. pyrenoidosa* for the synthesis of BioH_2 and biomethane. The microalgae biomass was initially pretreated by using steam and diluted acid in order to achieve impressive results.

In an attempt to reduce atmospheric CO_2 the implementation of anaerobic DF composite with wastewater photodegradation may help in CO_2 fixation. Until recently, the production of biohydrogen has served as the primary goal on the financial studies of employing enormous scale microalgae PBRs. Li et al. (2017) state that in order to attain an equilibrium of energy in microalgae-based technologies for biohydrogen, extremely improved technological advancements are required. This highlights the importance of continuously developing and advancing the area to maximise system reliability and reduce consumption of energy.

16.5.3 Genetic Engineering

The usage of CO_2 for biohydrogen production via microalgae may lower the carbon emission. The production of biohydrogen involves two main processes mainly, fermentation and biophotolysis. Glycogen is the precursor for the production of biohydrogen during the process of fermentation, whereas in biophotolysis, hydrogen is employed implicitly or explicitly (Anwar et al., 2019). The optimisation and investigation for the production of biohydrogen using microalgae-based technology still continues (Min Woon et al., 2023).

In order to improve the efficacy for the biohydrogen production, the genetic engineering techniques such as targeted and random mutagenesis, insertion of inducible genes, and overexpressing specific genes, including light-harvesting complexes (LHCs), are being investigated. The alteration of the hydrogenase enzyme which work as a vital enzyme is being focused (Wang et al., 2021).

Regardless, utilising genetic engineering in microalgae to make biohydrogen presents numerous difficulties that need to be resolved. The elevated cost of operations, demands for infrastructure, problems with maintenance, the flexibility of mass production, effective ways to extract are some of these challenges (Grama et al., 2022). To achieve viable levels, methods for boosting the yield and efficiency of biohydrogen generation should be developed.

The process of genetic engineering simulates the specific genes which play the primary role in maximising the biohydrogen production. The two-enzyme pyruvate formate lyase (PFL) and pyruvate ferredoxin oxidoreductase (PFOR) pathways are two pathways for biohydrogen generation (Peters and Sargent, 2023). The conversion of pyruvate into formate and acetyl-CoA is regulated by PFL. Further the enzyme formate hydrogen lyase (FHL) is crucial for converting formate into molecular hydrogen and CO_2.

In a study the microalgae *Clostridium* sp. was studied for the pyruvate:Fd oxidoreductase (PFOR) metabolic pathway by Dahiya et al. (2021). Ferredoxin (Fd) is reduced and acetyl-CoA is generated as a consequence of pyruvate oxidation by the action of PFOR. The diminution permits electrons from the reduced Fd to be transferred to protons by the action of the Hyd enzyme, ultimately resulting in the production of biohydrogen.

These results provide insight into the metabolic processes and enzymes that *Clostridium* sp. and other thermophilic bacteria use to produce biohydrogen. Gaining insight into these procedures is essential to clarifying the principles underlying biohydrogen synthesis and formulating plans to maximise and improve biohydrogen production in these microbes. However, focused approaches work better, and a number of ideas have been put up that use genetic engineering techniques to increase yields and circumvent existing limitations. To get the electrons ready for hydrogenase, Fd removes them from the electron transport chain, as previously mentioned. Even in the optimal circumstances for enzyme expression, other assimilatory pathways compete for the reductants, reducing the total output. Ferredoxin-NAD(P)+ reductase (FNR) is the main competitor for the electrons that it transports from Fd to CO_2 fixation. Fusion proteins were developed to get past this restriction. Investigating the fusion complex between Fd and hydrogenase in an in vivo culture of *C. reinhardtii* is one of the most current projects. In terms of production rates and oxygen tolerance, the combination performed better than the enzyme acting alone (Eilenberg et al., 2016). The D1 mutant's *C. reinhardtii* protein exhibited a comparable pattern of behaviour. A double amino acid mutation (L159I–N230Y) gave the mutant new characteristics, most notably increased oxygen tolerance compared to the wild strain. Compared to the wild type, the flavodiiron protein (FDP) mutants of *C. reinhardtii* generated more hydrogen during photosynthesis. This method once again demonstrates how the removal of a competitive pathway (such as the FDP-mediated O_2 photo-reduction pathway) directs electrons preferentially towards the synthesis of hydrogen. Furthermore, it has been discovered that these mutants' metabolism does not focus on CO_2 fixation even after extended exposure to light pulses. Thus, it is demonstrated how growth parameter adjustments and genetic engineering techniques combine to increase production. Therefore, even though genetic engineering has a lot of promise for increasing the generation of biohydrogen from microalgae, a lot of research and development work is still needed to get beyond the current obstacles and make this technology economically feasible. Various microalgae used for biohydrogen productions via genetic engineering techniques are summarised in Table 16.1.

TABLE 16.1 Microalgae biohydrogen production using genetic engineering approach

MICROALGAE	*MODE OF GENETIC ENGINEERING*	*REFERENCES*
Cyanothece sp.	Gene mutation	Chen et al. (2021)
Chlamydomonas reinhardtii	Inducible genes insertion	Zhang et al. (2023)
Chlamydomonas moewusii	RNA sequencing	Yang et al. (2019)
Chlamydomonas	Ultraviolet radiation	Sproles et al. (2021)
Phaeodactylum tricornutum	Genetic modification	Mutanda et al. (2020)
Chlamydomonas reinhardtii		
Auxenochlorella protothecoides	Metabolic pathway engineering	Kumar et al. (2020)

16.6 CHALLENGES AND FUTURE PROSPECT

The commercialisation potential of biohydrogen remains modest, even though generation from microalgal biomass is more sustainable and environmentally benign than production utilising the steam reforming process. The low concentration of microalgal cells in the growing medium, the high energy requirement of the process, and the low yield of hydrogen production are some of the obstacles causing this. Even though it is known that microalgae develop more quickly than terrestrial plants, it is believed that pretreatment of microalgae biomass is a crucial step in enhancing DF-based biohydrogen production. Breaking down the cell wall and releasing the organic materials from the cells is the main objective of the pretreatment procedures (Shobana et al., 2017). DF of microalgae biohydrogen could be significantly increased by appropriate pretreatment techniques to 50%–70% of theoretical values Additionally, preparation (such as heat treatment) may decrease methanogen activity. As a result, while hydrogen consumption has declined, hydrogen output has grown (Singh et al., 2022).

Genetic engineering has the potential to improve the production of microalgae biohydrogen in the future. It has become normal practice to modify microalgal strains via methods like mutation including gene knock-ins or knock-outs, in order to increase their functionality and capacity. Comparing these genetic alterations to native microalgal strains, biohydrogen production can be significantly increased. Researchers can optimise several elements of microalgae metabolism, like increasing enzyme, rerouting metabolic pathways towards enhanced hydrogen production, and introducing particular modifications (Go swami et al., 2021). By using focused genetic engineering, microalgae strains with significantly higher biohydrogen yields than their natural counterparts can be produced.

16.7 CONCLUSION

Microalgal biohydrogen has great promise as a carbon-negative, renewable way to address the looming fuel shortage problems. To optimise the processes of downstream gas harvesting and upstream microalgal cultivation, more study is still required. Although there has been a lot of work in creating methods to improve the production of hydrogen by microorganisms, many technical issues still need to be resolved before commercialisation is possible. Finding the best microalgal strains with high hydrogen productivity through selection and genetic engineering is a crucial research topic.

ACKNOWLEDGEMENT

Author MPR acknowledges the institutional financial support provided by MNNIT Allahabad [Seed Grant 379/R&C/2024-25].

REFERENCES

Ahmed, S.F., Rafa, N., Mofijur, M., Badruddin, I.A., Inayat, A., Ali, M.S., Farrok, O., Yunus Khan, T.M., 2021. Biohydrogen production from biomass sources: metabolic pathways and economic analysis. *Front. Energy Res.* 9. https://doi.org/10.3389/fenrg.2021.753878.

Anwar, M., Lou, S., Chen, L., Li, H., Hu, Z., 2019. Recent advancement and strategy on bio-hydrogen production from photosynthetic microalgae. *Bioresour. Technol.* 292, 121972. DOI: 10.1016/j.biortech.2019.121972.

Aydin, S., Yıldırım, E., Ince, O., Ince, B. 2017. Rumen anaerobic fungi create new opportunities for enhanced methane production from microalgae biomass. *Algal Res.* 23, 150–160. DOI: 10.1016/j.algal.2016.12.016

Barros, A.I., Gonçalves, A.L., Simões, M., Pires, J.C.M. 2015. Harvesting techniques applied to microalgae: a review. *Renew. Sust. Energ. Rev.* 41, 1489–1500. https://doi.org/10.1016/j.rser.2014.09.037

Batista, A.P., Ambrosano, L., Graça, S., Sousa, C., Marques, P.A.S.S., Ribeiro, B., Botrel, E. P., Castro Neto, P., Gouveia, L., 2015. Combining urban wastewater treatment with biohydrogen production – an integrated microalgae-based approach. *Bioresour. Technol.* 184, 230–235. https://doi.org/10.1016/j.biortech.2014.10.064.

Carrere, H., Antonopoulou, G., Affes, R., Passos, F., Battimelli, A., Lyberatos, G., Ferrer, I. 2016. Review of feedstock pretreatment strategies for improved anaerobic digestion: from lab-scale research to full-scale application. *Bioresour. Technol.* 199, 386–397. https://doi.org/10.1016/j.biortech.2015.09.007

Chandu, S.S., 2020. Production of hydrogen using microbial biomass and photobiological process. *Int. J. Res. Appl. Sci. Eng. Technol.* 8, 435–443. https://doi.org/10.22214/ijraset.2020.32539.

Chen, J., Li, Q., Wang, L., Fan, C., Liu, H., 2021. Advances in whole-cell photobiological hydrogen production. *Adv. NanoBiomed Res.* 1, 2000051. https://doi.org/10.1002/anbr.202000051.

Cho, S., Park, S., Seon, J., Yu, J., Lee, T. 2013. Evaluation of thermal, ultrasonic and alkali pretreatments on mixed-microalgal biomass to enhance anaerobic methane production. *Bioresour. Technol.* 143, 330–336. https://doi.org/10.1016/j.biortech.2013.06.017

Dahiya, S., Chatterjee, S., Sarkar, O., Mohan, S.V., 2021. Renewable hydrogen production by dark-fermentation: current status, challenges and perspectives. *Bioresour. Technol.* 321, 124354. https://doi.org/10.1016/j.biortech.2020.124354.

Dai, L., Fan, L., Duan, D., Ruan, R., Wang, Y., Liu, Y., Zhou, Y., Yu, Z., Liu, Y., Jiang, L., 2017. Production of hydrocarbon-rich bio-oil from soapstock via fast microwave-assisted catalytic pyrolysis. *J. Anal. Appl. Pyrolysis.* 125, 356–362. https://doi.org/10.1016/j.jaap.2017.03.003.

Eilenberg, H., Weiner, I., Ben-Zvi, O., Pundak, C., Marmari, A., Liran, O., Wecker, M.S., Milrad, Y., Yacoby, I., 2016. The dual effect of a ferredoxin-hydrogenase fusion protein in vivo: successful divergence of the photosynthetic electron flux towards hydrogen production and elevated oxygen tolerance. *Biotechnol. Biofuels.* 9, 182. https://doi.org/10.1186/s13068-016-0601-3.

Escapa, A., San-Martín, M.I., Mateos, R., Mor´an, A., 2015. Scaling-up of membraneless microbial electrolysis cells (MECs) for domestic wastewater treatment: ottlenecks and limitations. *Bioresour. Technol.* 180, 72–78. https://doi.org/10.1016/j.biortech.2014.12.096.

Giang, T.T., Lunprom, S., Liao, Q., Reungsang, A., Salakkam, A. 2019. Enhancing hydrogen production from *Chlorella* sp. biomass by pre-hydrolysis with simultaneous saccharification and fermentation (PSSF). *Energies.* 12(5), 908–908. https://doi.org/10.3390/en12050908

Grama, S.B., Liu, Z., Li, J., 2022. Emerging trends in genetic engineering of microalgae for commercial applications. *Mar. Drugs.* 20, 285. https://doi.org/10.3390/md20050285.

Guo, H., Daroch, M., Liu, L., Qiu, G., Geng, S., Wang, G. 2013. Biochemical features and bioethanol production of microalgae from coastal waters of Pearl River Delta. *Bioresour. Technol.* 127, 422–428. https://doi.org/10.1016/j.biortech.2012.10.006

Han, M., Zhang, C., Ho, S.-H., 2023. Immobilized microalgal system: an achievable idea for upgrading current microalgal wastewater treatment. *Environ. Sci. Ecotechnol.* 14, 100227. https://doi.org/10.1016/j.ese.2022.100227.

Harwood, C.S., 2020. Iron-only and vanadium nitrogenases: fail-safe enzymes or something more? *Annu. Rev. Microbiol.* 74, 247–266. https://doi.org/10.1146/annurev-micro-022620-014338.

Karishma, S., Saravanan, A., Senthil Kumar, P., Rangasamy, G., 2022. Sustainable production of biohydrogen from algae biomass: critical review on pretreatment methods, mechanism and challenges. *Bioresour. Technol.* 366, 128187. https://doi.org/10.1016/j.biortech.2022.128187.

Karstens, K., Wahlefeld, S., Horch, M., Grunzel, M., Lauterbach, L., Lendzian, F., Zebger, I., Lenz, O., 2015. Impact of the iron–sulfur cluster proximal to the active site on the catalytic function of an O_2-tolerant NAD+-reducing [NiFe]-hydrogenase. *Biochemistry.* 54, 389–403. https://doi.org/10.1021/bi501347u.

Kumar Sharma, A., Kumar Ghodke, P., Manna, S., Chen, W.-H., 2021. Emerging technologies for sustainable production of biohydrogen production from microalgae: a state-of-the-art review of upstream and downstream processes. *Bioresour. Technol.* 342, 126057. https://doi.org/10.1016/j.biortech.2021.126057.

Kumar, K., Roy, S., Das, D. 2013. Continuous mode of carbon dioxide sequestration by C. sorokiniana and subsequent use of its biomass for hydrogen production by *E. cloacae* iit-bt 08. *Bioresour. Technol.* 145, 116–122.

Lapponi, M.J., Mendez, M.B., Trelles, J.A., Rivero, C.W., 2022. Cell immobilization strategies for biotransformations. *Curr. Opin. Green Sustain. Chem.* 33, 100565. https://doi.org/10.1016/j.cogsc.2021.100565.

Li, T., Strous, M., Melkonian, M., 2017. Biofilm-based photobioreactors: their design and improving productivity through efficient supply of dissolved inorganic carbon. *FEMS Microbiol. Lett.* 364. https://doi.org/10.1093/femsle/fnx218.

Limongi, A.R., Viviano, E., De Luca, M., Radice, R.P., Bianco, G., Martelli, G., 2021. Biohydrogen from microalgae: production and applications. *Appl. Sci.* 11, 1616. https://doi.org/10.3390/app11041616.

Lunprom, S., Phanduang, O., Salakkam, A., Liao, Q., Reungsang, A. 2019. A sequential process of anaerobic solid-state fermentation followed by dark fermentation for biohydrogen production from *Chlorella* sp. *Int. J. Hydrogen Energy.* 44(6), 3306–3316. https://doi.org/10.1016/j.ijhydene.2018.06.012

Mahdy, A., Ballesteros, M., González-Fernández, C. 2016. Enzymatic pretreatment of *Chlorella vulgaris* for biogas production: influence of urban wastewater as a sole nutrient source on macromolecular profile and biocatalyst efficiency. *Bioresour. Technol.* 199, 319–325. https://doi.org/10.1016/j.biortech.2015.08.080

Mandotra, S.K., Sharma, C., Srivastava, N., Ahluwalia, A.S., Ramteke, P.W., 2021. Current prospects and future developments in algal bio-hydrogen production: a review. *Biomass Convers. Biorefinery.* 13, 8575–8592. https://doi.org/10.1007/s13399-021-01414-z.

Melitos, G., Voulkopoulos, X., Zabaniotou, A., 2021. Waste to sustainable biohydrogen production via photo-fermentation and biophotolysis A systematic review. *Renew. Energy Environ. Sustain.* 6, 45. https://doi.org/10.1051/rees/2021047.

Min Woon, J., Shiong Khoo, K., Akermi, M., Alanazi, M.M., Wei Lim, J., Jing Chan, Y., Sean Goh, P., Silas Chidi, B., Kee Lam, M., Zaini, J., Roil Bilad, M., Zhou, Y., Tasnim Sahrin, N., Musa Ardo, F., 2023. Reviewing biohydrogen production from microalgal cells through fundamental mechanisms, enzymes and factors that engendering new challenges and prospects. *Fuel* 346, 128312. https://doi.org/10.1016/j.fuel.2023.128312.

Mishra, V., Mudgal, N., Rawat, D., Poria, P., Mukherjee, P., Sharma, U., Kumria, P., Pani, B., Singh, M., Yadav, A., Farooqi, F., Sharma, R.S., 2023. Integrating microalgae into textile wastewater treatment processes: advancements and opportunities. *J. Water Process Eng.* 55, 104128. https://doi.org/10.1016/j.jwpe.2023.104128.

Mutanda, T., Naidoo, D., Bwapwa, J.K., Anandraj, A., 2020. Biotechnological applications of microalgal oleaginous compounds: current trends on microalgal bioprocessing of products. *Front. Energy Res.* 8. https://doi.org/10.3389/fenrg.2020.598803.

Nagarajan, D., Chang, J.S., Lee, D.J. 2020. Pretreatment of microalgal biomass for efficient biohydrogen production – recent insights and future perspectives. *Bioresour. Technol.* 302, 122871–122871. https://doi.org/10.1016/j.biortech.2020.122871

Naveena, B., Armshaw, P., Tony Pembroke, J. 2015. Ultrasonic intensification as a tool for enhanced microbial biofuel yields. *Biotechnol. Biofuels.* 8(1), 140–140. https://doi.org/10.1186/s13068-015-0321-0

Nazir, M.S., Mahdi, A.J., Bilal, M., Sohail, H.M., Ali, N., Iqbal, H.M.N. 2019. Environmental impact and pollution-related challenges of renewable wind energy paradigm – a review. *Sci. Total Environ.* 683, 436–444. https://doi.org/10.1016/j.scitotenv.2019.05.274

Oncel, S.S., Kose, A., Faraloni, C., Imamoglu, E., Elibol, M., Torzillo, G., Sukan, F.V., 2014. Biohydrogen production using mutant strains of Chlamydomonas reinhardtii: the effects of light intensity and illumination patterns. *Biochem. Eng. J.* 92, 47–52. https://doi.org/10.1016/j.bej.2014.06.019.

Olabi, A.G., Abdelkareem, M.A., Mahmoud, M., Mahmoud, M.S., Elsaid, K., Obaideen, K., Rezk, H., Eisa, T., Chae, K.-J., Sayed, E.T., 2024. Multiple-criteria decision-making for hydrogen production approaches based on economic, social, and environmental impacts. *Int. J. Hydrogen Energy* 52, 854–868. https://doi.org/10.1016/j.ijhydene.2023.10.293.

Passos, F., Felix, L., Rocha, H., Pereira, J.d.O., de Aquino, S. 2016. Reuse of microalgae grown in full-scale wastewater treatment ponds: thermochemical pretreatment and biogas production. *Bioresour. Technol.* 209, 305–312.

Patel, A.K., Choi, Y.Y., Sim, S.J., 2020. Emerging prospects of mixotrophic microalgae: way forward to sustainable bioprocess for environmental remediation and cost- effective biofuels. *Bioresour. Technol.* 300, 122741 https://doi.org/10.1016/j.biortech.2020.122741.

Peters, K., Sargent, F., 2023. Formatehydrogenlyase, formic acid translocation and hydrogen production: dynamic membrane biology during fermentation. *Biochim. Biophys. Acta Bioenerg.* 1864, 148919. https://doi.org/10.1016/j.bbabio.2022.148919.

Pfaffinger, C.E., Schone, ¨ D., Trunz, S., Lowe, ¨ H., Weuster-Botz, D., 2016. Model-based optimization of microalgae areal productivity in flat-plate gas-lift photobioreactors. *Algal Res.* 20, 153–163. https://doi.org/10.1016/j.algal.2016.10.002.

Phwan, C.K., Ong, H.C., Chen, W.H., Ling, T.C., Ng, E.P., Show, P.L. 2018. Overview: comparison of pretreatment technologies and fermentation processes of bioethanol from microalgae. *Energy Convers. Manag.* 173, 81–94. https://doi.org/10.1016/j.enconman.2018.07.054

Raheem, A., Prinsen, P., Vuppaladadiyam, A.K., Zhao, M., Luque, R. 2018. A review on sustainable microalgae based biofuel and bioenergy production: recent developments. *J. Clean. Produc.* 181, 42–59. https://doi.org/10.1016/j.jclepro.2018.01.125

Roy, S., Kumar, K., Ghosh, S., Das, D. 2014. Thermophilic biohydrogen production using pre-treated algal biomass as substrate. *Biomass Bioenerg.* 61, 157–166.

Ruiz-Marin, A., Canedo-Lopez, Y., Chavez-Fuentes, ´ P., 2020. Biohydrogen production by *Chlorella vulgaris* and *Scenedesmus obliquus* immobilized cultivated in artificial wastewater under different light quality. *Amb. Express.* 10, 191. https://doi.org/10.1186/s13568-020-01129-w.

Satheesh, S., Pugazhendi, A., Al-Mur, B.A., Balasubramani, R., 2023. Biohydrogen production coupled with wastewater treatment using selected microalgae. *Chemosphere.* 334, 138932. https://doi.org/10.1016/j.chemosphere.2023.138932.

Sharma, A.K., Sharma, A., Singh, Y., Chen, W.-H., 2021. Production of a sustainable fuel from microalgae *Chlorella minutissima* grown in a 1500 L open raceway ponds. *Biomass Bioenergy.* 149, 106073. https://doi.org/10.1016/j.biombioe.2021.106073.

Shobana, S., Saratale, G.D., Pugazhendhi, A., Arvindnarayan, S., Periyasamy, S., Kumar, G., Kim, S.-H., 2017. Fermentative hydrogen production from mixed and pure microalgae biomass: key challenges and possible opportunities. *Int. J. Hydrogen Energy.* 42, 26440–26453. https://doi.org/10.1016/j.ijhydene.2017.07.050.

Singh, H., Tomar, S., Qureshi, K.A., Jaremko, M., Rai, P.K., 2022. Recent advances in biomass pretreatment technologies for biohydrogen production. *Energies.* 15, 999. https://doi.org/10.3390/en15030999.

Sole-Bundó, M., Carrère, H., Garfí, M., Ferrer, I. 2017. Enhancement of microalgae anaerobic digestion by thermo-alkaline pretreatment with lime (CaO). *Algal Res.* 24, 199–206. https://doi.org/10.1016/j.algal.2017.03.025

Soomro, R.R., Ndikubwimana, T., Zeng, X., Lu, Y., Lin, L., Danquah, M.K. 2016. Development of a two-stage microalgae dewatering process – a life cycle assessment approach. *Front. Plant Sci.* 7(FEB2016), 1–12. doi: 10.3389/fpls.2016.00113

Sproles, A.E., Fields, F.J., Smalley, T.N., Le, C.H., Badary, A., Mayfield, S.P., 2021. Recent advancements in the genetic engineering of microalgae. *Algal Res.* 53, 102158. https://doi.org/10.1016/j.algal.2020.102158.

Sun, Y., He, J., Yang, G., Sun, G., Sage, V., 2019. A review of the enhancement of biohydrogen generation by chemicals addition. *Catalysts.* 9, 353. https://doi.org/10.3390/catal9040353.

Tan, J.S., Lee, S.Y., Chew, K.W., Lam, M.K., Lim, J.W., Ho, S.H., Show, P.L. 2020. A review on microalgae cultivation and harvesting, and their biomass extraction processing using ionic liquids. *Bioengineered.* 11(1), 116–129. https://doi.org/10.1080/21655979.2020.1711626

Wadjeam, P., Reungsang, A., Imai, T., Plangklang, P. 2019. Co-digestion of cassava starch wastewater with buffalo dung for bio-hydrogen production. *Int. J. Hydrogen Energy.* 44(29), 14694–14706. https://doi.org/10.1016/j.ijhydene.2019.04.138

Wang, M., Lee, E., Dilbeck, M.P., Liebelt, M., Zhang, Q., Ergas, S.J. 2017. Thermal pretreatment of microalgae for biomethane production: experimental studies, kinetics and energy analysis. *J. Chem. Technol. Biotechnol.* 92(2), 399–407. https://doi.org/10.1002/jctb.5018

Wang, Y., Yuan, H., Martinez, A., Hong, P., Xu, H., Bockmiller, F.R., 2021. Polymer electrolyte membrane fuel cell and hydrogen station networks for automobiles: status, technology, and perspectives. *Adv. Appl. Energy.* 2, 100011. https://doi.org/10.1016/j.adapen.2021.100011.

Wu, X., Cen, Q., Addy, M., Zheng, H., Luo, S., Liu, Y., Cheng, Y., Zhou, W., Chen, P., Ruan, R., 2019. A novel algal biofilm photobioreactor for efficient hog manure wastewater utilization and treatment. *Bioresour. Technol.* 292, 121925. https://doi.org/10.1016/j.biortech.2019.121925.

Xia, A., Cheng, J., Ding, L., Lin, R., Huang, R., Zhou, J., Cen, K. 2013. Improvement of the energy conversion efficiency of *Chlorella pyrenoidosa* biomass by a three-stage process comprising dark fermentation, photofermentation, and methanogenesis. *Bioresour. Technol.* 146, 436–443. https://doi.org/10.1016/j.biortech.2013.07.077

Yaashikaa, P.R., Keerthana Devi, M., Senthil Kumar, P., Rangasamy, G., Rajendran, S., Xiao, L., 2023. A review on pretreatment methods, photobioreactor design and metabolic engineering approaches of algal biomass for enhanced biohydrogen production. *Int. J. Hydrogen Energy.* 48, 21110–21127. https://doi.org/10.1016/j.ijhydene.2022.10.092.

Yang, D.-W., Syn, J.-W., Hsieh, C.-H., Huang, C.-C., Chien, L.-F., 2019. Genetically engineered hydrogenases promote biophotocatalysis-mediated H_2 production in the green alga Chlorella sp. DT. *Int. J. Hydrogen Energy.* 44, 2533–2545. https://doi.org/10.1016/j.ijhydene.2018.11.088.

Yang, Z., Guo, R., Xu, X., Fan, X., Li, X. 2010. Enhanced hydrogen production from lipidextracted microalgal biomass residues through pretreatment. *Int. J. Hydrogen Energy.* 35(18), 9618–9623. https://doi.org/10.1002/jctb.2537

Yun, Y.M., Jung, K.W., Kim, D.H., Oh, Y.K., Cho, S.K., Shin, H.S. 2013. Optimization of dark fermentative H_2 production from microalgal biomass by combined (acid+ultrasonic) pretreatment. *Bioresour. Technol.* 141, 220–226. https://doi.org/10.1016/j.ijhydene.2018.11.088

Zhang, J., Xue, D., Wang, C., Fang, D., Cao, L., Gong, C., 2023. Genetic engineering for biohydrogen production from microalgae. *iScience.* 26, 107255. https://doi.org/10.1016/j.isci.2023.107255.

Index

For Product Safety Concerns and Information please contact our EU representative GPSR@taylorandfrancis.com Taylor & Francis Verlag GmbH, Kaufingerstraße 24, 80331 München, Germany